PRINTED TEST BANK

CHRISTINE HERMANN
Radford University

CHEMISTRY

A Molecular Approach

NIVALDO J. TRO

Third Edition

PEARSON

Boston Columbus Indianapolis New York San Francisco Upper Saddle River
Amsterdam Cape Town Dubai London Madrid Milan Munich Paris Montréal Toronto
Delhi Mexico City São Paulo Sydney Hong Kong Seoul Singapore Taipei Tokyo

Editor in Chief : Adam Jaworski
Senior Acquisitions Editor: Terry Haugen
Senior Marketing Manager: Jonathan Cottrell
Project Editor: Jessica Moro
Assistant Editor: Erin Kneuer
Managing Editor, Chemistry and Geosciences: Gina M. Cheselka
Production Project Manager: Nicole Jackson
Operations Specialist: Jeffrey Sargent
Cover Designer: Seventeenth Street Studios

Cover Image Credit: Nanotube sensor: A carbon nanotube treated with a capture agent, in yellow, can bind with and delect the purple-colored target protein—this changes the electrical resistance of the nanotube and creates a sensing device. Artist: Ethan Minot, in Nanotube technology leading to fast, lower-cost medical diagnostics, Oregon State University, @ 2012, 01 pp., *http://www.flickr.com/photos/oregonstateuniversity/6816133738/*

Credits and acknowledgments borrowed from other sources and reproduced, with permission, in this textbook appear on the appropriate page within the text.

1 2 3 4 5 6 7 8 9 10—EBM—16 15 14 13 12

PEARSON

www.pearsonhighered.com ISBN-10: 0-321-81367-7; ISBN-13: 978-0-321-81367-1

Contents

Chapter 1 Matter, Measurement, and Problem Solving

Multiple Choice Questions

1) Molecules can be described as
   A) mixtures of two or more pure substances.
   B) mixtures of two or more elements that has a specific ratio between components.
   C) two or more atoms chemically joined together.
   D) heterogeneous mixtures.
   E) homogeneous mixtures.

Answer: C
Diff: 1 *Page Ref: 1.1*

2) Give the composition of water.
   A) two hydrogen atoms and two oxygen atoms
   B) one hydrogen atom and one oxygen atom
   C) two hydrogen atoms and one oxygen atom
   D) one hydrogen atom and two oxygen atoms

Answer: C
Diff: 1 *Page Ref: 1.1*

3) Give the composition of hydrogen peroxide.
   A) two hydrogen atoms and two oxygen atoms
   B) one hydrogen atom and one oxygen atom
   C) two hydrogen atoms and one oxygen atom
   D) one hydrogen atom and two oxygen atoms

Answer: A
Diff: 1 *Page Ref: 1.1*

4) Which of the following represents a *hypothesis*?
   A) Sodium reacts with water to form sodium hydroxide and hydrogen gas.
   B) Nitrogen gas is a fairly inert substance.
   C) Nickel has a silvery sheen.
   D) When a substance combusts, it combines with air.
   E) When wood burns, heat is given off.

Answer: D
Diff: 1 *Page Ref: 1.2*

5) Which of the following represents a valid *hypothesis*?
   A) Neon does not react with oxygen.
   B) Sodium metal reacts violently with water.
   C) Lead is soft and malleable.
   D) Oxygen is a gas at room temperature.
   E) Metals tend to lose electrons.

Answer: E
Diff: 1 *Page Ref: 1.2*

6) The statement, "In a chemical reaction, matter is neither created nor destroyed" is called
A) the Law of Conservation of Mass.
B) Dalton's Atomic Theory.
C) the Scientific Method.
D) the Law of Multiple Proportions.
E) the Law of Definite Proportions.

Answer: A
Diff: 1 *Page Ref: 1.2*

7) Dalton's Atomic Theory states
A) that all elements have several isotopes.
B) that matter is composed of small indestructible particles.
C) that the properties of matter are determined by the properties of atoms.
D) that energy is neither created nor destroyed during a chemical reaction.
E) that an atom is predominantly empty space.

Answer: B
Diff: 1 *Page Ref: 1.2*

8) The Scientific Method
A) is just a theory.
B) is a strict set of rules and procedures that lead to inarguable fact.
C) isn't used much in modern chemistry.
D) is based on continued observation and experiment.
E) is a framework for proving an argument you know to be true.

Answer: D
Diff: 1 *Page Ref: 1.2*

9) Which of the following statements is TRUE?
A) A scientific law is fact.
B) Once a theory is constructed, it is considered fact.
C) A hypothesis is speculation that is difficult to test.
D) An observation explains why nature does something.
E) A scientific law summarizes a series of related observations.

Answer: E
Diff: 1 *Page Ref: 1.2*

10) Identify a liquid.
A) definite volume and definite shape
B) definite volume and no definite shape
C) no definite shape and definite volume
D) no definite shape and no definite volume

Answer: B
Diff: 1 *Page Ref: 1.3*

11) Identify a solid.
A) definite volume and definite shape
B) definite volume and no definite shape
C) no definite shape and definite volume
D) no definite shape and no definite volume

Answer: A
Diff: 1 *Page Ref: 1.3*

12) Identify a gas.
A) definite volume and definite shape
B) definite volume and no definite shape
C) no definite shape and definite volume
D) no definite shape and no definite volume

Answer: D
Diff: 1 *Page Ref: 1.3*

13) Which of the following statements about crystalline and amorphous solids is TRUE?
A) A crystalline solid is composed of atoms or molecules arranged with long-range repeating order.
B) An example of a crystalline solid is glass.
C) An example of an amorphous solid is table salt (NaCl).
D) An amorphous solid is composed of atoms or molecules with a majority of its volume empty.
E) All of the above statements are TRUE.

Answer: A
Diff: 1 *Page Ref: 1.3*

14) Which of the following statements about the phases of matter is TRUE?
A) In both solids and liquids, the atoms or molecules pack closely to one another.
B) Solids are highly compressible.
C) Gaseous substances have long-range repeating order.
D) There is only one type of geometric arrangement that the atoms or molecules in any solid can adopt.
E) Liquids have a large portion of empty volume between molecules.

Answer: A
Diff: 1 *Page Ref: 1.3*

15) A substance that can't be chemically broken down into simpler substances is
A) a homogeneous mixture.
B) an element.
C) a heterogeneous mixture.
D) a compound.
E) an electron.

Answer: B
Diff: 1 *Page Ref: 1.3*

16) A substance composed of two or more elements in a fixed, definite proportion is
A) a homogeneous mixture.
B) a heterogeneous mixture.
C) a compound.
D) a solution.
E) an alloy.

Answer: C
Diff: 1 *Page Ref: 1.3*

17) Decanting is
A) a process in which the more volatile liquid is boiled off.
B) dissolving a solid into a liquid.
C) separating a solid from a liquid by pouring off the liquid.
D) pouring a mixture through a filter paper to separate the solid from the liquid.
E) heating a mixture of two solids to fuse them together.

Answer: C
Diff: 1 *Page Ref: 1.3*

18) Distillation is
A) a process in which the more volatile liquid is boiled off.
B) dissolving a solid into a liquid.
C) separating a solid from a liquid by pouring off the liquid.
D) pouring a mixture through a filter paper to separate the solid from the liquid.
E) heating a mixture of two solids to fuse them together.

Answer: A
Diff: 1 *Page Ref: 1.3*

19) Filtration is
A) a process in which the more volatile liquid is boiled off.
B) dissolving a solid into a liquid.
C) separating a solid from a liquid by pouring off the liquid.
D) pouring a mixture through a filter paper to separate the solid from the liquid.
E) heating a mixture of two solids to fuse them together.

Answer: D
Diff: 1 *Page Ref: 1.3*

20) Two or more substances in variable proportions, where the composition is constant throughout are
A) a compound.
B) an element.
C) a heterogeneous mixture.
D) a homogeneous mixture.
E) a crystalline solid.

Answer: D
Diff: 1 *Page Ref: 1.3*

21) Two or more substances in variable proportions, where the composition is variable throughout are
- A) a solution.
- B) a homogeneous mixture.
- C) a compound.
- D) an amorphous solid..
- E) a heterogeneous mixture.

Answer: E
Diff: 1 *Page Ref: 1.3*

22) A physical change
- A) occurs when iron rusts.
- B) occurs when sugar is heated into caramel.
- C) occurs when glucose is converted into energy within your cells.
- D) occurs when water is evaporated.
- E) occurs when propane is burned for heat.

Answer: D
Diff: 1 *Page Ref: 1.4*

23) A chemical change
- A) occurs when methane gas is burned.
- B) occurs when paper is shredded.
- C) occurs when water is vaporized.
- D) occurs when salt is dissolved in water.
- E) occurs when powdered lemonade is stirred into water.

Answer: A
Diff: 1 *Page Ref: 1.4*

24) Which of thc following represents a physical property?
- A) Sodium metal is extremely reactive with chlorine gas.
- B) Mercury is a silvery liquid at room temperature.
- C) Aluminum has a tendency to "rust."
- D) Butane is highly flammable.
- E) Argon has an unreactive nature.

Answer: B
Diff: 1 *Page Ref: 1.4*

25) Which of the following represents a chemical property of hydrogen gas?
- A) It is gaseous at room temperature.
- B) It is less dense than air.
- C) It reacts explosively with oxygen.
- D) It is colorless.
- E) It is tasteless.

Answer: C
Diff: 1 *Page Ref: 1.4*

26) Which of the following statements about energy is FALSE?
A) Energy can be converted from one type to another.
B) The total energy of a system remains constant.
C) Kinetic energy is the energy associated with its position or composition.
D) Energy is the capacity to do work.
E) Systems tend to change in order to lower their potential energy.

Answer: C
Diff: 1 *Page Ref: 1.5*

27) Define thermal energy.
A) energy associated with the temperature of an object
B) energy associated with the motion of an object
C) energy associated with the force of an object
D) energy associated with the gravity of an object
E) energy associated with the position or composition of an object

Answer: A
Diff: 1 *Page Ref: 1.5*

28) Define kinetic energy.
A) energy associated with the temperature of an object
B) energy associated with the motion of an object
C) energy associated with the force of an object
D) energy associated with the gravity of an object
E) energy associated with the position or composition of an object

Answer: B
Diff: 1 *Page Ref: 1.5*

29) Define potential energy.
A) energy associated with the temperature of an object
B) energy associated with the motion of an object
C) energy associated with the force of an object
D) energy associated with the gravity of an object
E) energy associated with the position or composition of an object

Answer: E
Diff: 1 *Page Ref: 1.5*

30) All of the following are SI base units of measurement, EXCEPT
A) meter B) gram C) second D) kelvin E) mole

Answer: B
Diff: 1 *Page Ref: 1.6*

31) The outside temperature is 35°C, what is the temperature in K?
A) –238 K B) 308 K C) 95 K D) 31 K E) 63 K

Answer: B
Diff: 2 *Page Ref: 1.6*

32) Determine the density of an object that has a mass of 149.8 g and displaces 12.1 mL of water when placed in a graduated cylinder.
A) 8.08 g/mL
B) 1.38 g/mL
C) 12.4 g/mL
D) 18.1 g/mL
E) 11.4 g/mL

Answer: C
Diff: 2 *Page Ref: 1.6*

33) Determine the volume of an object that has a mass of 455.6 g and a density of 19.3 g/cm ^{3}.
A) 87.9 mL B) 42.4 mL C) 18.5 mL D) 23.6 mL E) 31.2 mL

Answer: D
Diff: 2 *Page Ref: 1.6*

34) A student performs an experiment to determine the density of a sugar solution. She obtains the following results: 1.11 g/mL, 1.81 g/mL, 1.95 g/mL, 1.75 g/mL. If the actual value for the density of the sugar solution is 1.75 g/mL, which statement below best describes her results?
A) Her results are precise, but not accurate.
B) Her results are accurate, but not precise.
C) Her results are both precise and accurate
D) Her results are neither precise nor accurate.
E) It isn't possible to determine with the information given.

Answer: D
Diff: 1 *Page Ref: 1.7*

35) A student performs an experiment to determine the density of a sugar solution. She obtains the following results: 1.71 g/mL, 1.73 g/mL, 1.67 g/mL, 1.69 g/mL. If the actual value for the density of the sugar solution is 1.40 g/mL, which statement below best describes her results?
A) Her results are precise, but not accurate.
B) Her results are accurate, but not precise.
C) Her results are both precise and accurate
D) Her results are neither precise nor accurate.
E) It isn't possible to determine with the information given.

Answer: A
Diff: 1 *Page Ref: 1.7*

36) A student performs an experiment to determine the density of a sugar solution. She obtains the following results: 1.79 g/mL, 1.81 g/mL, 1.80 g/mL, 1.81 g/mL. If the actual value for the density of the sugar solution is 1.80 g/mL, which statement below best describes her results?
A) Her results are precise, but not accurate.
B) Her results are accurate, but not precise.
C) Her results are both precise and accurate
D) Her results are neither precise nor accurate.
E) It isn't possible to determine with the information given.

Answer: C
Diff: 1 *Page Ref: 1.7*

37) Systematic error is defined as
A) error that tends to be too high or too low.
B) error that has equal probability of being too high and too low.
C) error that averages out with repeated trials.
D) error that is random.

Answer: A
Diff: 1 *Page Ref: 1.7*

38) Read the water level with the correct number of significant figures.

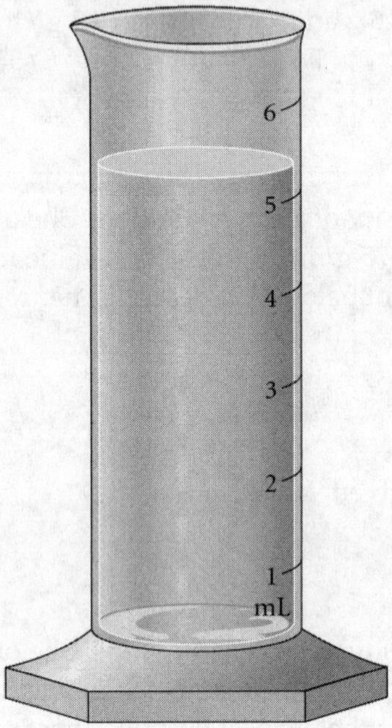

A) 5 mL
B) 5.3 mL
C) 5.32 mL
D) 5.320 mL
E) 5.3200 mL

Answer: B
Diff: 2 *Page Ref: 1.7*

39) Read the temperature with the correct number of significant figures.

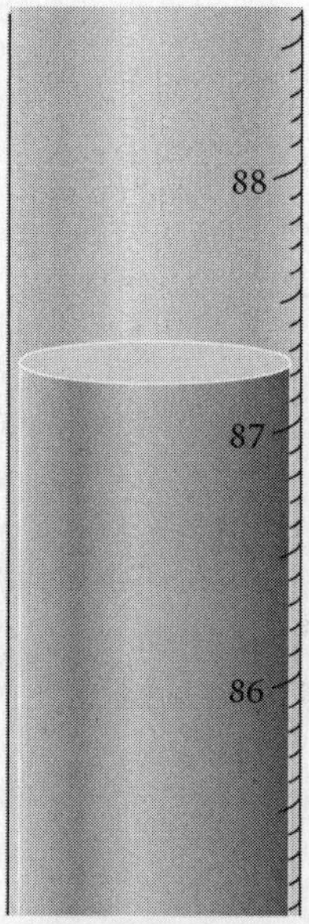

A) 87°C
B) 87.2°C
C) 87.20°C
D) 87.200°C
E) 87.2000°C

Answer: C
Diff: 2 *Page Ref: 1.7*

40) Read the length of the metal bar with the correct number of significant figures.

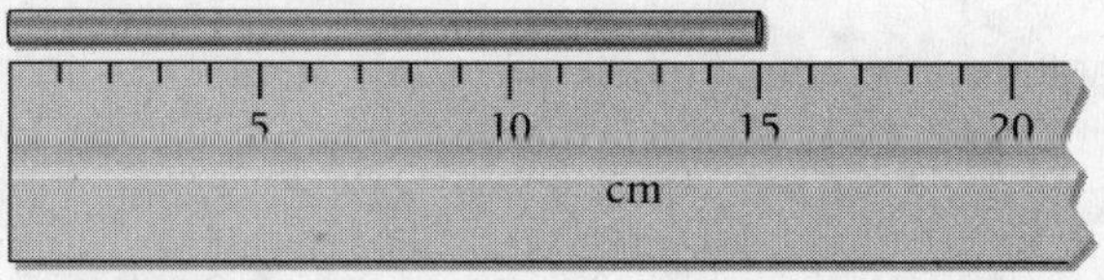

A) 20 cm
B) 15 cm
C) 15.0 cm
D) 15.00 cm
E) 15.000 cm

Answer: C
Diff: 2 *Page Ref: 1.7*

41) Read the length of the metal bar with the correct number of significant figures.

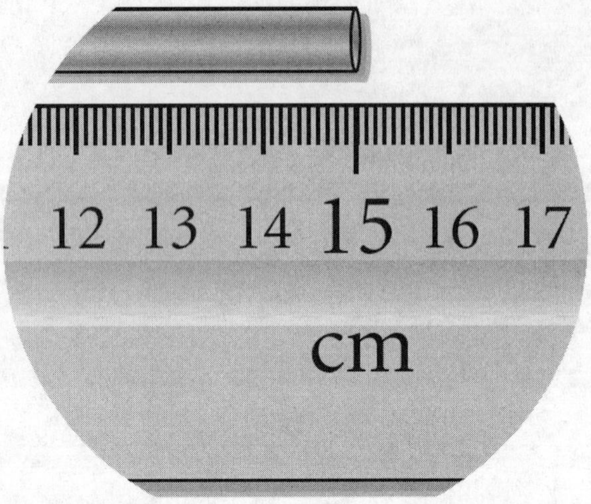

A) 20 cm
B) 15 cm
C) 15.0 cm
D) 15.00 cm
E) 15.000 cm

Answer: D
Diff: 2 *Page Ref: 1.7*

42) Identify the exact number

A) 2 B) 2.0 C) 2.00 D) 2.000 E) 2.0000

Answer: A
Diff: 1 *Page Ref: 1.7*

43) How many significant figures are in the measurement, 20.300 m?

A) 3 B) 4 C) 5 D) 1 E) 2

Answer: C
Diff: 2 *Page Ref: 1.7*

44) What answer should be reported, with the correct number of significant figures, for the following calculation? (433.621 – 333.9) × 11.900

A) 1.19×10^3
B) 1.187×10^3
C) 1.1868×10^3
D) 1.18680×10^3
E) 1.186799×10^3

Answer: A
Diff: 2 *Page Ref: 1.7*

45) What answer should be reported, with the correct number of significant figures, for the following calculation? (249.362 + 41) / 63.498

A) 4.6 B) 4.57 C) 4.573 D) 4.5728 E) 4.57277

Answer: B
Diff: 3 *Page Ref: 1.7*

46) What answer should be reported, with the correct number of significant figures, for the following calculation? (965.43 × 3.911) + 9413.4136

A) 13189
B) 13189.2
C) 1.32×10^4
D) 1.3×10^4
E) 1.319×10^4

Answer: A
Diff: 3 *Page Ref: 1.7*

47) How many mm are in 3.20 cm?

A) 3.20×10^1 mm
B) 3.20×10^{-1} mm
C) 3.20×10^{-2} mm
D) 3.20×10^2 mm
E) 3.20×10^3 mm

Answer: A
Diff: 2 *Page Ref: 1.8*

48) If an object has a density of 8.65 g/cm^3, what is its density in units of kg/m^3?

A) 8.65×10^{-3} kg/m^3
B) 8.65×10^{-7} kg/m^3
C) 8.65×10^3 kg/m^3
D) 8.65×10^1 kg/m^3
E) 8.65×10^{-1} kg/m^3

Answer: C
Diff: 3 *Page Ref: 1.8*

49) If the walls in a room are 955 square feet in area, and a gallon of paint covers 15 square yards, how many gallons of paint are needed for the room? (3 ft = 1 yd)

A) 47 gallons
B) 21 gallons
C) 7.1 gallons
D) 24 gallons
E) 2.3 gallons

Answer: C
Diff: 4 *Page Ref: 1.8*

50) Gas is sold for $1.399 per liter in Toronto, Canada. Your car needs 12.00 gallons. How much will your credit card be charged in Canadian dollars?

A) $16.79 B) $67.15 C) $4.44 D) $63.54

Answer: D
Diff: 5 Page Ref: 1.8

Algorithmic Questions

1) In the blood, hemoglobin is the carrier of

A) oxygen
B) carbon monoxide
C) nitrogen
D) hydrogen
E) helium

Answer: A
Diff: 1 Page Ref: 1.1

2) Identify the crystalline solid.

A) plastic B) glass C) table salt D) water E) bleach

Answer: C
Diff: 1 Page Ref: 1.3

3) Choose the pure substance from the list below.

A) tea
B) a casserole
C) carbon monoxide
D) sugar water
E) pomegranate juice

Answer: C
Diff: 1 Page Ref: 1.3

4) A blueberry scone is an example of

A) a compound. B) an element.
C) a heterogeneous mixture. D) a homogeneous mixture.

Answer: C
Diff: 2 Page Ref: 1.3

5) Hydrochloric acid is an example of

A) a compound. B) an element.
C) a heterogeneous mixture. D) a homogeneous mixture.

Answer: D
Diff: 2 Page Ref: 1.3

6) Gold is an example of

A) a compound. B) an element.
C) a heterogeneous mixture. D) a homogeneous mixture.

Answer: B
Diff: 2 Page Ref: 1.3

7) Ammonia is an example of
A) a compound.
B) an element.
C) a heterogeneous mixture.
D) a homogeneous mixture.

Answer: A
Diff: 2 *Page Ref: 1.3*

8) Identify a solid.
A) copper B) oxygen C) water D) helium E) air

Answer: A
Diff: 1 *Page Ref: 1.3*

9) Identify a liquid.
A) oxygen B) copper C) salt D) mercury E) sugar

Answer: D
Diff: 1 *Page Ref: 1.3*

10) Choose the pure substance from the list below.
A) sea water B) diamond C) air D) tea E) milk

Answer: B
Diff: 1 *Page Ref: 1.3*

11) Choose the element from the list below.
A) sodium chloride
B) water
C) hydrogen peroxide
D) helium
E) rust

Answer: D
Diff: 1 *Page Ref: 1.3*

12) Choose the compound from the list below.
A) gold B) butane C) neon D) helium E) lithium

Answer: B
Diff: 1 *Page Ref: 1.3*

13) Choose the heterogeneous mixture from the list below.
A) sports drink
B) chlorine gas
C) beer
D) lasagna
E) carbon (graphite)

Answer: D
Diff: 1 *Page Ref: 1.3*

14) Choose the homogeneous mixture from the list below.
A) cola B) mud C) ice water D) dirt E) salsa

Answer: A
Diff: 1 *Page Ref: 1.3*

15) Choose the homogeneous mixture from the list below.
A) carbonated soda
B) air
C) chocolate chip cookies
D) trail mix
E) blood

Answer: B
Diff: 2 *Page Ref: 1.3*

16) Which of the following are examples of physical change?
A) sugar is dissolved in water
B) tea is brewed
C) dry ice sublimes
D) water freezes
E) All of these are examples of physical change.

Answer: E
Diff: 1 *Page Ref: 1.4*

17) Which of the following are examples of physical change?
A) dew forms on a blade of grass
B) a halloween light stick glows after shaking
C) browning meat
D) hydrogen balloon explodes when contacted with a flame
E) None of the above are physical changes.

Answer: A
Diff: 1 *Page Ref: 1.4*

18) Which of the following are examples of a chemical change?
A) copper building materials develop a green patina over time
B) a candle burns
C) rubbing alcohol evaporates
D) Both A and B are examples of chemical change.
E) All of the above are examples of chemical change.

Answer: D
Diff: 1 *Page Ref: 1.4*

19) Which of the following are examples of a chemical change?
A) coffee brewing
B) water freezing
C) leaves turning color in the fall
D) salt dissolves in water
E) None of the above are chemical changes.

Answer: C
Diff: 1 *Page Ref: 1.4*

20) Identify the unit of measurement which is a SI base unit of measurement.
A) second B) Celsius C) quart D) ton E) kilometer

Answer: A
Diff: 1 *Page Ref: 1.6*

21) Kilogram is a measure of
A) mass
B) time
C) length
D) temperature
E) volume

Answer: A
Diff: 1 *Page Ref: 1.6*

22) What symbol is used to represent the factor 10^3?
A) M B) k C) μ D) n

Answer: B
Diff: 1 *Page Ref: 1.6*

23) Which of the following is the **smallest** volume?
A) 22 cm^3 B) 0.50 dL C) 2.8×10^3 mL D) 5.0×10^7 nL

Answer: A
Diff: 1 *Page Ref: 1.6*

24) What symbol is used to represent the factor 10^{-1}?
A) M B) m C) μ D) d

Answer: D
Diff: 1 *Page Ref: 1.6*

25) The factor 10 corresponds to which prefix?
A) centi B) deci C) deka D) milli

Answer: C
Diff: 1 *Page Ref: 1.6*

26) The factor 10^{-3} corresponds to which prefix?
A) deka B) deci C) milli D) centi

Answer: C
Diff: 1 *Page Ref: 1.6*

27) What decimal power does the abbreviation f represent?
A) 1×10^6 B) 1×10^3 C) 1×10^{-1} D) 1×10^{-15} E) 1×10^{-12}

Answer: D
Diff: 1 *Page Ref: 1.6*

28) What decimal power does the abbreviation pico represent?
A) 1×10^6 B) 1×10^9 C) 1×10^{-1} D) 1×10^{-12} E) 1×10^{-15}

Answer: D
Diff: 1 *Page Ref: 1.6*

29) Which of the following are examples of intensive properties?
A) density
B) volume
C) mass
D) None of the above are examples of intensive properties.
E) All of the above are examples of intensive properties.

Answer: A
Diff: 1 *Page Ref: 1.6*

30) Which of the following are examples of extensive properties?
A) mass
B) color
C) density
D) temperature
E) odor

Answer: A
Diff: 1 *Page Ref: 1.6*

31) Identify the common substance that has the highest density.
A) sugar
B) ice
C) glass
D) mercury
E) aluminum

Answer: D
Diff: 1 *Page Ref: 1.6*

32) Identify the common substance that has the lowest density.
A) ethanol
B) aluminum
C) titanium
D) table salt
E) sugar

Answer: A
Diff: 1 *Page Ref: 1.6*

33) What is the volume (in cm^3) of a 43.6 g piece of metal with a density of 2.71 g/cm^3?
A) 16.1
B) 19.5
C) .425
D) 6.65
E) none of the above

Answer: A
Diff: 2 *Page Ref: 1.6*

34) A piece of metal ore weighs 8.25 g. When a student places it into a graduated cylinder containing water, the liquid level rises from 21.25 mL to 26.47 mL. What is the density of the ore?
A) 0.312 g/mL B) 0.633 g/mL C) 1.58 g/mL D) 3.21 g/mL

Answer: C
Diff: 2 *Page Ref: 1.6*

35) A mass of mercury occupies 0.950 L. What volume would an equal mass of ethanol occupy? The density of mercury is 13.546 g/mL and the density of ethanol is 0.789 g/mL.
A) 0.0553 L B) 0.0613 L C) 16.3 L D) 18.1 L

Answer: C
Diff: 2 *Page Ref: 1.6*

36) If the melting point of vanadium metal is 1910°C, what is its melting point in Kelvin?
A) 1029 K B) 1637 K C) 2183 K D) 3470 K

Answer: C
Diff: 3 *Page Ref: 1.6*

37) If the temperature is 128°F, what is the temperature in degrees celsius?
A) 262.4°C B) 401°C C) 53.3°C D) –145°C E) 328°C

Answer: C
Diff: 2 *Page Ref: 1.6*

38) If a solution has a temperature of 255 K, what is its temperature in degrees celsius?
A) 491°C B) 528°C C) 123.9°C D) 355°C E) –18°C

Answer: E
Diff: 2 *Page Ref: 1.6*

39) Determine the mass of an object that has a volume of 88.6 mL and a density of 9.77 g/mL.
A) 98.37 g B) 0.110 g C) 9.07 g D) 78.8 g E) 866 g

Answer: E
Diff: 2 *Page Ref: 1.6*

40) The outside air temperature is 40°F, what is the temperature in Kelvin?
A) 313 K B) 377 K C) 281 K D) 277 K

Answer: D
Diff: 3 *Page Ref: 1.6*

41) The outside temperature is 55°C, what is the temperature in °F?
A) 131°F B) 12.8°F C) 23°F D) 77°F E) 99°F

Answer: A
Diff: 3 *Page Ref: 1.6*

42) Which of the following is the **lowest** temperature?
A) 42°C B) 57°F
C) 318 K D) All of these temperatures are all equal.

Answer: B
Diff: 3 *Page Ref: 1.6*

43) How many significant figures are in the measurement 5.34 g?
A) 1 B) 2 C) 4 D) 3 E) 5

Answer: D
Diff: 1 *Page Ref:* 1.7

44) How many significant figures are in 0.00523980 mL?
A) 3 B) 4 C) 5 D) 6 E) 7

Answer: D
Diff: 2 *Page Ref:* 1.7

45) How many significant figures are in 3.876×10^4 m?
A) 3 B) 4 C) 5 D) 7 E) 8

Answer: B
Diff: 2 *Page Ref:* 1.7

46) How many significant figures are in the measurement, 463.090 m?
A) 2 B) 3 C) 4 D) 5 E) 6

Answer: E
Diff: 2 *Page Ref:* 1.7

47) How many significant figures are in the measurement, 0.005890 g?
A) 4 B) 5 C) 6 D) 7 E) 8

Answer: A
Diff: 2 *Page Ref:* 1.7

48) The correct answer (reported to the proper number of significant figures) to the following is ________.

$11.5 \times 8.78 =$ ________

Answer: 101
Diff: 2 *Page Ref:* 1.7

49) Round the following number to four significant figures and express the result in standard exponential notation: 0.00222755
A) 0.2228×10^{-2}
B) 0.002228
C) 2.228×10^{3}
D) 2.228×10^{-3}
E) 22.28×10^{2}

Answer: D
Diff: 2 *Page Ref:* 1.7

50) Which of the following numbers has the greatest number of significant figures?
A) 0.5070 B) 0.201 C) 418000 D) 6.02×10^{24}

Answer: A
Diff: 2 *Page Ref:* 1.7

51) How many of the following numbers contain 3 significant figures?

0.408   9.040   0.0400 9.05×10^{24}

A) one   B) two   C) three   D) four

Answer: C

Diff: 2   *Page Ref: 1.7*

52) How many significant figures are there in the answer to the following problem?

$(9.992 \times 3.200) + 0.610 = ?$

A) one   B) two   C) three   D) four

Answer: C

Diff: 2   *Page Ref: 1.7*

53) How many significant figures are there in the answer for the following problem?

$56.4 + 0.8822 + 21 = ?$

A) one   B) two   C) three   D) four

Answer: B

Diff: 2   *Page Ref: 1.7*

54) How many significant figures are there in the answer for the following problem?

$$\frac{[(143.7 - 121) \times 2.06]}{0.600} = ?$$

A) one   B) two   C) three   D) four

Answer: B

Diff: 2   *Page Ref: 1.7*

55) An acetylene molecule contains 2 atoms of carbon. The number 2 represents how many significant figures?

A) one   B) two   C) three   D) infinite

Answer: D

Diff: 2   *Page Ref: 1.7*

56) Round off 00907506 to four significant figures.

A) 0091   B) 9076   C) 9100   D) 9.075×10^{5}

Answer: D

Diff: 2   *Page Ref: 1.7*

57) The width, length, and height of a large, custom-made shipping crate are 1.22 m, 3.22 m, and 0.83 m, respectively. The volume of the box using the correct number of significant figures is ________ m^3.

A) 3.26057   B) 3.3   C) 3.26   D) 3.261   E) 3.2606

Answer: B

Diff: 2   *Page Ref: 1.7*

58) The correct answer (reported to the proper number of significant figures) to the following is ________.

$(1815 - 1806) \times (9.11 \times 7.92) =$ ________

Answer: 600
Diff: 4 *Page Ref: 1.7*

59) Without using a calculator, solve the following problem:

$$\frac{[(1 \times 10^{4}) \times (1 \times 10^{3})]^{2}}{(1 \times 10^{-9})}$$

A) 1×10^{-4} B) 1×10^{5} C) 1×10^{23} D) 1×10^{32}

Answer: C
Diff: 1 *Page Ref: 1.8*

60) Without using a calculator, solve the following problem:

$$\frac{[(1 \times 10^{-8}) \times (1 \times 10^{5})]^{2}}{(1 \times 10^{6})}$$

A) 1×10^{6} B) 1×10^{0} C) 1×10^{-12} D) 1×10^{-18}

Answer: C
Diff: 1 *Page Ref: 1.8*

61) Which of the following is the greatest mass?
A) 1000 μg
B) 1.000×10^{-4} kg
C) 1.000×10^{-4} cg
D) 1.000×10^{-8} Mg

Answer: B
Diff: 2 *Page Ref: 1.8*

62) The mass of a proton is 1.67×10^{-27} kg. What is the mass of a proton in picograms?
A) 1.67×10^{-18} pg
B) 1.67×10^{-15} pg
C) 1.67×10^{-12} pg
D) 1.67×10^{-9} pg

Answer: C
Diff: 2 *Page Ref: 1.8*

63) The mass of a single zinc atom is 1.086×10^{-22} g. This is the same mass as
A) 1.086×10^{-16} mg.
B) 1.086×10^{-25} kg.
C) 1.086×10^{-28} μg.
D) 1.086×10^{-31} ng.

Answer: B
Diff: 2 *Page Ref: 1.8*

64) A student weighed 30.00 μg of sulfur in the lab. This is the same mass as
A) 3.000×10^{-8} g.
B) 3.000×10^{-5} kg.
C) 3.000×10^{-5} mg.
D) 3.000×10^{4} ng.

Answer: D
Diff: 2 *Page Ref: 1.8*

65) Convert 2 μm to meters.

A) 2×10^{-9} m B) 2×10^{-6} m C) 2×10^{-3} m D) 2×10^{6} m

Answer: B
Diff: 2 *Page Ref: 1.8*

66) The average distance between nitrogen and oxygen atoms is 115 pm in a compound called nitric oxide. What is this distance in millimeters?

A) 1.15×10^{-8} mm B) 1.15×10^{-7} mm
C) 1.15×10^{13} mm D) 1.15×10^{17} mm

Answer: B
Diff: 2 *Page Ref: 1.8*

67) The diameter of an atom is approximately 1×10^{-10} m. What is the diameter in millimeters?

A) 1×10^{-16} mm B) 1×10^{-13} mm C) 1×10^{-7} mm D) 1×10^{-4} mm

Answer: C
Diff: 2 *Page Ref: 1.8*

68) Which of the following volumes is equal to 40 mL?

A) 40 cm^3 B) 40 dm^3 C) 0.40 L D) 0.00040 kL

Answer: A
Diff: 2 *Page Ref: 1.8*

69) Convert 10 cm^3 to m^3.

A) 1×10^{-5} m^3 B) 1×10^{-1} m^3 C) 1×10^{3} m^3 D) 1×10^{7} m^3

Answer: A
Diff: 2 *Page Ref: 1.8*

70) Convert 35 m^3 to liters.

A) 3.5×10^{-2} L B) 3.5 L C) 3.5×10^{2} L D) 3.5×10^{4} L

Answer: D
Diff: 2 *Page Ref: 1.8*

71) What wavelength of light would you report in units of nm, if the light had a wavelength of 9.90×10^{-10} m?

A) 9.90×10^{-3} nm
B) 9.90×10^{-19} nm
C) 9.90 nm
D) 0.990 nm
E) 990 nm

Answer: D
Diff: 2 *Page Ref: 1.8*

72) How many mg does a 690 kg sample contain?

A) 6.90×10^{-4} mg
B) 6.90×10^{7} mg
C) 6.90×10^{-3} mg
D) 6.90×10^{6} mg
E) 6.90×10^{8} mg

Answer: E
Diff: 2 *Page Ref: 1.8*

73) How many kL does a 8.55×10^{8} cL sample contain?

A) 8.55×10^{3} kL
B) 8.55×10^{13} kL
C) 8.55×10^{4} kL
D) 8.55×10^{12} kL
E) 8.55×10^{2} kL

Answer: A
Diff: 2 *Page Ref: 1.8*

74) How many cm^3 are contained in 3.77×10^{4} mm^3?

A) 3.77×10^{4} cm^3
B) 3.77×10^{1} cm^3
C) 3.77×10^{-10} cm^3
D) 3.77×10^{20} cm^3
E) 3.77×10^{6} cm^3

Answer: B
Diff: 2 *Page Ref: 1.8*

75) How many mL are in 7.59 L?

A) 7.59×10^{-3} mL
B) 7.59×10^{1} mL
C) 7.59×10^{3} mL
D) 7.59×10^{-1} mL
E) 7.59×10^{2} mL

Answer: C
Diff: 2 *Page Ref: 1.8*

76) 38.325 lb = ________ grams. (1 lb = 454 g)

Answer: 17400
Diff: 4 *Page Ref: 1.8*

77) If 1.4% of the mass of a human body is calcium, how many kilograms of calcium are there in a 165-pound man?

A) 1.0 kg Ca
B) 5.1 kg Ca
C) 1.0×10^{2} kg Ca
D) 5.1×10^{2} kg

Answer: A
Diff: 5 *Page Ref: 1.8*

78) A fishing boat accidentally spills 3.0 barrels of diesel oil into the ocean. Each barrel contains 42 gallons. If the oil film on the ocean is 2.5×10^2 nm thick, how many square meters will the oil slick cover?

A) 1.9×10^{-3} m^2 B) 1.9×10^6 m^2 C) 1.9×10^7 m^2 D) none of these

Answer: B
Diff: 5 *Page Ref: 1.8*

79) Because of the high heat and low humidity in the summer in Death Valley, California, a visitor requires about one quart of water for every two miles traveled on foot. Calculate the approximate number of liters required for a person to walk 10. kilometers in Death Valley.

A) 2.9 L B) 12 L C) 30 L D) 47 L

Answer: A
Diff: 5 *Page Ref: 1.8*

80) The estimated costs for remodelling the interior of an apartment are: three 1-gallon cans of paint at $13.22 each, two paint brushes at $9.53 each, and $135 for a helper. The total estimated cost with the appropriate significant figures is $________.

A) 193.72 B) 1.9×10^2 C) 194 D) 2×10^2 E) 193.7

Answer: C
Diff: 5 *Page Ref: 1.8*

81) How many liters of wine can be held in a wine barrel whose capacity is 22.0 gal?
1 gal = 4 qt = 3.7854 L.

A) 1.72×10^{-4}
B) 0.172
C) 83.3
D) 5.81×10^3
E) 5.81

Answer: C
Diff: 5 *Page Ref: 1.8*

82) The recommended adult dose of Elixophyllin®, a drug used to treat asthma, is 6.00 mg/kg of body mass. Calculate the dose in milligrams for a 115-lb person. 1 lb = 453.59 g.

A) 24 B) 1,521 C) 1.5 D) 313 E) 3.1×10^5

Answer: D
Diff: 5 *Page Ref: 1.8*

83) The density of air under ordinary conditions at 25°C is 1.19 g/L. How many kilograms of air are in a room that measures 11.0 ft × 11.0 ft and has an 10.0 ft ceiling? 1 in. = 2.54 cm (exactly); 1 L = 10^3 cm^3

A) 3.66 B) 0.152 C) 4.08×10^4 D) 0.0962 E) 40.8

Answer: E
Diff: 5 *Page Ref: 1.8*

84) How many liters of air are in a room that measures 10.0 ft × 11.0 ft and has an 8.00 ft ceiling?
1 in. = 2.54 cm (exactly); 1 L = 10^3 cm^3
A) 2.49×10^4
B) 92.8
C) 26.8
D) 2.68×10^7
E) 8.84×10^5
Answer: A
Diff: 5 *Page Ref: 1.8*

85) A recipe requires 1.66 liters of milk for a soup base. How many quarts are needed?
A) 1.57 qts B) 1.75 qts C) 1.66 qts D) 3.65 qts E) 0.754 qts
Answer: B
Diff: 2 *Page Ref: 1.8*

86) If a room requires 25.4 square yards of carpeting, what is the area of the floor in units of ft^2?
(3 ft = 1 yd)
A) 76.2 ft^2 B) 8.47 ft^2 C) 152.4 ft^2 D) 229 ft^2 E) 64.5 ft^2
Answer: D
Diff: 3 *Page Ref: 1.8*

87) A person is 68.00 inches tall. How tall is she in cm?
A) 172.7 cm B) 26.77 cm C) 27.20 cm D) 204.0 cm E) 136.0 cm
Answer: A
Diff: 3 *Page Ref: 1.8*

88) A person weighs 77.1 kg. What is his weight in pounds?
A) 154 pounds
B) 170 pounds
C) 35.0 pounds
D) 30.8 pounds
E) 193 pounds
Answer: B
Diff: 3 *Page Ref: 1.8*

89) A person is 2.000 yards tall. How tall is he in cm?
A) 182.9 cm
B) 28.35 cm
C) 0.1411 cm
D) 5.080 cm
E) 15.24 cm
Answer: A
Diff: 3 *Page Ref: 1.8*

90) An alligator is 152.4 cm long. How long is he in feet?
A) 5.00 feet B) 60.0 ft C) 12.7 ft D) 32.2 ft E) 720.0 ft
Answer: A
Diff: 3 *Page Ref: 1.8*

91) How many cm are in 23.2 ft?
A) 9.14 cm B) 278 cm C) 58.9 cm D) 707 cm E) 0.761 cm

Answer: D
Diff: 3 *Page Ref: 1.8*

Matching Questions

Match the following.

1) kilo
Diff: 1 *Page Ref: 1.6*

2) centi
Diff: 1 *Page Ref: 1.6*

3) milli
Diff: 1 *Page Ref: 1.6*

4) nano
Diff: 1 *Page Ref: 1.6*

5) micro
Diff: 1 *Page Ref: 1.6*

6) deci
Diff: 1 *Page Ref: 1.6*

7) mega
Diff: 1 *Page Ref: 1.6*

A) 10^{-2}
B) 10^{6}
C) 10^{-1}
D) 10^{-6}
E) 10^{-3}
F) 10^{-9}
G) 10^{3}

Answers: 1) G 2) A 3) E 4) F 5) D
6) C 7) B

Short Answer Questions

1) Define an atom.
Answer: An atom is the submicroscopic particle that constitutes the fundamental building block of ordinary matter.
Diff: 1 *Page Ref: 1.1*

2) Define matter.
Answer: Matter is anything that occupies space and has mass.
Diff: 1 *Page Ref: 1.3*

3) A sample of liquid isopropyl alcohol is placed in a sealed container. Some of the volatile isopropyl alcohol vaporizes. Does the mass of the sealed container and its contents change during the vaporization? Explain.

Answer: No. The vaporized isopropyl alcohol is just in a different physical state. It still has mass and therefore the gas plus the remaining liquid and container have the same total mass after the vaporization of some of the isopropyl alcohol.
Diff: 1 *Page Ref: 1.4*

4) What is the difference between a physical property and a chemical property? Give an example of each.

Answer: A physical property is something that can be observed without changing the chemical identity of the substance, such as color or scent. A chemical property can only be observed while the chemical identity of a substance is changing, such as sodium metals tendency to react with water to form hydrogen gas and sodium hydroxide.
Diff: 1 *Page Ref: 1.4*

5) Define energy.

Answer: Energy is the capacity to do work.
Diff: 1 *Page Ref: 1.5*

6) Define the law of the conservation of energy.

Answer: Energy is neither created or destroyed.
Diff: 1 *Page Ref: 1.5*

7) A flash drive contains 4 gigabytes. How many bytes does it contain?

Answer: 4,000,000,000 bytes or 4,292,967,296 bytes, if someone is computer literate
Diff: 1 *Page Ref: 1.6*

8) Describe the difference between an intensive and extensive property using examples.

Answer: An intensive property does NOT depend on the amount of the substance present, such as color or density. An extensive property is one that does depend on the amount of the substance, such as mass or volume.
Diff: 1 *Page Ref: 1.6*

9) What happens to the density of a sample of iron metal as it is heated from room temperature to 100°C? (This is below the melting point of iron.)

Answer: Since the mass of the iron stays constant, but the volume increases as the temperature is raised, the density of the iron decreases upon heating.
Diff: 1 *Page Ref: 1.6*

10) What does it mean to be an exact number? Give an example of an exact number.

Answer: An exact number has an infinite number of significant figures even though we typically don't write many of them out. If there are 26 people in a classroom, there are exactly 26.00000.... people in that room. There is no possibility of a half person, so this is an exact whole number with no ambiguity.
Diff: 1 *Page Ref: 1.7*

11) Define random error.

Answer: Random error has an equal probability of being too high or too low.
Diff: 1 *Page Ref: 1.7*

Chapter 2 Atoms and Elements

Multiple Choice Questions

1) In a chemical reaction, matter is neither created or destroyed. Which law does this refer to?
   A) Law of Definite Proportions
   B) Law of the Conservation of Mass
   C) Law of Modern Atomic Theory
   D) Law of Multiple Proportions
   E) First Law of Thermodynamics

Answer: B
Diff: 1 *Page Ref: 2.3*

2) All samples of a given compound, regardless of their source or how they were prepared, have the same proportions of their constituent elements. Which law does this refer to?
   A) Law of Definite Proportions
   B) Law of the Conservation of Mass
   C) Law of Modern Atomic Theory
   D) Law of Multiple Proportions
   E) First Law of Thermodynamics

Answer: A
Diff: 1 *Page Ref: 2.3*

3) When two elements form two different compounds, the masses of element B that combine with 1 g of element A can be expressed as a ratio of small whole numbers. Which law does this refer to?
   A) Law of Definite Proportions
   B) Law of the Conservation of Mass
   C) Law of Modern Atomic Theory
   D) Law of Multiple Proportions
   E) First Law of Thermodynamics

Answer: D
Diff: 1 *Page Ref: 2.3*

4) Which of the following is an example of the law of multiple proportions?
   A) A sample of chlorine is found to contain three times as much Cl–35 as Cl–37.
   B) Two different compounds formed from carbon and oxygen have the following mass ratios:
   1.33 g O: 1 g C and 2.66 g O: 1 g C.
   C) Two different samples of table salt are found to have the same ratio of sodium to chlorine.
   D) The atomic mass of bromine is found to be 79.90 amu.
   E) Nitrogen dioxide always has a mass ratio of 2.28 g O: 1 g N.

Answer: B
Diff: 1 *Page Ref: 2.3*

5) Which of the following statements is FALSE according to Dalton's Atomic Theory?
A) Atoms combine in simple whole number ratios to form compounds.
B) All atoms of chlorine have identical properties that distinguish them from other elements.
C) One carbon atom will combine with one oxygen atom to form a molecule of carbon monoxide.
D) Atoms of sodium do not change into another element during chemical reaction with chlorine.
E) An atom of nitrogen can be broken down into smaller particles that will still have the unique properties of nitrogen.

Answer: E
Diff: 1 *Page Ref: 2.3*

6) Identify the description of an atom.
A) neutrons and electrons in nucleus; protons in orbitals
B) neutrons in nucleus; protons and electrons in orbitals
C) protons and neutrons in nucleus; electrons in orbitals
D) protons and electrons in nucleus; neutrons in orbitals
E) electrons in nucleus; protons and neutrons in orbitals

Answer: C
Diff: 1 *Page Ref: 2.5*

7) Identify the charges of the protons, neutrons, and electrons.
A) protons +1, neutrons 0, electrons −1
B) protons 0, neutrons −1, electrons +1
C) protons −1, neutrons 0, electrons +1
D) protons 0, neutrons +1, electrons −1
E) protons +1, neutrons −1, electrons 0

Answer: A
Diff: 1 *Page Ref: 2.6*

8) The mass number is equal to
A) the sum of the number of the electrons and protons.
B) the sum of the number of the neutrons and electrons.
C) the sum of the number of protons, neutrons, and electrons.
D) the sum of the number of protons and neutrons.

Answer: D
Diff: 1 *Page Ref: 2.6*

9) What does "X" represent in the following symbol?

$^{80}_{35}X$

A) mercury
B) chlorine
C) scandium
D) bromine
E) selenium

Answer: D
Diff: 1 *Page Ref: 2.6*

10) What does "X" represent in the following symbol?

$^{28}_{14}X$

A) silicon
B) sulfur
C) zinc
D) ruthenium
E) nickel

Answer: A
Diff: 1 *Page Ref: 2.6*

11) Determine the number of protons, neutrons and electrons in the following:

$^{40}_{18}X$

A) $p^+ = 18$ $n° = 18$ $e^- = 22$
B) $p^+ = 18$ $n° = 22$ $e^- = 18$
C) $p^+ = 22$ $n° = 18$ $e^- = 18$
D) $p^+ = 18$ $n° = 22$ $e^- = 40$
E) $p^+ = 40$ $n° = 22$ $e^- = 18$

Answer: B
Diff: 1 *Page Ref: 2.6*

12) Determine the number of protons, neutrons and electrons in the following:

$^{25}_{12}X$

A) $p^+ = 12$ $n° = 25$ $e^- = 12$
B) $p^+ = 12$ $n° = 12$ $e^- = 13$
C) $p^+ = 12$ $n° = 13$ $e^- = 12$
D) $p^+ = 25$ $n° = 12$ $e^- = 13$
E) $p^+ = 12$ $n° = 13$ $e^- = 25$

Answer: C
Diff: 1 *Page Ref: 2.6*

13) Determine the number of protons, neutrons and electrons in the following:

$^{65}_{29}X$

A) $p^+ = 36$ $n° = 29$ $e^- = 36$
B) $p^+ = 29$ $n° = 29$ $e^- = 36$
C) $p^+ = 36$ $n° = 36$ $e^- = 29$
D) $p^+ = 29$ $n° = 36$ $e^- = 29$
E) $p^+ = 29$ $n° = 36$ $e^- = 36$

Answer: D
Diff: 1 *Page Ref: 2.6*

14) What element is defined by the following information?

$p^+ = 11$ $n° = 12$ $e^- = 11$

A) sodium   B) vanadium   C) magnesium   D) titanium

Answer: A
Diff: 1   *Page Ref: 2.6*

15) What element is defined by the following information?

$p^+ = 20$ $n° = 20$ $e^- = 20$

A) zirconium
B) calcium
C) potassium
D) neon
E) argon

Answer: B
Diff: 1   *Page Ref: 2.6*

16) What element is defined by the following information?
$p^+ = 17$ $n° = 20$ $e^- = 17$

A) calcium   B) rubidium   C) chlorine   D) neon   E) oxygen

Answer: C
Diff: 1   *Page Ref: 2.6*

17) Which of the following statements about subatomic particles is TRUE?
A) A neutral atom contains the same number of protons and electrons.
B) Protons have about the same mass as electrons.
C) Electrons make up most of the mass of an atom.
D) Protons and neutrons have opposite, but equal in magnitude, charges.
E) Neutrons and electrons are found in the nucleus of an atom.

Answer: A
Diff: 1   *Page Ref: 2.6*

18) Which of the following statements about isotopes is TRUE?
A) Isotopes of the same element differ only in the number of electrons they contain.
B) An isotope of an atom with a larger number of neutrons is larger than an isotope of the same atom that contains fewer neutrons.
C) Isotopes of the same element have the same mass.
D) Isotopes of the same element don't usually have the same properties.
E) Some elements have 3 or more naturally occurring isotopes.

Answer: E
Diff: 1   *Page Ref: 2.6*

19) Give the symbol for silver.
A) S   B) Si   C) Ar   D) Ag   E) Sl

Answer: D
Diff: 2   *Page Ref: 2.6*

20) Ions differ in the number of
A) electrons.
B) neutrons.
C) protons.
D) neutrons and protons.
E) electrons and protons.
Answer: A
Diff: 2 *Page Ref:* 2.6

21) What species is represented by the following information?

$p^+ = 12$ $n^\circ = 14$ $e^- = 10$

A) Si^{4+} B) Mg C) Ne D) Si E) Mg^{2+}
Answer: E
Diff: 2 *Page Ref:* 2.6

22) What species is represented by the following information?

$p^+ = 47$ $n^\circ = 62$ $e^- = 46$

A) Ag^+ B) Nd C) Pd D) Ag E) Pd^+
Answer: A
Diff: 2 *Page Ref:* 2.6

23) What species is represented by the following information?

$p^+ = 17$ $n^\circ = 18$ $e^- = 18$

A) Cl B) Cl^- C) Ar D) Ar^+ E) Kr
Answer: B
Diff: 2 *Page Ref:* 2.6

24) Identify the largest atom or ion of carbon.
A) $p^+ = 6$ $n^\circ = 6$ $e^- = 6$
B) $p^+ = 6$ $n^\circ = 7$ $e^- = 6$
C) $p^+ = 6$ $n^\circ = 6$ $e^- = 7$
D) $p^+ = 6$ $n^\circ = 6$ $e^- = 5$
Answer: C
Diff: 2 *Page Ref:* 2.6

25) Predict the charge that an aluminum ion would have.
A) 5– B) 1+ C) 1– D) 2+ E) 3+
Answer: E
Diff: 2 *Page Ref:* 2.7

26) Predict the charge that a calcium ion would have.
A) 6– B) 2– C) 3+ D) 2+ E) 1+
Answer: D
Diff: 2 *Page Ref:* 2.7

27) Predict the charge that an ion formed from sulfur would have.
A) 1-　B) 6+　C) 3-　D) 4+　E) 2-

Answer: E
Diff: 2　*Page Ref:* 2.7

28) Predict the charge that an ion formed from sodium would have.
A) 1-　B) 6+　C) 3-　D) 1+　E) 2-

Answer: D
Diff: 2　*Page Ref:* 2.7

29) Which of the following elements is NOT a metal?
A) Ba　B) Mg　C) Xe　D) Pb　E) Ga

Answer: C
Diff: 1　*Page Ref:* 2.7

30) Which of the following elements is a lanthanide?
A) Ce　B) Na　C) U　D) Ar　E) Rg

Answer: A
Diff: 1　*Page Ref:* 2.7

31) Which of the following elements is a actinide?
A) Ce　B) Na　C) U　D) Ar　E) Rg

Answer: C
Diff: 1　*Page Ref:* 2.7

32) Which of the following statements is FALSE?
A) Halogens are very reactive elements.
B) The alkali metals are fairly unreactive.
C) Sulfur is a main group element.
D) Noble gases do not usually form ions.
E) Zn is a transition metal.

Answer: B
Diff: 1　*Page Ref:* 2.7

33) Which of the following does NOT describe a metal?
A) good conductor of heat
B) good conductor of electricity
C) tends to gain electrons
D) forms ionic compounds with nonmetals
E) found on the left side of the periodic table.

Answer: C
Diff: 1　*Page Ref:* 2.7

34) Which of the following does NOT describe a nonmetal?
A) tend to gain electrons
B) found in the upper right hand corner of the periodic table
C) poor conductor of electricity
D) nonmetals are generally unreactive
E) poor conductor of heat

Answer: D
Diff: 1 *Page Ref: 2.7*

35) Semiconductors are
A) metalloids. B) noble gases. C) nonmetals. D) metals.

Answer: A
Diff: 1 *Page Ref: 2.7*

36) Which of the following statements is FALSE?
A) Anions are usually larger than their corresponding atom.
B) Metals tend to form cations.
C) Atoms are usually larger than their corresponding cation.
D) The halogens tend to form 1+ ions.
E) Nonmetals tend to gain electrons.

Answer: D
Diff: 1 *Page Ref: 2.7*

37) Identify the green–yellowish gas that is used as a disinfecting agent.
A) chlorine B) bromine C) iodine D) fluorine

Answer: A
Diff: 1 *Page Ref: 2.7*

38) Identify the instrument that is used to determine the mass of a molecule.
A) mass spectrometer
B) nuclear magnetic resonance spectrometer
C) infrared spectrometer
D) gas chromatograph
E) ultraviolet spectrophotometer

Answer: A
Diff: 1 *Page Ref: 2.8*

39) The atomic mass for cadmium is
A) 48 B) 112.41 C) 40.08 D) 20

Answer: B
Diff: 1 *Page Ref: 2.8*

40) Calculate the atomic mass of silver if silver has 2 naturally occurring isotopes with the following masses and natural abundances:

Ag–107	106.90509 amu	51.84%
Ag–109	108.90476 amu	48.46%

A) 107.90 amu
B) 108.00 amu
C) 107.79 amu
D) 108.32 amu
E) 108.19 amu

Answer: E
Diff: 2 *Page Ref: 2.8*

41) Calculate the atomic mass of gallium if gallium has 2 naturally occurring isotopes with the following masses and natural abundances:

Ga–69	68.9256 amu	60.11%
Ga–71	70.9247 amu	39.89%

A) 69.72 amu
B) 69.93 amu
C) 70.00 amu
D) 69.80 amu
E) 70.68 amu

Answer: A
Diff: 2 *Page Ref: 2.8*

42) Silver has an atomic mass of 107.868 amu. The Ag–109 isotope (108.905 amu) is 48.16%. What is the amu of the other isotope?
A) 106.905 amu B) 106.908 amu C) 106.903 amu D) 106.911 amu

Answer: A
Diff: 3 *Page Ref: 2.8*

43) Gallium has an atomic mass of 69.723 amu. The Ga–69 (68.926 amu) is 60.11%. What is the amu of the other isotope?
A) 70.924 amu B) 70.928 amu C) 70.932 amu D) 70.920 amu

Answer: A
Diff: 3 *Page Ref: 2.8*

44) Calculate the atomic mass of element "X", if it has 2 naturally occurring isotopes with the following masses and natural abundances:

X-45	44.8776 amu	32.88%
X-47	46.9443 amu	67.12%

A) 46.26 amu
B) 45.91 amu
C) 46.34 amu
D) 46.84 amu
E) 44.99 amu

Answer: A
Diff: 3 *Page Ref: 2.8*

45) What mass (in mg) does 2.63 moles of nickel have?
A) 44.8 mg
B) 2.23×10^4 mg
C) 129 mg
D) 3.56×10^5 mg
E) 1.54×10^5 mg

Answer: E
Diff: 3 *Page Ref: 2.9*

46) How many moles of Kr are contained in 398 mg of Kr?
A) 4.75×10^{-3} moles Kr
B) 33.4 moles Kr
C) 2.11×10^{-4} moles Kr
D) 2.99×10^{-3} moles Kr
E) 1.19×10^{-4} moles Kr

Answer: A
Diff: 3 *Page Ref: 2.9*

47) How many moles of Cs are contained in 595 kg of Cs?
A) 2.23×10^2 moles Cs
B) 4.48×10^3 moles Cs
C) 7.91×10^4 moles Cs
D) 1.26×10^3 moles Cs
E) 5.39×10^2 moles Cs

Answer: B
Diff: 3 *Page Ref: 2.9*

48) How many iron atoms are contained in 354 g of iron?

A) 2.62×10^{25} Fe atoms
B) 2.13×10^{26} Fe atoms
C) 4.69×10^{24} Fe atoms
D) 3.82×10^{24} Fe atoms
E) 9.50×10^{22} Fe atoms

Answer: D
Diff: 3 *Page Ref: 2.9*

49) How many phosphorus atoms are contained in 158 kg of phosphorus?

A) 3.07×10^{27} phosphorus atoms
B) 2.95×10^{27} phosphorus atoms
C) 3.25×10^{28} phosphorus atoms
D) 1.18×10^{24} phosphorus atoms
E) 8.47×10^{24} phosphorus atoms

Answer: A
Diff: 3 *Page Ref: 2.9*

50) Calculate the mass (in kg) of 4.87×10^{25} atoms of Zn.

A) 5.29 kg B) 1.89 kg C) 8.09 kg D) 1.24 kg E) 1.09 kg

Answer: A
Diff: 4 *Page Ref: 2.9*

51) Calculate the mass (in ng) of 2.33×10^{20} atoms of oxygen.

A) 6.19×10^{6} ng
B) 1.62×10^{7} ng
C) 2.25×10^{3} ng
D) 3.73×10^{6} ng
E) 4.69×10^{7} ng

Answer: A
Diff: 4 *Page Ref: 2.9*

52) How many xenon atoms are contained in 2.36 moles of xenon?

A) 3.92×10^{24} xenon atoms
B) 2.55×10^{23} xenon atoms
C) 1.42×10^{24} xenon atoms
D) 7.91×10^{25} xenon atoms
E) 1.87×10^{26} xenon atoms

Answer: C
Diff: 2 *Page Ref: 2.9*

53) How many argon atoms are contained in 7.66×10^5 mmol of argon?

A) 4.61×10^{26} Ar atoms

B) 1.84×10^{28} Ar atoms

C) 1.15×10^{28} Ar atoms

D) 7.86×10^{20} Ar atoms

E) 3.24×10^{26} Ar atoms

Answer: A

Diff: 2 *Page Ref: 2.9*

Algorithmic Questions

1) Identify the element that has an atomic number of 15.

A) sulfur B) oxygen C) phosphorus D) silicon

Answer: C

Diff: 1 *Page Ref: 2.6*

2) Give the symbol for fluorine.

A) F B) Fl C) Fo D) Fu E) Fr

Answer: A

Diff: 2 *Page Ref: 2.6*

3) An atom of ^{131}I contains ________ protons.

A) 53 B) 184 C) 78 D) 124 E) 131

Answer: A

Diff: 1 *Page Ref: 2.6*

4) An atom of ^{131}Xe contains ________ electrons.

A) 131 B) 185 C) 77 D) 123 E) 54

Answer: E

Diff: 1 *Page Ref: 2.6*

5) The atomic number of an atom of ^{80}Br is ________.

A) 115 B) 35 C) 45 D) 73 E) 80

Answer: B

Diff: 1 *Page Ref: 2.6*

6) How many electrons are in nickel?

A) 28 B) 30 C) 31 D) 30.7 E) 58.7

Answer: A

Diff: 1 *Page Ref: 2.6*

7) How many neutrons are in nickel?

A) 28 B) 30 C) 31 D) 30.7 E) 58.7

Answer: C

Diff: 1 *Page Ref: 2.6*

8) How many protons are in magnesium?

A) 12 B) 13 C) 14 D) 12.3 E) 24.3

Answer: A
Diff: 1 *Page Ref: 2.6*

9) An ion has 8 protons, 9 neutrons, and 10 electrons. The symbol for the ion is ________.

A) $^{17}O^{2-}$ B) $^{17}O^{2+}$ C) $^{19}F^{+}$ D) $^{19}F^{-}$ E) $^{17}Ne^{2+}$

Answer: A
Diff: 1 *Page Ref: 2.6*

10) Isotopes differ in the number of

A) beta particles
B) protons
C) electrons
D) neutrons
E) neutrons and protons

Answer: D
Diff: 1 *Page Ref: 2.6*

11) Identify a cation.

A) An atom that has lost an electron.
B) An atom that has gained an electron.
C) An atom that has lost a proton and a neutron.
D) An atom that has gained a neutron.

Answer: A
Diff: 2 *Page Ref: 2.6*

12) Identify an anion.

A) An atom that has lost an electron.
B) An atom that has gained an electron.
C) An atom that has lost a neutron and a proton.
D) An atom that has gained a neutron.

Answer: B
Diff: 2 *Page Ref: 2.6*

13) How many electrons does the Al^{3+} ion possess?

A) 16 B) 10 C) 6 D) 0 E) 13

Answer: B
Diff: 1 *Page Ref: 2.6*

14) How many protons does the Br^{-} ion possess?

A) 34 B) 36 C) 6 D) 8 E) 35

Answer: E
Diff: 1 *Page Ref: 2.6*

15) What is the chemical symbol for titanium?

A) Th B) Ti C) Tl D) Tm

Answer: B
Diff: 2 *Page Ref: 2.6*

16) What is the chemical symbol for mercury?

A) Ag B) Au C) Hg D) Pb

Answer: C
Diff: 2 *Page Ref: 2.6*

17) What is the chemical symbol for copper?

A) Co B) Cr C) Cu D) C

Answer: C
Diff: 2 *Page Ref: 2.6*

18) Which element has the chemical symbol, Ru?

A) rubidium B) ruthenium C) rutherfordium D) rhodium

Answer: B
Diff: 2 *Page Ref: 2.6*

19) Which element has the chemical symbol, S?

A) selenium B) silicon C) sulfur D) scandium

Answer: C
Diff: 2 *Page Ref: 2.6*

20) Which are isotopes? An atom that has an atomic number of 20 and a mass number of 42 is an isotope of an atom that has

A) an atomic number of 21 and a mass number of 42.
B) an atomic number of 20 and a mass number of 40.
C) 22 neutrons and 20 protons.
D) 22 protons and 20 neutrons.

Answer: B
Diff: 2 *Page Ref: 2.6*

21) Which of the following represent isotopes?

A: ${}^{32}_{15}X$ B: ${}^{32}_{16}X$ C: ${}^{31}_{15}X$ D: ${}^{34}_{17}X$

A) A and B B) A and C C) A and D D) C and D

Answer: B
Diff: 2 *Page Ref: 2.6*

22) What does "X" represent in the following symbol?

${}^{81}_{35}X$

A) nickel B) krypton C) mercury D) thallium E) bromine

Answer: E
Diff: 1 *Page Ref: 2.6*

23) How many protons (p) and neutrons (n) are in an atom of ${}^{90}_{38}Sr$?

A) 38 p, 52 n B) 38 p, 90 n C) 52 p, 38 n D) 90 p, 38 n

Answer: A
Diff: 2 *Page Ref: 2.6*

24) How many protons (p) and neutrons (n) are in an atom of barium-130?

A) 56 p, 74 n B) 56 p, 130 n C) 74 p, 56 n D) 130 p, 56 n

Answer: A

Diff: 2 *Page Ref:* 2.6

25) What is the element symbol for an atom that has 5 protons and 6 neutrons?

A) B B) C C) H D) Na

Answer: A

Diff: 2 *Page Ref:* 2.6

26) How many electrons are in a neutral atom of bromine-81?

A) 1 B) 35 C) 36 D) 81

Answer: B

Diff: 2 *Page Ref:* 2.6

27) Identify the chemical symbol of element Q in $^{80}_{34}Q$.

A) Br B) Hg C) Pd D) Se

Answer: D

Diff: 2 *Page Ref:* 2.6

28) An atom of ^{118}Xe contains ________ neutrons.

A) 54 B) 172 C) 64 D) 110 E) 118

Answer: C

Diff: 2 *Page Ref:* 2.6

29) The mass number of an atom of ^{128}Xe is ________.

A) 54 B) 182 C) 74 D) 128 E) 120

Answer: D

Diff: 2 *Page Ref:* 2.6

30) What is the identity of element Q if the ion Q^{2+} contains 10 electrons?

A) C B) O C) Ne D) Mg

Answer: D

Diff: 2 *Page Ref:* 2.6

31) Give the number of neutrons in Al^{+3}.

A) 10 B) 16 C) 17 D) 13 E) 14

Answer: E

Diff: 2 *Page Ref:* 2.6

32) Give the number of electrons in P^{-3}.

A) 18 B) 12 C) 19 D) 15 E) 16

Answer: A

Diff: 2 *Page Ref:* 2.6

33) Give the number of protons in Na^{+1}.

A) 10 B) 13 C) 9 D) 11 E) 12

Answer: D

Diff: 2 *Page Ref: 2.6*

34) How many electrons are in the ion, Cu^{2+}?

A) 27 B) 29 C) 31 D) 64

Answer: A

Diff: 2 *Page Ref: 2.6*

35) How many electrons are in the ion, P^{3-}?

A) 12 B) 18 C) 28 D) 34

Answer: B

Diff: 2 *Page Ref: 2.6*

36) In which of the following sets do all species have the same number of electrons?

A) F^-, Ne, Mg^{2+} B) Ge, Se^{2-}, Br^- C) K^+, Rb^+, Cs^+ D) Br, Br^-, Br^+

Answer: A

Diff: 2 *Page Ref: 2.6*

37) In which of the following sets do all species have the same number of protons?

A) F^-, Ne, Mg^{2+} B) Ge, Se^{2-}, Br^- C) K^+, Rb^+, Cs^+ D) Br, Br^-, Br^+

Answer: D

Diff: 2 *Page Ref: 2.6*

38) Predict the charge that the ion formed from fluorine would have.

A) 1- B) 2+ C) 1+ D) 4+ E) 2-

Answer: A

Diff: 2 *Page Ref: 2.7*

39) Predict the charge of the most stable ion of bromine.

A) 2+ B) 1+ C) 3+ D) 1- E) 2-

Answer: D

Diff: 1 *Page Ref: 2.7*

40) Predict the charge of the most stable ion of potassium.

A) 3+ B) 1- C) 2+ D) 2- E) 1+

Answer: E

Diff: 1 *Page Ref: 2.7*

41) Which of the following elements is a metal?

A) As B) S C) Br D) Fe E) Kr

Answer: D

Diff: 1 *Page Ref: 2.7*

42) Which of the following elements is a metalloid?
A) As B) S C) Br D) Fe E) Kr

Answer: A
Diff: 1 *Page Ref: 2.7*

43) Which of the following elements is a noble gas?
A) As B) S C) Br D) Fe E) Kr

Answer: E
Diff: 1 *Page Ref: 2.7*

44) Which of the following elements is a halogen?
A) As B) S C) Br D) Fe E) Kr

Answer: C
Diff: 1 *Page Ref: 2.7*

45) Which of the following elements is a nonmetal?
A) Ce B) N C) Br D) K E) Be

Answer: B
Diff: 1 *Page Ref: 2.7*

46) Which of the following elements is a alkali metal?
A) Ce B) N C) Br D) K E) Be

Answer: D
Diff: 1 *Page Ref: 2.7*

47) Which of the following elements is an alkali earth metal?
A) Ce B) N C) Br D) K E) Be

Answer: E
Diff: 1 *Page Ref: 2.7*

48) Which of the following elements is a transition metal?
A) Ce B) N C) Br D) K E) Be

Answer: A
Diff: 1 *Page Ref: 2.7*

49) Cesium belongs to the ________ group of the periodic table.
A) alkali metal B) alkaline earth metal
C) halogen D) noble gas

Answer: A
Diff: 1 *Page Ref: 2.7*

50) Iodine belongs to the ________ group of the periodic table.
A) alkali metal B) alkaline earth metal
C) halogen D) noble gas

Answer: C
Diff: 1 *Page Ref: 2.7*

51) Argon belongs to the ________ group of the periodic table.
A) alkali metal
B) alkaline earth metal
C) halogen
D) noble gas

Answer: D
Diff: 1 *Page Ref: 2.7*

52) Barium belongs to the ________ group of the periodic table.
A) alkali metal
B) alkaline earth metal
C) halogen
D) noble gas

Answer: B
Diff: 1 *Page Ref: 2.7*

53) Which of the following elements has chemical properties similar to tellurium?
A) fluorine B) hydrogen C) nitrogen D) sulfur

Answer: D
Diff: 1 *Page Ref: 2.7*

54) Which of the following elements is a gas at room temperature?
A) bromine B) carbon C) helium D) sodium

Answer: C
Diff: 1 *Page Ref: 2.7*

55) Which of the following elements is **not** a solid at room temperature?
A) Ag B) Al C) Xe D) Fe

Answer: C
Diff: 1 *Page Ref: 2.7*

56) Which of the following elements is classified as a semimetal?
A) calcium B) germanium C) fluorine D) uranium

Answer: B
Diff: 1 *Page Ref: 2.7*

57) Which of the following elements is a good conductor of heat and electricity?
A) carbon B) iodine C) neon D) aluminum

Answer: D
Diff: 1 *Page Ref: 2.7*

58) Which one of the following elements is a **poor** conductor of heat and electricity?
A) copper B) fluorine C) iron D) lead

Answer: B
Diff: 1 *Page Ref: 2.7*

59) All of the following elements are nonmetals except
A) arsenic. B) nitrogen. C) helium. D) oxygen.

Answer: A
Diff: 1 *Page Ref: 2.7*

60) The atomic mass for potassium is
A) 30.97 B) 15 C) 39.10 D) 19

Answer: C
Diff: 1 Page Ref: 2.8

61) The atomic number for potassium is
A) 30.97 B) 15 C) 39.10 D) 19

Answer: D
Diff: 1 Page Ref: 2.8

62) Which of the following contains the MOST atoms? You shouldn't need to do a calculation here.
A) 10.0 g Mg
B) 10.0 g Ne
C) 10.0 g Ca
D) 10.0 g Rb
E) 10.0 g Cs

Answer: B
Diff: 1 Page Ref: 2.9

63) Which of the following contains the FEWEST atoms? You shouldn't need to do a calculation here.
A) 10.0 g Na B) 10.0 g Li C) 10.0 g Ar D) 10.0 g Kr E) 10.0 g Cs

Answer: E
Diff: 1 Page Ref: 2.9

64) How many Fe atoms are contained in 787 g of Fe?
A) 5.90×10^{25} Fe atoms
B) 7.09×10^{21} Fe atoms
C) 8.49×10^{24} Fe atoms
D) 4.27×10^{22} Fe atoms
E) 4.18×10^{24} Fe atoms

Answer: C
Diff: 3 Page Ref: 2.9

65) Calculate the mass (in g) of 2.1×10^{24} atoms of W.
A) 3.9×10^{2} g
B) 2.4×10^{2} g
C) 3.2×10^{2} g
D) 1.5×10^{2} g
E) 6.5×10^{2} g

Answer: E
Diff: 3 Page Ref: 2.9

66) How many silver atoms are contained in 3.75 moles of silver?

A) 1.23×10^{24} silver atoms
B) 2.26×10^{24} silver atoms
C) 1.61×10^{23} silver atoms
D) 2.44×10^{26} silver atoms
E) 6.50×10^{25} silver atoms

Answer: B
Diff: 2 *Page Ref: 2.9*

67) What mass (in g) does 3.99 moles of Kr have?

A) 334 g B) 476 g C) 211 g D) 240 g E) 144 g

Answer: A
Diff: 2 *Page Ref: 2.9*

68) How many moles of potassium are contained in 300 g of potassium?

A) 7.67 moles
B) 1.44 moles
C) 20.0 moles
D) 15.8 moles
E) 9.69 moles

Answer: A
Diff: 2 *Page Ref: 2.9*

69) How many moles are in 2.16×10^{24} atoms of silver?

A) 35.9 moles
B) 3.59 moles
C) 0.359 moles
D) 6.08 moles
E) 1.79 moles

Answer: B
Diff: 2 *Page Ref: 2.9*

70) How many atoms are in 2.50 moles of SO_2?

A) 4.52×10^{24} atoms
B) 1.52×10^{24} atoms
C) 5.02×10^{23} atoms
D) 3.01×10^{24} atoms
E) 7.53×10^{23} atoms

Answer: A
Diff: 3 *Page Ref: 2.9*

71) How many molecules are in 2.50 moles of SO_2?

A) 4.52×10^{24} atoms
B) 1.51×10^{24} atoms
C) 5.02×10^{23} atoms
D) 3.01×10^{24} atoms
E) 7.53×10^{23} atoms

Answer: B
Diff: 3 Page Ref: 2.9

72) How many atoms of nitrogen are in 2.50 moles of NO_2?

A) 4.52×10^{24} atoms
B) 1.51×10^{24} atoms
C) 5.02×10^{23} atoms
D) 3.01×10^{24} atoms
E) 7.53×10^{23} atoms

Answer: B
Diff: 3 Page Ref: 2.9

73) How many atoms of oxygen are in 2.50 moles of CO_2?

A) 4.52×10^{24} atoms
B) 1.51×10^{24} atoms
C) 5.02×10^{23} atoms
D) 3.01×10^{24} atoms
E) 7.53×10^{23} atoms

Answer: D
Diff: 3 Page Ref: 2.9

74) What mass (in kg) does 5.01 moles of iron have?

A) 0.352 kg B) 0.122 kg C) 0.820 kg D) 0.280 kg E) 0.632 kg

Answer: D
Diff: 3 Page Ref: 2.9

Matching Questions

Match the following.

1) magnesium
Diff: 1 *Page Ref: 2.6*

2) carbon
Diff: 1 *Page Ref: 2.6*

3) potassium
Diff: 1 *Page Ref: 2.6*

4) iron
Diff: 1 *Page Ref: 2.6*

5) silicon
Diff: 1 *Page Ref: 2.6*

A) Mg

B) Fe

C) K

D) Si

E) C

Answers: 1) A 2) E 3) C 4) B 5) D

Short Answer Questions

1) Describe an atom and what it is made up of according to modern atomic theory.

Answer: An atom is made up of a nucleus surrounded by electrons. The nucleus contains protons (positively charged particles) and neutrons (neutral particles) and is where most of the mass of an atom comes from, but is a tiny fraction of an atom's volume. The nucleus is surrounded by negatively charged electrons, the same number as there are protons in the nucleus. An atom is therefore neutral overall.
Diff: 2 *Page Ref: 2.3*

2) The atomic number is equal to the number of ________.

Answer: protons
Diff: 1 *Page Ref: 2.6*

3) Why do the isotopes of the same element have the same atomic size?

Answer: Isotopes only differ in the number of neutrons contained within the nucleus. Since the size of an atom is determined by the electrons, isotopes of the same element should be the same size.
Diff: 1 *Page Ref: 2.6*

4) Why doesn't a mass spectrum of silver have a peak at 107.9 amu?

Answer: The average atomic mass of silver is 107.9 amu, but there are no atoms of silver that weigh 107.9 amu. One isotope weighs more and another weighs less.
Diff: 1 *Page Ref: 2.6*

5) Are anions typically larger or smaller than their corresponding atom? Why?

Answer: Anions are larger than their corresponding atom because the anion contains more electrons than the atom. Since electrons repel one another AND determine the size of the atom or ion, adding electrons to the atom to form an anion makes it larger.

Diff: 1 *Page Ref: 2.6*

6) Give the name of the element whose symbol is Na.

Answer: sodium

Diff: 2 *Page Ref: 2.6*

7) Describe the difference between ions and isotopes.

Answer: Ions have the loss or gain of electrons; isotopes differ in the number of neutrons.

Diff: 2 *Page Ref: 2.6*

8) Give an example of an halogen.

Answer: F, Br, I, or Cl

Diff: 1 *Page Ref: 2.7*

9) What group of elements in the periodic table are the most unreactive and why?

Answer: The noble gases are the most unreactive since they do not combine with other elements to form compounds.

Diff: 1 *Page Ref: 2.7*

10) Why do elements in the same group tend to have similar chemical properties?

Answer: Since elements in the same group have the same number of valence electrons (similar electron configurations) they tend to have similar chemical reactivity, since chemical reactions typically involve valence electrons.

Diff: 1 *Page Ref: 2.7*

11) Give the name of the instrument that is used to measure masses of atoms and the percent abundance of isotopes.

Answer: mass spectrometer

Diff: 2 *Page Ref: 2.8*

12) The number 6.022×10^{23} is called ________.

Answer: Avogadro's number

Diff: 1 *Page Ref: 2.9*

Chapter 3 Molecules, Compounds, and Chemical Equations

Multiple Choice Questions

1) An ionic bond is best described as
   A) the sharing of electrons.
   B) the transfer of electrons from one atom to another.
   C) the attraction that holds the atoms together in a polyatomic ion.
   D) the attraction between 2 nonmetal atoms.
   E) the attraction between 2 metal atoms.

Answer: B
Diff: 1 *Page Ref: 3.2*

2) A covalent bond is best described as
   A) the sharing of electrons between atoms.
   B) the transfer of electrons.
   C) a bond between a metal and a nonmetal.
   D) a bond between a metal and a polyatomic ion.
   E) a bond between two polyatomic ions.

Answer: A
Diff: 1 *Page Ref: 3.2*

3) Give a possible molecular formula for C_3H_5ClO
   A) $C_6H_{10}ClO_2$
   B) $C_5H_{10}Cl_2O_2$
   C) $C_6H_{10}Cl_2O_2$
   D) $C_6H_{10}O_2$
   E) $C_6H_{12}Cl_2O_2$

Answer: C
Diff: 1 *Page Ref: 3.3*

4) What is the empirical formula for $Hg_2(NO_3)_2$?
   A) $Hg_2(NO_3)_2$
   B) $HgNO_3$
   C) $Hg(NO_3)_2$
   D) Hg_2NO_3
   E) $Hg_4(NO_3)_4$

Answer: B
Diff: 2 *Page Ref: 3.3*

5) Give the name for SnO.
   A) tin (I) oxide B) tin (II) oxide C) tin (III) oxide D) tin (IV) oxide

Answer: B
Diff: 2 *Page Ref: 3.5*

6) Determine the name for $TiCO_3$. Remember that titanium forms several ions.

A) titanium (II) carbonate
B) titanium carbide
C) titanium carbonite
D) titanium (II) carbonite
E) titanium (I) carbonate

Answer: A
Diff: 2 *Page Ref:* 3.5

7) Give the structure for sodium perchlorate.

A) $NaClO$ B) $NaClO_2$ C) $NaClO_3$ D) $NaClO_4$

Answer: D
Diff: 2 *Page Ref:* 3.5

8) Write the name for $Sn(SO_4)_2$. Remember that Sn forms several ions.

A) tin (I) sulfite
B) tin (IV) sulfate
C) tin sulfide
D) tin (II) sulfite
E) tin (I) sulfate

Answer: B
Diff: 2 *Page Ref:* 3.5

9) Give the name for $NaHSO_3$.

A) sodium sulfite
B) sodium sulfate
C) sodium hydrogen sulfate
D) sodium hydrogen sulfite
E) sodium sulfide

Answer: D
Diff: 2 *Page Ref:* 3.5

10) Give the name for $NaHSO_3$.

A) monosodium sulfite
B) monosodium sulfate
C) sodium bisulfate
D) sodium bisulfite
E) sodium bisulfide

Answer: D
Diff: 2 *Page Ref:* 3.5

11) Give the formula for calcium hydrogen sulfate.

A) $CaHSO_4$ B) $Ca_2(HSO_4)_2$ C) Ca_2HSO_4 D) $Ca(HSO_4)_2$

Answer: D
Diff: 2 *Page Ref:* 3.5

12) Give the formula for calcium bisulfate.

A) $CaHSO_4$ B) $Ca_2(HSO_4)_2$ C) Ca_2HSO_4 D) $Ca(HSO_4)_2$

Answer: D
Diff: 2 *Page Ref: 3.5*

13) Give the name for $KMnO_4$.

A) potassium manganese tetraoxide
B) potassium manganate
C) potassium permanganate
D) potassium permagnesium
E) potassium magnesate

Answer: C
Diff: 2 *Page Ref: 3.5*

14) Determine the name for $CoCl_2{\cdot}6H_2O$. Remember that Co forms several ions.

A) cobalt chloride hydrate
B) cobalt (I) chloride heptahydrate
C) cobalt (II) chloride heptahydrate
D) cobalt (II) chloride hexahydrate
E) cobalt (I) chloride

Answer: D
Diff: 2 *Page Ref: 3.5*

15) Give the correct formula for aluminum sulfate.

A) Al_2SO_4 B) $Al(SO_4)_3$ C) $Al_3(SO_4)_2$ D) $Al_2(SO_4)_3$

Answer: D
Diff: 3 *Page Ref: 3.5*

16) Write the formula for copper (II) sulfate pentahydrate.

A) $Cu_2SO_3{\cdot}H_5$
B) $Cu_2S{\cdot}H_2O$
C) $CuS{\cdot}5H_2O$
D) $(CuSO_4)_5$
E) $CuSO_4{\cdot}5H_2O$

Answer: E
Diff: 3 *Page Ref: 3.5*

17) Determine the name for H_2CO_3.

A) carbonous acid
B) dihydrogen carbonate
C) carbonic acid
D) hydrocarbonic acid
E) hydrocarbide acid

Answer: C
Diff: 2 *Page Ref: 3.6*

18) Give the formula for sulfurous acid.

A) H_2SO_3 B) HSO_3 C) H_2SO_4 D) HSO_4

Answer: A

Diff: 2 *Page Ref: 3.6*

19) Give the name for H_2SO_4.

A) sulfuric acid
B) persulfurous acid
C) sulfurous acid
D) hyposulfurous acid
E) persulfuric acid

Answer: A

Diff: 2 *Page Ref: 3.6*

20) Give the name for HNO_3.

A) nitric acid
B) nitrous acid
C) hydrogen nitrate
D) hydrogen nitrite
E) hydrogen nitride

Answer: A

Diff: 2 *Page Ref: 3.6*

21) Determine the name for P_4O_{10}.

A) phosphorus (IV) oxide
B) diphosphorus pentoxide
C) phosphorus oxide
D) phosphorus (II) oxide
E) tetraphosphorus decoxide

Answer: E

Diff: 3 *Page Ref: 3.6*

22) Determine the name for N_2O_5.

A) dinitrogen pentoxide
B) nitrogen oxide
C) nitrogen (IV) oxide
D) nitrogen (II) oxide
E) nitrogen tetroxide

Answer: A

Diff: 3 *Page Ref: 3.6*

23) Calculate the molar mass for $Mg(ClO_4)_2$.

A) 223.21 g/mol
B) 123.76 g/mol
C) 119.52 g/mol
D) 247.52 g/mol
E) 75.76 g/mol

Answer: A

Diff: 2 *Page Ref: 3.8*

24) Calculate the molar mass of $Al(C_2H_3O_2)_3$.

A) 86.03 g/mol
B) 204.13 g/mol
C) 56.00 g/mol
D) 258.09 g/mol
E) 139.99 g/mol

Answer: B
Diff: 2 *Page Ref: 3.8*

25) Calculate the molar mass of $Ca_3(PO_4)_2$.

A) 87.05 g/mol
B) 215.21 g/mol
C) 310.18 g/mol
D) 279.21 g/mol
E) 246.18 g/mol

Answer: C
Diff: 2 *Page Ref: 3.8*

26) Calculate the molar mass of H_2CO_3.

A) 62.03 g/mol
B) 29.02 g/mol
C) 61.02 g/mol
D) 60.01 g/mol
E) 74.04 g/mol

Answer: A
Diff: 2 *Page Ref: 3.8*

27) The molecular weight of urea ($(NH_2)_2CO$), a compound used as a nitrogen fertilizer, is ________ amu (rounded to one decimal place).

A) 44.0 B) 43.0 C) 60.1 D) 8.0 E) 32.0

Answer: C
Diff: 3 *Page Ref: 3.8*

28) How many millimoles of $Ca(NO_3)_2$ contain 4.78×10^{22} formula units of $Ca(NO_3)_2$? The molar mass of $Ca(NO_3)_2$ is 164.10 g/mol.

A) 12.6 mmol $Ca(NO_3)_2$
B) 13.0 mmol $Ca(NO_3)_2$
C) 20.7 mmol $Ca(NO_3)_2$
D) 79.4 mmol $Ca(NO_3)_2$
E) 57.0 mmol $Ca(NO_3)_2$

Answer: D
Diff: 2 *Page Ref: 3.8*

29) How many molecules of N_2O_4 are in 76.3 g N_2O_4? The molar mass of N_2O_4 is 92.02 g/mol.

A) 5.54×10^{25} N_2O_4 molecules

B) 7.26×10^{23} N_2O_4 molecules

C) 1.38×10^{24} N_2O_4 molecules

D) 4.59×10^{25} N_2O_4 molecules

E) 4.99×10^{23} N_2O_4 molecules

Answer: E
Diff: 3 *Page Ref: 3.8*

30) How many moles of N_2O_4 are in 76.3 g N_2O_4? The molar mass of N_2O_4 is 92.02 g/mol.

A) 7.02×10^3 moles
B) 1.42×10^{-4} moles
C) 1.00 mole
D) 1.21 moles
E) 0.829 moles

Answer: E
Diff: 3 *Page Ref: 3.8*

31) How many C_2H_4 molecules are contained in 45.8 mg of C_2H_4? The molar mass of C_2H_4 is 28.05 g/mol.

A) 9.83×10^{20} C_2H_4 molecules

B) 7.74×10^{26} C_2H_4 molecules

C) 2.71×10^{20} C_2H_4 molecules

D) 3.69×10^{23} C_2H_4 molecules

E) 4.69×10^{23} C_2H_4 molecules

Answer: A
Diff: 3 *Page Ref: 3.8*

32) What is the mass (in kg) of 6.89×10^{25} molecules of CO_2? The molar mass of CO_2 is 44.01 g/mol.

A) 3.85 kg B) 5.04 kg C) 2.60 kg D) 3.03 kg E) 6.39 kg

Answer: B
Diff: 3 *Page Ref: 3.8*

33) Calculate the mass percent composition of sulfur in $Al_2(SO_4)_3$.

A) 28.12 % B) 9.372 % C) 42.73 % D) 21.38 % E) 35.97 %

Answer: A
Diff: 2 *Page Ref: 3.9*

34) Calculate the mass percent composition of lithium in Li_3PO_4.

A) 26.75 % B) 17.98 % C) 30.72 % D) 55.27 % E) 20.82 %

Answer: B
Diff: 2 *Page Ref: 3.9*

35) How many atoms of oxygen are contained in 47.6 g of $Al_2(CO_3)_3$? The molar mass of $Al_2(CO_3)_3$ is 233.99 g/mol.

A) 1.23×10^{23} O atoms
B) 2.96×10^{24} O atoms
C) 2.87×10^{25} O atoms
D) 1.10×10^{24} O atoms
E) 3.68×10^{23} O atoms

Answer: D
Diff: 4 *Page Ref: 3.9*

36) How many atoms of carbon are contained in 47.6 g of $Al_2(CO_3)_3$? The molar mass of $Al_2(CO_3)_3$ is 233.99 g/mol.

A) 1.23×10^{23} C atoms
B) 2.96×10^{24} C atoms
C) 2.87×10^{25} C atoms
D) 1.10×10^{24} C atoms
E) 3.68×10^{23} C atoms

Answer: E
Diff: 4 *Page Ref: 3.9*

37) How many sodium ions are contained in 99.6 mg of Na_2SO_3? The molar mass of Na_2SO_3 is 126.05 g/mol.

A) 1.52×10^{27} sodium ions
B) 4.76×10^{20} sodium ions
C) 2.10×10^{21} sodium ions
D) 1.05×10^{21} sodium ions
E) 9.52×10^{20} sodium ions

Answer: E
Diff: 4 *Page Ref: 3.9*

38) How many SO_3^{2-} ions are contained in 99.6 mg of Na_2SO_3? The molar mass of Na_2SO_3 is 126.05 g/mol.

A) 1.52×10^{27} SO_3 ions
B) 4.76×10^{20} SO_3 ions
C) 2.10×10^{21} SO_3 ions
D) 1.05×10^{21} SO_3 ions
E) 9.52×10^{20} SO_3 ions

Answer: B
Diff: 4 *Page Ref: 3.9*

39) Determine the volume of hexane that contains 5.33×10^{22} molecules of hexane. The density of hexane is 0.6548 g/mL and its molar mass is 86.17 g/mol.

A) 8.59 mL B) 13.5 mL C) 7.40 mL D) 12.4 mL E) 11.6 mL

Answer: E
Diff: 4 *Page Ref: 3.9*

40) How many molecules of butane are contained in 25.0 mL of butane? The density of butane is 0.6011 g/mL and the molar mass is 58.12 g/mol.

A) 2.59×10^{23} molecules butane
B) 1.46×10^{27} molecules butane
C) 6.87×10^{23} molecules butane
D) 1.56×10^{23} molecules butane
E) 7.14×10^{25} molecules butane

Answer: D
Diff: 4 *Page Ref: 3.9*

41) Determine the molecular formula of a compound that has a molar mass of 183.2 g/mol and an empirical formula of $C_2H_5O_2$.

A) $C_2H_5O_2$
B) $C_6H_{15}O_6$
C) $C_3H_7O_3$
D) $C_4H_{10}O_4$
E) $C_8H_{20}O_8$

Answer: B
Diff: 2 *Page Ref: 3.10*

42) Determine the molecular formula of a compound that has a molar mass of 92.0 g/mol and an empirical formula of NO_2.

A) N_2O_3 B) N_3O_6 C) N_2O_4 D) NO_2 E) N_2O_5

Answer: C
Diff: 2 *Page Ref: 3.10*

43) Determine the empirical formula for a compound that is 36.86% N and 63.14% O by mass.

A) NO B) N_2O C) NO_2 D) N_2O_3 E) NO_3

Answer: D
Diff: 3 *Page Ref: 3.10*

44) Determine the empirical formula for a compound that is found to contain 10.15 mg P and 34.85 mg Cl.

A) P_3Cl B) PCl C) PCl_2 D) P_2Cl_3 E) PCl_3

Answer: E
Diff: 3 *Page Ref: 3.10*

45) Determine the empirical formula for a compound that contains C, H and O. It contains 52.14% C and 34.73% O by mass.

A) C_2H_6O
B) CHO
C) $C_4H_{13}O_2$
D) CH_4O_3
E) CH_3O

Answer: A
Diff: 3 *Page Ref:* 3.10

46) Determine the molecular formula for a compound that is 70.79% carbon, 8.91% hydrogen, 4.59% nitrogen, and 15.72% oxygen.

A) $C_{18}H_{27}NO_3$ B) $C_{18}H_{27}NO_2$ C) $C_{17}H_{27}NO_3$ D) $C_{17}H_{26}NO_3$

Answer: A
Diff: 5 *Page Ref:* 3.10

47) Determine the molecular formula of a compound that is 49.48% carbon, 5.19% hydrogen, 28.85% nitrogen, and 16.48% oxygen. The molecular weight is 194.19 g/mol.

A) $C_8H_{12}N_4O_2$ B) $C_4H_5N_2O$ C) $C_8H_{10}N_4O_2$ D) $C_8H_{10}N_2O$

Answer: C
Diff: 5 *Page Ref:* 3.10

48) Combustion analysis of 63.8 mg of a C, H and O containing compound produced 145.0 mg of CO_2 and 59.38 mg of H_2O. What is the empirical formula for the compound?

A) C_5H_2O B) CHO C) C_3H_6O D) C_3H_7O E) C_6HO_3

Answer: C
Diff: 5 *Page Ref:* 3.10

49) Write a **balanced** equation to show the reaction of sulfurous acid with lithium hydroxide to form water and lithium sulfite.

A) $H_2SO_4(aq) + LiOH(aq) \rightarrow H_2O(l) + Li_2SO_4(aq)$
B) $H_2SO_3(aq) + 2\ LiOH(aq) \rightarrow 2\ H_2O(l) + Li_2SO_3(aq)$
C) $HSO_3(aq) + LiOH(aq) \rightarrow H_2O(l) + LiSO_3(aq)$
D) $HSO_4(aq) + LiOH(aq) \rightarrow H_2O(l) + LiSO_4(aq)$
E) $H_2S(aq) + 2\ LiOH(aq) \rightarrow 2\ H_2O(l) + Li_2S(aq)$

Answer: B
Diff: 3 *Page Ref:* 3.11

50) Write a **balanced** equation to show the reaction of gaseous ethane with gaseous oxygen to form carbon monoxide gas and water vapor.

A) $2\ C_2H_6(g) + 7\ O_{2(g)} \rightarrow 4\ CO_2(g) + 6\ H_2O(g)$
B) $C_2H_6(g) + 5\ O_{(g)} \rightarrow 2\ CO(g) + 3\ H_2O(g)$
C) $2\ C_2H_6(g) + 5\ O_{2(g)} \rightarrow 4\ CO(g) + 6\ H_2O(g)$
D) $C_2H_6(g) + 7\ O_{(g)} \rightarrow 2\ CO_2(g) + 3\ H_2O(g)$
E) $2\ CH_3(g) + 5\ O_{(g)} \rightarrow 2\ CO(g) + 3\ H_2O(g)$

Answer: C
Diff: 3 *Page Ref:* 3.11

51) Write a **balanced** equation to show the reaction of aqueous aluminum acetate with aqueous ammonium phosphate to form solid aluminum phosphate and aqueous ammonium acetate.

A) $Al(C_2H_3O_2)_2(aq) + (NH_4)_2PO_4(aq) \rightarrow AlPO_{4(s)} + 2\ NH_4C_2H_3O_2(aq)$

B) $Al(C_2H_3O_2)_2(aq) + (NH_3)_2PO_4(aq) \rightarrow AlPO_{4(s)} + 2\ NH_3C_2H_3O_2(aq)$

C) $Al(CO_3)_2(aq) + (NH_3)_2PO_4(aq) \rightarrow AlPO_{4(s)} + 2\ NH_3CO_3(aq)$

D) $Al(C_2H_3O_2)_3(aq) + (NH_4)_3PO_4(aq) \rightarrow AlPO_{4(s)} + 3\ NH_4C_2H_3O_2(aq)$

E) $Al(CO_2)_3(aq) + (NH_4)_3PO_3(aq) \rightarrow AlPO_{3(s)} + 3\ NH_4CO_2(aq)$

Answer: D
Diff: 3 *Page Ref: 3.11*

52) Which of the following is one possible form of pentane?

A) $CH_3CH_2CH_2CH_2CH_3$

B) $CH_3CH{=}CHCH_2CH_3$

C) $CH_3CH_2CH_2CH_2CH_2CH_3$

D) $CH_3CH_2CH_2CH_2CH_2NH_2$

E) $CH_3CH_2\text{-}O\text{-}CH_2CH_2CH_3$

Answer: A
Diff: 2 *Page Ref: 3.12*

53) Which of the following compounds is ethanol?

A) C_2H_6

B) C_2H_5OH

C) CH_3CO_2H

D) $CH_3CO_2CH_3$

E) CH_3OCH_3

Answer: B
Diff: 2 *Page Ref: 3.12*

54) Which of the following compounds is nail polish remover?

A) $(CH_3)_2C{=}O$

B) C_2H_5OH

C) CH_3CO_2H

D) $CH_3CO_2CH_3$

E) CH_3OCH_3

Answer: A
Diff: 2 *Page Ref: 3.12*

Algorithmic Questions

1) Identify the compound with ionic bonds.

A) Ne B) CO C) N_2 D) H_2O E) KBr

Answer: E
Diff: 1 *Page Ref: 3.2*

2) Identify the compound with covalent bonds.

A) CH_4 B) Kr C) KBr D) Li E) NaCl

Answer: A
Diff: 1 *Page Ref: 3.2*

3) Which of the following contains BOTH ionic and covalent bonds?

A) $CaCl_2$
B) COS
C) $BaSO_4$
D) SF_6
E) None of the above contain both ionic and covalent bonds.

Answer: C
Diff: 2 *Page Ref: 3.2*

4) What is the empirical formula for $C_{12}H_{24}O_6$?

A) C_2H_5O B) CHO C) C_2H_4O D) CHO_2 E) CH_2O

Answer: C
Diff: 1 *Page Ref: 3.3*

5) Write a possible molecular formula for C_4H_4O.

A) $C_{32}H_{32}O_8$ B) $C_{12}H_{12}O_2$ C) C_2H_2O D) C_8H_8O

Answer: A
Diff: 1 *Page Ref: 3.3*

6) Which of the following is an ionic compound?

A) SF_2
B) $Mg_3(PO_4)_2$
C) Cl_2O
D) CH_2O
E) PBr_5

Answer: B
Diff: 1 *Page Ref: 3.4*

7) Which of the following is a molecular compound?

A) KCN B) LiOH C) SrI_2 D) ZnS E) P_4O_{10}

Answer: E
Diff: 1 *Page Ref: 3.4*

8) In which set do all elements tend to form anions in binary ionic compounds?

A) C, S, Pb B) K, Fe, Br C) Li, Na, K D) N, O, I

Answer: D
Diff: 1 *Page Ref: 3.4*

9) What type of bonding is found in the compound OF_2?

A) covalent bonding
B) hydrogen bonding
C) ionic bonding
D) metallic bonding

Answer: A
Diff: 1 *Page Ref: 3.4*

10) Which one of the following compounds contains ionic bonds?

A) SrO
B) HBr
C) PBr_3
D) SiO_2

Answer: A
Diff: 1 *Page Ref: 3.4*

11) Which of the following is the correct chemical formula for a molecule of astatine?

A) At
B) At^-
C) At^+
D) At_2

Answer: D
Diff: 1 *Page Ref: 3.4*

12) Which of the compounds, Li_3P , PH_3, C_2H_6, IBr_3 are ionic compounds?

A) only C_2H_6
B) only Li_3P
C) Li_3P and PH_3
D) PH_3, C_2H_6, and IBr_3

Answer: B
Diff: 1 *Page Ref: 3.4*

13) Which of the compounds C_4H_{10}, $BaCl_2$, $Ni(NO_3)_2$, SF_6 are expected to exist as molecules?

A) only C_4H_{10}
B) C_4H_{10} and SF_6
C) C_4H_{10}, $Ni(NO_3)_2$, and SF_6
D) $BaCl_2$ and $Ni(NO_3)_2$

Answer: B
Diff: 1 *Page Ref: 3.4*

14) Which of the following is an atomic element?

A) chlorine
B) hydrogen
C) nitrogen
D) oxygen
E) calcium

Answer: E
Diff: 1 *Page Ref: 3.4*

15) Which of the following is a molecular element?

A) neon
B) lithium
C) selenium
D) magnesium
E) titanium

Answer: C
Diff: 1 *Page Ref: 3.4*

16) Which of the following is a molecular element?

A) barium
B) argon
C) xenon
D) iodine
E) potassium

Answer: D
Diff: 1 *Page Ref: 3.4*

17) Which of the following is a molecular compound?

A) $CuCl_2$ B) NaI C) $NaNO_3$ D) CH_3Cl E) RbBr

Answer: D
Diff: 1 *Page Ref: 3.4*

18) Which of the following exists as a diatomic molecule?

A) hydrogen
B) carbon
C) phosphorus
D) lithium
E) krypton

Answer: A
Diff: 1 *Page Ref: 3.4*

19) Which of the following exists as a polyatomic molecule?

A) neon
B) carbon
C) phosphorus
D) lithium
E) krypton

Answer: C
Diff: 1 *Page Ref: 3.4*

20) Which of the following is an ionic compound?

A) NaBr B) CO_2 C) PCl_3 D) CCl_4 E) $SeBr_2$

Answer: A
Diff: 1 *Page Ref: 3.4*

21) Which of the following is a molecular compound?

A) CuI_2 B) N_2O_4 C) $NaNO_3$ D) $SrSO_3$ E) RbI

Answer: B
Diff: 1 *Page Ref: 3.4*

22) Which of the following elements has the **least** tendency to form an ion?

A) Ca B) K C) Kr D) Se

Answer: C
Diff: 1 *Page Ref: 3.4*

23) In which set do all elements tend to form cations in binary ionic compounds?
A) K, Ga, O B) Sr, Ni, Hg C) N, P, Bi D) O, Br, I

Answer: B
Diff: 3 *Page Ref:* 3.5

24) What is the charge on the Sc ions in Sc_2O_3?
A) 3– B) 1+ C) 2+ D) 3+

Answer: D
Diff: 2 *Page Ref:* 3..5

25) Write the formula for the compound formed between potassium and sulfur.
A) KS B) KS_2 C) K_2S D) K_2SO_3 E) K_3S_2

Answer: C
Diff: 2 *Page Ref:* 3.5

26) Write the formula for barium nitrite.
A) Ba_3N_2
B) $BaNO_3$
C) BaN
D) $Ba(NO_2)_2$
E) $Ba(NO_2)_3$

Answer: D
Diff: 2 *Page Ref:* 3.5

27) Write the formula for barium nitride.
A) Ba_3N_2
B) $Ba(NO_3)_2$
C) BaN
D) Ba_2N_3
E) $Ba(NO_2)_2$

Answer: A
Diff: 2 *Page Ref:* 3.5

28) Give the structure for cesium chlorate.
A) CsClO B) $CsClO_2$ C) $CsClO_3$ D) $CsClO_4$

Answer: C
Diff: 2 *Page Ref:* 3.5

29) Give the structure for lithium hypochlorite.
A) LIClO B) $LIClO_2$ C) $LIClO_3$ D) $LIClO_4$

Answer: A
Diff: 2 *Page Ref:* 3.5

30) Write the name for $Sr_3(PO_4)_2$.
A) strontium (III) phosphite
B) strontium (II) phosphite
C) strontium phosphate
D) tristrontium phosphorustetraoxide
E) strontium phosphite

Answer: C
Diff: 3 *Page Ref:* 3.5

31) Write the name for PbS.
A) lead (I) sulfate
B) lead (I) sulfide
C) lead (II) sulfide
D) lead (II) sulfate
E) lead sulfide

Answer: C
Diff: 3 *Page Ref:* 3.5

32) Write the name for CoS.
A) cobaltous sulfate
B) cobaltous sulfide
C) cobaltic sulfide
D) cobaltic sulfate
E) cobalt sulfide

Answer: B
Diff: 3 *Page Ref:* 3.5

33) Give the formula for stannous nitrate
A) $Sn(NO_2)_3$
B) $Sn(NO_3)_3$
C) $Sn(NO_3)_2$
D) $Sn(NO_2)_2$
E) $Sn_2(NO_2)$

Answer: C
Diff: 3 *Page Ref:* 3.5

34) The solid compound, K_2SO_4, contains
A) K^+, S^{6+}, and O^{2-} ions.
B) K^+ and $SO_4{}^{-2}$ ions.
C) $K_2{}^+$ and $SO_4{}^{-2}$ ions.
D) K_2SO_4 molecules.

Answer: B
Diff: 3 *Page Ref:* 3.5

35) What is the chemical formula for iron(III) sulfate?
A) Fe_3S
B) Fe_3SO_4
C) Fe_2S_3
D) $Fe_2(SO_4)_3$

Answer: D
Diff: 3 *Page Ref:* 3.5

36) Rb_2S is named

A) rubidium disulfide.
B) rubidium sulfide.
C) rubidium(II) sulfide.
D) rubidium sulfur.

Answer: B
Diff: 3 *Page Ref: 3.5*

37) What is the chemical formula for calcium hydroxide?

A) CaH_2
B) CaOH
C) $CaOH_2$
D) $Ca(OH)_2$

Answer: D
Diff: 3 *Page Ref: 3.5*

38) What is the chemical formula for magnesium hydride?

A) MgH_2
B) MgOH
C) $MgOH_2$
D) $Mg(OH)_2$

Answer: A
Diff: 3 *Page Ref: 3.5*

39) The chemical formula for lithium peroxide is

A) LiOH.
B) LiO_2.
C) Li_2O.
D) Li_2O_2.

Answer: D
Diff: 3 *Page Ref: 3.5*

40) The compound, $Cu(IO_3)$, is named

A) copper iodate(II).
B) copper(I) iodate.
C) copper(I) iodate(II).
D) copper(II) iodate.

Answer: D
Diff: 3 *Page Ref: 3.5*

41) The chemical formula for calcium nitride is

A) $Ca(NO_3)_2$.
B) $Ca(NO_2)_2$.
C) Ca_3N_2.
D) CaN_2.

Answer: C
Diff: 3 *Page Ref: 3.5*

42) The chemical formula for the sulfite ion is

A) S^-.
B) S^{2-}.
C) SO_3^{2-}.
D) SO_4^{2-}.

Answer: C
Diff: 3 *Page Ref: 3.5*

43) The ion, IO_2^-, is named

A) iodate ion.
B) iodite ion.
C) iodine dioxide ion.
D) iodine(II) oxide ion.

Answer: B
Diff: 3 *Page Ref: 3.5*

44) The compound, NO_2, is named

A) nitrate.
B) nitrite.
C) nitrogen dioxide.
D) nitrogen (IV) oxide.

Answer: C
Diff: 3 *Page Ref: 3.6*

45) An aqueous solution of H_2S is named

A) hydrosulfuric acid. B) hydrosulfurous acid.
C) sulfuric acid. D) sulfurous acid.

Answer: A
Diff: 3 *Page Ref: 3.6*

46) The chemical formula for nitrous acid is

A) $H_3N(aq)$. B) $HNO_2(aq)$. C) $HNO_3(aq)$. D) $H_2N_2O_6(aq)$.

Answer: B
Diff: 3 *Page Ref: 3.6*

47) Determine the name for aqueous HF.

A) fluoric acid
B) fluorous acid
C) hydrofluorous acid
D) hydrogen fluorate
E) hydrofluoric acid

Answer: E
Diff: 2 *Page Ref: 3.6*

48) Determine the name for Cl_2O.

A) chlorine oxide
B) dichlorine monoxide
C) chlorine (I) oxide
D) chlorine (II) oxide
E) chlorate

Answer: B
Diff: 3 *Page Ref: 3.6*

49) Give the name for PBr_3.

A) phosphorus tribromide
B) potassium tribromide
C) phosphorus (III) bromide
D) phosphorus (II) bromide
E) phosphorus bromide

Answer: A
Diff: 3 *Page Ref: 3.6*

50) Determine the name for $HBrO_3$.

A) hydrobromic acid
B) hydrobromus acid
C) bromate acid
D) bromic acid
E) perbromic acid

Answer: D
Diff: 3 *Page Ref: 3.6*

51) What is the molar mass of nitrogen gas?

A) 14.0 g/mol
B) 28.0 g/mol
C) 6.02×10^{23} g/mol
D) 1.20×10^{23} g/mol

Answer: B
Diff: 3 *Page Ref: 3.8*

52) What is the mass of a single fluorine molecule, F_2?

A) 3.155×10^{-23} g
B) 6.310×10^{-23} g
C) 19.00 g
D) 38.00 g

Answer: B
Diff: 3 *Page Ref: 3.8*

53) What is the mass of 0.500 mol of dichlorodifluoromethane, CCl_2F_2?

A) 4.14×10^{-3} g
B) 60.5 g
C) 121 g
D) 242 g

Answer: B
Diff: 3 *Page Ref: 3.8*

54) How many moles are there in 3.00 g of ethanol, CH_3CH_2OH?

A) 0.00725 mol
B) 0.0652 mol
C) 15.3 mol
D) 138 mol

Answer: B
Diff: 3 *Page Ref: 3.8*

55) How many moles of C_8H_{18} contain 9.25×10^{24} molecules of C_8H_{18} ?

A) 65.1 moles C_8H_{18}
B) 28.6 moles C_8H_{18}
C) 34.9 moles C_8H_{18}
D) 46.2 moles C_8H_{18}
E) 15.4 moles C_8H_{18}

Answer: E
Diff: 2 *Page Ref: 3.8*

56) What is the mass of 8.50×10^{22} molecules of NH_3?

A) 0.00829 g
B) 0.417 g
C) 2.40 g
D) 121 g

Answer: C
Diff: 3 *Page Ref: 3.8*

57) What is the mass of 3.91×10^{24} molecules of SeO_2? The molar mass of SeO_2 is 110.96 g/mol.

A) 205 g
B) 294 g
C) 721 g
D) 341 g
E) 685 g

Answer: C
Diff: 3 *Page Ref: 3.8*

58) What is the molar mass of 1-butene if 5.38×10^{16} molecules of 1-butene weigh 5.00 μg?

A) 56.0 g/mol
B) 178 g/mol
C) 224 g/mol
D) 447 g/mol

Answer: A
Diff: 3 *Page Ref: 3.8*

59) What mass of carbon dioxide, CO_2, contains the same number of molecules as 3.00 g of trichlorofluoromethane, CCl_3F?

A) 0.106 g B) 0.961 g C) 1.04 g D) 9.37 g

Answer: B
Diff: 3 *Page Ref: 3.8*

60) What mass of phosphorus pentafluoride, PF_5, has the same number of fluorine atoms as 25.0 g of oxygen difluoride, OF_2?

A) 0.933 g B) 10.0 g C) 23.3 g D) 146 g

Answer: C
Diff: 3 *Page Ref: 3.8*

61) How many anions are there in 2.50 g of $MgBr_2$?

A) 8.18×10^{21} anions
B) 1.64×10^{22} anions
C) 4.43×10^{25} anions
D) 8.87×10^{25} anions

Answer: B
Diff: 3 *Page Ref: 3.8*

62) Which of the following has the greatest mass?

A) 3.88×10^{22} molecules of O_2
B) 1.00 g of O_2
C) 0.0312 mol of O_2
D) All of the above have the same mass.

Answer: A
Diff: 3 *Page Ref: 3.8*

63) Which of the following has the **smallest** mass?

A) 3.50×10^{23} molecules of I_2
B) 85.0 g of Cl_2
C) 2.50 mol of F_2
D) 0.050 kg of Br_2

Answer: D
Diff: 3 *Page Ref: 3.8*

64) Give the mass percent of oxygen in $C_{14}H_{19}NO_2$.

A) 13.72% B) 72.07% C) 8.21% D) 6.00%

Answer: A
Diff: 3 *Page Ref: 3.9*

65) How many moles of NCl_3 contain 2.55×10^{24} chlorine atoms?

A) 1.41 moles NCl_3
B) 4.23 moles NCl_3
C) 12.7 moles NCl_3
D) 7.87 moles NCl_3
E) 2.82 moles NCl_3

Answer: A
Diff: 2 *Page Ref: 3.9*

66) How many moles of PBr_3 contain 3.68×10^{25} bromine atoms?

A) 61.1 moles PBr_3

B) 20.4 moles PBr_3

C) 16.4 moles PBr_3

D) 54.5 moles PBr_3

E) 49.1 moles PBr_3

Answer: B
Diff: 3 *Page Ref:* 3.9

67) How many moles of C_3H_8 contain 4.95×10^{24} hydrogen atoms?

A) 8.22 moles C_3H_8

B) 6.58 moles C_3H_8

C) 1.03 moles C_3H_8

D) 9.73 moles C_3H_8

E) 3.09 moles C_3H_8

Answer: C
Diff: 3 *Page Ref:* 3.9

68) A sample of pure lithium nitrate contains 10.1% lithium by mass. What is the % lithium by mass in a sample of pure lithium carbonate that has twice the mass of the first sample?

A) 5.05% B) 10.1% C) 20.2% D) 40.4%

Answer: B
Diff: 2 *Page Ref:* 3.9

69) How many Fe(II) ions are there in 20.0 g of $FeSO_4$?

A) 2.19×10^{-25} iron(II) ions B) 7.92×10^{22} iron(II) ions

C) 4.57×10^{24} iron(II) ions D) 1.82×10^{27} iron(II) ions

Answer: B
Diff: 4 *Page Ref:* 3.9

70) A sample of pure calcium fluoride with a mass of 15.0 g contains 7.70 g of calcium. How much calcium is contained in 40.0 g of calcium fluoride?

A) 2.27 g B) 7.70 g C) 15.0 g D) 20.5 g

Answer: D
Diff: 2 *Page Ref:* 3.9

71) Which one of the following contains 39% carbon by mass?

A) C_2H_2 B) CH_4 C) CH_3NH_2 D) CO_2

Answer: C
Diff: 2 *Page Ref:* 3.9

72) Determine the mass percent (to the hundredths place) of H in sodium bicarbonate ($NaHCO_3$).

Answer: 1.20
Diff: 2 *Page Ref:* 3.9

73) How many oxygen atoms are there in 7.00 g of sodium dichromate, $Na_2Cr_2O_7$?

A) 0.187 oxygen atoms
B) 2.30×10^{21} oxygen atoms
C) 1.60×10^{22} oxygen atoms
D) 1.13×10^{23} oxygen atoms

Answer: D
Diff: 4 *Page Ref: 3.9*

74) How many chloride ions are there in 4.50 mol of aluminum chloride?

A) 3.00 chloride ions
B) 13.5 chloride ions
C) 2.71×10^{24} chloride ions
D) 8.13×10^{24} chloride ions

Answer: D
Diff: 4 *Page Ref: 3.9*

75) How many cations are there in 10.0 g of sodium phosphate?

A) 3.67×10^{22} cations
B) 1.10×10^{23} cations
C) 9.87×10^{24} cations
D) 2.96×10^{25} cations

Answer: B
Diff: 4 *Page Ref: 3.9*

76) What is the empirical formula of a compound that is 62.0% C, 10.4% H, and 27.5% O by mass?

A) C_3HO
B) C_6HO_3
C) $C_6H_{12}O_2$
D) $C_5H_{10}O_2$
E) C_3H_6O

Answer: E
Diff: 4 *Page Ref: 3.10*

77) What is the empirical formula of a substance that contains 2.64 g of C, 0.444 g of H, and 3.52 g of O?

A) CH_2O
B) $C_2H_4O_2$
C) $C_2H_4O_3$
D) $C_3H_4O_4$

Answer: A
Diff: 3 *Page Ref: 3.10*

78) Which one of the following is **not** an empirical formula?

A) CHO
B) CH_2O
C) C_2H_4O
D) $C_2H_4O_2$

Answer: D
Diff: 3 *Page Ref: 3.10*

79) Methane and oxygen react to form carbon dioxide and water. What mass of water is formed if 0.80 g of methane reacts with 3.2 g of oxygen to produce 2.2 g of carbon dioxide?

A) 1.8 g
B) 2.2 g
C) 3.7 g
D) 4.0 g

Answer: A
Diff: 4 *Page Ref: 3.10*

80) Combustion analysis of an unknown compound containing only carbon and hydrogen produced 0.2845 g of CO_2 and 0.1451 g of H_2O. What is the empirical formula of the compound?

A) CH_2 B) C_2H_5 C) C_4H_{10} D) C_5H_2

Answer: B
Diff: 4 *Page Ref: 3.10*

81) Combustion analysis of 1.200 g of an unknown compound containing carbon, hydrogen, and oxygen produced 2.086 g of CO_2 and 1.134 g of H_2O. What is the empirical formula of the compound?

A) C_2H_5O B) $C_2H_5O_2$ C) $C_2H_{10}O_3$ D) $C_3H_8O_2$

Answer: D
Diff: 4 *Page Ref: 3.10*

82) A certain alcohol contains only three elements, carbon, hydrogen, and oxygen. Combustion of a 50.00 gram sample of the alcohol produced 95.50 grams of CO_2 and 58.70 grams of H_2O. What is the empirical formula of the alcohol?

Answer: C_2H_6O
Diff: 4 *Page Ref: 3.10*

83) What is the stoichiometric coefficient for oxygen when the following equation is balanced using the lowest, whole-number coefficients?

$$\_\_\_\_ C_2H_6O(l) + \_\_\_\_ O_2(g) \rightarrow \_\_\_\_ CO_2(g) + \_\_\_\_ H_2O(l)$$

A) 9 B) 7 C) 5 D) 3

Answer: D
Diff: 5 *Page Ref: 3.11*

84) Aluminum metal reacts with aqueous iron(II) chloride to form aqueous aluminum chloride and iron metal. What is the stoichiometric coefficient for aluminum when the chemical equation is balanced using the lowest, whole-number stoichiometric coefficients?

A) 1 B) 2 C) 3 D) 4

Answer: B
Diff: 5 *Page Ref: 3.11*

85) Calcium phosphate reacts with sulfuric acid to form calcium sulfate and phosphoric acid. What is the coefficient for sulfuric acid when the equation is balanced using the lowest, whole-numbered coefficients?

A) 1 B) 2 C) 3 D) none of these

Answer: C
Diff: 5 *Page Ref: 3.11*

86) Identify an amine.

A) $CH_3CH_2OCH_2CH_3$
B) CH_3CH_2OH
C) $CH_3CH_2NH_2$
D) $CH_3CH_2CH_2CH_3$
E) CH_3COOH

Answer: C
Diff: 2 *Page Ref: 3.12*

87) Identify an ether.

A) $CH_3CH_2OCH_2CH_3$
B) CH_3CH_2I
C) $CH_3CH_2NH_2$
D) $CH_3CH_2CH_2CH_3$
E) CH_3COOH

Answer: A
Diff: 2 *Page Ref: 3.12*

88) Identify a carboxylic acid.

A) $CH_3CH_2SCH_2CH_3$
B) CH_3CH_2Br
C) $CH_3CH_2NH_2$
D) $CH_3CH_2CH_2CH_3$
E) CH_3COOH

Answer: E
Diff: 2 *Page Ref: 3.12*

89) Identify a hydrocarbon.

A) $CH_3CH_2SCH_2CH_3$
B) CH_3CH_2I
C) $CH_3CH_2PH_2$
D) $CH_3CH_2CH_2CH_3$
E) CH_3COOH

Answer: D
Diff: 2 *Page Ref: 3.12*

90) Which of the following compounds is vinegar?

A) C_2H_6
B) C_2H_5Br
C) CH_3CO_2H
D) $CH_3CO_2CH_3$
E) CH_3NHCH_3

Answer: C
Diff: 2 *Page Ref: 3.12*

Matching Questions

Match the following.

1) oxygen
Diff: 1 *Page Ref: 3.4*

2) chlorine
Diff: 1 *Page Ref: 3.4*

3) neon
Diff: 1 *Page Ref: 3.4*

4) calcium
Diff: 1 *Page Ref: 3.4*

5) iodine
Diff: 1 *Page Ref: 3.4*

6) hydrogen
Diff: 1 *Page Ref: 3.4*

A) $H_2(g)$

B) $O(g)$

C) $I_2(s)$

D) $Cl_2(g)$

E) $Cl(g)$

F) $I(s)$

G) $Ne_2(g)$

H) $O_2(g)$

I) $H(g)$

J) $Ca(s)$

K) $Ne(g)$

L) $Ca_2(s)$

Answers: 1) H 2) D 3) K 4) J 5) C
6) A

Short Answer Questions

1) How can one compound contain both ionic and covalent bonds? Give an example.

Answer: An ionic compound that contains a polyatomic ion, such as $NaNO_3$, has both ionic bonds (that hold the sodium and nitrate ions together) as well as covalent bonds (that hold the atoms within the nitrate ion together).

Diff: 1 *Page Ref: 3.2*

2) Describe the difference between a molecular formula and an empirical formula. Give an example.

Answer: A molecular formula is the exact number of each type of atom necessary to build a specific molecule. An empirical formula is simply the smallest whole number ratio between atoms in a compound. For example, C_2H_4 is the molecular formula for ethene. The empirical formula for ethene is CH_2, the smallest whole number ratio between the elements.

Diff: 1 *Page Ref: 3.3*

3) Define empirical formula.

Answer: An empirical formula gives relative numbers of atoms of each element per formula unit.
Diff: 2 *Page Ref: 3.3*

4) What is the structure of the covalent compound formed by nitrogen and oxygen? Is this the only possibility? Explain.

Answer: Since nitrogen and oxygen are both nonmetals, they combine by sharing electrons. This can be done in multiple different ways. Some possible compounds are N_2O, N_2O_3, NO_2.
Diff: 2 *Page Ref: 3.3*

5) Describe a structural formula

Answer: In a structural formula, lines are used to represent covalent bonds to show how the atoms in the molecule are connected to each other.
Diff: 2 *Page Ref: 3.3*

6) Describe the difference between an atomic element and a molecular element.

Answer: Atomic elements exist with a single atom at their basic unit; molecular elements have more than one atom in their basic unit.
Diff: 1 *Page Ref: 3.4*

7) Describe the difference between ionic and molecular compounds. Give an example of each.

Answer: An ionic compound is formed between a metal and a nonmetal (or polyatomic ions) and is held together through the attraction of opposite charges. An example is NaCl. A molecular compound is usually formed between 2 or more nonmetals and is held together through the sharing of electrons between atoms. An example is CO_2.
Diff: 1 *Page Ref: 3.4*

8) Why aren't prefixes used in naming ionic compounds?

Answer: The charges on the ions dictate how many must be present to form a neutral unit. Molecular compounds do not have such constraints and therefore must use prefixes to denote the number of atoms present.
Diff: 1 *Page Ref: 3.5*

9) Give the name for HNO_2.

Answer: nitrous acid
Diff: 2 *Page Ref: 3.6*

10) Calculate the mass percent composition of oxygen in $Al_2(SO_4)_3$.

Answer: 56.11%
Diff: 2 *Page Ref: 3.9*

11) Balance the following equation.

$$\_\_\_\_ C_{10}H_{12} + \_\_\_\_ O_2 \rightarrow \_\_\_\_ H_2O + \_\_\_\_ CO_2$$

Answer: $C_{10}H_{12} + 13\ O_2 \rightarrow 6\ H_2O + 10\ CO_2$
Diff: 5 *Page Ref: 3.11*

12) Balance the following equation.

_____ C_9H_{20} + _____ O_2 → _____ H_2O + _____ CO_2

Answer: $C_9H_{20} + 14\ O_2 \rightarrow 10\ H_2O + 9\ CO_2$
Diff: 5 *Page Ref: 3.11*

Chapter 4 Chemical Quantities and Aqueous Reactions

Multiple Choice Questions

1) Global warming is thought to be caused by the increase of one particular gas. Name the gas.

A) oxygen
B) carbon monoxide
C) carbon dioxide
D) nitrogen
E) helium

Answer: C
Diff: 1 *Page Ref: 4.1*

2) How many grams of Li_3N can be formed from 1.75 moles of Li? Assume an excess of nitrogen.

$$6\ Li(s) + N_2(g) \rightarrow 2\ Li_3N(s)$$

A) 18.3 g Li_3N
B) 20.3 g Li_3N
C) 58.3 g Li_3N
D) 61.0 g Li_3N
E) 15.1 g Li_3N

Answer: B
Diff: 3 *Page Ref: 4.2*

3) How many moles of oxygen are formed when 58.6 g of KNO_3 decomposes according to the following reaction? The molar mass of KNO_3 is 101.11 g/mol.

$$4\ KNO_3(s) \rightarrow 2\ K_2O(s) + 2\ N_2(g) + 5\ O_2(g)$$

A) 0.290 mol O_2
B) 0.580 mol O_2
C) 18.5 mol O_2
D) 0.724 mol O_2
E) 1.73 mol O_2

Answer: D
Diff: 3 *Page Ref: 4.2*

4) How many moles of nitrogen are formed when 58.6 g of KNO_3 decomposes according to the following reaction? The molar mass of KNO_3 is 101.11 g/mol.

$$4\ KNO_3(s) \rightarrow 2\ K_2O(s) + 2\ N_2(g) + 5\ O_2(g)$$

A) 0.290 mol N_2
B) 0.580 mol N_2
C) 18.5 mol N_2
D) 0.724 mol N_2
E) 1.73 mol N_2

Answer: A
Diff: 3 *Page Ref: 4.2*

5) Two samples of calcium fluoride are decomposed into their constituent elements. The first sample produced 0.154 g of calcium and 0.146 g of fluorine. If the second sample produced 294 mg of fluorine, how many g of calcium were formed?
A) 0.280 g
B) 3.09×10^2 g
C) 3.13 g
D) 0.309 g
E) 2.80×10^2 g

Answer: D
Diff: 3 *Page Ref: 4.2*

6) Two samples of potassium iodide are decomposed into their constituent elements. The first sample produced 13.0 g of potassium and 42.3 g of iodine. If the second sample produced 24.4 kg of potassium, how many kg of iodine were produced?
A) 13.3 kg B) 22.5 kg C) 79.4 kg D) 44.4 kg E) 92.4 kg

Answer: C
Diff: 3 *Page Ref: 4.2*

7) A 14.01 g sample of N_2 reacts with 3.02 g of H_2 to form ammonia (NH_3). If ammonia is the only product, what mass of ammonia is formed?
A) 17.01 g B) 1.10 g C) 14.01 g D) 3.02 g E) 23.07 g

Answer: A
Diff: 3 *Page Ref: 4.2*

8) Consider the following balanced reaction. How many grams of water are required to form 75.9 g of HNO_3? Assume that there is excess NO_2 present. The molar masses are as follows: H_2O = 18.02 g/mol, HNO_3 = 63.02 g/mol.

$$3\ NO_2(g) + H_2O(l) \rightarrow 2\ HNO_3(aq) + NO(g)$$

A) 38.0 g H_2O
B) 21.7 g H_2O
C) 43.4 g H_2O
D) 10.9 g H_2O
E) 26.5 g H_2O

Answer: D
Diff: 4 *Page Ref: 4.2*

9) Consider the following reaction. How many moles of oxygen are required to produce 2.33 moles of water? Assume that there is excess C_3H_7SH present.

$$C_3H_7SH(l) + 6\ O_2(g) \rightarrow 3\ CO_2(g) + SO_2(g) + 4\ H_2O(g)$$

A) 1.55 moles O_2
B) 3.50 moles O_2
C) 2.33 moles O_2
D) 4.14 moles O_2
E) 6.21 moles O_2

Answer: B
Diff: 2 *Page Ref: 4.2*

10) Consider the following balanced reaction. What mass (in g) of CO_2 can be formed from 288 mg of O_2? Assume that there is excess C_3H_7SH present.

$$C_3H_7SH(l) + 6\ O_2(g) \rightarrow 3\ CO_2(g) + SO_2(g) + 4\ H_2O(g)$$

A) 0.396 g CO_2
B) 0.209 g CO_2
C) 0.792 g CO_2
D) 0.126 g CO_2
E) 0.198 g CO_2

Answer: E
Diff: 4 *Page Ref: 4.2*

11) According to the following reaction, how many grams of sulfur are formed when 37.4 g of water are formed?

$$2\ H_2S(g) + SO_2(g) \rightarrow 3\ S(s) + 2H_2O(l)$$

A) 99.8 g S B) 66.6 g S C) 56.1 g S D) 44.4 g S E) 14.0 g S

Answer: A
Diff: 4 *Page Ref: 4.2*

12) How many molecules of H_2S are required to form 79.0 g of sulfur according to the following reaction? Assume excess SO_2.

$$2\ H_2S(g) + SO_2(g) \rightarrow 3\ S(s) + 2H_2O(l)$$

A) 1.48×10^{24} molecules H_2S
B) 9.89×10^{23} molecules H_2S
C) 5.06×10^{25} molecules H_2S
D) 3.17×10^{25} molecules H_2S
E) 2.44×10^{23} molecules H_2S

Answer: B
Diff: 5 *Page Ref: 4.2*

13) Give the theoretical yield, in moles, of CO_2 from the reaction of 4.00 moles of C_8H_{18} with 4.00 moles of O_2.

$$2\ C_8H_{18} + 25\ O_2 \rightarrow 16\ CO_2 + 18\ H_2O$$

A) 0.640 moles B) 64.0 moles C) 2.56 moles D) 16.0 moles

Answer: C
Diff: 3 *Page Ref: 4.3*

14) A 12.39 g sample of phosphorus reacts with 42.54 g of chlorine to form only phosphorus trichloride (PCl_3). If it is the only product, what mass of PCl_3 is formed?

A) 30.15 g B) 54.93 g C) 140.01 g D) 79.71 g E) 91.86 g

Answer: B
Diff: 3 *Page Ref: 4.3*

15) Determine the limiting reactant (LR) and the mass (in g) of nitrogen that can be formed from 50.0 g N_2O_4 and 45.0 g N_2H_4. Some possibly useful molar masses are as follows: N_2O_4 = 92.02 g/mol, N_2H_4 = 32.05 g/mol.

$$N_2O_4(l) + 2\ N_2H_4(l) \rightarrow 3\ N_2(g) + 4\ H_2O(g)$$

A) LR = N_2H_4, 59.0 g N_2 formed
B) LR = N_2O_4, 105 g N_2 formed
C) LR = N_2O_4, 45.7 g N_2 formed
D) LR = N_2H_4, 13.3 g N_2 formed
E) No LR, 45.0 g N_2 formed

Answer: C
Diff: 5 *Page Ref: 4.3*

16) Determine the theoretical yield of HCl if 60.0 g of BCl_3 and 37.5 g of H_2O are reacted according to the following balanced reaction. A possibly useful molar mass is BCl_3 = 117.16 g/mol.

$$BCl_3(g) + 3\ H_2O(l) \rightarrow H_3BO_3(s) + 3\ HCl(g)$$

A) 75.9 g HCl
B) 132 g HCl
C) 187 g HCl
D) 56.0 g HCl
E) 25.3 g HCl

Answer: D
Diff: 5 *Page Ref: 4.3*

17) According to the following reaction, what amount of Al_2S_3 remains when 20.00 g of Al_2S_3 and 2.00 g of H_2O are reacted? A few of the molar masses are as follows: Al_2S_3 = 150.17 g/mol, H_2O = 18.02 g/mol.

$$Al_2S_3(s) + 6\ H_2O(l) \rightarrow 2\ Al(OH)_3(s) + 3\ H_2S(g)$$

A) 28.33 g B) 14.00 g C) 8.33 g D) 19.78 g E) 17.22 g

Answer: E
Diff: 5 *Page Ref: 4.3*

18) Give the percent yield when 28.16 g of CO_2 are formed from the reaction of 4.000 moles of C_8H_{18} with 4.000 moles of O_2.

$$2\ C_8H_{18} + 25\ O_2 \rightarrow 16\ CO_2 + 18\ H_2O$$

A) 20.00% B) 25.00% C) 50.00% D) 12.50%

Answer: B
Diff: 5 *Page Ref: 4.3*

19) Give the percent yield when 28.16 g of CO_2 are formed from the reaction of 4.000 moles of C_8H_{18} with 8.000 moles of O_2.

$$2\ C_8H_{18} + 25\ O_2 \rightarrow 16\ CO_2 + 18\ H_2O$$

A) 20.00% B) 25.00% C) 50.00% D) 12.50%

Answer: D
Diff: 5 *Page Ref: 4.3*

20) Give the percent yield when 28.16 g of CO_2 are formed from the reaction of 8.000 moles of C_8H_{18} with 4.000 moles of O_2.

$$2\ C_8H_{18} + 25\ O_2 \rightarrow 16\ CO_2 + 18\ H_2O$$

A) 20.00% B) 25.00% C) 50.00% D) 12.50%

Answer: B
Diff: 5 *Page Ref: 4.3*

21) Determine the percent yield of a reaction that produces 28.65 g of Fe when 50.00 g of Fe_2O_3 react with excess Al according to the following reaction.

$$Fe_2O_3(s) + 2\ Al(s) \rightarrow Al_2O_3(s) + 2\ Fe(s)$$

A) 61.03 % B) 28.65 % C) 57.30 % D) 20.02 % E) 81.93 %

Answer: E
Diff: 5 *Page Ref: 4.3*

22) Determine the molarity of a solution formed by dissolving 97.7 g LiBr in enough water to yield 750.0 mL of solution.

A) 1.50 M B) 1.18 M C) 0.130 M D) 0.768 M E) 2.30 M

Answer: A
Diff: 3 *Page Ref: 4.4*

23) Determine the molarity of a solution formed by dissolving 468 mg of MgI_2 in enough water to yield 50.0 mL of solution.

A) 0.0297 M B) 0.0337 M C) 0.0936 M D) 0.0107 M E) 0.0651 M

Answer: B
Diff: 3 *Page Ref: 4.4*

24) How many molecules of sucrose ($C_{12}H_{22}O_{11}$, molar mass = 342.30 g/mol) are contained in 14.3 mL of 0.140 M sucrose solution?

A) 8.29×10^{22} molecules $C_{12}H_{22}O_{11}$

B) 1.21×10^{21} molecules $C_{12}H_{22}O_{11}$

C) 6.15×10^{22} molecules $C_{12}H_{22}O_{11}$

D) 1.63×10^{23} molecules $C_{12}H_{22}O_{11}$

E) 5.90×10^{24} molecules $C_{12}H_{22}O_{11}$

Answer: B
Diff: 3 *Page Ref: 4.4*

25) According to the following reaction, how many moles of $Fe(OH)_2$ can form from 175.0 mL of 0.227 M LiOH solution? Assume that there is excess $FeCl_2$.

$$FeCl_2(aq) + 2\ LiOH(aq) \rightarrow Fe(OH)_2(s) + 2\ LiCl(aq)$$

A) 3.97×10^{-2} moles
B) 2.52×10^{-2} moles
C) 1.99×10^{-2} moles
D) 5.03×10^{-2} moles
E) 6.49×10^{-2} moles

Answer: C
Diff: 3 *Page Ref: 4.4*

26) According to the following reaction, what volume of 0.244 M KCl solution is required to react exactly with 50.0 mL of 0.210 M $Pb(NO_3)_2$ solution?

$$2\ KCl(aq) + Pb(NO_3)_2(aq) \rightarrow PbCl_2(s) + 2\ KNO_3(aq)$$

A) 97.4 mL B) 116 mL C) 43.0 mL D) 86.1 mL E) 58.1 mL

Answer: D
Diff: 4 *Page Ref: 4.4*

27) According to the following reaction, what mass of $PbCl_2$ can form from 235 mL of 0.110 M KCl solution? Assume that there is excess $Pb(NO_3)_2$.

$$2\ KCl(aq) + Pb(NO_3)_2(aq) \rightarrow PbCl_2(s) + 2\ KNO_3(aq)$$

A) 7.19 g B) 3.59 g C) 1.80 g D) 5.94 g E) 1.30 g

Answer: B
Diff: 4 *Page Ref: 4.4*

28) Determine the number of grams H_2 formed when 250.0 mL of 0.743 M HCl solution reacts with 3.41×10^{23} atoms of Fe according to the following reaction.

$$2\ HCl(aq) + Fe(s) \rightarrow H_2(g) + FeCl_2(aq)$$

A) 0.374 g B) 1.33 g C) 1.14 g D) 0.187 g E) 1.51 g

Answer: D
Diff: 4 *Page Ref: 4.4*

29) What volume of 0.305 M $AgNO_3$ is required to react exactly with 155.0 mL of 0.274 M Na_2SO_4 solution? Hint: You will want to write a balanced reaction.

A) 581 mL B) 173 mL C) 345 mL D) 139 mL E) 278 mL

Answer: E
Diff: 5 *Page Ref: 4.4*

30) What mass (in g) of AgCl is formed from the reaction of 75.0 mL of a 0.078 M $AgC_2H_3O_2$ solution with 55.0 mL of 0.109 M $MgCl_2$ solution?

$$2\ AgC_2H_3O_2(aq) + MgCl_2(aq) \rightarrow 2\ AgCl(s) + Mg(C_2H_3O_2)_2(aq)$$

A) 0.838 g B) 1.72 g C) 0.859 g D) 2.56 g E) 1.70 g

Answer: A
Diff: 5 *Page Ref: 4.4*

31) Identify acetic acid.
A) strong electrolyte, weak acid
B) weak electrolyte, weak acid
C) strong electrolyte, strong acid
D) weak electrolyte, strong acid
E) nonelectrolyte

Answer: B
Diff: 1 *Page Ref: 4.5*

32) Identify HCl.
A) strong electrolyte, weak acid
B) weak electrolyte, weak acid
C) strong electrolyte, strong acid
D) weak electrolyte, strong acid
E) nonelectrolyte

Answer: C
Diff: 1 *Page Ref: 4.5*

33) Choose the statement below that is TRUE.
A) A weak acid solution consists of mostly nonionized acid molecules.
B) The term "strong electrolyte" means that the substance is extremely reactive.
C) A strong acid solution consists of only partially ionized acid molecules.
D) The term "weak electrolyte" means that the substance is inert.
E) A molecular compound that does not ionize in solution is considered a strong electrolyte.

Answer: A
Diff: 1 *Page Ref: 4.5*

34) Give the **complete ionic equation** for the reaction (if any) that occurs when aqueous solutions of lithium sulfide and copper (II) nitrate are mixed.
A) $Li^+(aq) + SO_4^{2-}(aq) + Cu^+(aq) + NO_3^-(aq) \rightarrow CuS(s) + Li^+(aq) + NO_3^-(aq)$
B) $Li^+(aq) + S^-(aq) + Cu^+(aq) + NO_3^-(aq) \rightarrow CuS(s) + LiNO_3(aq)$
C) $2\ Li^+(aq) + S^{2-}(aq) + Cu^{2+}(aq) + 2\ NO_3^-(aq) \rightarrow Cu^{2+}(aq) + S^{2-}(aq) + 2\ LiNO_3(s)$
D) $2\ Li^+(aq) + S^{2-}(aq) + Cu^{2+}(aq) + 2\ NO_3^-(aq) \rightarrow CuS(s) + 2\ Li^+(aq) + 2\ NO_3^-(aq)$
E) No reaction occurs.

Answer: D
Diff: 2 *Page Ref: 4.7*

35) Give the **net ionic equation** for the reaction (if any) that occurs when aqueous solutions of K_2S and $Fe(NO_3)_2$ are mixed.

A) $K^+(aq) + NO_3^-(aq) \rightarrow KNO_3(s)$

B) $Fe^{2+}(aq) + S^{2-}(aq) + 2\ K^+(aq) + 2\ NO_3^-(aq) \rightarrow FeS(s) + 2\ K^+(aq) + 2\ NO_3^-(aq)$

C) $Fe^{2+}(aq) + S^{2-}(aq) + 2\ K^+(aq) + 2\ NO_3^-(aq) \rightarrow Fe^{2+}(aq) + S^{2-}(aq) + 2\ KNO_3(s)$

D) $Fe^{2+}(aq) + S^{2-}(aq) \rightarrow FeS(s)$

E) No reaction occurs.

Answer: D

Diff: 2 *Page Ref:* 4.7

36) Give the **net ionic equation** for the reaction (if any) that occurs when aqueous solutions of H_2SO_4 and KOH are mixed.

A) $H^+(aq) + OH^-(aq) \rightarrow H_2O(l)$

B) $2\ K^+(aq) + SO_4^{2-}(aq) \rightarrow K_2SO_4(s)$

C) $H^+(aq) + OH^-(aq) + 2\ K^+(aq) + SO_4^{2-}(aq) \rightarrow H_2O(l) + K_2SO_4(s)$

D) $H_2^{2+}(aq) + OH^-(aq) \rightarrow H_2(OH)_2(l)$

E) No reaction occurs.

Answer: A

Diff: 2 *Page Ref:* 4.7

37) Give the **net ionic equation** for the reaction (if any) that occurs when aqueous solutions of $Al(C_2H_3O_2)_3$ and $LiNO_3$ are mixed.

A) $Al^{3+}(aq) + 3\ NO_3^-(aq) \rightarrow Al(NO_3)_3(s)$

B) $Li^+(aq) + C_2H_3O_2^-(aq) \rightarrow LiC_2H_3O_2(s)$

C) $Al^{3+}(aq) + 3\ NO_3^-(aq) + Li^+(aq) + C_2H_3O_2^-(aq) \rightarrow Al(NO_3)_3(aq) + LiC_2H_3O_2(s)$

D) $3\ Li^+(aq) + (C_2H_3O_2)_3^{3-}(aq) \rightarrow Li_3(C_2H_3O_2)_3(s)$

E) No reaction occurs.

Answer: E

Diff: 2 *Page Ref:* 4.7

38) Give the **net ionic equation** for the reaction (if any) that occurs when aqueous solutions of Na_2CO_3 and HCl are mixed.

A) $2\ H^+(aq) + CO_3^{2-}(aq) \rightarrow H_2CO_3(s)$

B) $2\ Na^+(aq) + CO_3^{2-}(aq) + 2\ H^+(aq) + 2\ Cl^-(aq) \rightarrow H_2CO_3(s) + 2\ NaCl(aq)$

C) $2\ H^+(aq) + CO_3^{2-}(aq) \rightarrow H_2O(l) + CO_2(g)$

D) $2\ Na^+(aq) + CO_3^{2-}(aq) + 2\ H^+(aq) + 2\ Cl^-(aq) \rightarrow H_2CO_3(s) + 2\ Na^+(aq) + 2\ Cl^-(aq)$

E) No reaction occurs.

Answer: C

Diff: 2 *Page Ref:* 4.7

39) Give the **complete ionic equation** for the reaction (if any) that occurs when aqueous solutions of $MgSO_3$ and HI are mixed.

A) $2\ H^+(aq) + SO_3^{2-}(aq) \rightarrow H_2SO_3(s)$

B) $Mg^{2+}(aq) + 2\ I^-(aq) \rightarrow MgI_2(s)$

C) $2\ H^+(aq) + SO_3^{2-}(aq) + Mg^{2+}(aq) + 2\ I^-(aq) \rightarrow H_2SO_3(s) + MgI_2(aq)$

D) $2\ H^+(aq) + SO_3^{2-}(aq) \rightarrow H_2O(l) + SO_2(g)$

E) No reaction occurs.

Answer: D
Diff: 2 *Page Ref: 4.7*

40) Which of the following is an acid-base reaction?

A) $C(s) + O_2(g) \rightarrow CO_2(g)$

B) $2\ HClO_4(aq) + Ca(OH)_2(aq) \rightarrow 2\ H_2O(l) + Ca(ClO_4)_2(aq)$

C) $Fe(s) + 2\ AgNO_3(aq) \rightarrow 2\ Ag(s) + Fe(NO_3)_2(aq)$

D) $MgSO_4(aq) + Ba(NO_3)_2(aq) \rightarrow Mg(NO_3)_2(aq) + BaSO_4(s)$

E) None of the above are acid base reactions.

Answer: B
Diff: 1 *Page Ref: 4.8*

41) Which of the following is a gas-evolution reaction?

A) $2\ C_2H_6(l) + 7\ O_2(g) \rightarrow 4\ CO_2(g) + 6\ H_2O(g)$

B) $2\ H_2(g) + O_2(g) \rightarrow 2\ H_2O(g)$

C) $LiCl(aq) + NaNO_3(aq) \rightarrow LiNO_3(aq) + NaCl(g)$

D) $NH_4Cl(aq) + KOH(aq) \rightarrow KCl(aq) + NH_3(g) + H_2O(l)$

E) None of the above are gas-evolution reactions.

Answer: D
Diff: 1 *Page Ref: 4.8*

42) The titration of 25.0 mL of an unknown concentration H_2SO_4 solution requires 83.6 mL of 0.12 M LiOH solution. What is the concentration of the H_2SO_4 solution (in M)?

A) 0.20 M B) 0.40 M C) 0.10 M D) 0.36 M E) 0.25 M

Answer: A
Diff: 4 *Page Ref: 4.8*

43) The titration of 80.0 mL of an unknown concentration H_3PO_4 solution requires 126 mL of 0.218 M KOH solution. What is the concentration of the H_3PO_4 solution (in M)?

A) 1.03 M B) 0.343 M C) 0.114 M D) 0.138 M E) 0.0461 M

Answer: C
Diff: 5 *Page Ref: 4.8*

44) Choose the reaction that represents the combustion of $C_6H_{12}O_2$.

A) $C_6H_{12}O_2(l) + 8\ O_2(g) \rightarrow 6\ CO_2(g) + 6\ H_2O(g)$
B) $Mg(s) + C_6H_{12}O_2(l) \rightarrow MgC_6H_{12}O_2(aq)$
C) $6\ C(s) + 6\ H_2(g) + O_2(g) \rightarrow C_6H_{12}O_2(l)$
D) $C_6H_{12}O_2(l) \rightarrow 6\ C(s) + 6\ H_2(g) + O_2(g)$
E) None of the above represent the combustion of $C_6H_{12}O_2$.

Answer: A
Diff: 1 *Page Ref: 4.9*

45) Determine the oxidation state of P in PO_3^{3-}.

A) +3 B) +6 C) +2 D) 0 E) -3

Answer: A
Diff: 2 *Page Ref: 4.9*

46) Determine the oxidation state of C in CO_3^{-2}.

A) +4 B) +2 C) -2 D) -4 E) +6

Answer: A
Diff: 2 *Page Ref: 4.9*

47) Determine the oxidation state of nitrogen in NO.

A) +5 B) +3 C) 0 D) +2 E) +4

Answer: D
Diff: 2 *Page Ref: 4.9*

48) Determine the oxidation state of nitrogen in NO_2.

A) +5 B) +3 C) 0 D) +2 E) +4

Answer: E
Diff: 2 *Page Ref: 4.9*

49) What element is undergoing oxidation (if any) in the following reaction?

$$CH_4(g) + 2\ O_2(g) \rightarrow CO_2(g) + 2\ H_2O(g)$$

A) O
B) H
C) C
D) both C and H
E) None of the elements is undergoing oxidation.

Answer: C
Diff: 2 *Page Ref: 4.9*

50) What element is undergoing reduction (if any) in the following reaction?

$$Zn(s) + 2\ AgNO_3(aq) \rightarrow Zn(NO_3)_2(aq) + 2\ Ag(s)$$

A) Zn
B) N
C) O
D) Ag
E) This is not an oxidation–reduction reaction.

Answer: D
Diff: 2 *Page Ref:* 4.9

51) What element is undergoing oxidation (if any) in the following reaction?

$$Zn(s) + 2\ AgNO_3(aq) \rightarrow Zn(NO_3)_2(aq) + 2\ Ag(s)$$

A) Zn
B) N
C) O
D) Ag
E) This is not an oxidation–reduction reaction.

Answer: A
Diff: 2 *Page Ref:* 4.9

52) Determine the reducing agent in the following reaction.

$$2\ Li(s) + Fe(C_2H_3O_2)_2(aq) \rightarrow 2\ LiC_2H_3O_2(aq) + Fe(s)$$

A) O B) H C) C D) Fe E) Li

Answer: E
Diff: 2 *Page Ref:* 4.9

53) Determine the oxidizing agent in the following reaction.

$$Ni(s) + 2\ AgClO_4(aq) \rightarrow Ni(ClO_4)_2(aq) + 2\ Ag(s)$$

A) Ag
B) Ni
C) Cl
D) O
E) This is not an oxidation–reduction reaction.

Answer: A
Diff: 2 *Page Ref:* 4.9

Algorithmic Questions

1) According to the following balanced reaction, how many moles of NO are formed from 8.44 moles of NO_2 if there is plenty of water present?

$$3\ NO_2(g) + H_2O(l) \rightarrow 2\ HNO_3(aq) + NO(g)$$

A) 25.32 moles NO
B) 12.66 moles NO
C) 8.44 moles NO
D) 5.63 moles NO
E) 2.82 moles NO

Answer: E
Diff: 2 *Page Ref: 4.2*

2) According to the following balanced reaction, how many moles of HNO_3 are formed from 8.44 moles of NO_2 if there is plenty of water present?

$$3\ NO_2(g) + H_2O(l) \rightarrow 2\ HNO_3(aq) + NO(g)$$

A) 25.32 moles NO
B) 12.66 moles NO
C) 8.44 moles NO
D) 5.63 moles NO
E) 2.82 moles NO

Answer: D
Diff: 2 *Page Ref: 4.2*

3) Consider the following reaction. How many moles of oxygen are required to produce 10.0 moles of water? Assume that there is excess C_3H_7SH present.

$$C_3H_7SH(l) + 6\ O_2(g) \rightarrow 3\ CO_2(g) + SO_2(g) + 4\ H_2O(g)$$

A) 6.67 moles O_2
B) 15.0 moles O_2
C) 10.0 moles O_2
D) 40.0 moles O_2
E) 2.50 moles O_2

Answer: B
Diff: 2 *Page Ref: 4.2*

4) According to the following balanced reaction, how many moles of KO are required to exactly react with 5.44 moles of H_2O?

$$4\ KO(s) + 2\ H_2O(l) \rightarrow 4\ KOH(s) + O_2(g)$$

A) 5.44 moles H_2O
B) 10.9 moles H_2O
C) 21.8 moles H_2O
D) 2.72 moles H_2O
E) 1.36 moles H_2O

Answer: B
Diff: 2 *Page Ref: 4.2*

5) According to the following balanced reaction, how many moles of KOH are required to exactly react with 5.44 moles of H_2O?

$$4\ KO(s) + 2\ H_2O(l) \rightarrow 4\ KOH(s) + O_2(g)$$

A) 5.44 moles H_2O
B) 10.9 moles H_2O
C) 21.8 moles H_2O
D) 2.72 moles H_2O
E) 1.36 moles H_2O

Answer: B
Diff: 2 *Page Ref: 4.2*

6) How many grams of oxygen are formed when 6.21 moles of KOH are formed?

$$4\ KO(s) + 2\ H_2O(l) \rightarrow 4\ KOH(s) + O_2(g)$$

A) 6.21 g O_2 B) 24.8 g O_2 C) 49.7 g O_2 D) 199 g O_2 E) 99.4 g O_2

Answer: C
Diff: 3 *Page Ref: 4.2*

7) How many molecules of HCl are formed when 80.0 g of water reacts according to the following balanced reaction? Assume excess ICl_3.

$$2\ ICl_3 + 3\ H_2O \rightarrow ICl + HIO_3 + 5\ HCl$$

A) 4.46×10^{24} molecules HCl
B) 2.68×10^{24} molecules HCl
C) 8.04×10^{24} molecules HCl
D) 5.36×10^{24} molecules HCl
E) 4.46×10^{25} molecules HCl

Answer: A
Diff: 5 *Page Ref: 4.2*

8) Carbonic acid can form water and carbon dioxide upon heating. How much carbon dioxide is formed from 24.8 g of carbonic acid?

$$H_2CO_3 \rightarrow H_2O + CO_2$$

A) 17.6 g B) 34.9 g C) 24.8 g D) 7.20 g E) 85.4 g

Answer: A

Diff: 5 *Page Ref: 4.2*

9) Lithium and nitrogen react to produce lithium nitride:

$$6Li(s) + N_2(g) \rightarrow 2Li_3N(s)$$

How many moles of N_2 are needed to react with 0.500 mol of lithium?

A) 3.00 B) 0.500 C) 0.167 D) 1.50 E) 0.0833

Answer: E

Diff: 2 *Page Ref: 4.2*

10) Lithium and nitrogen react to produce lithium nitride:

$$6Li(s) + N_2(g) \rightarrow 2Li_3N(s)$$

How many moles of lithium nitride are produced when 0.450 mol of lithium react in this fashion?

A) 0.150 B) 0.900 C) 0.0750 D) 1.35 E) 0.225

Answer: A

Diff: 2 *Page Ref: 4.2*

11) Lithium and nitrogen react in a combination reaction to produce lithium nitride:

$$6Li(s) + N_2(g) \rightarrow 2Li_3N(s)$$

How many moles of lithium are needed to produce 0.60 mol of Li_3N when the reaction is carried out in the presence of excess nitrogen?

A) 0.30 B) 1.8 C) 0.20 D) 0.40 E) 3.6

Answer: B

Diff: 2 *Page Ref: 4.2*

12) Automotive air bags inflate when sodium azide decomposes explosively to its constituent elements:

$$2NaN_3(s) \rightarrow 2Na(s) + 3N_2(g)$$

How many moles of N_2 are produced by the decomposition of 2.88 mol of sodium azide?

A) 1.92 B) 8.64 C) 4.32 D) 0.960 E) 1.44

Answer: C

Diff: 2 *Page Ref: 4.2*

13) Magnesium burns in air with a dazzling brilliance to produce magnesium oxide:

$$2Mg(s) + O_2(g) \rightarrow 2MgO(s)$$

How many moles of O_2 are consumed when 0.770 mol of magnesium burns?

A) 0.0317 B) 2.60 C) 0.770 D) 1.54 E) 0.385

Answer: E

Diff: 2 *Page Ref:* 4.2

14) Automotive air bags inflate when sodium azide decomposes explosively to its constituent elements:

$$2NaN_3(s) \rightarrow 2Na(s) + 3N_2(g)$$

How many grams of sodium azide are required to produce 33.0 g of nitrogen?

A) 1.77 B) 0.785 C) 76.6 D) 51.1 E) 114.9

Answer: D

Diff: 3 *Page Ref:* 4.2

15) How many moles of CuO can be produced from 0.900 mol of Cu_2O in the following reaction?

$$2\ Cu_2O(s) + O_2(g) \rightarrow 4\ CuO(s)$$

A) 0.450 mol B) 0.900 mol C) 1.80 mol D) 3.60 mol

Answer: C

Diff: 5 *Page Ref:* 4.2

16) How many moles of BCl_3 are needed to produce 10.0 g of HCl(*aq*) in the following reaction?

$$BCl_3(g) + 3\ H_2O(l) \rightarrow 3\ HCl(aq) + B(OH)_3(aq)$$

A) 0.0914 mol B) 0.274 mol C) 0.823 mol D) 10.9 mol

Answer: A

Diff: 5 *Page Ref:* 4.2

17) How many grams of calcium chloride are needed to produce 1.50 g of potassium chloride?

$$CaCl_2(aq) + K_2CO_3(aq) \rightarrow 2\ KCl(aq) + CaCO_3(aq)$$

A) 0.896 g B) 1.12 g C) 2.23 g D) 4.47 g

Answer: B

Diff: 5 *Page Ref:* 4.2

18) Balance the chemical equation given below, and determine the number of moles of iodine that reacts with 30.0 g of aluminum.

_____ Al(s) + _____ I_2(s) → _____ Al_2I_6(s)

A) 0.741 mol B) 1.67 mol C) 2.22 mol D) 3.33 mol

Answer: B
Diff: 5 *Page Ref: 4.2*

19) Balance the chemical equation given below, and determine the number of grams of MgO are needed to produce 10.0 g of Fe_2O_3.

_____ MgO(s) + _____ Fe(s) → _____ Fe_2O_3(s) + _____ Mg(s)

A) 0.312 g B) 0.841 g C) 2.52 g D) 7.57 g

Answer: D
Diff: 5 *Page Ref: 4.2*

20) Dinitrogen monoxide gas decomposes to form nitrogen gas and oxygen gas. How many grams of oxygen are formed when 10.0 g of dinitrogen monoxide decomposes?
A) 0.275 g B) 3.64 g C) 7.27 g D) 14.5 g

Answer: B
Diff: 5 *Page Ref: 4.2*

21) If the density of ethanol, C_2H_5OH, is 0.789 g/mL. How many milliliters of ethanol are needed to produce 15.0 g of CO_2 according to the following chemical equation?

$C_2H_5OH(l) + 3\ O_2(g) \rightarrow 2\ CO_2(g) + 3\ H_2O(l)$

A) 6.19 mL B) 9.95 mL C) 19.9 mL D) 39.8 mL

Answer: B
Diff: 5 *Page Ref: 4.2*

22) Magnesium burns in air with a dazzling brilliance to produce magnesium oxide:

$2Mg(s) + O_2(g) \rightarrow 2MgO(s)$

When 4.00 g of magnesium burns, the theoretical yield of magnesium oxide is ________ g.
A) 4.00 B) 6.63 C) 0.165 D) 3.32 E) 13.3

Answer: B
Diff: 3 *Page Ref: 4.3*

23) If the percent yield for the following reaction is 65.0%, how many grams of $KClO_3$ are needed to produce 32.0 g of O_2?

$$2\ KClO_3(s) \rightarrow 2\ KCl(s) + 3\ O_2(g)$$

A) 53.1 g B) 81.7 g C) 126 g D) 283 g

Answer: C
Diff: 5 *Page Ref: 4.3*

24) If the percent yield for the following reaction is 75.0%, and 45.0 g of NO_2 are consumed in the reaction, how many grams of nitric acid, $HNO_3(aq)$ are produced?

$$3\ NO_2(g) + H_2O(l) \rightarrow 2\ HNO_3(aq) + NO(g)$$

A) 30.8 g B) 41.1 g C) 54.8 g D) 69.3 g

Answer: A
Diff: 5 *Page Ref: 4.3*

25) Determine the theoretical yield of H_2S (in moles) if 64 mol Al_2S_3 and 64 mol H_2O are reacted according to the following balanced reaction. A possibly useful molar mass is Al_2S_3 = 150.17 g/mol.

$$Al_2S_3(s) + 6\ H_2O(l) \rightarrow 2\ Al(OH)_3(s) + 3\ H_2S(g)$$

A) 192 mol H_2S
B) 64 mol H_2S
C) 128 mol H_2S
D) 96 mol H_2S
E) 32 mol H_2S

Answer: E
Diff: 3 *Page Ref: 4.3*

26) Lithium and nitrogen react in a combination reaction to produce lithium nitride:

$$6Li(s) + N_2(g) \rightarrow 2Li_3N(s)$$

In a particular experiment, 3.50-g samples of each reagent are reacted. The theoretical yield of lithium nitride is ________ g.

A) 3.52 B) 2.93 C) 17.6 D) 5.85 E) 8.7

Answer: D
Diff: 3 *Page Ref: 4.3*

27) Calcium oxide reacts with water in a combination reaction to produce calcium hydroxide:

$$CaO(s) + H_2O(l) \rightarrow Ca(OH)_2(s)$$

A 4.50-g sample of CaO is reacted with 4.34 g of H_2O. How many grams of water remain after the reaction is complete?

A) 0.00 B) 0.00892 C) 2.90 D) 1.04 E) 0.161

Answer: C

Diff: 4 *Page Ref: 4.3*

28) If 294 grams of FeS_2 is allowed to react with 176 grams of O_2 according to the following unbalanced equation, how many grams of Fe_2O_3 are produced?

$$FeS_2 + O_2 \rightarrow Fe_2O_3 + SO_2$$

Answer: 160

Diff: 4 *Page Ref: 4.3*

29) Calcium oxide reacts with water in a combination reaction to produce calcium hydroxide:

$$CaO(s) + H_2O(l) \rightarrow Ca(OH)_2(s)$$

In a particular experiment, a 5.00-g sample of CaO is reacted with excess water and 6.11 g of $Ca(OH)_2$ is recovered. What is the percent yield in this experiment?

A) 122 B) 1.22 C) 7.19 D) 92.5 E) 81.9

Answer: D

Diff: 4 *Page Ref: 4.3*

30) Sodium metal and water react to form hydrogen and sodium hydroxide. If 5.98 g of sodium react with water to form 0.26 g of hydrogen and 10.40 g of sodium hydroxide, what mass of water was involved in the reaction?

A) 4.68 g B) 5.98 g C) 10.14 g D) 10.66 g

Answer: A

Diff: 5 *Page Ref: 4.3*

31) Which substance is the limiting reactant when 2.0 g of sulfur reacts with 3.0 g of oxygen and 4.0 g of sodium hydroxide according to the following chemical equation
:

$$2\ S(s) + 3\ O_2(g) + 4\ NaOH(aq) \rightarrow 2\ Na_2SO_4(aq) + 2\ H_2O(l)$$

A) S(*s*)
B) $O_2(g)$
C) NaOH(*aq*)
D) None of these substances is the limiting reactant.

Answer: C

Diff: 5 *Page Ref: 4.3*

32) When 7.00×10^{22} molecules of ammonia react with 6.00×10^{22} molecules of oxygen according to the chemical equation shown below, how many grams of nitrogen gas are produced?

$$4\ NH_3(g) + 3\ O_2(g) \rightarrow 2\ N_2(g) + 6\ H_2O(g)$$

A) 1.63 g B) 1.86 g C) 4.19 g D) 6.51 g

Answer: A
Diff: 5 *Page Ref: 4.3*

33) When silver nitrate reacts with barium chloride, silver chloride and barium nitrate are formed. How many grams of silver chloride are formed when 10.0 g of silver nitrate reacts with 15.0 g of barium chloride?

A) 8.44 g B) 9.40 g C) 11.9 g D) 18.8 g

Answer: A
Diff: 5 *Page Ref: 4.3*

34) When 11.0 g of calcium metal is reacted with water, 5.00 g of calcium hydroxide is produced. Using the following balanced equation, calculate the percent yield for the reaction?

$$Ca(s) + 2\ H_2O(l) \rightarrow Ca(OH)_2(aq) + H_2(g)$$

A) 12.3% B) 24.6% C) 45.5% D) 84.0%

Answer: B
Diff: 5 *Page Ref: 4.3*

35) 7.0 g of nitrogen is reacted with 5.0 g of hydrogen to produce ammonia according to the chemical equation shown below. Which one of the following statements is **false**?

$$N_2(g) + 3\ H_2(g) \rightarrow 2\ NH_3(g)$$

A) 3.5 g of hydrogen are left over.
B) Hydrogen is the excess reactant.
C) Nitrogen is the limiting reactant.
D) The theoretical yield of ammonia is 15 g.

Answer: D
Diff: 5 *Page Ref: 4.3*

36) 5.0 g of iron is reacted with 5.0 g of water according to the chemical equation shown below. Which one of the following statements is **false**?

$$3\ Fe(s) + 4\ H_2O(l) \rightarrow Fe_3O_4(s) + 4\ H_2(g)$$

A) 6.91 g of Fe_3O_4 are produced.
B) 2.85 g of H_2O are left over.
C) Mass is conserved in this reaction.
D) Water is the limiting reactant.

Answer: D
Diff: 5 *Page Ref: 4.3*

37) Which of the following solutions will have the highest concentration of chloride ions?
A) 0.10 M LiCl
B) 0.10 M $MgCl_2$
C) 0.10 M $AlCl_3$
D) 0.05 M $CaCl_2$
E) All of these solutions have the same concentration of chloride ions.
Answer: C
Diff: 1 Page Ref: 4.4

38) Which of the following solutions will have the highest concentration of chloride ions?
A) 0.20 M LiCl
B) 0.40 M $MgCl_2$
C) 0.60 M $AlCl_3$
D) 0.40 M $CaCl_2$
E) All of these solutions have the same concentration of chloride ions.
Answer: C
Diff: 1 Page Ref: 4.4

39) How many milliliters of a 0.184 M $CsNO_3$ solution contain 0.113 moles of $CsNO_3$?
A) 543 mL B) 163 mL C) 614 mL D) 885 mL E) 326 mL
Answer: C
Diff: 2 Page Ref: 4.4

40) How many moles of CH_3CH_2F are contained in 548 mL of 0.0788 M CH_3CH_2F solution?
A) 4.32×10^{-2} mol
B) 2.32×10^{-2} mol
C) 6.95×10^{-2} mol
D) 1.44×10^{-2} mol
E) 5.26×10^{-2} mol
Answer: A
Diff: 2 Page Ref: 4.4

41) How many milliliters of a 0.266 M $NaNO_3$ solution are required to make 150.0 mL of 0.075 M $NaNO_3$ solution?
A) 53.2 mL B) 42.3 mL C) 18.8 mL D) 23.6 mL E) 35.1 mL
Answer: B
Diff: 2 Page Ref: 4.4

42) What volume (in mL) of 0.0887 M MgI_2 solution is needed to make 275.0 mL of 0.0224 M MgI_2 solution?
A) 72.3 mL B) 91.8 mL C) 10.9 mL D) 69.4 mL E) 14.4 mL
Answer: D
Diff: 2 Page Ref: 4.4

43) How many liters of a 0.0550 M NaBr solution contain 0.163 moles of NaBr?

A) 3.37 L
B) 1.48 L
C) 8.97 L
D) 2.96 L
E) 1.12 L

Answer: D
Diff: 2 *Page Ref:* 4.4

44) How many moles of LiF are contained in 258.6 mL of 0.0296 M LiF solution?

A) 1.31×10^{-3} mol
B) 8.74×10^{-3} mol
C) 1.14×10^{-3} mol
D) 3.67×10^{-3} mol
E) 7.65×10^{-3} mol

Answer: E
Diff: 2 *Page Ref:* 4.4

45) How many moles of KBr are required to make 250 mL of a 3.00 M solution?

A) 750 moles
B) 0.750 moles
C) 3 moles
D) 0.250 moles

Answer: B
Diff: 2 *Page Ref:* 4.4

46) What is the concentration of nitrate ions in a 0.125 M $Mg(NO_3)_2$ solution?

A) 0.125 M
B) 0.0625 M
C) 0.375 M
D) 0.250 M
E) 0.160 M

Answer: D
Diff: 2 *Page Ref:* 4.4

47) What is the concentration of magnesium ions in a 0.125 M $MgBr_2$ solution?

A) 0.125 M
B) 0.0625 M
C) 0.375 M
D) 0.250 M
E) 0.160 M

Answer: A
Diff: 2 *Page Ref:* 4.4

48) Determine the concentration of a solution prepared by diluting 25.0 mL of a stock 0.188 M $Ca(NO_3)_2$ solution to 150.0 mL.

A) 1.13 M
B) 0.0887 M
C) 0.0313 M
D) 0.0199 M
E) 0.0501 M

Answer: C
Diff: 2 *Page Ref:* 4.4

49) Determine the concentration of a solution prepared by diluting 20.0 mL of a 0.200 M CsCl to 250.0 mL.

A) 0.160 M
B) 0.0320 M
C) 2.50 M
D) 0.00800 M
E) 0.0160 M

Answer: E
Diff: 2 *Page Ref:* 4.4

50) Calculate the number of grams of solute in 500.0 mL of 0.189 M KOH.
A) 148
B) 1.68
C) 5.30×10^3
D) 5.30
E) 1.68×10^{-3}

Answer: D
Diff: 2 *Page Ref: 4.4*

51) How many grams of NaCl are required to make 500.0 mL of a 1.500 M solution?
A) 58.40 g B) 175.3 g C) 14.60 g D) 43.83 g

Answer: D
Diff: 3 *Page Ref: 4.4*

52) What is the concentration of $FeCl_3$ in a solution prepared by dissolving 20.0 g of $FeCl_3$ in enough water to make 275 mL of solution?
A) 4.48×10^{-4} M B) 0.448 M C) 2.23 M D) 2.23×10^3 M

Answer: B
Diff: 3 *Page Ref: 4.4*

53) How many grams of $AgNO_3$ are needed to make 250. mL of a solution that is 0.135 M?
A) 0.0917 g B) 0.174 g C) 5.73 g D) 91.7 g

Answer: C
Diff: 3 *Page Ref: 4.4*

54) What volume of a 0.540 M NaOH solution contains 11.5 g of NaOH?
A) 0.155 L B) 0.532 L C) 1.88 L D) 6.44 L

Answer: B
Diff: 3 *Page Ref: 4.4*

55) What is the concentration (M) of sodium ions in 1.57 L of a .398 M Na_3P solution?

Answer: 1.19
Diff: 3 *Page Ref: 4.4*

56) What is the concentration (M) of CH_3OH in a solution prepared by dissolving 16.8 g of C H_3OH in sufficient water to give exactly 230 mL of solution?

Answer: 2.28
Diff: 3 *Page Ref: 4.4*

57) How many grams of H_3PO_4 are in 265 mL of a 1.50 M solution of H_3PO_4?

Answer: 39.0
Diff: 3 *Page Ref: 4.4*

58) What is the concentration (M) of a NaCl solution prepared by dissolving 7.2 g of NaCl in sufficient water to give 425 mL of solution?

Answer: 0.29
Diff: 3 *Page Ref: 4.4*

59) How many grams of NaOH (MW = 40.0) are there in 250.0 mL of a 0.275 M NaOH solution?

Answer: 2.75
Diff: 3 *Page Ref: 4.4*

60) How many grams of CH_3OH must be added to water to prepare 150mL of a solution that is 2.0 M CH_3OH?

Answer: 9.6
Diff: 3 *Page Ref: 4.4*

61) What is the concentration of HCl in the **final** solution when 65 mL of a 12 M HCl solution is diluted with pure water to a total volume of 0.15 L?

A) 2.8×10^{-2} M B) 5.2 M C) 28 M D) 5.2×10^{3} M

Answer: B
Diff: 3 *Page Ref: 4.4*

62) How many milliliters of a 9.0 M H_2SO_4 solution are needed to make 0.35 L of a 3.5 M solution?

A) 0.14 mL B) 0.90 mL C) 140 mL D) 900 mL

Answer: C
Diff: 3 *Page Ref: 4.4*

63) A student prepared a stock solution by dissolving 10.0 g of KOH in enough water to make 150. mL of solution. She then took 15.0 mL of the stock solution and diluted it with enough water to make water to make 65.0 mL of a final solution. What is the concentration of KOH for the final solution?

A) 0.274 M B) 0.356 M C) 2.81 M D) 3.65 M

Answer: A
Diff: 3 *Page Ref: 4.4*

64) How many milliliters of a stock solution of 11.1 M HNO_3 would be needed to prepare 0.500 L of 0.500 M HNO_3?

A) 0.0444 B) 22.5 C) 2.78 D) 44.4 E) 0.0225

Answer: B
Diff: 3 *Page Ref: 4.4*

65) A stock solution of HNO_3 is prepared and found to contain 13.5 M of HNO_3. If 25.0 mL of the stock solution is diluted to a final volume of 0.500 L, the concentration of the diluted solution is ________ M.

A) 0.270 B) 1.48 C) 0.675 D) 675 E) 270

Answer: C
Diff: 3 *Page Ref: 4.4*

66) A $FeCl_3$ solution is 0.175 M. How many mL of a 0.175 M $FeCl_3$ solution are needed to make 450. mL of a solution that is 0.300 M in Cl^- ion?

A) 0.771 mL
B) 257 mL
C) 771 mL
D) It is not possible to make a more concentrated solution from a less concentrated solution.

Answer: B
Diff: 3 *Page Ref: 4.4*

67) Pure acetic acid ($HC_2H_3O_2$) is a liquid and is known as glacial acetic acid. Calculate the molarity of a solution prepared by dissolving 10.00 mL of glacial acetic acid at 25 °C in sufficient water to give 500.0 mL of solution. The density of glacial acetic acid at 25 °C is 1.05 g/mL.

A) 1.26×10^3 M
B) 21.0 M
C) 0.0210 M
D) 0.350 M
E) 3.50×10^{-4} M

Answer: D
Diff: 4 *Page Ref: 4.4*

68) What is the concentration of NO_3^- ions in a solution prepared by dissolving 25.0 g of $Ca(NO_3)_2$ in enough water to produce 300. mL of solution?

A) 0.254 M B) 0.508 M C) 0.672 M D) 1.02 M

Answer: D
Diff: 4 *Page Ref: 4.4*

69) If the reaction of phosphate ion with water is ignored, what is the **total** concentration of ions in a solution prepared by dissolving 3.00 g of K_3PO_4 in enough water to make 350. mL of solution?

A) 0.0101 M B) 0.0404 M C) 0.162 M D) 0.323 M

Answer: C
Diff: 4 *Page Ref: 4.4*

70) What is the concentration of an $AlCl_3$ solution if 150. mL of the solution contains 450. mg of Cl^- ion?

A) 2.82×10^{-2} M B) 6.75×10^{-2} M C) 8.46×10^{-2} M D) 2.54×10^{-1} M

Answer: A
Diff: 4 *Page Ref: 4.4*

71) How many fluoride ions are present in 65.5 mL of 0.210 M AlF_3 solution?

A) 4.02×10^{23} fluoride ions
B) 5.79×10^{24} fluoride ions
C) 2.48×10^{22} fluoride ions
D) 8.28×10^{21} fluoride ions
E) 1.21×10^{22} fluoride ions

Answer: C
Diff: 3 Page Ref: 4.4

72) How many aluminum ions are present in 65.5 mL of 0.210 M AlF_3 solution?

A) 4.02×10^{23} aluminum ions
B) 5.79×10^{24} aluminum ions
C) 2.48×10^{22} aluminum ions
D) 8.28×10^{21} aluminum ions
E) 1.21×10^{22} aluminum ions

Answer: D
Diff: 3 Page Ref: 4.4

73) How many ions are present in 30.0 mL of 0.600 M K_2CO_3 solution?

A) 3.25×10^{22} ions
B) 2.17×10^{22} ions
C) 1.08×10^{22} ions
D) 5.42×10^{22} ions
E) 3.61×10^{21} ions

Answer: A
Diff: 3 Page Ref: 4.4

74) How many lithium ions are present in 30.0 mL of 0.600 M Li_2CO_3 solution?

A) 3.25×10^{22} ions
B) 2.17×10^{22} ions
C) 1.08×10^{22} ions
D) 5.42×10^{22} ions
E) 3.61×10^{21} ions

Answer: B
Diff: 3 Page Ref: 4.4

75) How many carbonate ions are present in 30.0 mL of 0.600 M K_2CO_3 solution?

A) 3.25×10^{22} ions
B) 2.17×10^{22} ions
C) 1.08×10^{22} ions
D) 5.42×10^{22} ions
E) 3.61×10^{21} ions

Answer: C
Diff: 3 Page Ref: 4.4

76) A student dissolved 4.00 g of $Co(NO_3)_2$ in enough water to make 100. mL of stock solution. He took 4.00 mL of the stock solution and then diluted it with water to give 275. mL of a final solution. How many grams of NO_3^- ion are there in the final solution?

A) 0.0197 g B) 0.0394 g C) 0.0542 g D) 0.108 g

Answer: D
Diff: 4 *Page Ref: 4.4*

77) There are ________ mol of bromide ions in 0.900 L of a 0.500M solution of $AlBr_3$.

Answer: 1.35
Diff: 4 *Page Ref: 4.4*

78) How many moles of Co^{2+} are present in 0.150 L of a 0.200 M solution of CoI_2?

Answer: 0.0300
Diff: 4 *Page Ref: 4.4*

79) What is the molar concentration of sodium ions in a 0.450 M Na_3PO_4 solution?

A) 0.150 M B) 0.450 M C) 1.35 M D) 1.80 M

Answer: C
Diff: 4 *Page Ref: 4.4*

80) Calculate the concentration (M) of sodium ions in a solution made by diluting 40.0 mL of a 0.474 M solution of sodium sulfide to a total volume of 300 mL.

Answer: 0.126
Diff: 4 *Page Ref: 4.4*

81) Balance the chemical equation given below, and calculate the volume of nitrogen monoxide gas produced when 8.00 g of ammonia is reacted with 12.0 g of oxygen at 25°C? The density of nitrogen monoxide at 25°C is 1.23 g/L.

_____ $NH_3(g)$ + _____ $O_2(g)$ → _____ $NO(g)$ + _____ $H_2O(l)$

A) 7.32 L B) 11.1 L C) 11.5 L D) 17.3 L

Answer: A
Diff: 5 *Page Ref: 4.4*

82) How many milliliters of 0.260 M Na_2S are needed to react with 40.00 mL of 0.315 M $AgNO_3$?

$$Na_2S(aq) + 2\ AgNO_3(aq) \rightarrow 2\ NaNO_3(aq) + Ag_2S(s)$$

A) 24.2 mL B) 48.5 mL C) 66.0 mL D) 96.9 mL

Answer: A
Diff: 5 *Page Ref: 4.4*

83) How many grams of $CaCl_2$ are formed when 15.00 mL of 0.00237 M $Ca(OH)_2$ reacts with excess Cl_2 gas?

$$2\ Ca(OH)_2(aq) + 2\ Cl_2(g) \rightarrow Ca(OCl)_2(aq) + CaCl_2(s) + 2\ H_2O(l)$$

A) 0.00197 g B) 0.00394 g C) 0.00789 g D) 0.0507 g

Answer: A
Diff: 5 *Page Ref: 4.4*

84) When 31.2 mL of 0.500 M $AgNO_3$ is added to 25.0 mL of 0.300 M NH_4Cl, how many grams of AgCl are formed?

$$AgNO_3(aq) + NH_4Cl(aq) \rightarrow AgCl(s) + NH_4NO_3(aq)$$

A) 1.07 g B) 2.24 g C) 3.31 g D) 6.44 g

Answer: A
Diff: 5 *Page Ref: 4.4*

85) How many milliliters of 0.200 M $FeCl_3$ are needed to react with an excess of Na_2S to produce 1.38 g of Fe_2S_3 if the percent yield for the reaction is 65.0%?

$$3\ Na_2S(aq) + 2\ FeCl_3(aq) \rightarrow Fe_2S_3(s) + 6\ NaCl(aq)$$

A) 25.5 mL B) 43.1 mL C) 51.1 mL D) 102 mL

Answer: D
Diff: 5 *Page Ref: 4.4*

86) If 100. mL of 0.400 M Na_2SO_4 is added to 200. mL of 0.600 M NaCl, what is the concentration of Na^+ ions in the final solution? Assume that the volumes are additive.

A) 0.534 M B) 0.667 M C) 1.00 M D) 1.40 M

Answer: B
Diff: 5 *Page Ref: 4.4*

87) How many milliliters of 0.550 M hydriodic acid are needed to react with 15.00 mL of 0.217 M CsOH?

$$HI(aq) + CsOH(aq) \rightarrow CsI(aq) + H_2O(l)$$

A) 0.0263 mL B) 0.169 mL C) 5.92 mL D) 38.0 mL

Answer: C
Diff: 5 *Page Ref: 4.4*

88) Balance the chemical equation given below, and determine the number of milliliters of 0.00300 M phosphoric acid required to neutralize 45.00 mL of 0.00150 M calcium hydroxide.

_____ $Ca(OH)_2(aq)$ + _____ $H_3PO_4(aq)$ → _____ $Ca_3(PO_4)_2(aq)$ + _____ $H_2O(l)$

A) 3.04 mL B) 15.0 mL C) 22.5 mL D) 33.8 mL

Answer: B
Diff: 5 *Page Ref: 4.4*

89) According to the balanced equation shown below, 4.00 moles of oxalic acid, $H_2C_2O_4$, reacts with ________ moles of permanganate, MnO_4^-.

5 $H_2C_2O_4(aq)$ + 2 $MnO_4^-(aq)$ + 6 $H^+(aq)$ → 10 $CO_2(g)$ + $Mn^{2+}(aq)$ + 8 $H_2O(l)$

A) 1.60 B) 4.00 C) 8.00 D) 9.00

Answer: A
Diff: 5 *Page Ref: 4.4*

90) Based on the balanced chemical equation shown below, what volume of 0.250 M $K_2S_2O_3(aq)$, is needed to completely react with 24.88 mL of 0.125 M $KI_3(aq)$, according to the following chemical equation.

2 $S_2O_3^{2-}(aq)$ + $I_3^-(aq)$ → $S_4O_6^{2-}(aq)$ + 3 $I^-(aq)$

A) 6.22 mL B) 12.4 mL C) 24.9 mL D) 99.5 mL

Answer: C
Diff: 5 *Page Ref: 4.4*

91) What is the molarity of a potassium triiodide solution, $KI_3(aq)$, if 30.00 mL of the solution is required to completely react with 25.00 mL of a 0.200 M thiosulfate solution, $K_2S_2O_3(aq)$? The chemical equation for the reaction is

2 $S_2O_3^{2-}(aq)$ + $I_3^-(aq)$ → $S_4O_6^{2-}(aq)$ + 3 $I^-(aq)$.

A) 0.0833 M B) 0.120 M C) 0.167 M D) 0.333 M

Answer: A
Diff: 5 *Page Ref: 4.4*

92) Based on the balanced chemical equation shown below, determine the molarity of a solution containing $Fe^{2+}(aq)$, if 40.00 mL of the $Fe^{2+}(aq)$ solution is required to completely react with 30.00 mL of a 0.250 M potassium bromate, $KBrO_3(aq)$, solution. The chemical equation for the reaction is

6 $Fe^{2+}(aq)$ + $BrO_3^-(aq)$ + 6 $H^+(aq)$ → 6 $Fe^{3+}(aq)$ + $Br^-(aq)$ + 3 $H_2O(l)$.

A) 0.0312 M B) 0.188 M C) 1.12 M D) 2.00 M

Answer: C
Diff: 5 *Page Ref: 4.4*

93) Based on the balanced chemical equation shown below, determine the mass percent of Fe^{3+} in a 0.7450 g sample of iron ore, if 22.40 mL of a 0.1000 M stannous chloride, $SnCl_2(aq)$, solution is required to completely react with the Fe^{3+} present in the ore sample. The chemical equation for the reaction is

$$2\ Fe^{3+}(aq) + Sn^{2+}(aq) \rightarrow 2\ Fe^{2+}(aq) + Sn^{4+}(aq).$$

A) 5.365%   B) 8.396%   C) 16.79%   D) 33.58%

Answer: D
Diff: 5   *Page Ref: 4.4*

94) How many grams of H_2 gas can be produced by the reaction of 54.0 grams of Al(s) with an excess of dilute hydrochloric acid in the reaction shown below?

$$2\ Al(s) + 6\ HCl(aq) \rightarrow 2\ AlCl_3(aq) + 3\ H_2(g)$$

A) 2.68 g   B) 4.04 g   C) 6.05 g   D) 12.1 g

Answer: C
Diff: 5   *Page Ref: 4.4*

95) Which of the following solutions will have the highest electrical conductivity?
A) 0.045 M $Al_2(SO_4)_3$
B) 0.050 M $(NH_4)_2CO_3$
C) 0.10 M NaBr
D) 0.10 M NaI
E) 0.10 M NaF

Answer: A
Diff: 1   *Page Ref: 4.5*

96) Which one of the following compounds behaves as an acid when dissolved in water?
A) RaO   B) C_4H_{10} C) HI   D) RbOH

Answer: C
Diff: 1   *Page Ref: 4.5*

97) Which of the compounds of $H_2C_2O_4$, $Ca(OH)_2$, KOH, and HI, behave as acids when they are dissolved in water?
A) $Ca(OH)_2$ and KOH   B) $H_2C_2O_4$ and HI
C) only HI   D) only KOH

Answer: B
Diff: 1   *Page Ref: 4.5*

98) Which of the compounds $H_2C_2O_4$, $Ca(OH)_2$, KOH, and HI, behave as bases when they are dissolved in water?
A) $Ca(OH)_2$ and KOH   B) $H_2C_2O_4$ and HI
C) only HI   D) only KOH

Answer: A
Diff: 1   *Page Ref: 4.5*

99) When dissolved in water, KOH behaves as

A) an acid that forms K^+ and OH^- ions.
B) an acid that forms KO^- and H^+ ions.
C) a base that forms K^+ and OH^- ions.
D) a base that forms KO^- and H^+ ions.

Answer: C
Diff: 1 *Page Ref: 4.5*

100) H Cl , HI , H_2SO_4, Li Cl , and KI are all classified as

A) acids.
B) nonelectrolytes.
C) strong electrolytes.
D) weak electrolytes.

Answer: C
Diff: 1 *Page Ref: 4.5*

101) Which of the following is **NOT** a strong electrolyte?

A) NaOH
B) $CaCl_2$
C) $CaCO_3$
D) $NaC_2H_3O_2$
E) K_2SO_4

Answer: C
Diff: 1 *Page Ref: 4.5*

102) Which of the following is considered a **STRONG** electrolyte?

A) NH_4NO_3
B) $C_{12}H_{22}O_{11}$
C) $PbBr_2$
D) $HC_2H_3O_2$
E) CH_3SH

Answer: A
Diff: 1 *Page Ref: 4.5*

103) Identify water.

A) strong electrolyte, weak acid
B) weak electrolyte, weak acid
C) strong electrolyte, strong acid
D) weak electrolyte, strong acid
E) nonelectrolyte

Answer: E
Diff: 1 *Page Ref: 4.5*

104) Identify LiBr.

A) weak acid
B) weak electrolyte
C) strong acid
D) strong electrolyte
E) nonelectrolyte

Answer: D
Diff: 1 *Page Ref: 4.5*

105) Which one of the following compounds is insoluble in water?
A) Ca Cl_2 B) $NaNO_3$ C) Pb Cl_2 D) K_2CO_3

Answer: C
Diff: 3 *Page Ref: 4.5*

106) Which one of the following compounds is soluble in water?
A) $Cu_3(PO_4)_2$ B) CoS C) $Pb(NO_3)_2$ D) $ZnCO_3$

Answer: C
Diff: 3 *Page Ref: 4.5*

107) Which pair of compounds is soluble in water?
A) Ag Cl and AgBr B) CoS and K_2S
C) NaI and $Cu(NO_3)_2$ D) NH_4NO_3 and $BaCO_3$

Answer: C
Diff: 3 *Page Ref: 4.5*

108) How many of the following compounds are **soluble** in water?

$Cu(OH)_2$ $NaNO_3$ NH_4Cl Li_2S

A) 0 B) 1 C) 2 D) 3 E) 4

Answer: D
Diff: 1 *Page Ref: 4.5*

109) Which of the following compounds is **soluble** in water?
A) CaS
B) $MgCO_3$
C) $PbBr_2$
D) $PbSO_4$
E) None of these compounds is soluble in water.

Answer: A
Diff: 1 *Page Ref: 4.5*

110) Which of the following compounds is **insoluble** in water?
A) Hg_2Cl_2
B) $MgSO_4$
C) $(NH_4)_2CO_3$
D) SrS
E) All of these compounds are soluble in water.

Answer: A
Diff: 1 *Page Ref: 4.5*

111) How many of the following compounds are **insoluble** in water?

$NaC_2H_3O_2$ $SrSO_4$ BaS $AlPO_4$

A) 0 B) 1 C) 2 D) 3 E) 4

Answer: C
Diff: 3 *Page Ref: 4.5*

112) What precipitate is most likely formed from a solution containing Ba^{+2}, Li^{+1}, OH^{-1}, and CO_3^{-2}.

A) LiOH B) $BaCO_3$ C) Li_2CO_3 D) $Ba(OH)_2$

Answer: B
Diff: 2 *Page Ref: 4.6*

113) The mixing of which pair of reactants will result in a precipitation reaction?

A) CsI(*aq*) + NaOH(*aq*)
B) HCl(*aq*) + $Ca(OH)_2$(*aq*)
C) K_2SO_4(*aq*) + $Hg_2(NO_3)_2$(*aq*)
D) $NaNO_3$(*aq*) + NH_4Cl(*aq*)

Answer: C
Diff: 3 *Page Ref: 4.6*

114) The mixing of which pair of reactants will result in a precipitation reaction?

A) $Ba(NO_3)_2$(*aq*) + Na_2CO_3(*aq*)
B) K_2SO_4(*aq*) + $Cu(NO_3)_2$(*aq*)
C) $NaClO_4$(*aq*) + $(NH_4)_2S$(*aq*)
D) NH_4Br(*aq*) + NH_4I(*aq*)

Answer: A
Diff: 3 *Page Ref: 4.6*

115) What reagent could be used to separate Br^- from $CH_3CO_2^-$ when added to an aqueous solution containing both?

A) $AgNO_3$(aq) B) $Ba(OH)_2$(aq) C) $CuSO_4$(aq) D) NaI(aq)

Answer: A
Diff: 3 *Page Ref: 4.6*

116) What reagent could **not** be used to separate Br^- from CO_3^{2-} when added to an aqueous solution containing both?

A) $AgNO_3$(aq) B) $Ca(NO_3)_2$(aq) C) $Cu(NO_3)_2$(aq) D) $Fe(NO_3)_2$(*aq*)

Answer: A
Diff: 3 *Page Ref: 4.6*

117) Which of the following is a precipitation reaction?

A) Zn(s) + 2 $AgNO_3$(aq) → 2 Ag(s) + $Zn(NO_3)_2$(aq)
B) NaCl(aq) + LiI(aq) → NaI(aq) + LiCl(aq)
C) 2 KI(aq) + $Hg_2(NO_3)_2$(aq) → Hg_2I_2(s) + 2 KNO_3(aq)
D) HI(aq) + NaOH(aq) → NaI(aq) + H_2O(l)
E) None of the above are precipitation reactions.

Answer: C
Diff: 1 *Page Ref: 4.6*

118) Which of the following pairs of aqueous solutions will form a precipitate when mixed?
A) HI+ LiOH
B) Li_2S + HF
C) K_2CO_3 + HNO_3
D) $MgCl_2$ + KOH
E) All of these solution pairs will produce a precipitate.

Answer: D
Diff: 2 Page Ref: 4.6

119) Which of the following pairs of aqueous solutions will form a precipitate when mixed?
A) K_2CO_3 + NaBr
B) Na_2 SO_4 + KOH
C) CaS + Na_2SO_4
D) None of these solution pairs will produce a precipitate.
E) All of these solution pairs will produce a precipitate.

Answer: C
Diff: 2 Page Ref: 4.6

120) Which of the following pairs of aqueous solutions will form a precipitate when mixed?
A) NH_4NO_3 + Li_2CO_3
B) $Hg_2(NO_3)_2$ + LiCl
C) NaCl + Li_3PO_4
D) $AgC_2H_3O_2$ + $Cu(NO_3)_2$
E) None of the above solution pairs will produce a precipitate.

Answer: B
Diff: 2 Page Ref: 4.6

121) Which of the following pairs of aqueous solutions will form a precipitate when mixed?
A) KOH + Li_2S
B) $(NH_4)_2SO_4$ + KCl
C) $Sr(C_2H_3O_2)_2$ + Na_2SO_4
D) KNO_3 + LiOH
E) None of the above solution pairs will produce a precipitate.

Answer: C
Diff: 2 Page Ref: 4.6

122) Lead ions can be precipitated from aqueous solutions by the addition of aqueous iodide:

$$Pb^{2+}(aq) + 2I^-(aq) \rightarrow PbI_2(s)$$

Lead iodide is virtually insoluble in water so that the reaction appears to go to completion. How many milliliters of 3.550 M HI(aq) must be added to a solution containing 0.700 mol of $Pb(NO_3)_2(aq)$ to completely precipitate the lead?

A) 2.54×10^{-3} mL
B) 394 mL
C) 197 mL
D) 0.197 mL
E) 0.394 mL

Answer: B
Diff: 4 *Page Ref: 4.6*

123) Silver ions can be precipitated from aqueous solutions by the addition of aqueous chloride:

$$Ag^+(aq) + Cl^-(aq) \rightarrow AgCl(s)$$

Silver chloride is virtually insoluble in water so that the reaction appears to go to completion. How many grams of solid NaCl must be added to 25.0 mL of 0.366 M $AgNO_3$ solution to completely precipitate the silver?

A) 9.15×10^{-3} g
B) 1.57×10^{-4} g
C) 0.535 g
D) 0.157 g
E) 6.39×10^3 g

Answer: C
Diff: 4 *Page Ref: 4.6*

124) Identify the spectator ions in the following molecular equation.

$$LiCl(aq) + AgNO_3(aq) \rightarrow AgCl(s) + LiNO_3(aq)$$

A) Ag^+ and Cl^-
B) Li^+ and NO_3^-
C) Li^+ and Cl^-
D) Ag^+ and NO_3^-
E) There are no spectator ions in this reaction.

Answer: B
Diff: 1 *Page Ref: 4.7*

125) Identify the polyprotic acid.

A) H_2SO_4 B) HCl C) LiCl D) LiOH E) $Ca(OH)_2$

Answer: A
Diff: 2 *Page Ref: 4.8*

126) How many H+ ions can the acid, H_2SO_4 , donate per molecule?
A) 0 B) 1 C) 2 D) 3

Answer: C
Diff: 1 *Page Ref: 4.8*

127) Which of the following compounds is **not** an Arrhenius acid?
A) CH_3CO_2H B) $CH_3CH_2NH_2$ C) HNO_2 D) H_2SO_4

Answer: B
Diff: 1 *Page Ref: 4.8*

128) Which of the following compounds is an Arrhenius base?
A) $C_6H_{12}O_6$ B) HOCl C) H_2SO_4 D) $C_6H_5NH_2$

Answer: D
Diff: 1 *Page Ref: 4.8*

129) Which of the following compounds is an Arrhenius base?
A) CH_3OH B) CH_3CO_2H C) HOCl D) CsOH

Answer: D
Diff: 1 *Page Ref: 4.8*

130) In an acid–base neutralization reaction 38.74 mL of 0.500 M potassium hydroxide reacts with 50.00 mL of sulfuric acid solution. What is the concentration of the H_2SO_4 solution?
A) 0.194 M B) 0.387 M C) 0.775 M D) 1.29 M

Answer: A
Diff: 4 *Page Ref: 4.8*

131) When 280. mL of 1.50×10^{-4} M hydrochloric acid is added to 135 mL of 1.75×10^{-4} M $Mg(OH)_2$, the resulting solution will be
A) acidic.
B) basic
C) neutral.
D) It is impossible to tell from the information given.

Answer: B
Diff: 4 *Page Ref: 4.8*

132) A solution is prepared by mixing 50.0 mL of 0.100 M HCl and 10.0 mL of 0.200 M NaCl. What is the molarity of chloride ion in this solution?
A) 0.183 B) 8.57 C) 3.50 D) 0.0500 E) 0.117

Answer: E
Diff: 4 *Page Ref: 4.8*

133) A solution is prepared by adding 1.60 g of solid NaCl to 50.0 mL of 0.100 M $CaCl_2$. What is the molarity of chloride ion in the final solution? Assume that the volume of the final solution is 50.0 mL.
A) 0.747 B) 0.647 C) 0.132 D) 0.232 E) 0.547

Answer: A
Diff: 4 *Page Ref: 4.8*

134) What is the molarity of a NaOH solution if 28.2 mL of a 0.355 M H_2SO_4 solution is required to neutralize a 25.0-mL sample of the NaOH solution?

A) 0.801 B) 0.315 C) 0.629 D) 125 E) 0.400

Answer: A
Diff: 4 *Page Ref: 4.8*

135) How many milliliters of 0.132 M $HClO_4$ solution are needed to neutralize 50.00 mL of 0.0789 M NaOH?

A) 0.521 B) 0.0120 C) 83.7 D) 0.0335 E) 29.9

Answer: E
Diff: 4 *Page Ref: 4.8*

136) What is the oxidation number of the sulfur atom in K_2SO_4 ?

A) -2 B) +2 C) +4 D) +6

Answer: D
Diff: 2 *Page Ref: 4.9*

137) What is the oxidation number of the chromium atom in K_2CrO_4 ?

A) -2 B) +2 C) +6 D) +7

Answer: C
Diff: 2 *Page Ref: 4.9*

138) Determine the oxidation state of nitrogen in KNO_3.

A) +5 B) +3 C) 0 D) +2 E) +4

Answer: A
Diff: 2 *Page Ref: 4.9*

139) Determine the oxidation state of nitrogen in KNO_2.

A) +5 B) +3 C) 0 D) +2 E) +4

Answer: B
Diff: 2 *Page Ref: 4.9*

140) Determine the oxidation state of hydrogen in H_2.

A) +5 B) +3 C) 0 D) +2 E) +4

Answer: C
Diff: 2 *Page Ref: 4.9*

141) What is the oxidation number of the oxygen atom in Na_2O_2 ?

A) -2 B) -1 C) +1 D) +2

Answer: B
Diff: 2 *Page Ref: 4.9*

142) What is the oxidation number change for the bromine atom in the following unbalanced reduction half reaction:

$$BrO_3^-(aq) + H^+(aq) \rightarrow Br^-(aq) + H_2O(l)?$$

A) -7 B) -6 C) +6 D) +7

Answer: B
Diff: 2 *Page Ref: 4.9*

143) Determine the oxidation state of Cr in $Cr(SO_4)_2$.

A) +2 B) +4 C) +6 D) 0 E) −2

Answer: B

Diff: 2 *Page Ref: 4.9*

144) Determine the oxidation state of S in Na_2SO_4.

A) −4 B) +2 C) +4 D) +6 E) −2

Answer: D

Diff: 2 *Page Ref: 4.9*

145) Identify the oxidation state of Ba in Ba(s).

$$Ba(s) + 2HF(aq) \rightarrow BaF_2(aq) + H_2(g)$$

A) +1 B) +2 C) 0 D) −1 E) −2

Answer: C

Diff: 2 *Page Ref: 4.9*

146) Identify the oxidation state of H in HF(aq).

$$Mg(s) + 2HF(aq) \rightarrow MgF_2(aq) + H_2(g)$$

A) +1 B) +2 C) 0 D) −1 E) −2

Answer: A

Diff: 2 *Page Ref: 4.9*

147) Identify the oxidation state of Ba in $BaCl_2(aq)$.

$$Ba(s) + 2HCl(aq) \rightarrow BaCl_2(aq) + H_2(g)$$

A) +1 B) +2 C) 0 D) −1 E) −2

Answer: B

Diff: 2 *Page Ref: 4.9*

148) Identify the oxidation state of H in $H_2(g)$.

$$Mg(s) + 2HI(aq) \rightarrow MgI_2(aq) + H_2(g)$$

A) +1 B) +2 C) 0 D) −1 E) −2

Answer: C

Diff: 2 *Page Ref: 4.9*

149) Which of the following is an oxidation-reduction reaction?

A) $HI(aq) + LiOH(aq) \rightarrow LiI(aq) + H_2O(l)$

B) $NaCl(aq) + AgNO_3(aq) \rightarrow AgCl(s) + NaNO_3(aq)$

C) $Pb(C_2H_3O_2)_2(aq) + 2\ NaCl(aq) \rightarrow PbCl_2(s) + 2\ NaC_2H_3O_2(aq)$

D) $Mg(s) + 2\ HCl(aq) \rightarrow MgCl_2(aq) + H_2(g)$

E) All of the above are oxidation-reduction reactions.

Answer: D

Diff: 2 *Page Ref: 4.9*

150) What are the coefficients in front of $NO_3^-(aq)$ and Cu(s) when the following redox equation is balanced in an acidic solution:

___ $NO_3^-(aq)$ + ___ Cu(s) → ___ NO(g) + ___ $Cu^{2+}(aq)$?

A) 2, 3 B) 2, 6 C) 3, 4 D) 3, 6

Answer: A

Diff: 2 *Page Ref: 4.9*

Matching Questions

Match the following.

1) $CH_4(g) + 2\ O_2(g) \rightarrow CO_2(g) + 2\ H_2O(g)$

Diff: 1 *Page Ref: 4.9*

2) $H_2SO_4(aq) + 2\ LiOH(aq) \rightarrow 2\ H_2O(l) + Li_2SO_4(aq)$

Diff: 1 *Page Ref: 4.8*

3) $Mg(s) + Cu(NO_3)_2(aq) \rightarrow Mg(NO_3)_2(aq) + Cu(s)$

Diff: 1 *Page Ref: 4.9*

4) $HCl(aq) + (NH_4)_2S(aq) \rightarrow H_2S(g) + 2NH_4Cl(aq)$

Diff: 1 *Page Ref: 4.8*

5) $NaCl(aq) + AgNO_3(aq) \rightarrow AgCl(s) + NaNO_3(aq)$

Diff: 1 *Page Ref: 4.6*

A) acid–base

B) gas evolution

C) combustion

D) precipitation

E) oxidation reduction

Answers: 1) C 2) A 3) E 4) B 5) D

Short Answer Questions

1) Describe the greenhouse effect.

Answer: Greenhouse gases allow visible light to pass through and warm the Earth's surface, but prevent heat energy from being radiated out into space.

Diff: 2 *Page Ref: 4.1*

2) Define a limiting reagent.

Answer: A limiting reagent is a reactant that is completely consumed in a chemical reaction and limits the amount of product.

Diff: 3 *Page Ref: 4.3*

3) How would the concentration change if a 1.0 L flask of 1.0 M NaCl were left uncapped on a laboratory bench for several days. Why?

Answer: The concentration would slowly increase as water from the solution evaporated. This is because the amount of NaCl in the flask would remain constant while the amount of water decreases.

Diff: 1 *Page Ref: 4.4*

4) Define an electrolyte.

Answer: A substance that dissolves in water to form a solution that conducts electricity.

Diff: 1 *Page Ref: 4.5*

5) Explain the difference between a strong and weak electrolyte. Give an example of each.

Answer: A strong electrolyte is either an ionic compound that is soluble in water or a molecular compound that ionizes completely in water. Possible examples are NaCl or HCl. A weak electrolyte is only slightly soluble or does not ionize to any great extent in water. Possible examples are AgCl or $HC_2H_3O_2$.

Diff: 1 *Page Ref: 4.5*

6) What causes a precipitation reaction to occur between two soluble compounds?

Answer: A precipitate, or insoluble compound, forms from a pair of the ions present. This "traps" some of the ions into a newly formed compound causing reaction.

Diff: 1 *Page Ref: 4.6*

7) Describe the difference between complete ionic and net ionic equations.

Answer: A complete ionic equation shows all of the substances present in the reaction mixture. A net ionic equation only shows the substances that are participating in the reaction. The spectator ions are absent in the net ionic equation.

Diff: 1 *Page Ref: 4.7*

8) Define a spectator ion.

Answer: An ion that does not participate in the reaction.

Diff: 1 *Page Ref: 4.7*

9) List the spectator ions in the following reaction.

$$Ba^{+2}(aq) + 2\,OH^{-}(aq) + 2\,Na^{+}(aq) + CO_3{}^{-2}(aq) \rightarrow 2\,Na^{+}(aq) + 2\,OH^{-}(aq) + BaCO_3(s)$$

Answer: Na^{+} and OH^{-}

Diff: 3 *Page Ref: 4.7*

10) How can you tell if a reaction is an oxidation–reduction reaction?

Answer: At least one species is losing electrons while at least one other species is gaining electrons. This can be evaluated by assigning oxidation numbers to each element present and looking to see if these change from reactant to product.

Diff: 1 *Page Ref: 4.9*

11) Determine the oxidation state of Mn in $KMnO_4$.

Answer: +7

Diff: 2 *Page Ref: 4.9*

Chapter 5 Gases

Multiple Choice Questions

1) An instrument used to atmospheric pressure is called a
   A) barometer
   B) manometer
   C) sphygmomanometer
   D) spectrophotometer
   E) spectrometer

Answer: A
Diff: 1 *Page Ref: 5.2*

2) An instrument used to measure the pressure of a gas in a laboratory is called a
   A) barometer
   B) manometer
   C) sphygmomanometer
   D) spectrophotometer
   E) spectrometer

Answer: B
Diff: 1 *Page Ref: 5.2*

3) An instrument used to measure blood pressure is called a
   A) barometer
   B) manometer
   C) sphygmomanometer
   D) spectrophotometer
   E) spectrometer

Answer: C
Diff: 1 *Page Ref: 5.2*

4) The volume of a gas is proportional to the temperature of a gas is known as
   A) Avogadro's Law
   B) Ideal Gas Law
   C) Charles's Law
   D) Boyle's Law
   E) Dalton's Law

Answer: C
Diff: 1 *Page Ref: 5.3*

5) The volume of a gas is proportional to number of moles of a gas is known as
   A) Avogadro's Law
   B) Ideal Gas Law
   C) Charles's Law
   D) Boyle's Law
   E) Dalton's Law

Answer: A
Diff: 1 *Page Ref: 5.3*

6) The volume of a gas is inversely proportional to the pressure of a gas is known as
A) Avogadro's Law
B) Ideal Gas Law
C) Charles's Law
D) Boyle's Law
E) Dalton's Law

Answer: D
Diff: 1 *Page Ref: 5.3*

7) A scuba diver experiences ________ as she descends into the ocean water.
A) decrease of 1 atm for each 20 m of depth
B) decrease of 1 atm for each 10 m of depth
C) increase of 1 atm for each 20 m of depth
D) increase of 1 atm for each 10 m of depth
E) no difference in pressure

Answer: D
Diff: 1 *Page Ref: 5.3*

8) What volume (in mL) will a sample of F_2 gas occupy in a syringe at 5.5 atm, if the F_2 has a volume of 25.0 mL at 1.2 atm?
A) 11 mL B) 17 mL C) 3.8 mL D) 5.5 mL E) 7.6 mL

Answer: D
Diff: 2 *Page Ref: 5.3*

9) If a sample of 0.29 moles of Ar occupies 3.8 L under certain conditions, what volume will 0.66 moles occupy under the same conditions?
A) 12 L B) 8.6 L C) 17 L D) 5.0 L E) 15 L

Answer: B
Diff: 2 *Page Ref: 5.3*

10) A syringe contains 0.65 moles of He gas that occupy 750.0 mL. What volume (in L) of gas will the syringe hold if 0.35 moles of Ne is added?
A) 0.87 L B) 4.9 L C) 1.2 L D) 2.1 L E) 1.9 L

Answer: C
Diff: 2 *Page Ref: 5.3*

11) To what temperature must a balloon, initially at 25°C and 2.00 L, be heated in order to have a volume of 6.00 L?
A) 993 K B) 403 K C) 75 K D) 655 K E) 894 K

Answer: E
Diff: 3 *Page Ref: 5.3*

12) A large balloon is initially filled to a volume of 25.0 L at 353 K and a pressure of 2575 mm Hg. What volume of gas will the balloon contain at 1.35 atm and 253 K?
A) 22.2 L B) 87.5 L C) 11.4 L D) 45.0 L E) 58.6 L

Answer: D
Diff: 3 *Page Ref: 5.3*

13) A syringe initially holds a sample of gas with a volume of 285 mL at 355 K and 1.88 atm. To what temperature must the gas in the syringe be heated/cooled in order to have a volume of 435 mL at 2.50 atm?

A) 139 K B) 572 K C) 175 K D) 466 K E) 721 K

Answer: E
Diff: 3 *Page Ref: 5.4*

14) A sample of gas initially has a volume of 859 mL at 565 K and 2.20 atm. What pressure will the sample have if the volume changes to 268 mL while the temperature is increased to 815 K?

A) 10.2 atm B) 9.83 atm C) 15.3 atm D) 6.53 atm E) 1.05 atm

Answer: A
Diff: 3 *Page Ref: 5.4*

15) What pressure will 14.0 g of CO exert in a 3.5 L container at 75°C?

A) 4.1 atm B) 5.0 atm C) 6.4 atm D) 1.1 atm E) 2.3 atm

Answer: A
Diff: 3 *Page Ref: 5.4*

16) What mass of NO_2 is contained in a 13.0 L tank at 4.58 atm and 385 K?

A) 18.8 g B) 53.1 g C) 24.4 g D) 86.7 g E) 69.2 g

Answer: D
Diff: 3 *Page Ref: 5.4*

17) What is the volume of 5.60 g of O_2 at 7.78 atm and 415K?

A) 1.53 L B) 565 L C) 24.5 L D) 25.0 L E) 0.766 L

Answer: E
Diff: 3 *Page Ref: 5.4*

18) Using the graph below, determine the gas that has the lowest density at STP.

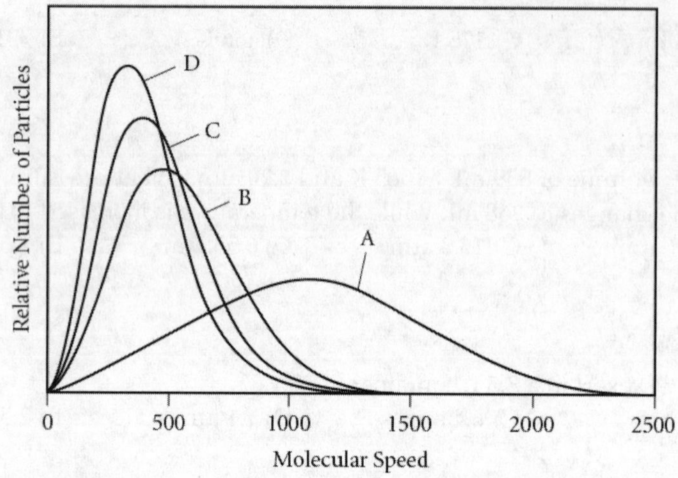

A) A
B) B
C) C
D) D
E) All of the gases have the same density at STP.

Answer: A
Diff: 1 *Page Ref:* 5.5

19) Using the graph below, determine the gas that has the highest density at STP.

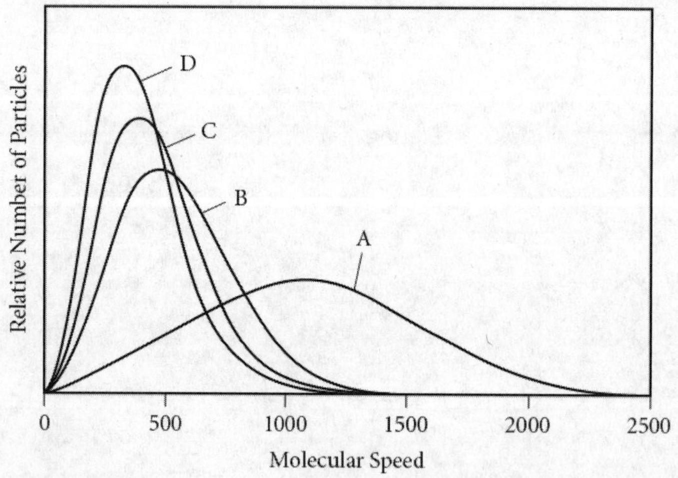

A) A
B) B
C) C
D) D
E) All of the gases have the same density at STP.

Answer: D
Diff: 1 *Page Ref:* 5.5

20) Which of the following will cause the volume of an ideal gas to triple in value?
A) Raising the temperature from 25°C to 75°C at constant pressure.
B) Lowering the absolute temperature by a factor of 3 at constant pressure.
C) Raising the absolute temperature by a factor of 3 while increasing the pressure by a factor of 3.
D) Lowering the absolute temperature by a factor of 3 while increasing the pressure by a factor of 3.
E) Lowering the pressure by a factor of 3 while the temperature stays constant.

Answer: E
Diff: 1 *Page Ref: 5.5*

21) Determine the density of CO_2 gas at STP.
A) 1.96 g/L B) 1.80 g/L C) 2.24 g/L D) 4.46 g/L E) 5.10 g/L

Answer: A
Diff: 2 *Page Ref: 5.5*

22) Determine the density of NH_3 gas at 435 K and 1.00 atm.
A) 2.10 g/L B) 0.477 g/L C) 0.321 g/L D) 2.24 g/L E) 0.851 g/L

Answer: B
Diff: 3 *Page Ref: 5.5*

23) A 0.334 g sample of an unknown halogen occupies 109 mL at 398 K and 1.41 atm. What is the identity of the halogen?
A) Br_2 B) F_2 C) Cl_2 D) I_2 E) Ge

Answer: C
Diff: 3 *Page Ref: 5.5*

24) A 0.465 g sample of an unknown compound occupies 245 mL at 298 K and 1.22 atm. What is the molar mass of the unknown compound?
A) 26.3 g/mol
B) 33.9 g/mol
C) 12.2 g/mol
D) 38.0 g/mol
E) 81.8 g/mol

Answer: D
Diff: 3 *Page Ref: 5.5*

25) A compound is found to be 30.45% N and 69.55 % O by mass. If 1.63 g of this compound occupy 389 mL at 0.00°C and 775 mm Hg, what is the molecular formula of the compound?
A) NO_2 B) N_2O C) N_4O_2 D) N_2O_5 E) N_2O_4

Answer: E
Diff: 4 *Page Ref: 5.5*

26) Define hypoxia.
A) oxygen starvation
B) increased oxygen concentration in body tissues
C) increased nitrogen concentration in body tissues and fluids
D) nitrogen starvation

Answer: A
Diff: 1 *Page Ref: 5.6*

27) Define vapor pressure.
A) partial pressure of water in a liquid mixture
B) partial pressure of water in a gaseous mixture
C) condensation of water
D) water dissolved in a liquid
E) water molecules

Answer: B
Diff: 1 *Page Ref: 5.6*

28) The total pressure of a gas mixture is the sum of the partial pressure of its components is known as
A) Avogadro's Law
B) Ideal Gas Law
C) Charles's Law
D) Boyle's Law
E) Dalton's Law

Answer: E
Diff: 1 *Page Ref: 5.6*

29) A gas mixture consists of N_2, O_2, and Ne, where the mole fraction of N_2 is 0.55 and the mole fraction of Ne is 0.25. If the mixture is at STP in a 5.0 L container, how many molecules of O_2 are present?
A) 4.5×10^{22} molecules O_2
B) 2.7×10^{22} molecules O_2
C) 3.7×10^{23} molecules O_2
D) 1.1×10^{23} molecules O_2
E) 9.3×10^{24} molecules O_2

Answer: B
Diff: 3 *Page Ref: 5.6*

30) A mixture of 10.0 g of Ne and 10.0 g Ar have a total pressure of 1.6 atm. What is the partial pressure of Ne?
A) 1.1 atm B) 0.80 atm C) 0.54 atm D) 0.40 atm E) 1.3 atm

Answer: A
Diff: 3 *Page Ref: 5.6*

31) The following reaction is used to generate hydrogen gas in the laboratory. If 243 mL of gas is collected at 25°C and has a total pressure of 745 mm Hg, what mass of hydrogen is produced? A possibly useful table of water vapor pressures is provided below.

$Mg(s) + 2\ HCl(aq) \rightarrow MgCl_2(aq) + H_2(g)$

T (°C)	P (mm Hg)
20	17.55
25	23.78
30	31.86

A) 0.0196 g H_2
B) 0.0717 g H_2
C) 0.0190 g H_2
D) 0.0144 g H_2
E) 0.0449 g H_2

Answer: C
Diff: 3 *Page Ref: 5.6*

32) A syringe contains 589 mL of CO at 325 K and 1.2 atm pressure. A second syringe contains 473 mL of N_2 at 298 K and 2.6 atm. What is the final pressure if the contents of these two syringes are injected into a 1.00 L container at STP?

A) 0.59 atm B) 1.1 atm C) 1.7 atm D) 1.9 atm E) 3.8 atm

Answer: C
Diff: 5 *Page Ref: 5.6*

33) What pressure would a gas mixture in a 10.0 L tank exert if it were composed of 48.5 g He and 94.6 g CO_2 at 398 K?

A) 39.6 atm B) 7.02 atm C) 32.6 atm D) 46.6 atm E) 58.7 atm

Answer: D
Diff: 5 *Page Ref: 5.6*

34) Determine the volume of O_2 (at STP) formed when 50.0 g of $KClO_3$ decomposes according to the following reaction. The molar mass for $KClO_3$ is 122.55 g/mol.

$$2\ KClO_3(s) \rightarrow 2\ KCl(s) + 3\ O_2(g)$$

A) 9.14 L B) 8.22 L C) 12.3 L D) 13.7 L E) 14.6 L

Answer: D
Diff: 4 *Page Ref: 5.7*

35) Determine the volume of H_2S (at 375 K and 1.20 atm) needed to produce 55.0 g of S. Assume that there is excess SO_2 present.

$$2\ H_2S(g) + SO_2(g) \rightarrow 3\ S(s) + 2\ H_2O(g)$$

A) 44.0 L B) 29.3 L C) 22.7 L D) 34.1 L E) 66.0 L

Answer: B
Diff: 4 *Page Ref: 5.7*

36) How many molecules of XeF_6 are formed from 12.9 L of F_2 (at 298 K and 2.60 atm) according to the following reaction? Assume that there is excess Xe.

$$Xe(g) + 3\,F_2(g) \rightarrow XeF_6(g)$$

A) 1.21×10^{23} molecules XeF_6

B) 8.25×10^{23} molecules XeF_6

C) 2.75×10^{23} molecules XeF_6

D) 7.29×10^{23} molecules XeF_6

E) 1.37×10^{23} molecules XeF_6

Answer: C
Diff: 4 *Page Ref: 5.7*

37) Determine the theoretical yield and the percent yield if 21.8 g of K_2CO_3 is produced from reacting 27.9 g KO_2 with 29.0 L of CO_2 (at STP). The molar mass of KO_2 = 71.10 g/mol and K_2CO_3 = 138.21 g/mol.

$$4\,KO_2(s) + 2\,CO_2(g) \rightarrow 2\,K_2CO_3(s) + 3\,O_2(g)$$

A) 27.1 g, 80.4 % yield
B) 179 g, 12.2 % yield
C) 91.7 g, 23.8 % yield
D) 206 g, 10.6 % yield
E) 61.0 g, 35.7 % yield

Answer: A
Diff: 4 *Page Ref: 5.7*

38) Determine the volume of SO_2 (at STP) formed from the reaction of 96.7 g of FeS_2 and 55.0 L of O_2 (at 398 K and 1.20 atm). The molar mass of FeS_2 is 119.99 g/mol.

$$4\,FeS_2(s) + 11\,O_2(g) \rightarrow 2\,Fe_2O_3(s) + 8\,SO_2(g)$$

A) 36.1 L B) 45.3 L C) 18.1 L D) 27.6 L E) 32.9 L

Answer: E
Diff: 5 *Page Ref: 5.7*

39) Determine the mass of water formed when 12.5 L NH_3 (at 298 K and 1.50 atm) is reacted with 18.9 L of O_2 (at 323 K and 1.1 atm).

$$4\ NH_3(g) + 5\ O_2(g) \rightarrow 4\ NO(g) + 6\ H_2O(g)$$

A) 17.0 g H_2O
B) 20.7 g H_2O
C) 37.7 g H_2O
D) 13.8 g H_2O
E) 27.9 g H_2O

Answer: A
Diff: 5 *Page Ref: 5.7*

40) Determine the total volume of all gases (at STP) formed when 50.0 mL of TNT ($C_3H_5(NO_3)_3$, d = 1.60 g/mL, molar mass = 227.10 g/mol) reacts according to the following reaction.

$$4\ C_3H_5(NO_3)_3(l) \rightarrow 6\ N_2(g) + O_2(g) + 12\ CO_2(g) + 10\ H_2O(g)$$

A) 4.93 L B) 57.2 L C) 29.6 L D) 448 L E) 175 L

Answer: B
Diff: 5 *Page Ref: 5.7*

41) A mixture of 1.0 mol He and 1.0 mol Ne are at STP in a rigid container. Which of the following statements is TRUE?
A) Both gases have the same average kinetic energy.
B) Both gases contribute equally to the density of the mixture under these conditions.
C) Both gases have the same molecular speed.
D) The mixture has a volume of 22.4 L
E) All of the above are TRUE.

Answer: A
Diff: 1 *Page Ref: 5.8*

42) Which statement is TRUE about kinetic molecular theory?
A) A single particle does not move in a straight line.
B) The size of the particle is large compared to the volume.
C) The collisions of particles with one another is completely elastic.
D) The average kinetic energy of a particle is not proportional to the temperature.

Answer: C
Diff: 1 *Page Ref: 5.8*

43) Which of the gases in the graph below has the largest molar mass?

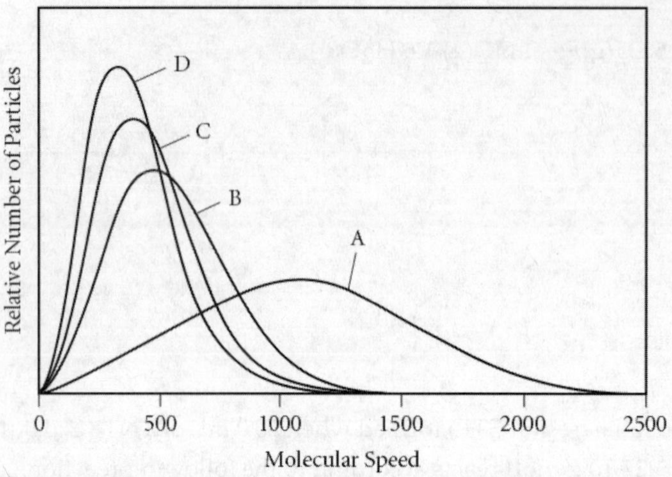

A) A
B) B
C) C
D) D
E) There is not enough information to determine.

Answer: D
Diff: 2 *Page Ref:* 5.8

44) Calculate the root mean square velocity of nitrogen molecules at 25°C.
A) 729 m/s B) 515 m/s C) 149 m/s D) 297 m/s

Answer: B
Diff: 3 *Page Ref:* 5.8

45) The rate of effusion of two different gases is known as
A) Avogadro's Law
B) Graham's Law
C) Charles's Law
D) Boyle's Law
E) Dalton's Law

Answer: B
Diff: 1 *Page Ref:* 5.9

46) Give the definition for diffusion.
A) gas molecules mix equally
B) gas molecules spread out in a concentration gradient
C) gas molecules escape from a container into a vacuum through a small hole
D) average distance between collisions
E) gas molecules mix unequally

Answer: B
Diff: 1 *Page Ref:* 5.9

47) The rate of effusion of oxygen to an unknown gas is 0.935. What is the other gas?
A) Ne B) Ar C) F_2 D) N_2

Answer: D
Diff: 3 *Page Ref: 5.9*

48) Which of the following statements is TRUE?
A) Particles of different masses have the same average speed at a given temperature.
B) The larger a molecule, the faster it will effuse.
C) At very high pressures, a gas will occupy a larger volume than predicted by the ideal gas law.
D) For a given gas, the lower the temperature, the faster it will effuse.
E) None of the above statements are true.

Answer: C
Diff: 1 *Page Ref: 5.10*

49) Which of the following statements is TRUE?
A) At a given temperature, lighter gas particles travel more slowly than heavier gas particles.
B) The smaller a gas particle, the slower it will effuse
C) The higher the temperature, the lower the average kinetic energy of the sample.
D) At low temperatures, intermolecular forces become important and the pressure of a gas will be lower than predicted by the ideal gas law.
E) None of the above statements are true.

Answer: D
Diff: 1 *Page Ref: 5.10*

50) Which of the following compounds will behave LEAST like an ideal gas at low temperatures?
A) He B) SO_2 C) H_2 D) N_2 E) F_2

Answer: B
Diff: 1 *Page Ref: 5.10*

51) This equation is used to calculate the properties of a gas under nonideal conditions.
A) Charles's Law
B) Avogadro's Law
C) Boyle's Law
D) van der Waals equation
E) Dalton's Law

Answer: D
Diff: 1 *Page Ref: 5.10*

Algorithmic Questions

1) Convert 1.50 atm to mm Hg.
A) 760 mm Hg
B) 875 mm Hg
C) 1140 mm Hg
D) 1000 mm Hg
E) 1520 mm Hg

Answer: C
Diff: 2 *Page Ref: 5.2*

2) Convert 1.50 atm to torr.
A) 760 torr B) 875 torr C) 1140 torr D) 1000 torr E) 1520 torr

Answer: C
Diff: 2 *Page Ref:* 5.2

3) Convert 1.75 atm to psi.
A) 52.4 psi B) 875 psi C) 25.7 psi D) 1000 psi E) 1330 psi

Answer: C
Diff: 2 *Page Ref:* 5.2

4) The atmospheric pressure is 700 mm Hg. What is the pressure in inches of Hg?
A) 16.0 in Hg
B) 0.921 in Hg
C) 13.5 in Hg
D) 27.6 in Hg
E) 32.5 in Hg

Answer: D
Diff: 3 *Page Ref:* 5.2

5) The atmospheric pressure is 715 mm Hg. What is the pressure in torr?
A) 715 torr B) 28.1 torr C) 13.8 torr D) 31.8 torr E) 760 torr

Answer: A
Diff: 2 *Page Ref:* 5.2

6) What is the pressure in a gas container that is connected to an open-end U-tube manometer if the pressure of the atmosphere is 742 torr and the level of mercury in the arm connected to the container is 8.60 cm higher than the level of mercury open to the atmosphere?
A) 656 mm Hg B) 733 mm Hg C) 751 mm Hg D) 828 mm Hg

Answer: A
Diff: 3 *Page Ref:* 5.2

7) If the pressure in a gas container that is connected to an open-end U-tube manometer is 116 kPa and the pressure of the atmosphere at the open end of the tube is 752 mm Hg, the level of mercury in the tube will
A) be 118 mm higher in the arm open to the atmosphere.
B) be 118 mm higher in the arm connected to the gas cylinder.
C) be 870 mm higher in the arm open to the atmosphere.
D) be 870 mm higher in the arm connected to the gas cylinder.

Answer: A
Diff: 3 *Page Ref:* 5.2

8) A container filled with gas is connected to an open-end manometer that is filled with mineral oil. The pressure in the gas container is 753 mm Hg and atmospheric pressure is 724 mm. How high will the level rise in the manometer if the densities of Hg and mineral oil are 13.6 g/mL and 0.822 g/mL respectively?
A) 1.75 mm B) 23.8 mm C) 29.0 mm D) 480 mm

Answer: D
Diff: 3 *Page Ref:* 5.2

9) What volume will a balloon occupy at 1.0 atm, if the balloon has a volume of 8.8 L at 4.4 atm?
A) 2.0 L B) 0.50 L C) 39 L D) 13 L E) 4.4 L

Answer: C
Diff: 2 *Page Ref: 5.3*

10) A gas occupies 4.23 L at 2.25 atm. What is the volume at 3.46 atm?
A) 6.50 L B) 1.84 L C) 2.75 L D) 32.9 L E) 0.364 L

Answer: C
Diff: 2 *Page Ref: 5.3*

11) A sample of 0.200 moles of nitrogen occupies 0.400 L. Under the same conditions, what number of moles occupies 1.200 L?
A) 0.600 moles B) 0.0667 moles C) 2.40 moles D) 0.0960 moles

Answer: A
Diff: 2 *Page Ref: 5.3*

12) To what volume will a sample of gas expand if it is heated from 50.0°C and 2.33 L to 500.0°C?
A) 5.58 L B) 23.3 L C) 0.233 L D) 0.97 L E) 0.184 L

Answer: A
Diff: 3 *Page Ref: 5.3*

13) A gas is at 35.0°C and 3.50 L. What is the temperature at 7.00 L?
A) 343°C B) 70.0°C C) 616°C D) 1.16°C E) 17.5°C

Answer: A
Diff: 3 *Page Ref: 5.3*

14) A fixed amount of gas at 25.0°C occupies a volume of 10.0 L when the pressure is 667 torr. Use Boyle's law to calculate the pressure (torr) when the volume is reduced to 7.88 L at a constant temperature of 25.0°C.
A) 846 torr
B) 0.118 torr
C) 5.26×10^4 torr
D) 526 torr
E) 1.11 torr

Answer: A
Diff: 2 *Page Ref: 5.3*

15) The volume of 350. mL of gas at 25°C is decreased to 135 mL at constant pressure. What is the final temperature of the gas?
A) -158°C B) 9.6°C C) 65°C D) 500°C

Answer: A
Diff: 3 *Page Ref: 5.3*

16) A fixed amount of gas at 25.0°C occupies a volume of 10.0 L when the pressure is 629 torr. Use Charles's law to calculate the volume (L) the gas will occupy when the temperature is increased to 121°C while maintaining the pressure at 629 torr.
A) 10.9 L B) 13.2 L C) 2.07 L D) 7.56 L E) 48.4 L

Answer: B
Diff: 2 *Page Ref: 5.3*

17) A basketball is inflated to a pressure of 1.90 atm in a 24.0°C garage. What is the pressure of the basketball outside where the temperature is –1.00°C?

A) 1.74 atm B) 1.80 atm C) 2.00 atm D) 2.08 atm

Answer: A
Diff: 3 *Page Ref: 5.4*

18) A balloon filled with helium gas at 20°C occupies 4.91 L at 1.00 atm. The balloon is immersed in liquid nitrogen at –196°C, while the pressure is raised to 5.20 atm. What is the volume of the balloon in the liquid nitrogen?

A) 0.25 L B) 3.6 L C) 6.7 L D) 97 L

Answer: A
Diff: 3 *Page Ref: 5.4*

19) What pressure (in atm) will 0.44 moles of CO_2 exert in a 2.6 L container at 25°C?

A) 0.35 atm B) 4.1 atm C) 4.7 atm D) 8.6 atm E) 3.6 atm

Answer: B
Diff: 2 *Page Ref: 5.4*

20) How many moles of Ar are contained in a 5.00 L tank at 155°C and 2.80 atm?

A) 0.399 moles
B) 1.10 moles
C) 2.51 moles
D) 0.455 moles
E) 0.289 moles

Answer: A
Diff: 2 *Page Ref: 5.4*

21) Calculate the temperature, in K, of 2.20 moles of gas occupying 3.30 L at 3.50 atm.

A) 64.0 K B) –209 K C) 337 K D) 28.0 K

Answer: A
Diff: 3 *Page Ref: 5.4*

22) What pressure will 2.6×10^{23} molecules of N_2 exert in a 3.9 L container at 45°C?

A) 5.7 atm B) 1.7 atm C) 2.9 atm D) 3.4 atm E) 4.6 atm

Answer: C
Diff: 3 *Page Ref: 5.4*

23) How many molecules of He are contained in a 10.0 L tank at 7.53 atm and 485 K?

A) 1.89×10^{24} molecules
B) 1.14×10^{24} molecules
C) 8.32×10^{24} molecules
D) 4.89×10^{24} molecules
E) 3.63×10^{24} molecules

Answer: B
Diff: 3 *Page Ref: 5.4*

24) What is the volume of 9.783×10^{23} atoms of Kr at 9.25 atm and 512K?
A) 7.38 L B) 3.69 L C) 1.85 L D) 15.4 L E) 30.8 L

Answer: A
Diff: 3 *Page Ref: 5.4*

25) How many molecules of N_2 are in a 400.0 mL container at 780 mm Hg and 135°C?
A) 7.01×10^{21} molecules B) 7.38×10^{21} molecules
C) 2.12×10^{22} molecules D) 2.23×10^{22} molecules

Answer: B
Diff: 3 *Page Ref: 5.4*

26) A 55.0-L steel tank at 20.0°C contains acetylene gas, C_2H_2, at a pressure of 1.39 atm. Assuming ideal behavior, how many grams of acetylene are in the tank?
A) 3.17 g B) 8.20 g C) 82.9 g D) 1210 g

Answer: C
Diff: 3 *Page Ref: 5.4*

27) What is the volume of 30.0 g of argon gas at 157°C and 2.50 kPa pressure?
A) 3.87 L B) 10.6 L C) 393 L D) 1070 L

Answer: D
Diff: 3 *Page Ref: 5.4*

28) What is the Celsius temperature of 100.0 g of chlorine gas in a 55.0-L container at 800 mm Hg?
A) -23°C B) 228°C C) 250°C D) 500°C

Answer: B
Diff: 3 *Page Ref: 5.4*

29) A gas bottle contains 0.250 mol of gas at 730 mm Hg pressure. If the final pressure is 1.15 atm, how many moles of gas were added to the bottle?
A) 0.0262 mol B) 0.0493 mol C) 0.276 mol D) 0.299 mol

Answer: B
Diff: 3 *Page Ref: 5.4*

30) A 2.75-L container filled with CO_2 gas at 25°C and 225 kPa pressure springs a leak. When the container is re-sealed, the pressure is 185 kPA and the temperature is 10°C. How many moles of gas were lost?
A) 0.0335 mol B) 0.728 mol C) 0.882 mol D) 3.39 mol

Answer: A
Diff: 3 *Page Ref: 5.4*

31) How many grams of F_2 gas are there in a 5.00-L cylinder at 4.00×10^3 mm Hg and 23°C?
A) 19.5 g B) 41.1 g C) 500 g D) 2.96×10^4 g

Answer: B
Diff: 3 *Page Ref: 5.4*

32) A 4.00-L flask contains nitrogen gas at 25°C and 1.00 atm pressure. What is the final pressure in the flask if an additional 2.00 g of N_2 gas is added to the flask and the flask cooled to -55°C?

A) 0.319 atm B) 1.05 atm C) 1.44 atm D) 1.96 atm

Answer: B
Diff: 4 *Page Ref: 5.4*

33) Three identical flasks contain three different gases at standard temperature and pressure. Flask A contains C_2H_4, flask B contains O_3, and flask C contains F_2. Which flask contains the largest number of molecules?

A) flask A
B) flask B
C) flask C
D) All contain same number of molecules.

Answer: D
Diff: 1 *Page Ref: 5.5*

34) The density of nitric oxide (NO) gas at 1.21 atm and 54.1°C is ________ g/L.

A) 0.0451 B) 0.740 C) 1.35 D) 0.273 E) 8.2

Answer: C
Diff: 3 *Page Ref: 5.5*

35) The density of krypton gas at 1.21 atm and 50.0°C is ________ g/L.

A) 0.0456 B) 0.262 C) 0.295 D) 3.82 E) 7.65

Answer: D
Diff: 3 *Page Ref: 5.5*

36) Give the temperature and pressure at STP.

A) 0°C and 1.00 atm
B) 0K and 1.00 atm
C) 25°C and 1.00 in Hg
D) 32K and 1 torr Hg
E) 0°C and 1 mm Hg

Answer: A
Diff: 1 *Page Ref: 5.5*

37) Which of the following gas samples would be most likely to behave ideally under the stated conditions?

A) Ar at STP
B) CO at 200 atm and 25°C
C) SO_2 at 2 atm and 0 K
D) N_2 at 1 atm and -70°C
E) H_2 at 400 atm and 25°C

Answer: A
Diff: 1 *Page Ref: 5.5*

38) Which of the following samples will have the greatest volume at STP?

A) 22 g Ne
B) 22 g He
C) 22 g O_2
D) 22 g Cl_2
E) All of these samples would have the same volume at STP.

Answer: B
Diff: 1 *Page Ref: 5.5*

39) Which of the following samples will have the lowest pressure if they are all at the same temperature and in identical containers (same volume)?

A) 15 g H_2
B) 15 g Ar
C) 15 g Kr
D) 15 g CO_2
E) All of these samples will have the same pressure.

Answer: C
Diff: 1 *Page Ref: 5.5*

40) Which of the following samples has the greatest density at STP?

A) NO_2
B) Xe
C) SO_2
D) SF_6
E) All of these samples have the same density at STP.

Answer: D
Diff: 1 *Page Ref: 5.5*

41) Place the following gases in order of increasing density at STP.

F_2 NH_3 N_2O_4 Ar

A) $N_2O_4 < Ar < F_2 < NH_3$
B) $Ar < N_2O_4 < F_2 < NH_3$
C) $F_2 < Ar < N_2O_4 < NH_3$
D) $NH_3 < F_2 < Ar < N_2O_4$
E) $Ar < F_2 < NH_3 < N_2O_4$

Answer: D
Diff: 1 *Page Ref: 5.5*

42) The density of a gas is 3.16 g/L at STP. What is the gas?

A) H_2 B) Br_2 C) Kr D) Cl_2 E) Ar

Answer: D
Diff: 2 *Page Ref: 5.5*

43) What volume will 4.91×10^{22} atoms of Ar occupy at STP?
A) 1.10 L B) 2.00 L C) 2.24 L D) 3.11 L E) 1.83 L

Answer: E
Diff: 3 *Page Ref: 5.5*

44) What volume will 0.780 moles of Xe occupy at STP?
A) 22.4 L B) 70.0 L C) 43.7 atm D) 17.5 L E) 15.6 L

Answer: D
Diff: 3 *Page Ref: 5.5*

45) A steel bottle contains argon gas at STP. What is the final pressure if the temperature is changed to 145°C?
A) 0.653 atm B) 0.713 atm C) 1.40 atm D) 1.53 atm

Answer: D
Diff: 3 *Page Ref: 5.5*

46) A gas occupies 22.4 L at STP and 19.0 L at 100°C and 1.50 atm pressure. How many moles of gas did the system gain or lose?
A) 0.08 moles gained B) 0.01 moles gained
C) 0.01 moles lost D) 0.07 moles lost

Answer: D
Diff: 3 *Page Ref: 5.5*

47) What is the density of hydrogen gas at STP?
A) 11.1 g/L B) 0.0450 g/L C) 0.0823 g/L D) 0.0899 g/L

Answer: D
Diff: 3 *Page Ref: 5.5*

48) Which of the following would have a density of 1.37 g/L at 7.0°C and 0.987 atm?
A) N_2 B) O_2 C) Kr D) Rn

Answer: B
Diff: 3 *Page Ref: 5.5*

49) The density of chlorine gas at 1.21 atm and 34.9°C is ________ g/L.
A) 0.0479 B) 0.295 C) 0.423 D) 1.70 E) 3.39

Answer: E
Diff: 3 *Page Ref: 5.5*

50) A 0.286-g sample of gas occupies 125 mL at 60. cm of Hg and 25 °C. What is the molar mass of the gas?
A) 5.9 g/mol B) 44 g/mol C) 59 g/mol D) 71 g/mol

Answer: D
Diff: 3 *Page Ref: 5.5*

51) Given three cylinders containing O_2 gas at the same volume and pressure. Cylinder A is at –15°C, cylinder B is at –5°F, cylinder C is at 255 K. Which cylinder contains the **largest** mass of oxygen?

A) cylinder A
B) cylinder B
C) cylinder C
D) All cylinders contain the same mass of O_2.

Answer: B
Diff: 3 *Page Ref: 5.5*

52) A 1.44-g sample of an unknown pure gas occupies a volume of 0.335 L at a pressure of 1.00 atm and a temperature of 100.0°C. The unknown gas is ________.

A) argon B) helium C) krypton D) neon E) xenon

Answer: E
Diff: 4 *Page Ref: 5.5*

53) Give the gas that is the lowest percent by volume in dry gas.

A) CO_2 B) He C) H_2 D) N_2

Answer: A
Diff: 1 *Page Ref: 5.6*

54) In a container containing CO, H_2, and O_2, what is the mole fraction of CO if the H_2 mole fraction is 0.22 and the O_2 mole fraction is 0.58?

A) 0.20 B) 0.30 C) 0.10 D) 0.50

Answer: A
Diff: 1 *Page Ref: 5.6*

55) A mixture of 0.220 moles CO, 0.350 moles H_2 and 0.640 moles He has a total pressure of 2.95 atm. What is the pressure of H_2?

A) 1.17 atm B) 0.853 atm C) 1.03 atm D) 0.969 atm E) 0.649 atm

Answer: B
Diff: 2 *Page Ref: 5.6*

56) A mixture of 0.220 moles CO, 0.350 moles N_2 and 0.640 moles Ne has a total pressure of 2.95 atm. What is the pressure of CO?

A) 1.86 atm B) 0.649 atm C) 0.536 atm D) 1.54 atm E) 0.955 atm

Answer: C
Diff: 2 *Page Ref: 5.6*

57) A mixture of N_2, O_2 and He have mole fractions of 0.25, 0.65, and 0.10, respectively. What is the pressure of N_2 if the total pressure of the mixture is 3.9 atm?

A) 2.5 atm B) 0.39 atm C) 0.67 atm D) 0.98 atm E) 1.33 atm

Answer: D
Diff: 2 *Page Ref: 5.6*

58) A mixture of Ar, Ne and He has a pressure of 7.85 atm. If the Ne has a mole fraction of 0.47 and He has a mole fraction of 0.23, what is the pressure of Ar?
A) 4.2 atm B) 3.7 atm C) 1.8 atm D) 5.5 atm E) 2.4 atm

Answer: E
Diff: 2 *Page Ref:* 5.6

59) A gas mixture contains CO, Ar and H_2. What is the total pressure of the mixture, if the mole fraction of H_2 is 0.350 and the pressure of H_2 is 0.480 atm?
A) 1.37 atm B) 0.168 atm C) 5.95 atm D) 0.729 atm E) 2.1 atm

Answer: A
Diff: 3 *Page Ref:* 5.6

60) The mole fraction of argon in dry air near sea level is 1.818×10^{-5} where the molar mass of argon is 39.948. The concentration of argon in the atmosphere is ________ ppm.
A) 1.07×10^8 B) 0.934 C) 9.34×10^6 D) 18.18 E) 9.34×10^9

Answer: D
Diff: 2 *Page Ref:* 5.6

61) A balloon contains 0.76 mol N_2, 0.18 mol O_2, 0.031 mol He and 0.026 mol H_2 at 749 mm Hg. What is the partial pressure of O_2?
A) 20 mm Hg B) 23 mm Hg C) 140 mm Hg D) 570 mm Hg

Answer: C
Diff: 3 *Page Ref:* 5.6

62) What is the total pressure in a 6.00-L flask which contains 0.127 mol of $H_2(g)$ and 0.288 mol of $N_2(g)$ at 20.0°C?
A) 0.510 atm B) 0.681 atm C) 1.16 atm D) 1.66 atm

Answer: D
Diff: 3 *Page Ref:* 5.6

63) The concentration of water vapor in a sample of air that has a partial pressure of water of 0.91 torr and a total pressure of air of 735 torr is ________ ppm.
A) 1.2×10^3 B) 1.2 C) 0.12 D) 8.1×10^{-4} E) 0.81

Answer: A
Diff: 3 *Page Ref:* 5.6

64) A sample of air from a home is found to contain 6.2 ppm of carbon monoxide. This means that if the total pressure is 695 torr, then the partial pressure of CO is ________ torr.
A) 4.3×10^3 B) 4.3×10^{-3} C) 4.3 D) 8.9×10^3 E) 1.1×10^8

Answer: B
Diff: 3 *Page Ref:* 5.6

65) The concentration of ozone in a sample of air that has a partial pressure of O_3 of 0.33 torr and a total pressure of air of 735 torr is ________ ppm.
A) 4.5×10^2 B) 0.45 C) 0.045 D) 2.2×10^{-3} E) 2.2

Answer: A
Diff: 3 *Page Ref:* 5.6

66) The mole fraction of carbon dioxide in dry air near sea level is 0.000375, where the molar mass of carbon dioxide is 44.010. The partial pressure of carbon dioxide when the total atmospheric pressure (dry air) is 97.5 kPa is ________ kPa.

A) 2.63×10^5
B) 5.97×10^3
C) 0.0370
D) 1.63
E) 8.40×10^{-4}

Answer: C
Diff: 3 *Page Ref:* 5.6

67) In the laboratory, hydrogen gas is usually made by the following reaction:

$$Zn(s) + 2\,HCl(aq) \rightarrow H_2(g) + ZnCl_2(aq)$$

How many liters of H_2 gas, collected over water at an atmospheric pressure of 752 mm Hg and a temperature of 21.0°C, can be made from 3.566 g of Zn and excess HCl? The partial pressure of water vapor is 18.65 mm Hg at 21.0°C.

A) 0.0975 L B) 1.30 L C) 1.33 L D) 1.36 L

Answer: D
Diff: 4 *Page Ref:* 5.6

68) A 0.600 g sample containing Ag_2O and inert material is heated, causing the silver oxide to decompose according to the following equation:

$$2\,Ag_2O(s) \rightarrow 4\,Ag(s) + O_2(g)$$

If 13.8 mL of gas are collected over water at 27°C and 1.00 atm external pressure, what is the percentage of silver oxide in the sample? The partial pressure of water is 26.7 mm Hg at 27 °C.

A) 10.4% B) 20.9% C) 41.8% D) 44.8%

Answer: C
Diff: 4 *Page Ref:* 5.6

69) Zinc reacts with aqueous sulfuric acid to form hydrogen gas:

$$Zn(s) + H_2SO_4(aq) \rightarrow ZnSO_4(aq) + H_2(g)$$

In an experiment, 225 mL of wet H_2 is collected over water at 27°C and a barometric pressure of 748 torr. How many grams of Zn have been consumed? The vapor pressure of water at 27°C is 26.74 torr.

A) 4.79×10^6 B) 0.567 C) 567 D) 431 E) 4.31×10^5

Answer: B
Diff: 4 *Page Ref:* 5.6

70) Zinc reacts with aqueous sulfuric acid to form hydrogen gas:

$$Zn(s) + H_2SO_4(aq) \rightarrow ZnSO_4(aq) + H_2(g)$$

In an experiment, 201 mL of wet H_2 is collected over water at 27°C and a barometric pressure of 733 torr. The vapor pressure of water at 27°C is 26.74 torr. The partial pressure of hydrogen in this experiment is ________ atm.

A) 0.929 B) 706 C) 0.964 D) 760 E) 1.00

Answer: A
Diff: 4 *Page Ref: 5.6*

71) The mole fraction of oxygen in dry air near sea level is 0.20948. The concentration of oxygen is ________ molecules per liter, assuming an atmospheric pressure of 739 torr and a temperature of 29.5°C.

A) 6.23
B) 0.00819
C) 4.93×10^{21}
D) 3.75×10^{24}
E) 5.07×10^{22}

Answer: C
Diff: 5 *Page Ref: 5.6*

72) The concentration of carbon monoxide in a sample of air is 9.2 ppm. There are ________ molecules of CO in 1.00 L of this air at 755 torr and 23°C.

A) 3.8×10^{-7} B) 2.2×10^{21} C) 2.9×10^{18} D) 1.7×10^{20} E) 2.3×10^{17}

Answer: E
Diff: 5 *Page Ref: 5.6*

73) The action of some commercial drain cleaners is based on the following reaction:

$$2\ NaOH(s) + 2\ Al(s) + 6\ H_2O(l) \rightarrow 2\ NaAl(OH)_4(s) + 3\ H_2(g)$$

What is the volume of H_2 gas formed at STP when 6.32 g of Al reacts with excess NaOH?

A) 3.50 L B) 5.25 L C) 7.87 L D) 8.59 L

Answer: C
Diff: 5 *Page Ref: 5.7*

74) How many liters of oxygen are needed to exactly react with 19.8 g of methane at STP?

$$CH_4(g) + 2\ O_2(g) \rightarrow CO_2(g) + 2\ H_2O(l)$$

A) 13.9 L B) 27.8 L C) 55.5 L D) 60.5 L

Answer: C
Diff: 5 *Page Ref: 5.7*

75) When 14.0 g of zinc metal reacts with excess HCl, how many liters of H_2 gas are produced at STP?

A) 0.208 L B) 0.416 L C) 4.80 L D) 9.60 L

Answer: C
Diff: 5 *Page Ref: 5.7*

76) How many grams of XeF_6 are required to react with 0.579 L of hydrogen gas at 6.46 atm and 45°C in the reaction shown below?

$XeF_6(s) + 3\,H_2(g) \rightarrow Xe(g) + 6\,HF(g)$

A) 9.58 g B) 11.7 g C) 35.2 g D) 106 g

Answer: B
Diff: 5 Page Ref: 5.7

77) A 1.000 kg sample of nitroglycerine, $C_3H_5N_3O_9$, explodes and releases gases with a temperature of 1985°C at 1.100 atm. What is the volume of gas produced?

$4\,C_3H_5N_3O_9(s) \rightarrow 12\,CO_2(g) + 10\,H_2O(g) + 6\,N_2(g) + O_2(g)$

A) 742.2 L B) 3525 L C) 4730 L D) 5378 L

Answer: D
Diff: 5 Page Ref: 5.7

78) Calcium hydride (CaH_2) reacts with water to form hydrogen gas:

$CaH_2(s) + 2H_2O(l) \rightarrow Ca(OH)_2(aq) + 2H_2(g)$

How many grams of CaH_2 are needed to generate 48.0 L of H_2 gas at a pressure of 0.888 atm and a temperature of 32°C?

A) 50.7 B) 0.851 C) 143 D) 35.8 E) 71.7

Answer: D
Diff: 5 Page Ref: 5.7

79) How many grams of zinc metal are required to produce 2.00 liters of hydrogen gas at STP according to the chemical equation shown below?

$Zn(s) + 2\,HCl(aq) \rightarrow ZnCl_2(aq) + H_2(g)$

A) 0.171 g B) 5.83 g C) 11.7 g D) 131 g

Answer: B
Diff: 5 Page Ref: 5.7

80) How many grams of calcium hydride are required to produce 4.56 L of hydrogen gas at 25.0°C and 0.975 atm pressure according to the chemical equation shown below?

$CaH_2(s) + 2\,H_2O(l) \rightarrow Ca(OH)_2(aq) + 2\,H_2(g)$

A) 3.82 g B) 7.64 g C) 15.3 g D) 45.6 g

Answer: A
Diff: 5 Page Ref: 5.7

81) How many liters of hydrogen gas can be generated by reacting 9.25 grams of barium hydride with water at 20°C and 755 mm Hg pressure according to the chemical equation shown below?

$BaH_2(s) + 2\,H_2O(l) \rightarrow Ba(OH)_2(aq) + 2\,H_2(g)$

A) 0.219 L B) 0.799 L C) 1.60 L D) 3.21 L

Answer: D
Diff: 5 Page Ref: 5.7

82) How many grams of water are required to produce 5.50 L of hydrogen gas at 25.0°C and 755 mm Hg pressure according to the chemical equation shown below?

$BaH_2(s) + 2\ H_2O(l) \rightarrow Ba(OH)_2(aq) + 2\ H_2(g)$

A) 2.01 g
B) 4.02 g
C) 4.07 g
D) 8.04 g

Answer: B
Diff: 5 *Page Ref:* 5.7

83) How many liters of hydrogen gas are needed to produce 100.0 grams of water at 25.0°C and 1.00 atm pressure according to the chemical equation shown below?

$2\ H_2(g) + O_2(g) \rightarrow 2\ H_2O(l)$

A) 67.9 L
B) 136 L
C) 203 L
D) 407 L

Answer: B
Diff: 5 *Page Ref:* 5.7

84) What is the **total** volume of the mixture of hydrogen gas and oxygen gas can be obtained from the electrolysis of 110.0 grams of water at 25.0°C and 1.00 atm pressure according to the chemical equation shown below?

$2\ H_2O(l) \rightarrow 2\ H_2(g) + O_2(g)$

A) 74.7 L
B) 149 L
C) 224 L
D) 447 L

Answer: C
Diff: 5 *Page Ref:* 5.7

85) How many mL of O_2 gas at 25°C and 755 mm Hg pressure can be produced from the thermal decomposition of 0.300 grams of $KClO_3(s)$ according to the chemical equation shown below?

$2\ KClO_3(s) \rightarrow 2\ KCl(s) + 3\ O_2(g)$

A) 30.1 mL
B) 40.2 mL
C) 90.4 mL
D) 181 mL

Answer: C
Diff: 5 *Page Ref:* 5.7

86) How many grams of $KCl(s)$ are produced from the thermal decomposition of $KClO_3(s)$ which produces 50.0 mL of $O_2(g)$ at 25°C and 1.00 atm pressure according to the chemical equation shown below?

$2\ KClO_3(s) \rightarrow 2\ KCl(s) + 3\ O_2(g)$?

A) 0.102 g
B) 0.152 g
C) 0.167 g
D) 0.304 g

Answer: A
Diff: 5 *Page Ref:* 5.7

87) How many liters of oxygen gas can be produced at STP from the decomposition of 0.250 L of 3.00 M H_2O_2 in the reaction according to the chemical equation shown below?

$2\ H_2O_2(l) \rightarrow 2\ H_2O(l) + O_2(g)$

A) 8.41 L
B) 11.2 L
C) 16.8 L
D) 33.6 L

Answer: A
Diff: 5 *Page Ref:* 5.7

88) What is the **total** volume of hydrogen gas and oxygen gas that can be produced from the thermal decomposition of 0.0425 grams of H_2O_2 at 700°C and 755 mm Hg according to the chemical equation shown below?

$H_2O_2(l) \rightarrow H_2(g) + O_2(g)$?

A) 50.2 mL B) 100 mL C) 151 mL D) 201 mL

Answer: D
Diff: 5 *Page Ref: 5.7*

89) How many liters of O_2 gas at 25°C and 1.00 atm pressure are needed to react with 60.25 grams of potassium hydride according to the chemical equation shown below?

$2\ KH(s) + O_2(g) \rightarrow H_2O(l) + K_2O(s)$

A) 9.19 L B) 18.4 L C) 36.8 L D) 73.5 L

Answer: B
Diff: 5 *Page Ref: 5.7*

90) How many milliliters of ozone gas at at 25.0°C and 1.00 atm pressure are needed to react with 45.00 mL of a 0.100 M aqueous solution of KI according to the chemical equation shown below?

$O_3(g) + 2\ I^-(aq) + H_2O(l) \rightarrow O_2(g) + I_2(s) + 2\ OH^-(aq)$

A) 55.0 mL B) 110. mL C) 165 mL D) 220. mL

Answer: A
Diff: 5 *Page Ref: 5.7*

91) Identify the gas particle that travels the fastest.

A) H_2 B) N_2 C) Ne D) Cl_2 E) CO

Answer: A
Diff: 1 *Page Ref: 5.8*

92) Identify the gas particle that travels the slowest.

A) H_2 B) N_2 C) Ne D) F_2 E) CO

Answer: E
Diff: 1 *Page Ref: 5.8*

93) What is the average speed (actually the root-mean-square speed) of a neon atom at 27°C?

A) 5.78 m/s B) 19.3 m/s C) 183 m/s D) 609 m/s

Answer: D
Diff: 3 *Page Ref: 5.8*

94) Which of the following gases has the **highest** average speed at 400K?

A) N_2 B) O_2 C) F_2 D) Cl_2

Answer: A
Diff: 3 *Page Ref: 5.8*

95) Which of the following gases has the **lowest** average speed at 25°C?

A) C_3H_8 B) Kr C) CH_3NH_2 D) SO_2

Answer: B
Diff: 3 *Page Ref: 5.8*

96) What is the temperature of NO_2 gas if the average speed (actually the root-mean-square speed) of the molecules is 750 m/s?

A) 1.38 K B) 1.04×10^3 K C) 1.38×10^3 K D) 1.04×10^6 K

Answer: B
Diff: 3 *Page Ref: 5.8*

97) If NO and NH_3 are allowed to effuse through a porous membrane under identical conditions, the rate of effusion for NH_3 will be ________ times that of NO.

A) 0.57 B) 0.75 C) 1.3 D) 1.8

Answer: C
Diff: 3 *Page Ref: 5.9*

98) Rank the following in order of decreasing rate of effusion.

F_2 SF_6 He Ar

A) SF_6 > Ar > F_2 > He
B) Ar > He > SF_6 > F_2
C) F_2 > Ar > He > SF_6
D) He > F_2 > Ar > SF_6
E) He > F_2 > SF_6 > Ar

Answer: D
Diff: 1 *Page Ref: 5.9*

99) Calculate the ratio of effusion rates of Cl_2 to O_2.

A) 47.7 B) 0.451 C) 2.22 D) 1.49 E) 0.671

Answer: E
Diff: 2 *Page Ref: 5.9*

100) A sample of N_2 effuses in 220 s. How long will the same size sample of Cl_2 take to effuse?

A) 350 s B) 138 s C) 558 s D) 86.8 s E) 388 s

Answer: A
Diff: 3 *Page Ref: 5.9*

101) Which of the following samples will have the greatest average speed at 355 K?

A) Kr
B) C_2H_4
C) Cl_2
D) CH_4
E) All of these samples will have the same average speed at the same temperature.

Answer: D
Diff: 1 *Page Ref: 5.9*

Matching Questions

Match the following.

1) barometer
Diff: 1 Page Ref: 5.2

2) manometer
Diff: 1 Page Ref: 5.2

3) sphygmomanometer
Diff: 1 Page Ref: 5.2

4) Boyle's Law
Diff: 1 Page Ref: 5.3

5) Charles's Law
Diff: 1 Page Ref: 5.3

6) Avogadro's Law
Diff: 1 Page Ref: 5.3

7) Ideal Gas Law
Diff: 1 Page Ref: 5.4

8) Dalton's Law
Diff: 1 Page Ref: 5.6

A) $P_T = P_A + P_B + P_C$ ...

B) measures blood pressure

C) $V_1/n_1 = V_2/n_2$

D) measures pressure of a gas in the laboratory

E) PV = nRT

F) measures atmospheric pressure

G) $V_1/T_1 = V_2/T_2$

H) $P_1V_1 = P_2V_2$

Answers: 1) F 2) D 3) B 4) H 5) G
6) C 7) E 8) A

Short Answer Questions

1) Define pressure.
Answer: Pressure is the force exerted per unit area by gas molecules as they strike the surfaces around them.
Diff: 1 Page Ref: 5.1

2) Consider a container of gas under a particular P, V, T set of conditions. Describe how the pressure would change if the volume were doubled while the absolute temperature was increasing by a factor of two.
Answer: The pressure would stay constant. The doubling of the volume would, by itself, decrease the pressure by a factor of 2. However, the temperature is also decreasing by a factor of 2. These two changing variables will cancel one another.
Diff: 2 Page Ref: 5.3

3) Why does HOT air rise?
Answer: As air is heated, it expands and becomes less dense (the same mass of air fills a larger volume). This lower density warm air will "float" to the top of cooler, more dense air.
Diff: 1 Page Ref: 5.5

4) Give the major gas in dry air.

Answer: nitrogen
Diff: 1 *Page Ref: 5.6*

5) Why doesn't Dalton's Law of Partial Pressures depend on the identity of the gases present?

Answer: It is assumed that all gas particles are behaving ideally and not interacting with one another. If that is true, the only difference between particles is associated with their mass. Since all of the gases are at the same temperature and have the same average kinetic energy, their masses don't have to be taken into account. All that matters is all of the different gas particles exert the same force on the walls of the container, no matter what their mass.
Diff: 1 *Page Ref: 5.6*

6) Given the equation $C_2H_6(g) + O_2(g) \rightarrow CO_2(g) + H_2O(g)$ (not balanced), determine the number of liters of CO_2 formed at STP when 240.0 grams of C_2H_6 is burned in excess oxygen gas.

Answer: 358
Diff: 5 *Page Ref: 5.7*

7) Given the equation $C_2H_6(g) + O_2(g) \rightarrow CO_2(g) + H_2O(g)$ (not balanced), determine the number of liters of O_2 consumed at STP when 270.0 grams of C_2H_6 is burned.

Answer: 706
Diff: 5 *Page Ref: 5.7*

8) Explain what the term "mean free path" describes. How does it change with decreasing pressure?

Answer: Mean free path describes the distance that a gas particle can travel before hitting another particle. As pressure decreases, the distance between particles increases, therefore increasing the mean free path of the particles.
Diff: 1 *Page Ref: 5.9*

9) Why does the rate of effusion increase with a decrease in the molar mass?

Answer: The rate of effusion depends on the speed of the gas particles. As the molar mass decreases at a given temperature, the average speed of the particles increases, increasing the rate of effusion.
Diff: 1 *Page Ref: 5.9*

10) Define effusion.

Answer: Effusion is the process by which a gas escapes from a container into a vacuum through a small hole.
Diff: 1 *Page Ref: 5.9*

Chapter 6 Thermochemistry

Multiple Choice Questions

1) Energy that is associated with the position or composition of an object is called
A) kinetic energy
B) thermal energy
C) potential energy
D) chemical energy

Answer: C
Diff: 1 *Page Ref: 6.2*

2) Energy that is associated with the motion of an object is called
A) kinetic energy
B) thermal energy
C) potential energy
D) chemical energy

Answer: A
Diff: 1 *Page Ref: 6.2*

3) Energy that is associated with the temperature of an object is called
A) kinetic energy
B) thermal energy
C) potential energy
D) chemical energy

Answer: B
Diff: 1 *Page Ref: 6.2*

4) Energy that is associated with the relative positions of electrons and nuclei in atoms and molecules is called
A) kinetic energy
B) thermal energy
C) potential energy
D) chemical energy

Answer: D
Diff: 1 *Page Ref: 6.2*

5) The law of ________ states that energy that can be neither created or destroyed.
A) kinetic energy
B) the consecration of energy
C) potential energy
D) the conservation of energy
E) thermochemistry

Answer: D
Diff: 1 *Page Ref: 6.2*

6) Which of the following signs on q and w represent a system that is doing work on the surroundings, as well as losing heat to the surroundings?
A) q = −, w = −
B) q = +, w = +
C) q = −, w = +
D) q = +, w = −
E) None of these represent the system referenced above.

Answer: A
Diff: 1 *Page Ref: 6.3*

7) Which of the following signs on q and w represent a system that is doing work on the surroundings, as well as gaining heat from the surroundings?

A) q = +, w = -
B) q = -, w = +
C) q = +, w = +
D) q = -, w = -
E) None of these represent the system referenced above.

Answer: A
Diff: 1 *Page Ref: 6.3*

8) For ΔE_{sys} to always be -, what must be true?

A) q = w B) +q > -w C) +w > -q D) -w > +q

Answer: D
Diff: 2 *Page Ref: 6.3*

9) Which of the following (with specific heat capacity provided) would show the smallest temperature change upon gaining 200.0 J of heat?

A) 50.0 g Al, C_{Al} = 0.903 J/g°C
B) 50.0 g Cu, C_{Cu} = 0.385 J/g°C
C) 25.0 g granite, $C_{granite}$ = 0.79 J/g°C
D) 25.0 g Au, C_{Au} = 0.128 J/g°C
E) 25.0 g Ag, C_{Ag} = 0.235 J/g°C

Answer: A
Diff: 2 *Page Ref: 6.4*

10) Determine the specific heat capacity of an alloy that requires 59.3 kJ to raise the temperature of 150.0 g alloy from 298 K to 398 K.

A) 4.38 J/g°C B) 2.29 J/g°C C) 3.95 J/g°C D) 2.53 J/g°C E) 1.87 J/g°C

Answer: C
Diff: 2 *Page Ref: 6.4*

11) A sample of copper absorbs 43.6 kJ of heat, resulting in a temperature rise of 75.0°C, determine the mass (in kg) of the copper sample if the specific heat capacity of copper is 0.385 J/g°C.

A) 1.51 kg B) 6.62 kg C) 1.26 kg D) 7.94 kg E) 3.64 kg

Answer: A
Diff: 2 *Page Ref: 6.4*

12) Determine the final temperature of a gold nugget (mass = 376 g) that starts at 398 K and loses 4.85 kJ of heat to a snowbank when it is lost. The specific heat capacity of gold is 0.128 J/g°C.

A) 133 K B) 398 K C) 187 K D) 297 K E) 377 K

Answer: D
Diff: 3 *Page Ref: 6.4*

13) Calculate the change in internal energy (ΔE) for a system that is giving off 25.0 kJ of heat and is changing from 12.00 L to 6.00 L in volume at 1.50 atm pressure. (Remember that 101.3 J = 1 L •atm)

A) +25.9 kJ B) -16.0 kJ C) -25.9 kJ D) -24.1 kJ E) 937 kJ

Answer: D
Diff: 2 *Page Ref: 6.4*

14) Calculate the change in internal energy (ΔE) for a system that is giving off 25.0 kJ of heat and is changing from 18.00 L to 15.00 L in volume at 1.50 atm pressure. (Remember that 101.3 J = 1 L •atm)

A) +25.5 kJ B) −16.0 kJ C) −25.5 kJ D) −24.5 kJ E) 456 kJ

Answer: D
Diff: 2 *Page Ref: 6.4*

15) A 6.55 g sample of aniline ($C_6H_5NH_2$, molar mass = 93.13 g/mol) was combusted in a bomb calorimeter. If the temperature rose by 32.9°C, use the information below to determine the heat capacity of the calorimeter.

$$4\ C_6H_5NH_2(l) + 35\ O_2(g) \rightarrow 24\ CO_2(g) + 14\ H_2O(g) + 4\ NO_2(g)$$

$$\Delta H°_{rxn} = -1.28 \times 10^4\ kJ$$

A) 97.3 kJ/°C B) 38.9 kJ/°C C) 5.94 kJ/°C D) 6.84 kJ/°C E) 12.8 kJ/°C

Answer: D
Diff: 3 *Page Ref: 6.5*

16) A 21.8 g sample of ethanol (C_2H_5OH) is burned in a bomb calorimeter, according to the following reaction. If the temperature rises from 25.0 to 62.3°C, determine the heat capacity of the calorimeter. The molar mass of ethanol is 46.07 g/mol.

$$C_2H_5OH(l) + 3\ O_2(g) \rightarrow 2\ CO_2(g) + 3\ H_2O(g) \qquad \Delta H°_{rxn} = -1235\ kJ$$

A) 4.99 kJ/°C B) 5.65 kJ/°C C) 63.7 kJ/°C D) 33.1 kJ/°C E) 15.7 kJ/°C

Answer: E
Diff: 3 *Page Ref: 6.5*

17) A 4.98 g sample of aniline ($C_6H_5NH_2$, molar mass = 93.13 g/mol) was combusted in a bomb calorimeter with a heat capacity of 4.25 kJ/°C. If the temperature rose from 29.5°C to 69.8°C, determine the value of $\Delta H°_{comb}$ for aniline.

A) $+7.81 \times 10^3$ kJ/mol
B) -3.20×10^3 kJ/mol
C) $+1.71 \times 10^3$ kJ/mol
D) -1.71×10^3 kJ/mol
E) -7.81×10^3 kJ/mol

Answer: B
Diff: 4 *Page Ref: 6.5*

18) The temperature rises from 25.00°C to 29.00°C in a bomb calorimeter when 3.50 g of sucrose undergoes combustion in a bomb calorimeter. Calculate ΔE_{rxn} for the combustion of sucrose in kJ/mol sucrose. The heat capacity of the calorimeter is 4.90 kJ/°C. The molar mass of sugar is 342.3 g/mol.

A) $- 1.92 \times 10^3$ kJ/mole
B) 1.92×10^3 kJ/mole
C) $- 1.23 \times 10^3$ kJ/mole
D) 2.35×10^4 kJ/mole

Answer: A
Diff: 4 *Page Ref: 6.5*

19) A 12.8 g sample of ethanol (C_2H_5OH) is burned in a bomb calorimeter with a heat capacity of 5.65 kJ/°C. Using the information below, determine the final temperature of the calorimeter if the initial temperature is 25.0°C. The molar mass of ethanol is 46.07 g/mol.

$$C_2H_5OH(l) + 3\,O_2(g) \rightarrow 2\,CO_2(g) + 3\,H_2O(g) \qquad \Delta H^\circ_{rxn} = -1235 \text{ kJ}$$

A) 53.4°C
B) 28.1°C
C) 111°C
D) 85.7°C
E) 74.2°C

Answer: D
Diff: 4 *Page Ref: 6.5*

20) A 35.6 g sample of ethanol (C_2H_5OH) is burned in a bomb calorimeter, according to the following reaction. If the temperature rose from 35.0 to 76.0°C and the heat capacity of the calorimeter is 23.3 kJ/°C, what is the value of ΔH°_{rxn}? The molar mass of ethanol is 46.07 g/mol.

$$C_2H_5OH(l) + 3\,O_2(g) \rightarrow 2\,CO_2(g) + 3\,H_2O(g) \qquad \Delta H^\circ_{rxn} = ?$$

A) -1.24×10^3 kJ/mol
B) $+1.24 \times 10^3$ kJ/mol
C) -8.09×10^3 kJ/mol
D) -9.55×10^3 kJ/mol
E) $+9.55 \times 10^3$ kJ/mol

Answer: A
Diff: 4 *Page Ref: 6.5*

21) A 6.55 g sample of aniline ($C_6H_5NH_2$, molar mass = 93.13 g/mol) was combusted in a bomb calorimeter with a heat capacity of 14.25 kJ/°C. If the initial temperature was 32.9°C, use the information below to determine the value of the final temperature of the calorimeter.

$$4\,C_6H_5NH_2(l) + 35\,O_2(g) \rightarrow 24\,CO_2(g) + 14\,H_2O(g) + 4\,NO_2(g)$$

$$\Delta H^\circ_{rxn} = -1.28 \times 10^4 \text{ kJ}$$

A) 257°C
B) 46.6°C
C) 48.7°C
D) 41.9°C
E) 931°C

Answer: C
Diff: 4 *Page Ref: 6.5*

22) Which statement is FALSE?

A) An exothermic reaction gives heat off heat to the surroundings.

B) Enthalpy is the sum of a system's internal energy and the product of pressure and volume.

C) ΔE_{rxn} is a measure of heat.

D) ΔH_{rxn} is the heat of reaction.

E) Endothermic has a positive ΔH.

Answer: C
Diff: 1 *Page Ref: 6.6*

23) Given w = 0, an endothermic reaction has the following.

A) $+\Delta H$ and $-\Delta E$ B) $-\Delta H$ and $+\Delta E$ C) $+\Delta H$ and $+\Delta E$ D) $-\Delta H$ and $-\Delta E$

Answer: C
Diff: 1 *Page Ref: 6.6*

24) How much energy is <u>required</u> to decompose 765 g of PCl_3, according to the reaction below? The molar mass of PCl_3 is 137.32 g/mol and may be useful.

$$4\,PCl_3(g) \rightarrow P_4(s) + 6\,Cl_2(g) \qquad \Delta H^\circ_{rxn} = +1207\text{ kJ}$$

A) 2.31×10^3 kJ

B) 4.33×103 kJ

C) 6.72×10^3 kJ

D) 1.68×10^3 kJ

E) 5.95×10^3 kJ

Answer: D
Diff: 3 *Page Ref: 6.6*

25) How much energy is <u>required</u> to decompose 612 g of PCl_3, according to the reaction below? The molar mass of PCl_3 is 137.32 g/mol and may be useful.

$$4\,PCl_3(g) \rightarrow P_4(s) + 6\,Cl_2(g) \qquad \Delta H^\circ_{rxn} = +1207\text{ kJ}$$

A) 1.85×10^3 kJ

B) 3.46×103 kJ

C) 5.38×10^3 kJ

D) 1.34×10^3 kJ

E) 4.76×10^3 kJ

Answer: D
Diff: 3 *Page Ref: 6.6*

26) How much energy is <u>evolved</u> during the formation of 98.7 g of Fe, according to the reaction below?

$$Fe_2O_3(s) + 2\ Al(s) \rightarrow Al_2O_3(s) + 2\ Fe(s) \qquad \Delta H^\circ_{rxn} = -852\ kJ$$

A) 753 kJ
B) 1.51×10^3 kJ
C) 4.20×10^3 kJ
D) 482 kJ
E) 241 kJ

Answer: A
Diff: 3 *Page Ref: 6.6*

27) How much energy is <u>evolved</u> during the formation of 197 g of Fe, according to the reaction below?

$$Fe_2O_3(s) + 2\ Al(s) \rightarrow Al_2O_3(s) + 2\ Fe(s) \qquad \Delta H^\circ_{rxn} = -852\ kJ$$

A) 1.52×10^3 kJ
B) 3.02×10^3 kJ
C) 8.40×10^3 kJ
D) 964 kJ
E) 482 kJ

Answer: A
Diff: 3 *Page Ref: 6.6*

28) Using the following thermochemical equation, determine the amount of heat produced per kg of CO_2 formed during the combustion of benzene (C_6H_6).

$$2\ C_6H_6(l) + 15\ O_2(g) \rightarrow 12\ CO_2(g) + 6\ H_2O(g) \qquad \Delta H^\circ_{rxn} = -6278\ kJ$$

A) 1.43×10^5 kJ/kg CO_2
B) 2.30×10^4 kJ/kg CO_2
C) 4.34×10^4 kJ/kg CO_2
D) 1.19×10^4 kJ/kg CO_2
E) 8.40×10^5 kJ/kg CO_2

Answer: D
Diff: 3 *Page Ref: 6.6*

29) Using the following equation for the combustion of octane, calculate the amount of moles of carbon dioxide formed from 100.0 g of octane. The molar mass of octane is 114.33 g/mole. The molar mass of carbon dioxide is 44.0095 g/mole.

$$2\ C_8H_{18} + 25\ O_2 \rightarrow 16\ CO_2 + 18\ H_2O \qquad \Delta H^\circ_{rxn} = -11018\ kJ$$

A) 18.18 moles
B) 6.997 moles
C) 14.00 moles
D) 8.000 moles
E) 10.93 moles

Answer: B
Diff: 4 *Page Ref: 6.6*

30) Using the following equation for the combustion of octane, calculate the amount of moles of oxygen that reacts with 100.0 g of octane. The molar mass of octane is 114.33 g/mole. The molar mass of carbon dioxide is 44.0095 g/mole.

$$2\ C_8H_{18} + 25\ O_2 \rightarrow 16\ CO_2 + 18\ H_2O \qquad \Delta H^\circ_{rxn} = -11018\ kJ$$

A) 18.18 moles
B) 6.997 moles
C) 14.00 moles
D) 8.000 moles
E) 10.93 moles

Answer: E
Diff: 4 *Page Ref: 6.6*

31) According to the following reaction, how much energy is evolved during the reaction of 32.5 g B_2H_6 and 72.5 g Cl_2? The molar mass of B_2H_6 is 27.67 g/mol.

$$B_2H_6(g) + 6\ Cl_2(g) \rightarrow 2\ BCl_3(g) + 6\ HCl(g) \qquad \Delta H^\circ_{rxn} = -1396\ kJ$$

A) 1640 kJ B) 238 kJ C) 1430 kJ D) 3070 kJ E) 429 kJ

Answer: B
Diff: 4 *Page Ref: 6.6*

32) According to the following reaction, how much energy is evolved during the reaction of 2.50 L B_2H_6 and 5.65 L Cl_2 (Both gases are initially at STP)? The molar mass of B_2H_6 is 27.67 g/mol.

$$B_2H_6(g) + 6\ Cl_2(g) \rightarrow 2\ BCl_3(g) + 6\ HCl(g) \qquad \Delta H^\circ_{rxn} = -1396\ kJ$$

A) 58.7 kJ B) 156 kJ C) 215 kJ D) 352 kJ E) 508 kJ

Answer: A
Diff: 4 *Page Ref: 6.6*

33) According to the following thermochemical equation, what mass of HF (in g) must react in order to produce 345 kJ of energy? Assume excess SiO_2.

$$SiO_2(s) + 4\ HF(g) \rightarrow SiF_4(g) + 2\ H_2O(l) \qquad \Delta H^\circ_{rxn} = -184\ kJ$$

A) 42.7 g B) 37.5 g C) 150. g D) 107 g E) 173 g

Answer: C
Diff: 4 *Page Ref: 6.6*

34) Using the following equation for the combustion of octane, calculate the amount of grams of carbon dioxide formed from 100.0 g of octane. The molar mass of octane is 114.33 g/mole. The molar mass of carbon dioxide is 44.0095 g/mole.

$$2\ C_8H_{18} + 25\ O_2 \rightarrow 16\ CO_2 + 18\ H_2O \qquad \Delta H^\circ_{rxn} = -11018\ kJ$$

A) 800.1 g B) 307.9 g C) 260.1 g D) 792.3 g

Answer: B
Diff: 5 *Page Ref: 6.6*

35) What volume of benzene (C_6H_6, d= 0.88 g/mL, molar mass = 78.11 g/mol) is required to produce 1.5×10^3 kJ of heat according to the following reaction?

$$2\ C_6H_6(l) + 15\ O_2(g) \rightarrow 12\ CO_2(g) + 6\ H_2O(g) \qquad \Delta H^\circ_{rxn} = -6278\ kJ$$

A) 75 mL B) 37 mL C) 21 mL D) 19 mL E) 42 mL

Answer: E
Diff: 5 *Page Ref: 6.6*

36) Which of the following statements is TRUE?
A) State functions do not depend on the path taken to arrive at a particular state.
B) ΔE_{rxn} can be determined using constant volume calorimetry.
C) Energy is neither created nor destroyed, excluding nuclear reactions.
D) ΔH_{rxn} can be determined using constant pressure calorimetry.
E) All of the above are true.

Answer: E
Diff: 1 *Page Ref: 6.7*

37) Two aqueous solutions are both at room temperature and are then mixed in a coffee cup calorimeter. The reaction causes the temperature of the resulting solution to fall below room temperature. Which of the following statements is TRUE?
A) The products have a lower potential energy than the reactants.
B) This type of experiment will provide data to calculate ΔE_{rxn}.
C) The reaction is exothermic.
D) Energy is leaving the system during reaction.
E) None of the above statements are true.

Answer: E
Diff: 2 *Page Ref: 6.7*

38) A piece of iron (mass = 25.0 g) at 398 K is placed in a styrofoam coffee cup containing 25.0 mL of water at 298 K. Assuming that no heat is lost to the cup or the surroundings, what will the final temperature of the water be? The specific heat capacity of iron = 0.449 J/g°C and water = 4.18 J/g°C.

A) 348 K B) 308 K C) 287 K D) 325 K E) 388 K

Answer: B
Diff: 5 Page Ref: 6.7

39) A student is preparing to perform a series of calorimetry experiments. She first wishes to determine the calorimeter constant (C_{cal}) for her coffee cup calorimeter. She pours a 50.0 mL sample of water at 345 K into the calorimeter containing a 50.0 mL sample of water at 298 K. She carefully records the final temperature of the water as 317 K. What is the value of C_{cal} for the calorimeter?

A) 19 J/K B) 28 J/K C) 99 J/K D) 21 J/K E) 76 J/K

Answer: C
Diff: 5 Page Ref: 6.7

40) A 100.0 mL sample of 0.300 M NaOH is mixed with a 100.0 mL sample of 0.300 M HNO_3 in a coffee cup calorimeter. If both solutions were initially at 35.00°C and the temperature of the resulting solution was recorded as 37.00°C, determine the $\Delta H°_{rxn}$ (in units of kJ/mol NaOH) for the neutralization reaction between aqueous NaOH and HCl. Assume 1) that no heat is lost to the calorimeter or the surroundings, and 2) that the density and the heat capacity of the resulting solution are the same as water.

A) −55.7 kJ/mol NaOH
B) −169 kJ/mol NaOH
C) −16.7 kJ/mol NaOH
D) −27.9 kJ/mol NaOH
E) − 34.4 kJ/mol NaOH

Answer: A
Diff: 5 Page Ref: 6.7

41) Two solutions, initially at 24.60°C, are mixed in a coffee cup calorimeter (C_{cal} = 15.5 J/°C). When a 100.0 mL volume of 0.100 M $AgNO_3$ solution is mixed with a 100.0 mL sample of 0.200 M NaCl solution, the temperature in the calorimeter rises to 25.30°C. Determine the $\Delta H°_{rxn}$ for the reaction as written below. Assume that the density and heat capacity of the solutions is the same as that of water.

$$NaCl(aq) + AgNO_3(aq) \rightarrow AgCl(s) + NaNO_3(aq) \qquad \Delta H°_{rxn} = ?$$

A) −35 kJ B) −69 kJ C) −250 kJ D) −16 kJ E) −140 kJ

Answer: B
Diff: 5 Page Ref: 6.7

42) Two solutions, initially at 24.69°C, are mixed in a coffee cup calorimeter (C_{cal} = 105.5 J/°C). When a 200.0 mL volume of 0.100 M $AgNO_3$ solution is mixed with a 100.0 mL sample of 0.100 M NaCl solution, the temperature in the calorimeter rises to 25.16°C. Determine the $\Delta H°_{rxn}$, in units of kJ/mol AgCl. Assume that the density and heat capacity of the solutions is the same as that of water. Hint: Write a balanced reaction for the process.

A) –32 kJ/mol AgCl
B) –78 kJ/mol AgCl
C) –64 kJ/mol AgCl
D) –25 kJ/mol AgCl
E) –59 kJ/mol AgCl

Answer: C
Diff: 5 *Page Ref: 6.7*

43) Use the standard reaction enthalpies given below to determine $\Delta H°_{rxn}$ for the following reaction:

$P_4(g) + 10\ Cl_2(g) \rightarrow 4PCl_5(s)$ $\Delta H°_{rxn}$ = ?

Given:

$PCl_5(s) \rightarrow PCl_3(g) + Cl_2(g)$ $\Delta H°_{rxn}$ = +157 kJ
$P_4(g) + 6\ Cl_2(g) \rightarrow 4\ PCl_3(g)$ $\Delta H°_{rxn}$ = –1207 kJ

A) –1835 kJ B) –1364 kJ C) –1050. kJ D) –1786 kJ E) –2100. kJ

Answer: A
Diff: 2 *Page Ref: 6.8*

44) Use the standard reaction enthalpies given below to determine $\Delta H°_{rxn}$ for the following reaction:

$2\ NO(g) + O_2(g) \rightarrow 2\ NO_2(g)$ $\Delta H°_{rxn}$ = ?

Given:

$N_2(g) + O_2(g) \rightarrow 2\ NO(g)$ $\Delta H°_{rxn}$ = +183 kJ
$1/2\ N_2(g) + O_2(g) \rightarrow NO_2(g)$ $\Delta H°_{rxn}$ = +33 kJ

A) –150. kJ B) –117 kJ C) –333 kJ D) +115 kJ E) +238 kJ

Answer: B
Diff: 2 *Page Ref: 6.8*

45) Use the standard reaction enthalpies given below to determine $\Delta H°_{rxn}$ for the following reaction:

$2\ S(s) + 3\ O_2(g) \rightarrow 2\ SO_3(g)$ $\Delta H°_{rxn}$ = ?

Given:

$SO_2(g) \rightarrow S(s) + O_2(g)$ $\Delta H°_{rxn}$ = +296.8 kJ
$2\ SO_2(g) + O_2(g) \rightarrow 2\ SO_3(g)$ $\Delta H°_{rxn}$ = –197.8 kJ

A) –494.6 kJ B) –692.4 kJ C) –791.4 kJ D) 1583 kJ E) –293.0 kJ

Answer: C
Diff: 2 *Page Ref: 6.8*

46) Use the standard reaction enthalpies given below to determine $\Delta H°_{rxn}$ for the following reaction:

$4\ SO_3(g) \rightarrow 4\ S(s) + 6\ O_2(g)$ $\Delta H°_{rxn} = ?$

Given:

$SO_2(g) \rightarrow S(s) + O_2(g)$ $\Delta H°_{rxn} = +296.8$ kJ

$2\ SO_2(g) + O_2(g) \rightarrow 2\ SO_3(g)$ $\Delta H°_{rxn} = -197.8$ kJ

A) -494.6 kJ B) -692.4 kJ C) -791.4 kJ D) 1583 kJ E) -293.0 kJ

Answer: D
Diff: 2 *Page Ref: 6.8*

47) Which of the following is not a standard state?
A) for a solid, it is 25°F
B) for a liquid, it is 25°C
C) for a gas, it is 1 atm
D) for a solution, it is 1 M
E) for a liquid, it is 1 atm

Answer: A
Diff: 1 *Page Ref: 6.9*

48) Which of the following is not a standard state?
A) for a liquid, it is 25°F
B) for a solid, it is 25°C
C) for a solid, it is 1 atm
D) for a solution, it is 1 M
E) for a liquid, it is 1 atm

Answer: A
Diff: 1 *Page Ref: 6.9*

49) Choose the thermochemical equation that illustrates $\Delta H°_f$ for Li_2SO_4.
A) $2\ Li^+(aq) + SO_4^{2-}(aq) \rightarrow Li_2SO_4(aq)$
B) $2\ Li(s) + 1/8\ S_8(s, \text{rhombic}) + 2\ O_2(g) \rightarrow Li_2SO_4(s)$
C) $Li_2SO_4(aq) \rightarrow 2\ Li^+(aq) + SO_4^{2-}(aq)$
D) $8\ Li_2SO_4(s) \rightarrow 16\ Li(s) + S_8(s, \text{rhombic}) + 16\ O_2(g)$
E) $16\ Li(s) + S_8(s, \text{rhombic}) + 16\ O_2(g) \rightarrow 8\ Li_2SO_4(s)$

Answer: B
Diff: 1 *Page Ref: 6.9*

50) Use the $\Delta H°_f$ information provided to calculate $\Delta H°_{rxn}$ for the following:

	$\Delta H°_f$ (kJ/mol)
$SO_2Cl_2(g)$	-364
$H_2O(l)$	-286
$HCl(g)$	-92
$H_2SO_4(l)$	-814

$SO_2Cl_2(g) + 2\ H_2O(l) \rightarrow 2\ HCl(g) + H_2SO_4(l)$ $\Delta H°_{rxn} = ?$

A) −256 kJ B) +161 kJ C) −62 kJ D) +800. kJ E) −422 kJ

Answer: C
Diff: 2 *Page Ref:* 6.9

51) Use the $\Delta H°_f$ and $\Delta H°_{rxn}$ information provided to calculate $\Delta H°_f$ for IF:

	$\Delta H°_f$ (kJ/mol)
$IF_7(g)$	-941
$IF_5(g)$	-840

$IF_7(g) + I_2(g) \rightarrow IF_5(g) + 2\ IF(g)$ $\Delta H°_{rxn} = -89$ kJ

A) 101 kJ/mol
B) −146 kJ/mol
C) −190. kJ/mol
D) −95 kJ/mol
E) 24 kJ/mol

Answer: D
Diff: 2 *Page Ref:* 6.9

52) Use the $\Delta H°_f$ and $\Delta H°_{rxn}$ information provided to calculate $\Delta H°_f$ for $SO_3(g)$:

	$\Delta H°_f$ (kJ/mol)
$SO_2(g)$	-297

$2\ SO_2(g) + O_2(g) \rightarrow 2\ SO_3(g)$ $\Delta H°_{rxn} = -198$ kJ

A) −792 kJ/mol
B) −248 kJ/mol
C) −495 kJ/mol
D) −578 kJ/mol
E) −396 kJ/mol

Answer: E
Diff: 2 *Page Ref:* 6.9

53) Use the information provided to determine ΔH°_{rxn} for the following reaction:

	ΔH°_f (kJ/mol)
$CH_4(g)$	-75
$CCl_4(g)$	-96
$HCl(g)$	-92

$CH_4(g) + 4\ Cl_2(g) \rightarrow CCl_4(g) + 4\ HCl(g)$ $\Delta H^\circ_{rxn} = ?$

A) −389 kJ B) −113 kJ C) +113 kJ D) −71 kJ E) +79 kJ

Answer: A
Diff: 2 *Page Ref:* 6.9

54) Use the information provided to determine ΔH°_{rxn} for the following reaction:

	ΔH°_f (kJ/mol)
$CH_4(g)$	-75
$CHCl_3(l)$	-134
$HCl(g)$	-92

$CH_4(g) + 3\ Cl_2(g) \rightarrow CHCl_3(l) + 3\ HCl(g)$ $\Delta H^\circ_{rxn} = ?$

A) −151 kJ B) −335 kJ C) +662 kJ D) +117 kJ E) −217 kJ

Answer: B
Diff: 2 *Page Ref:* 6.9

55) Use the information provided to determine ΔH°_{rxn} for the following reaction:

	ΔH°_f (kJ/mol)
$Fe_2O_3(s)$	-824
$Fe_3O_4(s)$	-1118
$CO(g)$	-111
$CO_2(g)$	-394

$3\ Fe_2O_3(s) + CO(g) \rightarrow 2\ Fe_3O_4(s) + CO_2(g)$ $\Delta H^\circ_{rxn} = ?$

A) +277 kJ B) −577 kJ C) −47 kJ D) +144 kJ E) −111 kJ

Answer: C
Diff: 2 *Page Ref:* 6.9

56) Which of the following statements is TRUE?

A) The burning of fossil fuels contributes to global warming.
B) Cars that run on hydrogen fuel cells are environmentally friendly.
C) The more energy produced per kg of CO_2 produced, the better the fuel.
D) Acid rain is one of the problems associated with the combustion of fossil fuels.
E) All of the above are true.

Answer: E
Diff: 1 *Page Ref:* 6.10

57) Which of the following is not a major contributor to energy consumption?
A) residential
B) commercial
C) transportation
D) industrial
E) atmospheric

Answer: E
Diff: 1 *Page Ref:* 6.10

58) Which of the following is the major contributor to energy consumption?
A) residential
B) commercial
C) transportation
D) industrial
E) atmospheric

Answer: D
Diff: 1 *Page Ref:* 6.10

Algorithmic Questions

1) For a process at constant pressure, 49,600 calories of heat are released. This quantity of heat is equivalent to
A) 4.82×10^{-6} J. B) 1.19×10^{4} J. C) 1.24×10^{4} J. D) 2.08×10^{5} J.

Answer: D
Diff: 2 *Page Ref:* 6.2

2) Which of the following is TRUE if $\Delta E_{sys} = -115$ J?
A) The system is gaining 115 J, while the surroundings are losing 115 J.
B) The system is losing 115 J, while the surroundings are gaining 115 J.
C) Both the system and the surroundings are gaining 115 J.
D) Both the system and the surroundings are losing 115 J.
E) None of the above are true.

Answer: B
Diff: 1 *Page Ref:* 6.3

3) Calculate the change internal energy (ΔE) for a system that is giving off 65.0 kJ of heat and is performing 855 J of work on the surroundings.
A) 64.1 kJ
B) -64.1 kJ
C) -65.9 kJ
D) 9.00×10^{2} kJ
E) -9.00×10^{2} kJ

Answer: C
Diff: 1 *Page Ref:* 6.3

4) Calculate the kinetic energy of a 150 g baseball moving at a speed of 39. m/s (87 mph).
A) 5.8 J B) 1.1×10^{2} J C) 5.8×10^{3} J D) 1.1×10^{5} J

Answer: B
Diff: 2 *Page Ref:* 6.3

5) Define heat capacity.
A) the quantity of heat required to lower the temperature of 1 mole of a substance by 1°C
B) the quantity of heat required to change a system's temperature by 1°C
C) the quantity of heat required to lower the temperature of 1 gram of a substance by 1°C
D) the quantity of heat required to raise the temperature of 1 g of a substance by 1°F
E) the quantity of heat required to lower the temperature of 1 liter of a substance by 1°C

Answer: B
Diff: 1 *Page Ref: 6.4*

6) Define specific heat capacity.
A) the quantity of heat required to lower the temperature of 1 mole of a substance by 1°C
B) the quantity of heat required to change a system's temperature by 1°C
C) the quantity of heat required to raise the temperature of 1 gram of a substance by 1°C
D) the quantity of heat required to lower the temperature of 1 gram of a substance by 1°F
E) the quantity of heat required to lower the temperature of 1 liter of a substance by 1°C

Answer: C
Diff: 1 *Page Ref: 6.4*

7) Define molar heat capacity.
A) the quantity of heat required to raise the temperature of 1 mole of a substance by 1°C
B) the quantity of heat required to change a system's temperature by 1°C
C) the quantity of heat required to raise the temperature of 1 gram of a substance by 1°C
D) the quantity of heat required to lower the temperature of 1 g of a substance by 1°F
E) the quantity of heat required to lower the temperature of 1 liter of a substance by 1°C

Answer: A
Diff: 1 *Page Ref: 6.4*

8) Give the units of heat capacity.
A) J/°C B) J/g°C C) Jmole/°C D) g/°C E) mole/°C

Answer: A
Diff: 1 *Page Ref: 6.4*

9) Give the units of specific heat capacity.
A) J/°C B) J/g°C C) Jmole °C D) g/°C E) gmole°C

Answer: B
Diff: 1 *Page Ref: 6.4*

10) Which of the following substances (with specific heat capacity provided) would show the greatest temperature change upon absorbing 100.0 J of heat?
A) 10.0 g Cu, C_{Cu} = 0.385 J/g°C
B) 10.0 g H_2O, C_{H_2O} = 4.18 J/g°C
C) 10.0 g ethanol, $C_{ethanol}$ = 2.42 J/g°C
D) 10.0 g Al, C_{Al} = 0.903 J/g°C
E) 10.0 g Pb, C_{Pb}= 0.128 J/g°C

Answer: E
Diff: 2 *Page Ref: 6.4*

11) A balloon is inflated from 0.0100 L to 0.400 L against an external pressure of 10.00 atm. How much work is done in joules? 101.3 J = 1 L x atm

A) -39.5 J B) 39.5 J C) 0.395 J D) -0.395 J E) -395 J

Answer: E
Diff: 3 *Page Ref:* 6.4

12) A piece of iron (C=0.449 J/g°C) and a piece of gold (C=0.128 J/g°C) have identical masses. If the iron has an initial temperature of 488 K and the gold has an initial temperature of 308 K, which of the following statements is TRUE of the outcome when the two metals are placed in contact with one another? Assume no heat is lost to the surroundings.

A) Since the two metals have the same mass, the final temperature of the two metals will be 398 K, exactly halfway between the two initial temperatures.
B) Since the two metals have the same mass, but the specific heat capacity of gold is much smaller than that of iron, the final temperature of the two metals will be closer to 308 K than to 488 K.
C) Since the two metals have the same mass, the thermal energy contained in the iron and gold after reaching thermal equilibrium will be the same.
D) Since the two metals have the same mass, the thermal energy contained in each metal after equilibrium will be the same.
E) None of the above are true.

Answer: D
Diff: 1 *Page Ref:* 6.4

13) Calculate the amount of heat (in kJ) required to raise the temperature of a 79.0 g sample of ethanol from 298.0 K to 385.0 K. The specific heat capacity of ethanol is 2.42 J/g°C.

A) 57.0 kJ B) 16.6 kJ C) 73.6 kJ D) 28.4 kJ E) 12.9 kJ

Answer: B
Diff: 2 *Page Ref:* 6.4

14) Calculate the amount of heat (in kJ) necessary to raise the temperature of 53.8 g benzene by 50.6 K. The specific heat capacity of benzene is 1.05 J/g°C

A) 1.61 kJ B) 16.6 kJ C) 2.59 kJ D) 2.86 kJ E) 3.85 kJ

Answer: D
Diff: 2 *Page Ref:* 6.4

15) Calculate the work, *w*, gained or lost by the system when a gas expands from 15 L to 40 L against a constant external pressure of 1.5 atm. 101.3 L • atm = 101 J

A) -6.1 kJ B) -3.8 kJ C) +3.8 kJ D) +6.1 kJ

Answer: B
Diff: 2 *Page Ref:* 6.4

16) The specific heat capacity of liquid mercury is 0.14 J/gK. How many joules of heat are needed to raise the temperature of 5.00 g of mercury from 15.0°C to 36.5°C?

A) 7.7×10^2 J
B) 15 J
C) 36 J
D) 0.0013 J
E) 1.7 J

Answer: B
Diff: 3 *Page Ref:* 6.4

17) The specific heat capacity of solid copper metal is 0.385 J/gK. How many joules of heat are needed to raise the temperature of a 1.55-kg block of copper from 33.0°C to 77.5°C?

A) 1.79×10^5 J
B) 26.6 J
C) 2.66×10^4 J
D) 5.58×10^{-6} J
E) 0.00558 J

Answer: C
Diff: 3 *Page Ref: 6.4*

18) A 5.00-g sample of liquid water at 25.0 C is heated by the addition of 84.0 J of energy. The final temperature of the water is ________°C. The specific heat capacity of liquid water is 4.18 J/gK.

A) 95.2 B) 25.2 C) -21.0 D) 29.0 E) 4.02

Answer: D
Diff: 3 *Page Ref: 6.4*

19) A 6.50-g sample of copper metal at 25.0°C is heated by the addition of 84.0 J of energy. The final temperature of the copper is ________°C. The specific heat capacity of copper is 0.38 J/gK.

A) 29.9 B) 25.0 C) 9.0 D) 59.0 E) 34.0

Answer: D
Diff: 3 *Page Ref: 6.4*

20) The specific heat capacity of liquid water is 4.18 J/g-K. How many joules of heat are needed to raise the temperature of 5.00 g of water from 25.1°C to 65.3°C?

A) 48.1 J
B) 840 J
C) 1.89×10^3 J
D) 2.08×10^{-2} J
E) 54.4 J

Answer: B
Diff: 3 *Page Ref: 6.4*

21) The specific heat capacity of methane gas is 2.20 J/g-K. How many joules of heat are needed to raise the temperature of 5.00 g of methane from 36.0°C to 75.0°C?

A) 88.6 J B) 429 J C) 1221 J D) 0.0113 J E) 22.9 J

Answer: B
Diff: 3 *Page Ref: 6.4*

22) The specific heat of copper is 0.385 J/(g • °C). If 34.2 g of copper, initially at 24.0°C, absorbs 4.689 kJ, what will be the final temperature of the copper?

A) 24.4°C B) 26.8°C C) 356°C D) 380°C

Answer: D
Diff: 3 *Page Ref: 6.4*

23) A sample of copper absorbs 43.6 kJ of heat, resulting in a temperature rise of 30.0°C, determine the mass (in kg) of the copper sample if the specific heat capacity of copper is 0.385 J/g°C.

A) 3.78 kg B) 9.93 kg C) 1.89 kg D) 11.9 kg E) 5.46 kg

Answer: A
Diff: 2 *Page Ref: 6.4*

24) It takes 11.2 kJ of energy to raise the temperature of 145 g of benzene from 23.0°C to 68.0°C. What is the specific heat of benzene?
A) 1.14 J/(g • °C) B) 1.72 J/(g • °C) C) 3.48 J/(g • °C) D) 5.25 J/(g • °C)

Answer: B
Diff: 3 *Page Ref: 6.4*

25) A 50.0-g sample of liquid water at 25.0°C is mixed with 29.0 g of water at 45.0°C. The final temperature of the water is ________ °C.
A) 102 B) 27.6 C) 35.0 D) 142 E) 32.3

Answer: E
Diff: 3 *Page Ref: 6.4*

26) What is the enthalpy change (in kJ) of a chemical reaction that raises the temperature of 250.0 mL of solution having a density of 1.25 g/ml by 7.80°C? (The specific heat of the solution is 3.74 J/g•K.)
A) -7.43 kJ
B) -12.51 kJ
C) 8.20 kJ
D) -9.12 kJ
E) 6.51 kJ

Answer: D
Diff: 4 *Page Ref: 6.4*

27) The combustion of titanium with oxygen produces titanium dioxide:

$$Ti(s) + O_2(g) \rightarrow TiO_2(s)$$

When 2.060 g of titanium is combusted in a bomb calorimeter, the temperature of the calorimeter increases from 25.00°C to 91.60°C. In a separate experiment, the heat capacity of the calorimeter is measured to be 9.84 kJ/K. The heat of reaction for the combustion of a mole of Ti in this calorimeter is ________ kJ/mol.
A) 14.3
B) 19.6
C) -311
D) -0.154
E) -1.52×10^4

Answer: E
Diff: 4 *Page Ref: 6.4*

28) Identify what a bomb calorimeter measures.
A) measures ΔH for aqueous solutions
B) measures ΔE for combustion reactions
C) measures ΔH for reduction solutions
D) measures ΔT for aqueous solutions
E) measures ΔE for oxidation reactions

Answer: B
Diff: 1 *Page Ref: 6.5*

29) Calculate the change in internal energy (ΔE) for a system that is absorbing 35.8 kJ of heat and is expanding from 8.00 to 24.0 L in volume at 1.00 atm. (Remember that 101.3 J = 1 L•atm)
A) +51.8 kJ B) -15.8 kJ C) -16.6 kJ D) -29.3 kJ E) +34.2 kJ

Answer: E
Diff: 2 *Page Ref: 6.5*

30) When 0.455 g of anthracene, $C_{14}H_{10}$, is combusted in a bomb calorimeter that has a water jacket containing 500.0 g of water, the temperature of the water increases by 8.63°C. Assuming that the specific heat of water is 4.18 J/(g • °C), and that the heat absorption by the calorimeter is negligible, estimate the enthalpy of combustion per mole of anthracene.
A) +39.7 kJ/mol B) -39.7 kJ/mol C) -7060 kJ/mol D) -8120 kJ/mol

Answer: C
Diff: 2 *Page Ref: 6.5*

31) Which of the following processes is endothermic?
A) the freezing of water
B) the combustion of butane
C) a hot cup of coffee (system) cools on a countertop
D) the chemical reaction in a "hot pack" often used to treat sore muscles
E) the vaporization of rubbing alcohol

Answer: E
Diff: 1 *Page Ref: 6.6*

32) Which of the following processes is exothermic?
A) a candle flame
B) baking bread
C) the chemical reaction in a "cold pack" often used to treat injuries
D) the vaporization of water
E) None of the above are exothermic.

Answer: A
Diff: 1 *Page Ref: 6.6*

33) Which of the following processes is endothermic?
A) mixing water and acid
B) rusting iron
C) photosynthesis
D) the electron affinity of a fluorine atom
E) None of the above processes are endothermic.

Answer: C
Diff: 1 *Page Ref: 6.6*

34) Using the following equation for the combustion of octane, calculate the heat of reaction for 400.0 g of octane. The molar mass of octane is 114.33 g/mole.

$$2\,C_8H_{18} + 25\,O_2 \rightarrow 16\,CO_2 + 18\,H_2O \qquad \Delta H^\circ_{rxn} = -11018\text{ kJ}$$

A) 19280 kJ B) 19.28 kJ C) 38560 kJ D) 50400 kJ

Answer: A
Diff: 3 *Page Ref: 6.6*

35) How much energy is evolved during the reaction of 51.2 g of Al, according to the reaction below? Assume that there is excess Fe_2O_3.

$$Fe_2O_3(s) + 2\ Al(s) \rightarrow Al_2O_3(s) + 2\ Fe(s) \qquad \Delta H^\circ_{rxn} = -852\ kJ$$

A) 51.2 kJ B) 448 kJ C) 224 kJ D) 1617 kJ E) 808 kJ

Answer: E
Diff: 3 *Page Ref: 6.6*

36) According to the following reaction, how much energy is required to decompose 59.0 kg of Fe_3O_4? The molar mass of Fe_3O_4 is 231.55 g/mol.

$$Fe_3O_4(s) \rightarrow 3\ Fe(s) + 2\ O_2(g) \qquad \Delta H^\circ_{rxn} = +1118\ kJ$$

A) 1.42×10^5 kJ
B) 1.13×10^4 kJ
C) 2.85×10^5 kJ
D) 5.70×10^5 kJ
E) 8.55×10^5 kJ

Answer: C
Diff: 3 *Page Ref: 6.6*

37) Using the following thermochemical equation, determine the amount of heat produced from the combustion of 24.3 g benzene (C_6H_6). The molar mass of benzene is 78.11 g/mole.

$$2\ C_6H_6(l) + 15\ O_2(g) \rightarrow 12\ CO_2(g) + 6\ H_2O(g) \qquad \Delta H^\circ_{rxn} = -6278\ kJ$$

A) 3910 kJ B) 1950 kJ C) 977 kJ D) 40.1 kJ E) 0.302 kJ

Answer: C
Diff: 3 *Page Ref: 6.6*

38) For a particular process that is carried out at constant pressure, $q = 145$ kJ and $w = -35$ kJ. Therefore,
A) $\Delta E = 110$ kJ and $\Delta H = 145$ kJ.
B) $\Delta E = 145$ kJ and $\Delta H = 110$ kJ.
C) $\Delta E = 145$ kJ and $\Delta H = 180$ kJ.
D) $\Delta E = 180$ kJ and $\Delta H = 145$ kJ.

Answer: A
Diff: 2 *Page Ref: 6.6*

39) When 5.00 mol of benzene is vaporized at a constant pressure of 1.00 atm and at its normal boiling point of 80.1°C, 169.5 kJ are absorbed and $P\Delta V$ for the vaporization process is equal to 14.5 kJ then
A) $\Delta E = 155.0$ kJ and $\Delta H = 169.5$ kJ.
B) $\Delta E = 184.0$ kJ and $\Delta H = 169.5$ kJ.
C) $\Delta E = 169.5$ kJ and $\Delta H = 184.0$ kJ.
D) $\Delta E = 169.5$ kJ and $\Delta H = 155.0$ kJ.

Answer: A
Diff: 4 *Page Ref: 6.6*

40) When 10.00 moles of $H_2(g)$ reacts with 5.000 mol of $O_2(g)$ to form 10.00 mol of $H_2O(l)$ at 25°C and a constant pressure of 1.00 atm. If 683.0 kJ of heat are released during this reaction, and $P\Delta V$ is equal to −37.00 kJ, then

A) $\Delta H° = +683.0$ kJ and $\Delta E° = +720.0$ kJ.
B) $\Delta H° = +683.0$ kJ and $\Delta E° = +646.0$ kJ.
C) $\Delta H° = -683.0$ kJ and $\Delta E° = -646.0$ kJ.
D) $\Delta H° = -683.0$ kJ and $\Delta E° = -720.0$ kJ.

Answer: C
Diff: 4 *Page Ref: 6.6*

41) At 1 atm pressure, the heat of sublimation of gallium is 277 kJ/mol and the heat of vaporization is 271 kJ/mol. To the correct number of significant figures, how much heat is required to melt 4.50 mol of gallium at 1 atm pressure?

A) 6 kJ
B) 27 kJ
C) 250 kJ
D) 274 kJ

Answer: B
Diff: 4 *Page Ref: 6.6*

42) According to the following thermochemical equation, what mass of H_2O (in g) must form in order to produce 488 kJ of energy?

$$SiO_2(s) + 4\ HF(g) \rightarrow SiF_4(g) + 2\ H_2O(l) \qquad \Delta H°_{rxn} = -184\ kJ$$

A) 34.0 g
B) 51 g
C) 27.1 g
D) 95.4 g
E) 47.8 g

Answer: D
Diff: 4 *Page Ref: 6.6*

43) How much heat is absorbed/released when 25.00 g of $NH_3(g)$ reacts in the presence of excess $O_2(g)$ to produce $NO(g)$ and $H_2O(l)$ according to the following chemical equation?

$$4\ NH_3(g) + 5\ O_2(g) \rightarrow 4\ NO(g) + 6\ H_2O(l) \qquad \Delta H° = 1168\ kJ$$

A) 428.6 kJ of heat are absorbed.
B) 428.6 kJ of heat are released.
C) 1715 kJ of heat are absorbed.
D) 1715 kJ of heat are released.

Answer: A
Diff: 4 *Page Ref: 6.6*

44) How much heat is absorbed when 45.00 g of $C(s)$ reacts in the presence of excess $SO_2(g)$ to produce $CS_2(l)$ and $CO(g)$ according to the following chemical equation?

$$5\ C(s) + 2\ SO_2(g) \rightarrow CS_2(l) + 4\ CO(g) \qquad \Delta H° = 239.9\ kJ$$

A) 179.8 kJ
B) 239.9 kJ
C) 898.5 kJ
D) 2158 kJ

Answer: A
Diff: 4 *Page Ref: 6.6*

45) At constant pressure, the combustion of 15.0 g of C_2H_6(*g*) releases 777 kJ of heat. What is Δ*H* for the reaction given below?

$$2\ C_2H_6(g) + 7\ O_2(g) \rightarrow 4\ CO_2(g) + 6\ H_2O(l)$$

A) –129 kJ B) –779 kJ C) –1560 kJ D) –3120 kJ

Answer: D
Diff: 4 *Page Ref: 6.6*

46) When 1.50 mol of CH_4(*g*) reacts with excess Cl_2(*g*) at constant pressure according to the chemical equation shown below, 1062 kJ of heat are released. Calculate the value of Δ*H* for this reaction, as written.

$$2\ CH_4(g) + 3\ Cl_2(g) \rightarrow 2\ CHCl_3(l) + 3\ H_2(g) \qquad \Delta H = ?$$

A) –1420 kJ B) –708 kJ C) +708 kJ D) +1420 kJ

Answer: A
Diff: 4 *Page Ref: 6.6*

47) Identify what a coffee cup calorimeter measures.
A) measures ΔH for aqueous solutions
B) measures ΔE for hydrolysis reactions
C) measures ΔH for reduction solutions
D) measures ΔT for combustion solutions
E) measures ΔE for oxidation reactions

Answer: A
Diff: 1 *Page Ref: 6.7*

48) In the presence of excess oxygen, methane gas burns in a constant-pressure system to yield carbon dioxide and water:

$$CH_4(g) + 2O_2(g) \rightarrow CO_2(g) + 2H_2O(l) \qquad \Delta H = -890.0\ kJ$$

Calculate the value of q (kJ) in this exothermic reaction when 1.70 g of methane is combusted at constant pressure.
A) –94.6 kJ
B) 0.0306 kJ
C) –0.0106 kJ
D) 32.7 kJ
E) -9.46×10^4 kJ

Answer: A
Diff: 3 *Page Ref: 6.7*

49) Hydrogen peroxide decomposes to water and oxygen at constant pressure by the following reaction:

$$2H_2O_2(l) \rightarrow 2H_2O(l) + O_2(g) \qquad \Delta H = -196 \text{ kJ}$$

Calculate the value of q (kJ) in this exothermic reaction when 4.00 g of hydrogen peroxide decomposes at constant pressure?
A) -23.1 kJ
B) -11.5 kJ
C) -0.0217 kJ
D) 1.44 kJ
E) -2.31×10^4 kJ

Answer: B
Diff: 3 *Page Ref: 6.7*

50) When 1.50 g of Ba(*s*) is added to 100.00 g of water in a container open to the atmosphere, the reaction shown below occurs and the temperature of the resulting solution rises from 22.00°C to 33.10°C. If the specific heat of the solution is 4.18 J/(g • °C), calculate ΔH for the reaction, as written.

$$Ba(s) + 2\ H_2O(l) \rightarrow Ba(OH)_2(aq) + H_2(g) \qquad \Delta H = ?$$

A) -431 kJ B) -3.14 kJ C) +3.14 kJ D) +431 kJ

Answer: A
Diff: 4 *Page Ref: 6.7*

51) Sodium metal reacts with water to produce hydrogen gas and sodium hydroxide according to the chemical equation shown below. When 0.025 mol of Na is added to 100.00 g of water, the temperature of the resulting solution rises from 25.00°C to 35.75°C. If the specific heat of the solution is 4.18 J/(g • °C), calculate ΔH for the reaction, as written.

$$2\ Na(s) + 2\ H_2O(l) \rightarrow 2\ NaOH(aq) + H_2(g) \qquad \Delta H = ?$$

A) -5.41 kJ B) -90.0 kJ C) -180 kJ D) -360 kJ

Answer: D
Diff: 4 *Page Ref: 6.7*

52) When 50.0 mL of 0.400 M $Ca(NO_3)_2$ is added to 50.0 mL of 0.800 M NaF, CaF_2 precipitates, as shown in the net ionic equation below. The initial temperature of both solutions is 23.0°C. Assuming that the reaction goes to completion, and that the resulting solution has a mass of 100.00 g and a specific heat of 4.18 J/(g • °C), calculate the final temperature of the solution.

$$Ca^{2+}(aq) + 2\ F^-(aq) \rightarrow CaF_2(s) \qquad \Delta H° = -11.5 \text{ kJ}$$

A) 22.45°C B) 23.55°C C) 24.10°C D) 24.65°C

Answer: B
Diff: 4 *Page Ref: 6.7*

53) Identify a substance that is not in its standard state.

A) CO B) Ba C) H_2 D) O_2 E) Ne

Answer: A
Diff: 1 *Page Ref: 6.9*

54) Identify a substance that is not in its standard state.

A) O_2 B) Zn C) H D) O_3 E) Xe

Answer: C
Diff: 1 *Page Ref: 6.9*

55) Which of the following processes is exothermic?

A) the ionization of a potassium atom
B) the breaking of a Br–Br bond
C) the sublimation of dry ice ($CO_2(s)$)
D) the reaction associated with $\Delta H°_f$ for an ionic compound
E) All of the above processes are exothermic.

Answer: D
Diff: 1 *Page Ref: 6.9*

56) Choose the reaction that illustrates $\Delta H°_f$ for $Mg(NO_3)_2$.

A) $Mg(s) + N_2(g) + 3O_2(g) \rightarrow Mg(NO_3)_2(s)$
B) $Mg^{2+}(aq) + 2\ NO_3^-(aq) \rightarrow Mg(NO_3)_2(aq)$
C) $Mg(s) + 2\ N(g) + 6\ O(g) \rightarrow Mg(NO_3)_2(s)$
D) $Mg(NO_3)_2(aq) \rightarrow Mg^{2+}(aq) + 2\ NO_3^-(aq)$
E) $Mg(NO_3)_2(s) \rightarrow Mg(s) + N_2(g) + 3O_2(g)$

Answer: A
Diff: 1 *Page Ref: 6.9*

57) Choose the reaction that illustrates $\Delta H°_f$ for $CsHCO_3$.

A) $Cs(s) + H_2(g) + C(s) + O_2(g) \rightarrow CsHCO_3(s)$
B) $Cs^+(aq) + HCO_3^{-1}(aq) \rightarrow CsHCO_3(s)$
C) $Cs^+(aq) + H_2O(l) + CO_2(g) \rightarrow CsHCO_3(s)$
D) $Cs(s) + 1/2\ H_2(g) + C(s) + 3/2\ O_2(g) \rightarrow CsHCO_3(s)$
E) $Cs(s) + 2\ H(g) + C(s) + 3\ O(g) \rightarrow CsHCO_3(s)$

Answer: D
Diff: 1 *Page Ref: 6.9*

58) Identify the greatest source of energy in the United States.

A) petroleum
B) natural gas
C) water
D) coal
E) nuclear power

Answer: A
Diff: 1 *Page Ref: 6.10*

59) Identify a greenhouse gas.
A) CO_2 B) H_2 C) CO D) NO E) NO_2

Answer: A
Diff: 1 *Page Ref: 6.10*

60) Identify an energy source that is not renewable.
A) solar
B) hydroelectric
C) nuclear energy
D) wind
E) water

Answer: C
Diff: 1 *Page Ref: 6.10*

Matching Questions

Match the following.

1) energy associated with the motion of an object
Diff: 1 *Page Ref: 6.2*

2) energy associated with the temperature of an object
Diff: 1 *Page Ref: 6.2*

3) energy associated with the position of an object
Diff: 1 *Page Ref: 6.2*

4) energy associated with the relative positions of electrons and nuclei in atoms and molecules
Diff: 1 *Page Ref: 6.2*

A) kinetic energy
B) potential energy
C) thermal energy
D) chemical energy

Answers: 1) A 2) C 3) B 4) D

Match the following.

5) $-\Delta E$
Diff: 1 *Page Ref: 6.3*

6) $+\Delta E$
Diff: 1 *Page Ref: 6.3*

7) $-w$
Diff: 1 *Page Ref: 6.3*

8) $+w$
Diff: 1 *Page Ref: 6.3*

9) $-q$
Diff: 1 *Page Ref: 6.3*

10) $+ q$
Diff: 1 *Page Ref: 6.3*

A) work done on the system by the surroundings

B) system gains thermal energy from the surroundings

C) system loses thermal energy to the surroundings

D) work done by the system on the surroundings

E) energy flows out of system into the surroundings

F) energy flows into the system from the surroundings

Answers: 5) E 6) F 7) D 8) A 9) C
10) B

Short Answer Questions

1) Describe the energy changes that occur when a book is held 6 ft off the floor and then dropped.

Answer: The book starts out with a certain amount of potential energy (stored energy due to position). When it begins to drop, that potential energy is then converted to kinetic energy (energy of motion).

Diff: 1 *Page Ref: 6.2*

2) Define chemical energy.

Answer: Chemical energy is the energy associated with the relative positions of electrons and nuclei in atoms and molecules.

Diff: 1 *Page Ref: 6.2*

3) Where does the energy absorbed during an endothermic reaction go?

Answer: During an endothermic reaction, the products are higher in potential energy than the reactants. The thermal (kinetic) energy absorbed during reaction is used in forming products with a higher potential energy.

Diff: 1 *Page Ref: 6.6*

4) Explain the difference between ΔH and ΔE.

Answer: Change in enthalpy (ΔH) only tracks the exchange of heat between a system and its surroundings. Internal energy change (ΔE) tracks both heat and work exchanged between a system and its surroundings.

Diff: 1 *Page Ref: 6.6*

5) Give the temperature and pressure for the standard state for a liquid.

Answer: 1 atmosphere and 25°C

Diff: 1 *Page Ref: 6.9*

6) Why are the standard heats of formation for elements in their most stable form assigned a value of "0"?

Answer: We can't measure absolute enthalpy, but we can measure changes in enthalpy. Assigning all elements in their most stable form a value of "0" gives us a starting point from which to measure all other enthalpy changes.

Diff: 1 *Page Ref: 6.9*

7) Give the equation with the elements in $MgSO_4$ in their standard state as the reactants and $MgSO_4$ as the product.

Answer: $Mg(s) + 1/8\ S_8\ (s, rhombic) + 2\ O_2(g) \rightarrow MgSO_4(s)$

Diff: 1 *Page Ref: 6.9*

8) Write the reaction illustrating $\Delta H°_f$ of $CaCO_3$ and draw the enthalpy diagram showing the relative positions of the reactants and products if it is an exothermic reaction.

Answer: $Ca(s) + C\ (s, graphite) + 3/2\ O_2(g) \rightarrow CaCO_3(s)$

The enthalpy diagram should show the reactants higher in enthalpy than the products.

Diff: 1 *Page Ref: 6.9*

Chapter 7 The Quantum–Mechanical Model of the Atom

Multiple Choice Questions

1) The vertical height of a wave is called
   A) wavelength
   B) amplitude
   C) frequency
   D) area
   E) median

Answer: B
Diff: 1 *Page Ref: 7.2*

2) The number of cycles that pass through a stationary point is called
   A) wavelength
   B) amplitude
   C) frequency
   D) area
   E) median

Answer: C
Diff: 1 *Page Ref: 7.2*

3) The distance between adjacent crests is called
   A) wavelength
   B) amplitude
   C) frequency
   D) area
   E) median

Answer: A
Diff: 1 *Page Ref: 7.2*

4) On the electromagnetic spectrum, visible light is immediately between two other wavelengths. Name them.
   A) infrared and x–ray
   B) radio and microwave
   C) gamma ray and ultraviolet
   D) microwave and x–ray
   E) infrared and ultraviolet

Answer: E
Diff: 1 *Page Ref: 7.2*

5) Which of the following visible colors of light has the highest frequency?
   A) green B) red C) blue D) yellow E) orange

Answer: C
Diff: 1 *Page Ref: 7.2*

6) Which of the following visible colors of light has the longest wavelength?
A) blue B) green C) yellow D) red E) violet

Answer: D
Diff: 1 *Page Ref: 7.2*

7) Which of the following colors of electromagnetic radiation has the shortest wavelength?
A) blue B) violet C) orange D) green E) yellow

Answer: B
Diff: 1 *Page Ref: 7.2*

8) Which of the following types of electromagnetic radiation has the lowest frequency?
A) yellow B) blue C) orange D) green E) purple

Answer: C
Diff: 1 *Page Ref: 7.2*

9) A sunburn is caused by overexposure to ________ radiation.
A) ultraviolet
B) gamma
C) microwave
D) x-ray
E) radio

Answer: A
Diff: 1 *Page Ref: 7.2*

10) ________ are used to image bones and internal organs.
A) Ultraviolet light
B) Gamma rays
C) Microwaves
D) X-rays
E) Radio waves

Answer: D
Diff: 1 *Page Ref: 7.2*

11) Food can be cooked by ________ radiation.
A) ultraviolet
B) gamma
C) microwave
D) x-ray
E) radio

Answer: C
Diff: 1 *Page Ref: 7.2*

12) When waves of equal amplitude from two sources are out of phase when they interact, it is called ________.
A) destructive interference
B) diffraction
C) constructive interference
D) effusion
E) amplitude

Answer: A
Diff: 1 *Page Ref: 7.2*

13) When waves of equal amplitude from two sources are in phase when they interact, it is called ________.
A) destructive interference
B) diffraction
C) constructive interference
D) effusion
E) amplitude

Answer: C
Diff: 1 *Page Ref: 7.2*

14) When a wave encounters an obstacle or a slit that is comparable in size to its wavelength, it bends around it. This characteristic is called ________.
A) destructive interference
B) diffraction
C) constructive interference
D) effusion
E) amplitude

Answer: B
Diff: 1 *Page Ref: 7.2*

15) Calculate the wavelength (in nm) of the blue light emitted by a mercury lamp with a frequency of 6.88×10^{14} Hz.
A) 229 nm B) 436 nm C) 206 nm D) 485 nm E) 675 nm

Answer: B
Diff: 2 *Page Ref: 7.2*

16) Which of the following occur as the energy of a photon increases?
A) the frequency decreases.
B) the speed increases.
C) the wavelength increases
D) the wavelength gets shorter.
E) None of the above occur as the energy of a photon increases.

Answer: D
Diff: 2 *Page Ref: 7.2*

17) Which of the following occur as the wavelength of a photon increases?
A) the frequency decreases
B) the energy increases
C) the speed decreases
D) Planck's constant decreases
E) None of the above occur as the wavelength of a photon increases.

Answer: A
Diff: 2 *Page Ref: 7.2*

18) Calculate the energy of the green light emitted, per photon, by a mercury lamp with a frequency of 5.49×10^{14} Hz.
A) 2.75×10^{-19} J
B) 3.64×10^{-19} J
C) 5.46×10^{-19} J
D) 1.83×10^{-19} J
E) 4.68×10^{-19} J

Answer: B
Diff: 2 *Page Ref: 7.2*

19) Calculate the energy of the orange light emitted, per photon, by a neon sign with a frequency of 4.89×10^{14} Hz.
A) 3.09×10^{-19} J
B) 6.14×10^{-19} J
C) 3.24×10^{-19} J
D) 1.63×10^{-19} J
E) 5.11×10^{-19} J

Answer: C
Diff: 2 *Page Ref: 7.2*

20) Calculate the frequency of the green light emitted by a hydrogen atom with a wavelength of 486.1 nm.
A) 1.46×10^{14} s^{-1}
B) 6.86×10^{14} s^{-1}
C) 4.33×10^{14} s^{-1}
D) 6.17×10^{14} s^{-1}
E) 1.62×10^{14} s^{-1}

Answer: D
Diff: 2 *Page Ref: 7.2*

21) Calculate the energy of the red light emitted by a neon atom with a wavelength of 703.2 nm.

A) 3.54×10^{-19} J

B) 4.27×10^{-19} J

C) 2.34×10^{-19} J

D) 6.45×10^{-19} J

E) 2.83×10^{-19} J

Answer: E

Diff: 3 *Page Ref:* 7.2

22) Calculate the energy of the violet light emitted by a hydrogen atom with a wavelength of 410.1 nm.

A) 4.85×10^{-19} J

B) 2.06×10^{-19} J

C) 1.23×10^{-19} J

D) 8.13×10^{-19} J

E) 5.27×10^{-19} J

Answer: A

Diff: 3 *Page Ref:* 7.2

23) How many photons are contained in a flash of green light (525 nm) that contains 189 kJ of energy?

A) 5.67×10^{23} photons

B) 2.01×10^{24} photons

C) 1.25×10^{31} photons

D) 4.99×10^{23} photons

E) 7.99×10^{30} photons

Answer: D

Diff: 3 *Page Ref:* 7.2

24) Determine the shortest frequency of light required to remove an electron from a sample of Ti metal, if the binding energy of titanium is 3.14×10^{3} kJ/mol.

A) 7.87×10^{15} Hz

B) 4.74×10^{15} Hz

C) 2.11×10^{15} Hz

D) 1.27×10^{15} Hz

E) 6.19×10^{15} Hz

Answer: A

Diff: 3 *Page Ref:* 7.2

25) What total energy (in kJ) is contained in 1.0 mol of photons, all with a frequency of 2.75×10^{14} Hz?

A) 182 kJ B) 219 kJ C) 457 kJ D) 326 kJ E) 110 kJ

Answer: E

Diff: 3 *Page Ref:* 7.2

26) Determine the longest wavelength of light required to remove an electron from a sample of potassium metal, if the binding energy for an electron in K is 1.76×10^3 kJ/mol.
A) 147 nm B) 68.0 nm C) 113 nm D) 885 nm E) 387 nm

Answer: B
Diff: 3 *Page Ref: 7.2*

27) Identify the color of a flame test for potassium.
A) violet B) red C) white D) yellow E) blue

Answer: A
Diff: 1 *Page Ref: 7.3*

28) Identify the color of a flame test for sodium.
A) violet B) red C) white D) yellow E) blue

Answer: D
Diff: 1 *Page Ref: 7.3*

29) Identify the color of a flame test for lithium.
A) violet B) red C) white D) yellow E) blue

Answer: B
Diff: 1 *Page Ref: 7.3*

30) Which of the following statements is TRUE?
A) The emission spectrum of a particular element is always the same and can be used to identify the element.
B) Part of the Bohr model proposed that electrons in the hydrogen atom are located in "stationary states" or particular orbits around the nucleus.
C) The uncertainty principle states that we can never know both the exact location and speed of an electron.
D) An orbital is the volume in which we are most likely to find an electron.
E) All of the above are true.

Answer: E
Diff: 1 *Page Ref: 7.4*

31) Calculate the wavelength of an electron ($m = 9.11 \times 10^{-28}$ g) moving at 3.66×10^6 m/s.
A) 1.99×10^{-10} m
B) 5.03×10^{-10} m
C) 1.81×10^{-10} m
D) 5.52×10^{-9} m
E) 2.76×10^{-9} m

Answer: A
Diff: 2 *Page Ref: 7.4*

32) Calculate the wavelength of a baseball (m = 155 g) moving at 32.5 m/s.

A) 7.60×10^{-36} m

B) 1.32×10^{-34} m

C) 2.15×10^{-32} m

D) 2.68×10^{-34} m

E) 3.57×10^{-32} m

Answer: B
Diff: 2 *Page Ref: 7.4*

33) Determine the velocity of a medicine ball (m = 10.0 kg) with a wavelength of 1.33×10^{-35} m.

A) 8.81 m/s B) 12.3 m/s C) 2.21 m/s D) 4.98 m/s E) 6.44 m/s

Answer: D
Diff: 2 *Page Ref: 7.4*

34) Determine the mass of a ball with a wavelength of 3.45×10^{-34} m and a velocity of 6.55 m/s.

A) 0.293 g B) 12.6 g C) 293 g D) 346 g E) 3.41 g

Answer: C
Diff: 2 *Page Ref: 7.4*

35) It is possible to determine the ionization energy for hydrogen using the Bohr equation. Calculate the ionization energy for an atom of hydrogen, making the assumption that ionization is the transition from $n = 1$ to $n = \infty$.

A) -2.18×10^{-18} J

B) $+2.18 \times 10^{-18}$ J

C) $+4.59 \times 10^{-18}$ J

D) -4.59×10^{-18} J

E) $+4.36 \times 10^{-18}$ J

Answer: B
Diff: 1 *Page Ref: 7.5*

36) For $n = 3$, what are the possible sublevels?

A) 0 B) 0, 1 C) 0, 1, 2 D) 0, 1, 2, 3

Answer: C
Diff: 1 *Page Ref: 7.5*

37) What are the possible orbitals for $n = 3$?

A) s, p, d B) s, p, d, f C) s D) s, p

Answer: A
Diff: 1 *Page Ref: 7.5*

38) What value of l is represented by a d orbital?

A) 1 B) 2 C) 0 D) 3

Answer: B
Diff: 1 *Page Ref: 7.5*

39) What value of l is represented by a f orbital?

A) 1 B) 2 C) 0 D) 3

Answer: D
Diff: 1 *Page Ref: 7.5*

40) How many orbitals are contained in the third principal level ($n = 3$) of a given atom?

A) 9 B) 3 C) 18 D) 7 E) 5

Answer: A
Diff: 1 *Page Ref: 7.5*

41) How many sublevels are contained in the second shell ($n = 2$) of a given atom?

A) 1 B) 2 C) 9 D) 4 E) 3

Answer: B
Diff: 1 *Page Ref: 7.5*

42) Which of the following statements is TRUE?

A) We can sometimes know the exact location and speed of an electron at the same time.
B) All orbitals in a given atom are roughly the same size.
C) Since electrons have mass, we must always consider them to have particle properties and never wavelike properties.
D) Atoms are roughly spherical because when all of the different shaped orbitals are overlapped, they take on a spherical shape.
E) All of the above are true.

Answer: D
Diff: 1 *Page Ref: 7.5*

43) Which of the following quantum numbers describes the shape of an orbital?

A) principal quantum number
B) magnetic quantum number
C) spin quantum number
D) Schrödinger quantum number
E) angular momentum quantum number

Answer: E
Diff: 1 *Page Ref: 7.5*

44) Which of the following quantum numbers describes the orientation of an orbital?

A) magnetic quantum number
B) principal quantum number
C) angular momentum quantum number
D) spin quantum number
E) Schrödinger quantum number

Answer: A
Diff: 1 *Page Ref: 7.5*

45) Which of the following quantum numbers describes the size and energy of an orbital?
A) magnetic quantum number
B) principal quantum number
C) angular momentum quantum number
D) spin quantum number
E) Schrödinger quantum number

Answer: B
Diff: 1 *Page Ref: 7.5*

46) How many different values of l are possible in the third principal level?
A) 1 B) 2 C) 3 D) 0 E) 4

Answer: C
Diff: 1 *Page Ref: 7.5*

47) If two electrons in the same atom have the same value of "l", they are
A) in the same sublevel, but not necessarily in the same level.
B) in the same level, but different sublevel.
C) in the same orbital.
D) in different levels and in different shaped orbitals.
E) none of the above.

Answer: A
Diff: 1 *Page Ref: 7.5*

48) Determine the energy change associated with the transition from $n = 2$ to $n = 5$ in the hydrogen atom.
A) -2.18×10^{-19} J
B) $+6.54 \times 10^{-19}$ J
C) $+4.58 \times 10^{-19}$ J
D) -1.53×10^{-19} J
E) $+3.76 \times 10^{-19}$ J

Answer: C
Diff: 2 *Page Ref: 7.5*

49) Determine the energy change associated with the transition from $n = 3$ to $n = 2$ in the hydrogen atom.
A) $+3.03 \times 10^{-19}$ J
B) -1.82×10^{-19} J
C) $+5.51 \times 10^{-19}$ J
D) -3.03×10^{-19} J
E) $+2.69 \times 10^{-19}$ J

Answer: D
Diff: 2 *Page Ref: 7.5*

50) Calculate the energy change associated with the transition from $n = 4$ to $n = 1$ in the hydrogen atom.

A) $+4.89 \times 10^{-18}$ J
B) $+1.64 \times 10^{-18}$ J
C) -6.12×10^{-18} J
D) $+3.55 \times 10^{-18}$ J
E) -2.04×10^{-18} J

Answer: E
Diff: 2 *Page Ref: 7.5*

51) Which of the following statements is TRUE?

A) The principal quantum number (n) describes the shape of an orbital.
B) The angular momentum quantum number (l) describes the the size and energy associated with an orbital.
C) The magnetic quantum number (m_l) describes the orientation of the orbital.
D) An orbital is the path that an electron follows during its movement in an atom.
E) All of the above are true.

Answer: C
Diff: 2 *Page Ref: 7.5*

52) It is possible to determine the ionization energy for hydrogen using the Bohr equation. Calculate the ionization energy (in kJ) for a mole of hydrogen atoms, making the assumption that ionization is the transition from $n = 1$ to $n = \infty$.

A) 7.62×10^{3} kJ
B) 2.76×10^{3} kJ
C) 1.31×10^{3} kJ
D) 3.62×10^{3} kJ
E) 5.33×10^{3} kJ

Answer: C
Diff: 2 *Page Ref: 7.5*

53) Calculate the wavelength of light associated with the transition from $n = 1$ to $n = 3$ in the hydrogen atom.

A) 103 nm B) 155 nm C) 646 nm D) 971 nm E) 136 nm

Answer: A
Diff: 3 *Page Ref: 7.5*

54) Calculate the frequency of light associated with the transition from $n = 2$ to $n = 3$ in the hydrogen atom.

A) 2.19×10^{14} s^{-1}
B) 5.59×10^{14} s^{-1}
C) 4.57×10^{14} s^{-1}
D) 1.79×10^{14} s^{-1}
E) 3.28×10^{14} s^{-1}

Answer: C
Diff: 3 *Page Ref: 7.5*

55) Determine the end (final) value of n in a hydrogen atom transition, if the electron starts in $n = 4$ and the atom emits a photon of light with a wavelength of 486 nm.

A) 1 B) 5 C) 3 D) 4 E) 2

Answer: E
Diff: 4 *Page Ref: 7.5*

56) Determine the end (final) value of n in a hydrogen atom transition, if the electron starts in $n = 2$ and the atom absorbs a photon of light with a frequency of 4.57×10^{14} Hz.

A) 3 B) 1 C) 4 D) 6 E) 7

Answer: A
Diff: 4 *Page Ref: 7.5*

57) Determine the end (final) value of n in a hydrogen atom transition, if the electron starts in n = 1 and the atom absorbs a photon of light with an energy of 2.044×10^{-18} J.

A) 3 B) 4 C) 2 D) 5 E) 6

Answer: B
Diff: 4 *Page Ref: 7.5*

58) Give the numbers for m_l for an s orbital.

A) 0 B) −1, 0, 1 C) 0, 1 D) 1

Answer: A
Diff: 1 *Page Ref: 7.5*

59) Give the numbers for m_l for a d orbital.

A) 0, 1, 2 B) −1, 0, 1 C) 1, 2, 3 D) −2, −1, 0, 1, 2

Answer: D
Diff: 1 *Page Ref: 7.5*

60) Give the numbers for m_l for a p orbital.

A) 0, 1 B) −1, 0, 1 C) 1, 2 D) −2, −1, 0, 1, 2

Answer: B
Diff: 1 *Page Ref: 7.5*

61) Give the numbers for m_l for an f orbital.

A) 0, 1, 2, 3 B) 1, 2, 3, 4
C) −3, −2, −1, 0, 1, 2, 3 D) −2, −1, 0, 1, 2

Answer: C
Diff: 1 *Page Ref: 7.5*

62) Describe the shape of a p orbital.

A) spherical
B) dumbbell shaped
C) three balls
D) four balls
E) eight balls

Answer: B
Diff: 1 *Page Ref: 7.6*

63) Describe the shape of a s orbital.
A) spherical
B) dumbbell shaped
C) three balls
D) four balls
E) eight balls

Answer: A
Diff: 1 *Page Ref: 7.6*

64) What is the maximum number of s orbitals that are possible?
A) 1 B) 3 C) 7 D) 5 E) 9

Answer: A
Diff: 1 *Page Ref: 7.6*

65) What is the maximum number of p orbitals that are possible?
A) 1 B) 3 C) 7 D) 5 E) 9

Answer: B
Diff: 1 *Page Ref: 7.6*

66) What is the maximum number of d orbitals that are possible?
A) 1 B) 3 C) 7 D) 5 E) 9

Answer: D
Diff: 1 *Page Ref: 7.6*

67) What is the maximum number of f orbitals that are possible?
A) 1 B) 3 C) 7 D) 5 E) 9

Answer: C
Diff: 1 *Page Ref: 7.6*

Algorithmic Questions

1) Electromagnetic radiation with a wavelength of 525 nm appears as green light to the human eye. The frequency of this light is ________ s^{-1}.
A) 5.71×10^{14}
B) 5.71×10^{5}
C) 1.58×10^{2}
D) 1.58×10^{11}
E) 1.75×10^{-15}

Answer: A
Diff: 1 *Page Ref: 7.2*

2) An FM radio station broadcasts electromagnetic radiation at a frequency of 100.6 MHz. The wavelength of this radiation is ________ m.

A) 2.982×10^6
B) 2.982
C) 3.018×10^{16}
D) 3.018×10^{10}
E) 0.3353

Answer: B
Diff: 1 *Page Ref: 7.2*

3) Place the following types of electromagnetic radiation in order of increasing wavelength.

visible light gamma rays microwaves

A) gamma rays< microwaves <visible light
B) microwaves < visible light <gamma rays
C) microwaves < gamma rays < visible light
D) visible light < gamma rays <microwaves
E) gamma rays < visible light < microwaves

Answer: E
Diff: 1 *Page Ref: 7.2*

4) Place the following types of electromagnetic radiation in order of increasing frequency.

x–rays microwaves gamma rays

A) microwaves < x–rays< gamma rays
B) gamma rays < x–rays < microwaves
C) microwaves < gamma rays < x–rays
D) gamma rays < microwaves < x–rays
E) x–rays < gamma rays < microwaves

Answer: A
Diff: 1 *Page Ref: 7.2*

5) Place the following types of electromagnetic radiation in order of decreasing energy.

gamma rays radio waves microwaves

A) radio waves > microwaves > gamma rays
B) gamma rays > microwaves > radio waves
C) radio waves > gamma rays > microwaves
D) gamma rays > radio waves >microwaves
E) microwaves > radio waves > gamma rays

Answer: B
Diff: 1 *Page Ref: 7.2*

6) Identify the color that has a wavelength of 680 nm.

A) blue B) green C) red D) yellow

Answer: C
Diff: 1 *Page Ref: 7.2*

7) Identify the color that has a wavelength of 460 nm.
A) blue B) green C) red D) yellow

Answer: A
Diff: 1 *Page Ref:* 7.2

8) Identify the color that has a wavelength of 540 nm.
A) blue B) green C) red D) yellow

Answer: B
Diff: 1 *Page Ref:* 7.2

9) Calculate the wavelength (in nm) of a the red light emitted by a neon sign with a frequency of 4.92×10^{14} Hz.
A) 610 nm B) 164 nm C) 492 nm D) 792 nm E) 192 nm

Answer: A
Diff: 2 *Page Ref:* 7.2

10) Calculate the frequency of the red light emitted by a neon sign with a wavelength of 690 nm.
A) $2.30 \times 10^{14}\ s^{-1}$
B) $3.90 \times 10^{14}\ s^{-1}$
C) $4.35 \times 10^{14}\ s^{-1}$
D) $5.05 \times 10^{14}\ s^{-1}$
E) $3.45 \times 10^{14}\ s^{-1}$

Answer: C
Diff: 2 *Page Ref:* 7.2

11) Electromagnetic radiation with a wavelength of 525 nm appears as green light to the human eye. The energy of one photon of this light is ________ J.
A) 1.04×10^{-31}
B) 3.79×10^{-28}
C) 3.79×10^{-19}
D) 1.04×10^{-22}
E) 2.64×10^{18}

Answer: C
Diff: 3 *Page Ref:* 7.2

12) How many photons are contained in a burst of yellow light (589 nm) from a sodium lamp that contains 630 kJ of energy?
A) 2.12×10^{13} photons
B) 3.06×10^{30} photons
C) 1.87×10^{24} photons
D) 4.03×10^{28} photons
E) 2.48×10^{25} photons

Answer: C
Diff: 3 *Page Ref:* 7.2

13) How much energy (in kJ) do 3.0 moles of photons, all with a wavelength of 665 nm, contain?
A) 160 kJ B) 360 kJ C) 406 kJ D) 540 kJ E) 246 kJ

Answer: D
Diff: 3 *Page Ref: 7.2*

14) Electromagnetic radiation with a wavelength of 531 nm appears as green light to the human eye. The energy of one photon of this light is 3.74×10^{-19} J. Thus, a laser that emits 1.3×10^{-2} J of energy in a pulse of light at this wavelength produces ________ photons in each pulse.
A) 2.9×10^{-17}
B) 9.2×10^{-24}
C) 1.8×10^{19}
D) 3.5×10^{16}
E) 6.5×10^{13}

Answer: D
Diff: 3 *Page Ref: 7.2*

15) The de Broglie wavelength of an electron with a velocity of 6.00×10^{6} m/s is ________ m. The mass of the electron is 9.11×10^{-28} g.
A) 8.25×10^{9}
B) 8.25×10^{12}
C) 1.21×10^{-16}
D) 1.21×10^{-13}
E) 1.21×10^{-10}

Answer: E
Diff: 2 *Page Ref: 7.4*

16) Determine the mass of a ball with a velocity of 35.0 m/s and a wavelength of 8.92×10^{-34} m.
A) 26.0 g B) 594 g C) 2.07 g D) 47.1 g E) 21.2 g

Answer: E
Diff: 2 *Page Ref: 7.4*

17) Determine the velocity of a marble (m = 8.66 g) with a wavelength of 3.46×10^{-33} m.
A) 45.2 m/s B) 2.21 m/s C) 22.1 m/s D) 38.8 m/s E) 52.9 m/s

Answer: C
Diff: 2 *Page Ref: 7.4*

18) Which of the following transitions (in a hydrogen atom) represent **emission** of the longest wavelength photon?
A) $n = 1$ to $n = 3$
B) $n = 3$ to $n = 2$
C) $n = 3$ to $n = 4$
D) $n = 4$ to $n = 2$
E) $n = 5$ to $n = 4$

Answer: E
Diff: 1 *Page Ref: 7.5*

19) Which of the following transitions (in a hydrogen atom) represent **absorption** of the smallest frequency photon?

A) $n = 5$ to $n = 6$
B) $n = 5$ to $n = 3$
C) $n = 2$ to $n = 1$
D) $n = 1$ to $n = 3$
E) $n = 1$ to $n = 2$

Answer: A
Diff: 1 *Page Ref: 7.5*

20) Choose the transition (in a hydrogen atom) below that represents the **absorption** of the shortest wavelength photon.

A) $n = 1$ to $n = 2$
B) $n = 2$ to $n = 3$
C) $n = 4$ to $n = 5$
D) $n = 6$ to $n = 3$
E) $n = 3$ to $n = 1$

Answer: A
Diff: 1 *Page Ref: 7.5*

21) Which of the following transitions represent the **emission** of a photon with the largest energy?

A) $n = 2$ to $n = 1$
B) $n = 3$ to $n = 1$
C) $n = 6$ to $n = 4$
D) $n = 1$ to $n = 4$
E) $n = 2$ to $n = 4$

Answer: B
Diff: 1 *Page Ref: 7.5*

22) What are the possible values of n and m_l for an electron in a 3d orbital?

A) $n = 1$, 2, or 3 and $m_l = 2$
B) $n = 1$, 2, or 3 and $m_l = -2, -1, 0, +1$, or $+2$
C) $n = 3$ and $m_l = 2$
D) $n = 4$ and $m_l = -2, -1, 0, +1, +2$

Answer: C
Diff: 1 *Page Ref: 7.5*

23) How much energy (in kJ) is required to ionize 2.78 moles of hydrogen atoms?

A) 1.83×10^3 kJ
B) 1.67×10^3 kJ
C) 6.06×10^3 kJ
D) 3.65×10^3 kJ
E) 5.89×10^3 kJ

Answer: D
Diff: 3 *Page Ref: 7.5*

24) How many different values of m_l are possible in the 5d sublevel?
A) 2 B) 1 C) 3 D) 5 E) 7

Answer: D
Diff: 1 *Page Ref: 7.5*

25) How many different values of m_l are possible in the 2p sublevel?
A) 4 B) 1 C) 3 D) 5 E) 7

Answer: C
Diff: 1 *Page Ref: 7.5*

26) Give all possible values of l for a 2 sublevel.
A) 2 B) -1/2 C) 0, 1 D) -1, 0, 1 E) 1, 2

Answer: C
Diff: 1 *Page Ref: 7.5*

27) Give the value of l for a 3p sublevel.
A) 1 B) -4 C) -1 D) 2 E) -2

Answer: A
Diff: 1 *Page Ref: 7.5*

28) Identify the correct values for a 1s sublevel.
A) $n = 3, l = 1, m_l = 0$
B) $n = 2, l = 1, m_l = 1$
C) $n = 1, l = 0, m_l = 0$
D) $n = 2, l = 0, m_l = -1$
E) $n = 4, l = -1, m_l = -2$

Answer: C
Diff: 1 *Page Ref: 7.5*

29) Identify the correct values for a 2p sublevel.
A) $n = 3, l = 1, m_l = 0$
B) $n = 2, l = 1, m_l = +2$
C) $n = 1, l = 0, m_l = 0$
D) $n = 2, l = 1, m_l = 0$
E) $n = 4, l = -1, m_l = 0$

Answer: D
Diff: 1 *Page Ref: 7.5*

30) Identify the correct values for a 3p sublevel.
A) $n = 3, l = 1, m_l = 0$
B) $n = 2, l = 1, m_l = +2$
C) $n = 1, l = 0, m_l = 0$
D) $n = 2, l = 0, m_l = 0$
E) $n = 4, l = -1, m_l = -2$

Answer: A
Diff: 1 *Page Ref: 7.5*

31) Identify the correct values for a 4f sublevel.

A) $n = 3, l = 1, m_l = 0$

B) $n = 2, l = 1, m_l = 2$

C) $n = 1, l = 0, m_l = 0$

D) $n = 2, l = 0, m_l = 0$

E) $n = 4, l = 3, m_l = -2$

Answer: E

Diff: 1 *Page Ref: 7.5*

32) In which orbital below would an electron (on average) be closest to the nucleus?

A) 2p B) 4s C) 2s D) 5d E) 2p

Answer: C

Diff: 1 *Page Ref: 7.5*

33) In which orbital below would an electron (on average) be farthest from the nucleus?

A) 1s B) 4f C) 3s D) 3d E) 2p

Answer: B

Diff: 1 *Page Ref: 7.5*

34) How many subshells are there in the shell with $n = 4$?

A) 3 B) 4 C) 6 D) 18

Answer: B

Diff: 1 *Page Ref: 7.5*

35) What are the possible values of l if $n = 6$?

A) 6

B) 0, 1, 2, 3, 4, or 5

C) −4, −3, −2, −1, 0, +1, +2, +3, or +4

D) −5, −4, −3, −2, −1, 0, +1, +2, +3, +4, or +5

Answer: B

Diff: 1 *Page Ref: 7.5*

36) How many orbitals are there in the fourth shell?

A) 3 B) 4 C) 6 D) 16

Answer: D

Diff: 1 *Page Ref: 7.5*

37) Each of the following sets of quantum numbers is supposed to specify an orbital. Which of the following sets of quantum numbers contains an error?

A) $n = 2, l = 1, m_l = +1$

B) $n = 4, l = 2, m_l = 0$

C) $n = 3, l = 3, m_l = -2$

D) $n = 1, l = 0, m_l = 0$

E) $n = 3, l = 0, m_l = 0$

Answer: C

Diff: 1 *Page Ref: 7.5*

38) Each of the following sets of quantum numbers is supposed to specify an orbital. Choose the one set of quantum numbers that does not contain an error.

A) $n = 2, l = 2, m_l = +1$

B) $n = 2, l = 2, m_l = 0$

C) $n = 3, l = 2, m_l = -3$

D) $n = 4, l = 3, m_l = -2$

E) $n = 4, l = 2, m_l = +4$

Answer: D
Diff: 1 *Page Ref: 7.5*

39) Each of the following sets of quantum numbers is supposed to specify an orbital. Choose the one set of quantum numbers that does not contain an error.

A) $n = 4, l = 4, m_l = 0$

B) $n = 3, l = 2, m_l = +3$

C) $n = 4, l = 0, m_l = -1$

D) $n = 3, l = 1, m_l = -2$

E) $n = 5, l = 3, m_l = -3$

Answer: E
Diff: 1 *Page Ref: 7.5*

40) How many different values of m_l are possible in the 5f sublevel?

A) 1 B) 7 C) 4 D) 6 E) 2

Answer: B
Diff: 1 *Page Ref: 7.5*

41) For a hydrogen atom, which electronic transition would result in the **emission** of a photon with the highest energy?

A) $2s \rightarrow 3p$ B) $2p \rightarrow 6d$ C) $6p \rightarrow 4s$ D) $7f \rightarrow 5d$

Answer: C
Diff: 2 *Page Ref: 7.5*

42) For hydrogen, what is the wavelength of the photon emitted when an electron drops from a 4*d* orbital to a 2*p* orbital in a hydrogen atom? The Rydberg constant is $1.097 \times 10^{-2}\ nm^{-1}$.

A) 656.3 nm B) 486.2 nm

C) 364.6 nm D) 2.057×10^{-3} nm

Answer: B
Diff: 3 *Page Ref: 7.5*

Matching Questions

Match the following.

1) $n = 1$ to $n = 2$
Diff: 3 *Page Ref: 7.5*

2) $n = 3$ to $n = 1$
Diff: 3 *Page Ref: 7.5*

3) $n = 2$ to $n = 3$
Diff: 3 *Page Ref: 7.5*

4) $n = 6$ to $n = 5$
Diff: 3 *Page Ref: 7.5*

5) $n = 5$ to $n = 3$
Diff: 3 *Page Ref: 7.5*

A) 122 nm

B) 1280 nm

C) 103 nm

D) 657 nm

E) 7460 nm

Answers: 1) A 2) C 3) D 4) E 5) B

Short Answer Questions

1) Define diffraction.

Answer: When a wave encounters an obstacle or a slit that is comparable in size to its wavelength, it bends around it.
Diff: 1 *Page Ref: 7.2*

2) Define constructive interference.

Answer: Waves combine that are in phase with each other.
Diff: 1 *Page Ref: 7.2*

3) What is the photoelectric effect?

Answer: Many metals emit electrons when light of high enough energy is shone on them. This observation brought the classical view of light into question.
Diff: 1 *Page Ref: 7.2*

4) Why do atoms only emit certain wavelengths of light when they are excited? (Why do line spectra exist?)

Answer: The energies of atoms are quantized. When an electron moves from one energy level to another during emission, a specific wavelength of light (with specific energy) is emitted. The electrons are not allowed "in between" quantized energy levels and thus only specific lines are observed.
Diff: 1 *Page Ref: 7.3*

5) Describe how a neon light works.

Answer: A neon sign contains glass tubes filled with neon gas. When an electric current is passed through the tube, the neon atoms absorb some of the energy and re-emits it as light.
Diff: 1 *Page Ref: 7.3*

6) Why don't we observe the wavelength of everyday macroscopic objects?

Answer: Due to the large mass of macroscopic objects, the deBroglie wavelength is extremely small. The wavelength is so small that it is impossible to detect compared to the size of the object.

Diff: 1 *Page Ref: 7.4*

7) How many orbitals are contained in the $n = 2$ level? Give the l and m_l values of each of them.

Answer: Four: the 2s and three 2p orbitals.
2s, $l = 0$, $m_l = 0$; 2p, $l = 1$, $m_l = -1$ and $l = 1$, $m_l = 0$ and $l = 1$, $m_l = +1$.

Diff: 1 *Page Ref: 7.5*

8) Sketch one of the 3p orbitals below. How are they different from the 2p orbitals?

Answer: They are larger in size and contain additional nodes.

Diff: 1 *Page Ref: 7.6*

9) Give an example of a d orbital.

Answer: d_{yz}, d_{xy}, d_{xz}, $d_{x^2-y^2}$, or d_{z^2}

Diff: 1 *Page Ref: 7.6*

10) Give an example of a p orbital.

Answer: p_x, p_y, or p_z

Diff: 1 *Page Ref: 7.6*

Chapter 8 Periodic Properties of the Elements

Multiple Choice Questions

1) No two electrons can have the same four quantum numbers is known as the
   - A) Pauli exclusion principle
   - B) Hund's rule
   - C) Aufbau principle
   - D) Heisenberg uncertainty principle

Answer: A
Diff: 1 *Page Ref: 8.3*

2) Only two electrons, with opposing spins, are allowed in each orbital is known as the
   - A) Pauli exclusion principle
   - B) Hund's rule
   - C) Aufbau principle
   - D) Heisenberg uncertainty principle

Answer: C
Diff: 1 *Page Ref: 8.3*

3) When filling degenerate orbitals, electrons fill them singly first, with parallel spins is known as
   - A) Pauli exclusion principle
   - B) Hund's rule
   - C) Aufbau principle
   - D) Heisenberg uncertainty principle

Answer: B
Diff: 1 *Page Ref: 8.3*

4) Choose the orbital diagram that represents the ground state of N.

A)

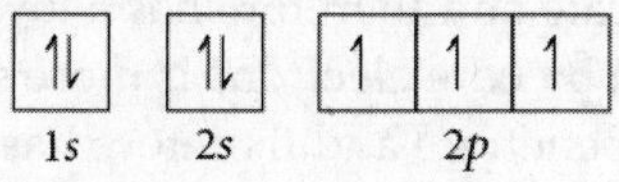

B)

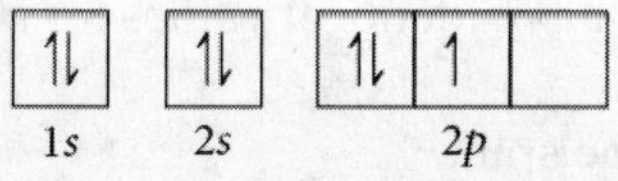

C)

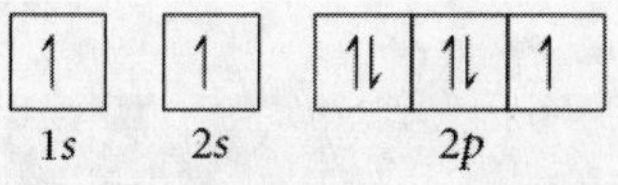

D)

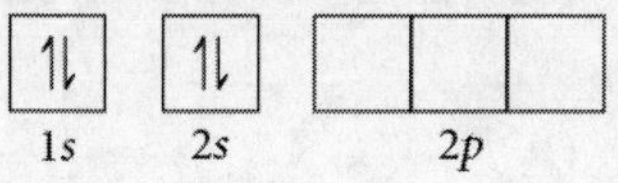

E)

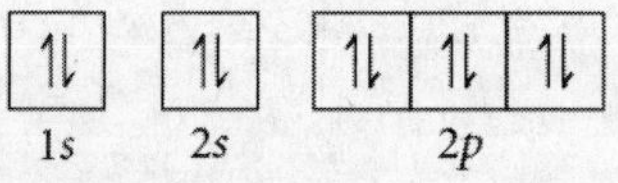

Answer: A
Diff: 2 *Page Ref: 8.3*

5) Choose the valence orbital diagram that represents the ground state of Zn.

A)

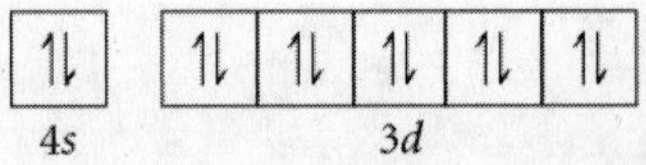

B)

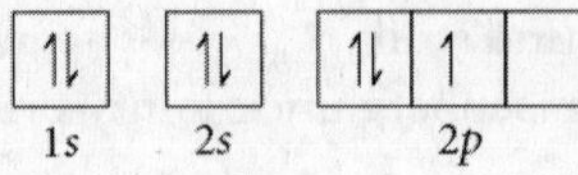

C)

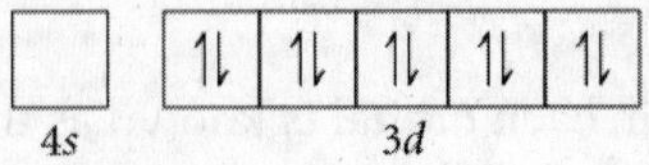

D)

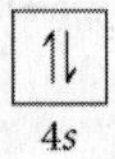

E)

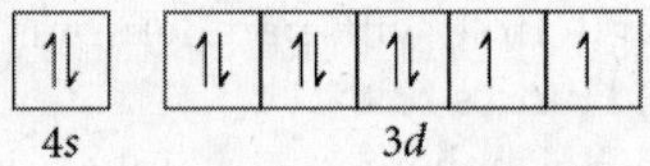

Answer: A
Diff: 2 *Page Ref:* 8.3

6) Which of the following statements is TRUE?
A) An orbital that penetrates into the region occupied by core electrons is less shielded from nuclear charge than an orbital that does not penetrate and therefore has a lower energy.
B) An orbital that penetrates into the region occupied by core electrons is more shielded from nuclear charge than an orbital that does not penetrate and therefore has a lower energy.
C) It is possible for two electrons in the same atom to have identical values for all four quantum numbers.
D) Two electrons in the same orbital can have the same spin.
E) None of the above are true.

Answer: A
Diff: 2 *Page Ref:* 8.3

7) The element that corresponds to the electron configuration $1s^2 2s^2 2p^6 3s^2 3p^6 4s^1 3d^5$ is ________.
A) titanium
B) vanadium
C) chromium
D) manganese
E) iron

Answer: C
Diff: 2 *Page Ref:* 8.3

8) Give the ground state electron configuration for Se.

A) $[Ar]4s^23d^{10}4p^4$
B) $[Ar]4s^24d^{10}4p^4$
C) $[Ar]4s^23d^{10}4p^6$
D) $[Ar]4s^23d^{10}$
E) $[Ar]3d^{10}4p^4$

Answer: A
Diff: 2 Page Ref: 8.3

9) Give the ground state electron configuration for I.

A) $[Kr]5s^24d^{10}5p^6$
B) $[Kr]5s^24d^{10}5p^5$
C) $[Kr]4d^{10}5p^6$
D) $[Kr]5s^25p^6$
E) $[Kr]5s^25d^{10}5p^6$

Answer: B
Diff: 2 Page Ref: 8.3

10) Give the ground state electron configuration for Sr.

A) $[Kr]5s^24d^2$
B) $[Kr]5s^24d^{10}5p^2$
C) $[Kr]5s^2$
D) $[Kr]5s^25d^{10}5p^2$
E) $[Kr]5s^24d^{10}$

Answer: C
Diff: 2 Page Ref: 8.3

11) Give the ground state electron configuration for Pb.

A) $[Xe]6s^26p^2$
B) $[Xe]6s^25d^{10}6p^2$
C) $[Xe]6s^25f^{14}6d^{10}6p^2$
D) $[Xe]6s^24f^{14}5d^{10}6p^2$
E) $[Xe]6s^24f^{14}5d^{10}6s^26p^2$

Answer: D
Diff: 2 Page Ref: 8.3

12) Give the ground state electron configuration for Cd.

A) $[Kr]5s^25d^{10}$
B) $[Kr]5s^24d^{10}5p^2$
C) $[Kr]4d^{10}$
D) $[Kr]5s^24d^8$
E) $[Kr]5s^24d^{10}$

Answer: E
Diff: 2 Page Ref: 8.3

13) Give the complete electronic configuration for Mn.

A) $1s^22s^22p^63s^23p^64s^24d^5$
B) $1s^22s^22p^63s^23p^64s^13d^6$
C) $1s^22s^22p^63s^23p^64s^23d^5$
D) $1s^22s^22p^63s^23p^64s^24p^5$

Answer: C
Diff: 3 *Page Ref: 8.3*

14) Identify the element that has a ground state electronic configuration of $[Kr]5s^24d^5$.

A) Tc
B) Mn
C) Nb
D) Ru

Answer: A
Diff: 1 *Page Ref: 8.3*

15) Give the number of core electrons for Cd.

A) 44
B) 48
C) 46
D) 45
E) 47

Answer: C
Diff: 2 *Page Ref: 8.4*

16) Give the number of valence electrons for Cd.

A) 8
B) 10
C) 12
D) 2
E) 6

Answer: D
Diff: 2 *Page Ref: 8.4*

17) How many valence electrons does an atom of Al possess?

A) 1
B) 2
C) 5
D) 3
E) 8

Answer: D
Diff: 2 *Page Ref: 8.4*

18) How many valence electrons do the halogens possess?

A) 5
B) 6
C) 2
D) 1
E) 7

Answer: E
Diff: 2 *Page Ref: 8.5*

19) How many valence electrons do the alkali earth metals possess?

A) 1
B) 2
C) 7
D) 6
E) 8

Answer: B
Diff: 2 *Page Ref: 8.5*

20) How many valence electrons do the alkali metals possess?

A) 1
B) 2
C) 7
D) 6
E) 8

Answer: A
Diff: 2 *Page Ref: 8.5*

21) How many valence electrons do the noble gases possess?

A) 1
B) 2
C) 7
D) 6
E) 8

Answer: E
Diff: 2 *Page Ref: 8.5*

22) Place the following elements in order of increasing atomic radius.

P Ba Cl

A) Ba < P < Cl
B) P < Cl < Ba
C) Cl < P < Ba
D) Cl < Ba < P
E) Ba < Cl < P

Answer: C
Diff: 1 *Page Ref: 8.6*

23) Place the following in order of increasing atomic radius.

As O Br

A) As < Br < O
B) O < As < Br
C) Br < As < O
D) As < O < Br
E) O < Br < As

Answer: E
Diff: 1 *Page Ref: 8.6*

24) A cation of +2 indicates that an element has
A) lost two electrons.
B) lost two protons.
C) lost two neutrons.
D) gained two electrons.
E) gained two protons.

Answer: A
Diff: 1 *Page Ref: 8.7*

25) Place the following in order of increasing radius.

Br^- Na^+ Rb^+

A) $Br^- < Rb^+ < Na^+$
B) $Na^+ < Rb^+ < Br^-$
C) $Rb^+ < Br^- < Na^+$
D) $Br^- < Na^+ < Rb^+$
E) $Rb^+ < Na^+ < Br^-$

Answer: B
Diff: 1 *Page Ref: 8.7*

26) Place the following in order of decreasing radius.

Te^{2-} F^{-} O^{2-}

A) $F^{-} > O^{2-} > Te^{2-}$
B) $F^{-} > Te^{2-} > O^{2-}$
C) $Te^{2-} > O^{2-} > F^{-}$
D) $Te^{2-} > F^{-} > O^{2-}$
E) $O^{2-} > F^{-} > Te^{2-}$

Answer: C
Diff: 1 *Page Ref: 8.7*

27) Choose the statement that is TRUE.
A) Outer electrons efficiently shield one another from nuclear charge.
B) Core electrons effectively shield outer electrons from nuclear charge.
C) Valence electrons are most difficult of all electrons to remove.
D) Core electrons are the easiest of all electrons to remove.
E) All of the above are true.

Answer: B
Diff: 1 *Page Ref: 8.7*

28) Which reaction below represents the first ionization of O?
A) $O^{+}(g) + e^{-} \rightarrow O(g)$
B) $O(g) + e^{-} \rightarrow O^{-}(g)$
C) $O^{-}(g) \rightarrow O(g) + e^{-}$
D) $O(g) \rightarrow O^{+}(g) + e^{-}$
E) $O^{-}(g) + e^{-} \rightarrow O^{2-}(g)$

Answer: D
Diff: 1 *Page Ref: 8.7*

29) Which reaction below represents the second ionization of Sr?
A) $Sr(g) \rightarrow Sr^{+}(g) + e^{-}$
B) $Sr^{2+}(g) + e^{-} \rightarrow Sr^{+}(g)$
C) $Sr^{+}(g) + e^{-} \rightarrow Sr(g)$
D) $Sr^{-}(g) + e^{-} \rightarrow Sr^{2-}(g)$
E) $Sr^{+}(g) \rightarrow Sr^{2+}(g) + e^{-}$

Answer: E
Diff: 1 *Page Ref: 8.7*

30) What period 3 element having the following ionization energies (all in kJ/mol)?

$IE_1 = 1012$ $IE_2 = 1900$ $IE_3 = 2910$ $IE_4 = 4960$ $IE_5 = 6270$ $IE_6 = 22{,}200$

A) Si B) S C) P D) Cl E) Mg

Answer: C
Diff: 1 *Page Ref: 8.7*

31) Place the following in order of increasing IE_1.

N F As

A) N < As < F
B) As < N < F
C) F < N < As
D) As < F < N
E) F < As < N

Answer: B
Diff: 1 *Page Ref: 8.7*

32) Place the following in order of decreasing IE_1.

Cs Mg Ar

A) Cs > Mg > Ar
B) Mg > Ar > Cs
C) Ar > Mg > Cs
D) Cs > Ar > Mg
E) Mg > Cs > Ar

Answer: C
Diff: 1 *Page Ref: 8.7*

33) Place the following in order of increasing IE_1.

K Ca Rb

A) Ca < K < Rb
B) Rb < Ca < K
C) Ca < Rb < K
D) Rb < K < Ca
E) K < Ca < Rb

Answer: D
Diff: 1 *Page Ref: 8.7*

34) Identify the species that has the smallest radius.
A) cation
B) anion
C) neutral
D) they are all the same size

Answer: A
Diff: 2 *Page Ref: 8.7*

35) Identify the species that has the smallest radius.
A) N^{-5} B) N^{-2} C) N^{0} D) N^{+1} E) N^{+3}

Answer: E
Diff: 2 *Page Ref: 8.7*

36) Identify the number of valence electrons in Cl^{-1}.

A) 6 B) 7 C) 8 D) 5 E) 4

Answer: C
Diff: 3 Page Ref: 8.7

37) Give the ground state electron configuration for Br^{-}.

A) $[Ar]4s^2 3d^{10} 4p^6$
B) $[Ar]4s^2 3d^{10} 4p^5$
C) $[Ar]4s^2 4p^6$
D) $[Ar]4s^2 4d^{10} 4p^6$
E) $[Ar]4s^2 3d^{10} 4p^4$

Answer: A
Diff: 3 Page Ref: 8.7

38) Give the ground state electron configuration for Mg^{2+}.

A) $1s^2 2s^2 2p^6 3s^2$
B) $1s^2 2s^2 2p^6$
C) $1s^2 2s^2 2p^6 3s^2 3p^2$
D) $1s^2 2s^2 2p^6 3s^2 3p^6$
E) $1s^2 2s^2 2p^6 3s^1$

Answer: B
Diff: 3 Page Ref: 8.7

39) Give the ground state electron configuration for Rb^{+}.

A) $[Ar]4s^2 4p^6$
B) $[Kr]5s^1$
C) $[Ar]4s^2 3d^{10} 4p^6$
D) $[Kr]5s^2$
E) $[Kr]5s^2 4d^2$

Answer: C
Diff: 3 Page Ref: 8.7

40) Give the ground state electron configuration for Se^{2-}.

A) $[Ar]4s^2 3d^{10} 4p^4$
B) $[Ar]4s^2 3d^{10} 4p^2$
C) $[Ar]4s^2 4p^6$
D) $[Ar]4s^2 3d^{10} 4p^6$
E) $[Ar]4s^2 3d^8 4p^6$

Answer: D
Diff: 3 Page Ref: 8.7

41) Give the ground state electron configuration for the **ion** of Ba.

A) [Kr]$5s^2 5p^6$
B) [Kr]$5s^2 4d^{10} 5p^6 6s^2 6p^2$
C) [Kr]$5s^2 4d^{10} 5p^6 6s^1$
D) [Kr]$5s^2 4d^{10} 5p^6 6s^2$
E) [Kr]$5s^2 4d^{10} 5p^6$

Answer: E
Diff: 3 *Page Ref: 8.7*

42) Choose the ground state electron configuration for Ti^{2+}.

A) [Ar]$3d^2$
B) [Ar]$4s^2$
C) [Ar]$4s^2 3d^2$
D) [Ar]$4s^2 3d^4$
E) [Ar]$3d^4$

Answer: A
Diff: 3 *Page Ref: 8.7*

43) Choose the ground state electron configuration for Zn^{2+}.

A) [Ar]$4s^2 3d^8$
B) [Ar]$3d^{10}$
C) [Ar]$4s^2 3d^6$
D) [Ar]
E) [Ar]$3d^8$

Answer: B
Diff: 3 *Page Ref: 8.7*

44) Choose the ground state electron configuration for Cr^{3+}.

A) [Ar]$4s^1 3d^2$
B) [Ar]
C) [Ar]$4s^2 3d^6$
D) [Ar]$3d^3$
E) [Ar]$4s^2 3d^1$

Answer: D
Diff: 3 *Page Ref: 8.7*

45) Choose the valence orbital diagram that represents the ground state of Sr^{2+}.

A)

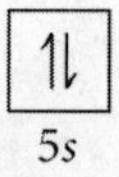

B)

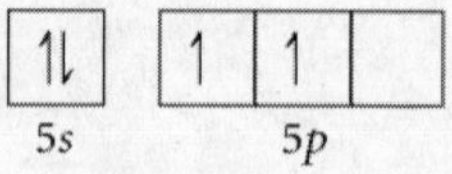

C)

D)

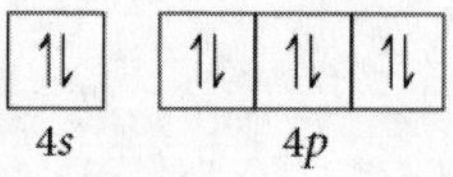

E)

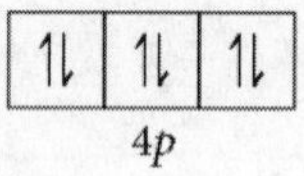

Answer: D

Diff: 3 *Page Ref: 8.7*

46) Choose the valence orbital diagram that represents the ground state of Se^{2-}.

A)

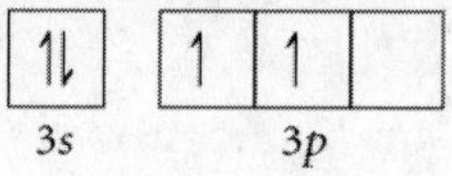

B)

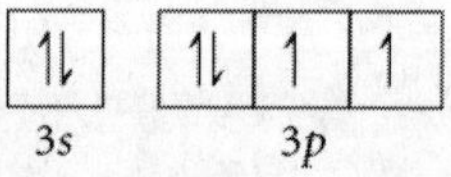

C)

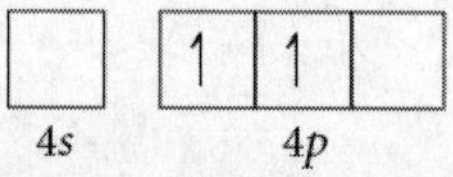

D)

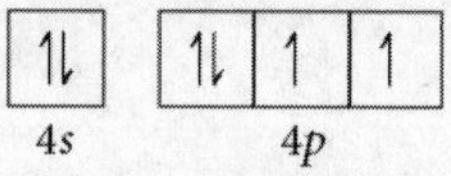

E)

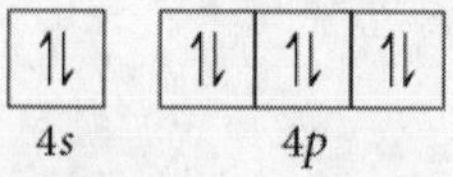

Answer: E

Diff: 3 *Page Ref: 8.7*

47) Which reaction below represents the **electron affinity** of Li?

A) $Li(g) + e^- \rightarrow Li^-(g)$
B) $Li(g) \rightarrow Li^+(g) + e^-$
C) $Li(g) + e^- \rightarrow Li^+(g)$
D) $Li^+(g) \rightarrow Li(g) + e^-$
E) $Li^+(g) + e^- \rightarrow Li(g)$

Answer: A
Diff: 1 *Page Ref: 8.8*

48) Which reaction below represents the **second electron affinity** of S?

A) $S(g) + e^- \rightarrow S^-(g)$
B) $S^-(g) + e^- \rightarrow S^{2-}(g)$
C) $S(g) \rightarrow S^+(g) + e^-$
D) $S^-(g) \rightarrow S(g) + e^-$
E) $S^{2-}(g) \rightarrow S^-(g) + e^-$

Answer: B
Diff: 1 *Page Ref: 8.8*

49) Place the following in order of decreasing metallic character.

P As K

A) P > As > K
B) As > P > K
C) K > P > As
D) As > K > P
E) K > As > P

Answer: E
Diff: 1 *Page Ref: 8.8*

50) Place the following in order of increasing metallic character.

Rb Cs K Na

A) K < Cs < Na < Rb
B) Na < K < Rb < Cs
C) Cs < Rb < K < Na
D) K < Cs < Rb < Na
E) Na < Rb < Cs < K

Answer: B
Diff: 1 *Page Ref: 8.8*

51) Describe the reaction of the noble gases with metals.

A) inert
B) vigorous
C) mild reaction
D) forms water
E) dissolves

Answer: A
Diff: 1 *Page Ref: 8.9*

52) Describe the reaction of the alkali metals with non-metals.
A) inert
B) vigorous
C) mild reaction
D) forms water
E) dissolves

Answer: B
Diff: 1 *Page Ref: 8.9*

53) Identify the most reactive alkali metal with water.
A) Cs B) Rb C) K D) Na E) Li

Answer: A
Diff: 1 *Page Ref: 8.9*

54) Identify a characteristic of halogens.
A) powerful reducing agents
B) forms water in reactions
C) powerful oxidizing agents
D) absorbs water in reactions
E) inert

Answer: C
Diff: 1 *Page Ref: 8.9*

55) Halogens can react with each other to form
A) covalent bonds.
B) ionic bonds.
C) salts.
D) hydrogen halides.
E) metal halides.

Answer: A
Diff: 1 *Page Ref: 8.9*

56) The U.S. Food and Drug Administration recommends the ingestion of recommended doses of ________ in the event of a nuclear emergency.
A) sodium bromide
B) sodiuim chloride
C) potassium iodide
D) potassium bromide
E) sodium iodide

Answer: C
Diff: 1 *Page Ref: 8.9*

Algorithmic Questions

1) Give the set of four quantum numbers that could represent the last electron added (using the Aufbau principle) to the Cl atom.

A) $n = 3, l = 1, m_l = 1, m_s = +\frac{1}{2}$

B) $n = 3, l = 0, m_l = 1, m_s = -\frac{1}{2}$

C) $n = 3, l = 2, m_l = 1, m_s = +\frac{1}{2}$

D) $n = 2, l = 1, m_l = 1, m_s = -\frac{1}{2}$

E) $n = 3, l = 1, m_l = 1, m_s = -\frac{1}{2}$

Answer: A
Diff: 2 *Page Ref:* 8.3

2) Give the set of four quantum numbers that could represent the last electron added (using the Aufbau principle) to the Ne atom.

A) $n = 2, l = 1, m_l = 1, m_s = +\frac{1}{2}$

B) $n = 3, l = 0, m_l = 1, m_s = +\frac{1}{2}$

C) $n = 3, l = 2, m_l = 1, m_s = +\frac{1}{2}$

D) $n = 2, l = 1, m_l = 1, m_s = -\frac{1}{2}$

E) $n = 3, l = 2, m_l = 1, m_s = -\frac{1}{2}$

Answer: D
Diff: 2 *Page Ref:* 8.3

3) Give the set of four quantum numbers that could represent the last electron added (using the Aufbau principle) to the Sr atom.

A) $n = 5, l = 0, m_l = 0, m_s = -\frac{1}{2}$

B) $n = 4, l = 1, m_l = 1\ m_s = -\frac{1}{2}$

C) $n = 5, l = 0, m_l = 0, m_s = +\frac{1}{2}$

D) $n = 4, l = 1, m_l = -1, m_s = +\frac{1}{2}$

E) $n = 5, l = 1, m_l = 1, m_s = +\frac{1}{2}$

Answer: A
Diff: 2 *Page Ref:* 8.3

4) Give the set of four quantum numbers that represent the last electron added (using the Aufbau principle) to the Zn atom.

A) $n = 4, l = 3, m_l = 3, m_s = +\frac{1}{2}$

B) $n = 3, l = 2, m_l = 2, m_s = -\frac{1}{2}$

C) $n = 3, l = 1, m_l = 1, m_s = -\frac{1}{2}$

D) $n = 3, l = 2, m_l = 2, m_s = +\frac{1}{2}$

E) $n = 4, l = 1\ m_l = 0, m_s = +\frac{1}{2}$

Answer: B
Diff: 2 *Page Ref:* 8.3

5) Identify the element that has a ground state electronic configuration of $[Ar]4s^2 3d^{10} 4p^1$.

A) Al B) In C) Ga D) B

Answer: C
Diff: 2 *Page Ref:* 8.3

6) Give the electron configuration for O.

A) $1s^2 2s^2 2p^4$
B) $1s^2 2p^4$
C) $1s^2 2s^2 2p^3$
D) $1s^2 2s^2 2p^5$
E) $1s^2 2s^1 2p^1$

Answer: A
Diff: 2 *Page Ref:* 8.3

7) The element that corresponds to the electron configuration $1s^2 2s^2 2p^6$ is _______.

A) sodium
B) magnesium
C) lithium
D) beryllium
E) neon

Answer: E
Diff: 2 *Page Ref:* 8.3

8) The element that corresponds to the electron configuration $1s^2 2s^2 2p^6 3s^2 3p^6 4s^2 3d^3$ is _______.

A) scandium
B) manganese
C) vanadium
D) iron
E) cobalt

Answer: C
Diff: 2 *Page Ref:* 8.3

9) The complete electron configuration of argon, element 18, is ________.

A) $1s^22s^22p^63s^23p^6$
B) $1s^22s^22p^{10}3s^23p^2$
C) $1s^42s^42p^63s^4$
D) $1s^42s^42p^{10}$
E) $1s^62s^62p^23s^4$

Answer: A
Diff: 2 *Page Ref:* 8.3

10) The complete electron configuration of gallium, element 31, is ________.

A) $1s^22s^22p^{10}3s^23p^{10}4s^23d^3$
B) $1s^22s^22p^63s^23p^63d^{10}4s^24p^1$
C) $1s^42s^42p^63s^43p^64s^43d^3$
D) $1s^42s^42p^{10}3s^43p^9$
E) $1s^42s^42p^83s^43p^84s^3$

Answer: B
Diff: 2 *Page Ref:* 8.3

11) The condensed electron configuration of silicon, element 14, is ________.

A) $[He]2s^42p^6$
B) $[Ne]2p^{10}$
C) $[Ne]3s^23p^2$
D) $[He]2s^4$
E) $[He]2s^62p^2$

Answer: C
Diff: 2 *Page Ref:* 8.3

12) The condensed electron configuration of krypton, element 36, is ________.

A) $[Kr]4s^23d^8$
B) $[Ar]4s^1$
C) $[Kr]4s^43d^8$
D) $[Ar]3d^{10}4s^24p^6$
E) $[Ar]4s^43d^4$

Answer: D
Diff: 2 *Page Ref:* 8.3

13) Which element has the ground-state electron configuration $[Xe]6s^2\ 4f^7$?

A) Re B) Ir C) Eu D) Gd

Answer: C
Diff: 2 *Page Ref:* 8.3

14) Give the number of valence electrons for Cl.

A) 8 B) 6 C) 5 D) 3 E) 7

Answer: E
Diff: 1 *Page Ref:* 8.4

15) Give the number of core electrons for Se.
A) 26 B) 30 C) 28 D) 32 E) 34

Answer: C
Diff: 1 *Page Ref: 8.4*

16) Give the number of valence electrons for Se.
A) 0 B) 8 C) 2 D) 4 E) 6

Answer: E
Diff: 1 *Page Ref: 8.4*

17) Give the number of valence electrons for Si.
A) 4 B) 2 C) 28 D) 12 E) 26

Answer: A
Diff: 2 *Page Ref: 8.4*

18) How many unpaired electrons are present in the ground state As atom?
A) 0 B) 3 C) 1 D) 2 E) 4

Answer: B
Diff: 2 *Page Ref: 8.4*

19) How many unpaired electrons are present in the ground state Kr atom?
A) 1 B) 2 C) 0 D) 3 E) 5

Answer: C
Diff: 2 *Page Ref: 8.4*

20) How many unpaired electrons are present in the ground state C atom?
A) 5 B) 3 C) 1 D) 2 E) 4

Answer: D
Diff: 2 *Page Ref: 8.4*

21) How many unpaired electrons are there in the ground state of Cl?
A) 5 B) 4 C) 3 D) 2 E) 1

Answer: E
Diff: 2 *Page Ref: 8.4*

22) How many of the following elements have 2 unpaired electrons in the ground state?

C Te Hf Si

A) 1 B) 2 C) 3 D) 4

Answer: D
Diff: 2 *Page Ref: 8.4*

23) How many of the following elements have 1 unpaired electron in the ground state?

B Al O F

A) 1 B) 2 C) 3 D) 4

Answer: C
Diff: 2 *Page Ref: 8.4*

24) Identify the number of valence electrons for Ti.
A) 8 B) 7 C) 4 D) 2

Answer: D
Diff: 2 *Page Ref: 8.4*

25) How many valence electrons does an atom of As have?
A) 3 B) 5 C) 8 D) 4 E) 6

Answer: B
Diff: 2 *Page Ref: 8.4*

26) How many valence electrons does an atom of Mg possess?
A) 2 B) 1 C) 8 D) 5 E) 3

Answer: A
Diff: 2 *Page Ref: 8.4*

27) How many valence electrons does an atom of Ti possess?
A) 2 B) 4 C) 6 D) 8 E) 0

Answer: A
Diff: 2 *Page Ref: 8.4*

28) How many valence electrons does an atom of Ag possess?
A) 2 B) 5 C) 11 D) 3 E) 1

Answer: E
Diff: 2 *Page Ref: 8.4*

29) How many valence shell electrons does an atom of indium have?
A) 1 B) 4 C) 3 D) 49 E) 2

Answer: C
Diff: 2 *Page Ref: 8.4*

30) An element that has the valence electron configuration $6s^2 6p^6$ belongs to which period and group?
A) period 6; group 6A
B) period 6; group 8A
C) period 7; group 6A
D) period 7; group 8A

Answer: B
Diff: 2 *Page Ref: 8.4*

31) Which of the following have their valence electrons in the same shell?
A) Li, N, F B) B, Si, As C) N, As, Bi D) He, Ne, F

Answer: A
Diff: 2 *Page Ref: 8.4*

32) Which of the following have the same number of valence electrons?
A) Rb, Sb, I B) Ga, Sn, Bi C) As, Sb, Bi D) Ar, Kr, Br

Answer: C
Diff: 2 *Page Ref: 8.4*

33) What is the general valence–electron ground–state electron configuration for neutral alkaline earth metals?

A) ns^1 B) ns^2 C) $1s^2 2s^1$ D) $1s^2 2s^2$

Answer: B

Diff: 2 *Page Ref: 8.4*

34) How many valence electrons does a neutral tellurium atom have?

A) 2 B) 4 C) 6 D) 52

Answer: C

Diff: 2 *Page Ref: 8.4*

35) Place the following elements in order of decreasing atomic radius.

Xe Rb Ar

A) Ar > Xe > Rb
B) Xe > Rb > Ar
C) Ar > Rb > Xe
D) Rb > Xe > Ar
E) Rb > Ar > Xe

Answer: D

Diff: 1 *Page Ref: 8.6*

36) Of the following, which atom has the largest atomic radius?

A) Rb B) I C) Cs D) At

Answer: C

Diff: 1 *Page Ref: 8.6*

37) Of the following, which atom has the smallest atomic radius?

A) K B) As C) Rb D) Sb

Answer: B

Diff: 1 *Page Ref: 8.6*

38) Which atom in each group (I and II) has the smallest atomic radius?

(I) Ba, Hf, At (II) As, Sb, Bi

A) Ba; As B) Ba; Bi C) At; As D) At; Bi

Answer: C

Diff: 1 *Page Ref: 8.6*

39) Give the number of core electrons for Cl^-.

A) 22 B) 30 C) 17 D) 12 E) 10

Answer: E

Diff: 3 *Page Ref: 8.7*

40) Give the number of valence electrons for Br^-.

A) 34 B) 36 C) 6 D) 8 E) 7

Answer: D

Diff: 3 *Page Ref: 8.7*

41) Choose the ground state electron configuration for Hf^{2+}.

A) $[Xe]6s^2$

B) $[Xe]6s^25d^2$

C) $[Xe]5d^2$

D) [Xe]

E) $[Xe]6s^25d^4$

Answer: C

Diff: 3 *Page Ref: 8.7*

42) How many electrons are in the outermost shell of the Ga^{3+} ion in its ground state?

A) 2 B) 3 C) 6 D) 18

Answer: D

Diff: 2 *Page Ref: 8.7*

43) Place the following in order of increasing radius.

Ba^{2+} Te^{2-} I^-

A) $Ba^{2+} < I^- < Te^{2-}$

B) $I^- < Ba^{2+} < Te^{2-}$

C) $Te^{2-} < I^- < Ba^{2+}$

D) $Ba^{2+} < Te^{2-} < I^-$

E) $I^- < Te^{2-} < Ba^{2+}$

Answer: A

Diff: 1 *Page Ref: 8.7*

44) Choose the paramagnetic species from below.

A) Ca B) O^{2-} C) Zn^{2+} D) Cd E) Nb^{3+}

Answer: E

Diff: 2 *Page Ref: 8.7*

45) Choose the diamagnetic species from below.

A) Sn^{2+}

B) I

C) N

D) Cr

E) None of the above are diamagnetic.

Answer: A

Diff: 2 *Page Ref: 8.7*

46) Choose the paramagnetic species from below.
A) Ti^{4+}
B) Se
C) Ar
D) All of the above are paramagnetic.
E) None of the above are paramagnetic.

Answer: B
Diff: 2 *Page Ref: 8.7*

47) How many of the following species are paramagnetic?

Sc^{3+} Cl^{-} Ba^{2+} Se

A) 0 B) 2 C) 1 D) 4 E) 3

Answer: C
Diff: 2 *Page Ref: 8.7*

48) How many of the following species are diamagnetic?

Fr Zr^{2+} Al^{3+} Hg^{2+}

A) 1 B) 3 C) 0 D) 2 E) 4

Answer: D
Diff: 2 *Page Ref: 8.7*

49) Which ion does **not** have a noble gas configuration in its ground state?
A) Sc^{3+} B) Al^{3+} C) Ga^{3+} D) As^{3-}

Answer: C
Diff: 2 *Page Ref: 8.7*

50) Of the following, which element has the highest first ionization energy?
A) beryllium B) boron C) carbon D) lithium

Answer: C
Diff: 1 *Page Ref: 8.7*

51) Of the following, which element has the highest first ionization energy?
A) Al B) Cl C) Na D) P

Answer: B
Diff: 1 *Page Ref: 8.7*

52) Of the following, which element has the highest first ionization energy?
A) barium B) lead C) cesium D) thallium

Answer: B
Diff: 1 *Page Ref: 8.7*

53) Of the following, which element has the highest first ionization energy?
A) Sr B) Rb C) Na D) Ca

Answer: D
Diff: 1 *Page Ref: 8.7*

54) Of the following, which element has the highest first ionization energy?

A) Sr B) Br C) K D) Te

Answer: B

Diff: 1 *Page Ref:* 8.7

55) Which ionization process requires the most energy?

A) $W(g) \rightarrow W^{+}(g) + e^{-}$
B) $W^{+}(g) \rightarrow W^{2+}(g) + e^{-}$
C) $W^{2+}(g) \rightarrow W^{3+}(g) + e^{-}$
D) $W^{3+}(g) \rightarrow W^{4+}(g) + e^{-}$

Answer: D

Diff: 1 *Page Ref:* 8.7

56) Which ionization process requires the most energy?

A) $O(g) \rightarrow O^{+}(g) + e^{-}$
B) $O^{+}(g) \rightarrow O^{2+}(g) + e^{-}$
C) $F(g) \rightarrow F^{+}(g) + e^{-}$
D) $F^{+}(g) \rightarrow F^{2+}(g) + e^{-}$

Answer: D

Diff: 1 *Page Ref:* 8.7

57) Which of the following species will have the highest ionization energy?

A) K^{+} B) Ar C) Cl^{-} D) S^{2-}

Answer: A

Diff: 1 *Page Ref:* 8.7

58) Which of the following represents the change in electronic configuration that is associated with the first ionization energy of strontium?

A) $[Kr]5s^{1}5p^{1} \rightarrow [Kr]5s^{1} + e^{-}$
B) $[Kr]5s^{2} \rightarrow [Kr]5s^{1}5p^{1}$
C) $[Kr]5s^{2} \rightarrow [Kr]5s^{1} + e^{-}$
D) $[Kr]5s^{2} + e^{-} \rightarrow [Kr]5s^{2}5p^{1}$

Answer: C

Diff: 1 *Page Ref:* 8.7

59) Give the set of four quantum numbers that could represent the electron gained to form the Br <u>**ION**</u> from the Br atom.

A) $n = 4, l = 1, m_l = 1, m_s = +\frac{1}{2}$

B) $n = 4, l = 0, m_l = -1, m_s = +\frac{1}{2}$

C) $n = 4, l = 1, m_l = 1, m_s = -\frac{1}{2}$

D) $n = 3, l = 2, m_l = 1, m_s = +\frac{1}{2}$

E) $n = 5, l = 1, m_l = -1, m_s = -\frac{1}{2}$

Answer: C

Diff: 3 *Page Ref:* 8.7

60) Give the set of four quantum numbers that could represent the electron lost to form the K **ION** from the K atom.

A) $n = 3, l = 1, m_l = 1, m_s = -\frac{1}{2}$

B) $n = 4, l = 0, m_l = 0, m_s = -\frac{1}{2}$

C) $n = 4, l = 4, m_l = 0, m_s = -\frac{1}{2}$

D) $n = 4, l = 0, m_l = 0, m_s = +\frac{1}{2}$

E) $n = 3, l = 0, m_l = -1, m_s = -\frac{1}{2}$

Answer: D
Diff: 3 *Page Ref: 8.7*

61) Give the set of four quantum numbers that could represent the electron lost to form the Rb **ION** from the Rb atom.

A) $n = 6, l = 0, m_l = -1\ m_s = -\frac{1}{2}$

B) $n = 4, l = 1, m_l = 1, m_s = +\frac{1}{2}$

C) $n = 5, l = 0, m_l = 0, m_s = -\frac{1}{2}$

D) $n = 4, l = 1, m_l = 0, m_s = -\frac{1}{2}$

E) $n = 5, l = 0, m_l = 0, m_s = +\frac{1}{2}$

Answer: E
Diff: 3 *Page Ref: 8.7*

62) Place the following in order of decreasing metallic character.

C O N F

A) O > N > F > C
B) F > O > N >C
C) C > N > O > F
D) F > N > O > C
E) N > C >O > F

Answer: C
Diff: 1 *Page Ref: 8.8*

63) Which element has the highest first electron affinity?

A) Na B) Mg C) O D) Ne

Answer: C
Diff: 1 *Page Ref: 8.8*

64) Identify the states of the halogens at room temperature.

A) Fluorine, chlorine, bromine, and iodine are solids.
B) Fluorine, chlorine, and bromine are solids, and iodine is a solid.
C) Fluorine and iodine are solids, chlorine and bromine are gases.
D) Fluorine is a gas, chlorine and bromine are liquids, and iodine is a solid.
E) Fluorine and chlorine are gases, bromine is a liquid, and iodine is a solid.

Answer: E
Diff: 1 *Page Ref:* 8.9

Matching Questions

Match the following.

1) valence electrons
Diff: 1 *Page Ref:* 8.4

2) core electrons
Diff: 1 *Page Ref:* 8.4

3) number of unpaired electrons in Na
Diff: 1 *Page Ref:* 8.4

4) 0
Diff: 1 *Page Ref:* 8.7

5) number of unpaired electrons in Ti^{2+}
Diff: 1 *Page Ref:* 8.7

A) electrons in completed shells

B) 1

C) electrons in the outermost shell

D) number of unpaired electrons in Zn^{2+}

E) 2

Answers: 1) C 2) A 3) B 4) D 5) E

Short Answer Questions

1) Why does the size of the transition elements stay roughly the same across a period?

Answer: The electrons added as the transition element increase in atomic number are NOT being added to the outermost shell. They are being added to an inner shell where they shield the outer electrons from nuclear charge. The number of outermost electrons are constant. For each electron added across the period, an additional proton is also added. This keeps the effective nuclear charge roughly constant as the transition elements increase in atomic number within a given period. Therefore, the size of the transition elements within a period stays roughly constant.
Diff: 2 *Page Ref:* 8.6

2) Define paramagnetic.

Answer: Paramagnetic is when an unpaired electron is attracted by an external magnetic field.
Diff: 1 *Page Ref:* 8.7

3) Define ionization energy.

Answer: Ionization energy is the energy required to remove an electron from an ion or an atom.
Diff: 1 *Page Ref: 8.7*

4) List the noble gas that has the highest ionization energy.

Answer: He
Diff: 1 *Page Ref: 8.7*

5) Below is a list of successive ionization energies (in kJ/mol) for a period 3 element. Identify the element and explain how you came to that conclusion.

$IE_2 = 2250$ $IE_3 = 3360$ $IE_4 = 4560$ $IE_5 = 7010$ $IE_6 = 8500$ $IE_7 = 27{,}100$

Answer: The large increase in ionization energy between IE_6 and IE_7 shows that there are 6 valence electrons in this element. The only period 3 element that contains 6 valence electrons is S.
Diff: 1 *Page Ref: 8.7*

6) Why do successive ionization energies increase?

Answer: After removing the first electron, the species possesses a positive charge. When trying to remove the second, third, fourth, etc... electrons you must add even more energy to overcome the attraction between the negative electron and the positively charged ion. As the ion becomes more positively charged, it becomes more difficult (higher energy cost) to separate an electron from the ion.
Diff: 1 *Page Ref: 8.7*

7) Why is the first ionization energy of sulfur smaller than the first ionization energy of phosphorus?

Answer: When sulfur loses one electron, it take on a particularly stable, half-filled p subshell. The removal of this first electron therefore requires less energy than the removal of an electron from phosphorus initially half-filled p subshell.
Diff: 2 *Page Ref: 8.7*

8) Give the ground state electron configuration for Cd^{+2}.

Answer: [Kr] $4d^{10}$
Diff: 3 *Page Ref: 8.7*

9) Give the number of core electrons for Mg^{2+}.

Answer: 2
Diff: 3 *Page Ref: 8.7*

10) Give the number of valence electrons for Mg^{2+}.

Answer: 8
Diff: 3 *Page Ref: 8.7*

11) Define electron affinity.

Answer: Electron affinity is the energy associated with the gaining of an electron by an atom in the gaseous state.
Diff: 1 *Page Ref: 8.8*

12) Why do Li, Na, and K have similar chemical properties?

Answer: They are in the same group (family) and all have the same number of valence electrons.
Diff: 1 *Page Ref: 8.9*

Chapter 9 Chemical Bonding I: Lewis Theory

Multiple Choice Questions

1) Which of the following statements is TRUE?
   A) A covalent bond is formed through the transfer of electrons from one atom to another.
   B) A pair of electrons involved in a covalent bond are sometimes referred to as "lone pairs."
   C) It is not possible for two atoms to share more than two electrons.
   D) Single bonds are shorter than double bonds.
   E) A covalent bond has a lower potential energy than the two separate atoms.

Answer: E
Diff: 2 *Page Ref:* 9.2

2) Which of the following represent the Lewis structure for N?

A) $\cdot\dot{\underset{\cdot}{N}}:$ B) $:\ddot{\underset{\cdot\cdot}{N}}:$ C) $\underset{\cdot}{N}:$ D) $\cdot\underset{\cdot}{N}:$ E) N·

Answer: A
Diff: 1 *Page Ref:* 9.3

3) Which of the following represent the Lewis structure for Cl?

A) $:\ddot{\underset{\cdot\cdot}{Cl}}:$ B) $:\dot{\underset{\cdot\cdot}{Cl}}:$ C) Cl· D) $\cdot\dot{\underset{\cdot}{Cl}}:$ E) $\cdot\dot{\underset{\cdot\cdot}{Cl}}:$

Answer: B
Diff: 1 *Page Ref:* 9.3

4) Which of the following represent the Lewis structure for Mg?

A) Mg· B) Mg C) Mg: D) $:\ddot{\underset{\cdot\cdot}{Mg}}:$ E) $\underset{\cdot\cdot}{Mg}:$

Answer: C
Diff: 1 *Page Ref:* 9.3

5) Which of the following represent the Lewis structure for Br^-?

A) $:\underset{\cdot\cdot}{Br}:^-$ B) $:\dot{\underset{\cdot\cdot}{Br}}:^-$ C) $\cdot\underset{\cdot\cdot}{Br}:^-$ D) $:\ddot{\underset{\cdot\cdot}{Br}}:^-$ E) $Br\cdot^-$

Answer: D
Diff: 1 *Page Ref:* 9.4

6) Which of the following represent the Lewis structure for Ca^{2+}?

A) $Ca:^{2+}$ B) Ca: C) $\underset{\cdot\cdot}{Ca}:^{2+}$ D) $:\ddot{\underset{\cdot\cdot}{Ca}}:^{2+}$ E) Ca^{2+}

Answer: E
Diff: 1 *Page Ref:* 9.4

7) Which of the following represent the Lewis structure for S^{2-}?

A) $:\ddot{\underset{\cdot\cdot}{S}}:^{2-}$ B) S^{2-} C) $:\underset{\cdot\cdot}{S}:^{2-}$ D) $\underset{\cdot\cdot}{S}:^{2-}$ E) $S:^{2-}$

Answer: A
Diff: 1 *Page Ref:* 9.4

8) Give the complete electronic configuration for Br^-.
A) $1s^22s^22p^63s^23p^64s^24p^6$
B) $1s^22s^22p^63s^23p^64s^24d^{10}4p^6$
C) $1s^22s^22p^63s^23p^64s^23d^{10}4p^6$
D) $1s^22s^22p^63s^23p^64s^23d^{10}4p^5$
E) $1s^22s^2p^63s^2p^64s^23d^{10}4p^6$

Answer: C
Diff: 3 *Page Ref: 9.4*

9) Give the complete electronic configuration for Ca^{2+}.
A) $1s^22s^22p^63s^24p^6$
B) $1s^22s^22p^63s^23p^6$
C) $1s^22s^22p^63s^23p^5$
D) $1s^22s^23p^64s^25p^6$
E) $1s^22s^2p^63s^2p^6$

Answer: B
Diff: 3 *Page Ref: 9.4*

10) Give the complete electronic configuration for S^{2-}.
A) $1s^22s^22p^63s^24p^6$
B) $1s^22s^22p^63s^23p^6$
C) $1s^22s^22p^63s^23p^5$
D) $1s^22s^23p^64s^25p^6$
E) $1s^22s^2p^63s^2p^6$

Answer: B
Diff: 3 *Page Ref: 9.4*

11) Which of the following statements is TRUE?
A) An ionic bond is much stronger than most covalent bonds.
B) An ionic bond is formed through the sharing of electrons.
C) Ionic compounds at room temperature typically conduct electricity.
D) Once dissolved in water, ionic compounds rarely conduct electricity.
E) None of the above are true.

Answer: A
Diff: 1 *Page Ref: 9.4*

12) Use Lewis theory to determine the chemical formula for the compound formed between Al and O.
A) Al_3O_2 B) Al_2O_3 C) AlO_2 D) Al_2O E) AlO

Answer: B
Diff: 2 *Page Ref: 9.4*

13) Which of the following reactions is associated with the lattice energy of Li_2O ($\Delta H°_{latt}$)?

A) $Li_2O(s) \rightarrow 2\ Li^+(g) + O^{2-}(g)$

B) $2\ Li^+(aq) + O^{2-}(aq) \rightarrow Li_2O(s)$

C) $2\ Li^+(g) + O^{2-}(g) \rightarrow Li_2O(s)$

D) $Li_2O(s) \rightarrow 2\ Li^+(aq) + O^{2-}(aq)$

E) $2\ Li(s) + \frac{1}{2}\ O_2(g) \rightarrow Li_2O(s)$

Answer: C
Diff: 2 *Page Ref:* 9.4

14) Which of the following reactions is associated with the lattice energy of CaS ($\Delta H°_{latt}$)?

A) $Ca(s) + S(s) \rightarrow CaS(s)$
B) $CaS(s) \rightarrow Ca(s) + S(s)$
C) $Ca^{2+}(aq) + S^{2-}(aq) \rightarrow CaS(s)$
D) $Ca^{2+}(g) + S^{2-}(g) \rightarrow CaS(s)$
E) $CaS(s) \rightarrow Ca^{2+}(aq) + S^{2-}(aq)$

Answer: D
Diff: 2 *Page Ref:* 9.4

15) Which of the following reactions is associated with the lattice energy of RbI ($\Delta H°_{latt}$)?

A) $Rb(s) + \frac{1}{2} I_2(g) \rightarrow RbI(s)$

B) $RbI(s) \rightarrow Rb^+(g) + I^-(g)$

C) $RbI(s) \rightarrow Rb(s) + \frac{1}{2} I_2(g)$

D) $RbI(s) \rightarrow Rb^+(aq) + I^-(aq)$
E) $Rb^+(g) + I^-(g) \rightarrow RbI(s)$

Answer: E
Diff: 2 *Page Ref:* 9.4

16) Place the following in order of **decreasing** magnitude of lattice energy.

K_2O Rb_2S Li_2O

A) $Li_2O > K_2O > Rb_2S$
B) $Li_2O > Rb_2S > K_2O$
C) $Rb_2S > K_2O > Li_2O$
D) $Rb_2S > Li_2O > K_2O$
E) $K_2O > Li_2O > Rb_2S$

Answer: A
Diff: 2 *Page Ref:* 9.4

17) Place the following in order of **decreasing** magnitude of lattice energy.

NaF RbBr KCl

A) RbBr > NaF > KCl
B) NaF > KCl > RbBr
C) KCl > NaF > RbBr
D) NaF > RbBr > KCl
E) RbBr > KCl > NaF

Answer: B
Diff: 2 *Page Ref:* 9.4

18) Place the following in order of **increasing** magnitude of lattice energy.

CaO MgO SrS

A) MgO < CaO < SrS
B) SrS < MgO < CaO
C) SrS < CaO < MgO
D) CaO < MgO < SrS
E) CaO < SrS < MgO

Answer: C
Diff: 2 *Page Ref:* 9.4

19) Use the data given below to construct a Born-Haber cycle to determine the lattice energy of CaO.

	ΔH°(kJ)
$Ca(s) \rightarrow Ca(g)$	193
$Ca(g) \rightarrow Ca^{+}(g) + e^{-}$	590
$Ca^{+}(g) \rightarrow Ca^{2+}(g) + e^{-}$	1010
$2\,O(g) \rightarrow O_2(g)$	-498
$O(g) + e^{-} \rightarrow O^{-}(g)$	-141
$O^{-}(g) + e^{-} \rightarrow O^{2-}(g)$	878
$Ca(s) + \frac{1}{2}O_2(g) \rightarrow CaO(s)$	-635

A) -3414 kJ B) +1397 kJ C) -2667 kJ D) +3028 kJ E) -2144 kJ

Answer: A
Diff: 3 *Page Ref:* 9.4

20) Use the data given below to construct a Born–Haber cycle to determine the electron affinity of Br.

	$\Delta H°(kJ)$
$K(s) \rightarrow K(g)$	89
$K(g) \rightarrow K^+(g) + e^-$	419
$Br_2(l) \rightarrow 2\ Br(g)$	193
$K(s) + \frac{1}{2} Br_2(g) \rightarrow KBr(s)$	-394
$KBr(s) \rightarrow K^+(g) + Br^-(g)$	674

A) -885 kJ B) -325 kJ C) +367 kJ D) -464 kJ E) +246 kJ

Answer: B
Diff: 3 *Page Ref: 9.4*

21) Use the data given below to construct a Born–Haber cycle to determine the heat of formation of KCl.

	$\Delta H°(kJ)$
$K(s) \rightarrow K(g)$	89
$K(g) \rightarrow K^+(g) + e^-$	418
$Cl_2(g) \rightarrow 2\ Cl(g)$	244
$Cl(g) + e^- \rightarrow Cl^-(g)$	-349
$KCl(s) \rightarrow K^+(g) + Cl^-(g)$	717

A) -1119 kJ B) -997 kJ C) -437 kJ D) +631 kJ E) +158 kJ

Answer: C
Diff: 3 *Page Ref: 9.4*

22) Use the data given below to construct a Born–Haber cycle to determine the bond energy of O_2.

	$\Delta H°(kJ)$
$Na(s) \rightarrow Na(g)$	107
$Na(g) \rightarrow Na^+(g) + e^-$	496
$O(g) + e^- \rightarrow O^-(g)$	-141
$O^-(g) + e^- \rightarrow O^{2-}(g)$	878
$2\ Na(s) + \frac{1}{2} O_2(g) \rightarrow Na_2O(s)$	-416
$2\ Na^+(g) + O^{2-}(g) \rightarrow Na_2O(s)$	-2608

A) 426 kJ B) 249 kJ C) 852 kJ D) 498 kJ E) 356 kJ

Answer: D
Diff: 3 *Page Ref: 9.4*

23) Use the data given below to construct a Born–Haber cycle to determine the second ionization energy of Ca.

	ΔH°(kJ)
$Ca(s) \rightarrow Ca(g)$	193
$Ca(g) \rightarrow Ca^{+}(g) + e^{-}$	590.
$2\,O(g) \rightarrow O_2(g)$	–498
$O(g) + e^{-} \rightarrow O^{-}(g)$	–141
$O^{-}(g) + e^{-} \rightarrow O^{2-}(g)$	878
$Ca(s) + \frac{1}{2}O_2(g) \rightarrow CaO(s)$	–635
$Ca^{2+}(g) + O^{2-}(g) \rightarrow CaO(s)$	–3414

A) 1010 kJ B) 1757 kJ C) 1508 kJ D) –3027 kJ E) –1514 kJ

Answer: A
Diff: 3 *Page Ref: 9.4*

24) A single covalent bond contains ________ of electrons.
A) 0 pairs B) 1 pair C) 2 pairs D) 3 pairs E) 4 pairs

Answer: B
Diff: 1 *Page Ref: 9.5*

25) A double covalent bond contains ________ of electrons.
A) 0 pairs B) 1 pair C) 2 pairs D) 3 pairs E) 4 pairs

Answer: C
Diff: 1 *Page Ref: 9.5*

26) A triple covalent bond contains ________ of electrons.
A) 0 pairs B) 1 pair C) 2 pairs D) 3 pairs E) 4 pairs

Answer: D
Diff: 1 *Page Ref: 9.5*

27) Identify the shortest bond.
A) single covalent bond
B) double covalent bond
C) triple covalent bond
D) all of the above bonds are the same length

Answer: C
Diff: 1 *Page Ref: 9.5*

28) Identify the weakest bond.
A) single covalent bond
B) double covalent bond
C) triple covalent bond
D) all of the above bonds are the same strength

Answer: A
Diff: 1 *Page Ref: 9.5*

29) Identify the strongest bond.
A) single covalent bond
B) double covalent bond
C) triple covalent bond
D) all of the above bonds are the same strength

Answer: C
Diff: 1 *Page Ref: 9.5*

30) Identify the number of bonding pairs and lone pairs of electrons in water.
A) 1 bonding pair and 1 lone pair
B) 1 bonding pair and 2 lone pairs
C) 2 bonding pairs and 2 lone pairs
D) 2 bonding pairs and 1 lone pair
E) 3 bonding pairs and 2 lone pairs

Answer: C
Diff: 1 *Page Ref: 9.5*

31) Identify the compound with the largest dipole moment in the gas phase.
A) Cl_2 B) ClF C) HF D) LiF

Answer: D
Diff: 1 *Page Ref: 9.6*

32) Identify the compound with the smallest dipole moment in the gas phase.
A) Cl_2 B) ClF C) HF D) LiF

Answer: A
Diff: 1 *Page Ref: 9.6*

33) Identify the compound with the highest percent ionic character.
A) HF B) IBr C) HCl D) LiF

Answer: D
Diff: 1 *Page Ref: 9.6*

34) Identify the compound with the smallest percent ionic character.
A) HF B) IBr C) HCl D) LiF

Answer: B
Diff: 1 *Page Ref: 9.6*

35) Choose the bond below that is **most** polar.
A) H–I B) H–Br C) H–F D) H–Cl E) C–H

Answer: C
Diff: 1 *Page Ref: 9.6*

36) Choose the bond below that is **least** polar.
A) P–F B) C–Br C) C–F D) C–I E) C–Cl

Answer: D
Diff: 1 *Page Ref: 9.6*

37) Using periodic trends, place the following bonds in order of **increasing** ionic character.

Si–P Si–Cl Si–S

A) Si–P < Si–Cl < Si–S
B) Si–P < Si–S < Si–Cl
C) Si–S < Si–Cl < Si–P
D) Si–Cl < Si–P < Si–S
E) Si–Cl < Si–S < Si–P

Answer: B
Diff: 1 *Page Ref: 9.6*

38) Choose the best Lewis structure for BeF_2.

A) 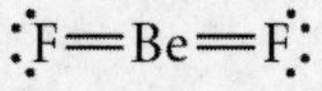
B) 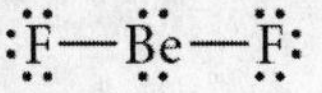
C) 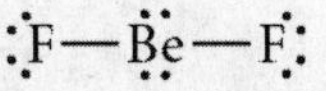
D) 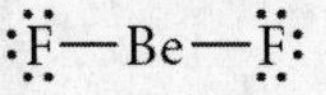
E) 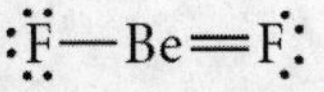

Answer: D
Diff: 2 *Page Ref: 9.7*

39) Choose the best Lewis structure for OCl_2.

A) 
B) 
C) 
D) 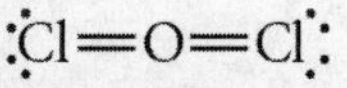
E) 

Answer: E
Diff: 2 *Page Ref: 9.7*

40) Choose the best Lewis structure for CH_2Cl_2.

A)

B)

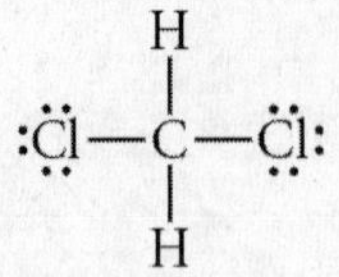

C)

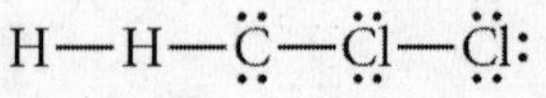

D)

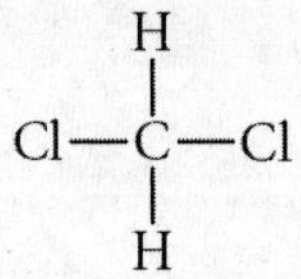

E)

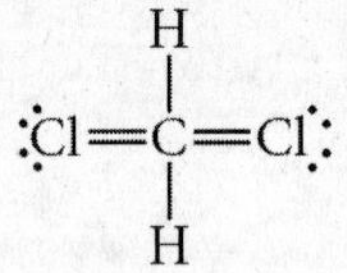

Answer: B
Diff: 2 *Page Ref:* 9.7

41) Give the number of valence electrons for CH_2Cl_2.

A) 16 B) 18 C) 20 D) 22 E) 12

Answer: C
Diff: 2 *Page Ref:* 9.7

42) Give the number of valence electrons for XeI_2.

A) 22 B) 20 C) 18 D) 24 E) 16

Answer: A
Diff: 2 *Page Ref:* 9.7

43) Choose the best Lewis structure for XeI_2.

A) 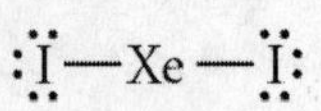

B) 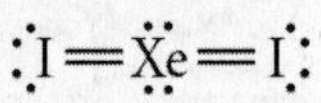

C) 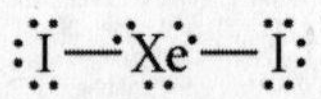

D) 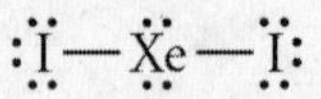

E) 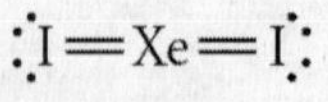

Answer: C
Diff: 2 *Page Ref:* 9.7

44) Give the number of valence electrons for ICl_5.

A) 36 B) 40 C) 42 D) 44 E) 46

Answer: C
Diff: 2 *Page Ref:* 9.7

45) Choose the best Lewis structure for ICl_5.

A)

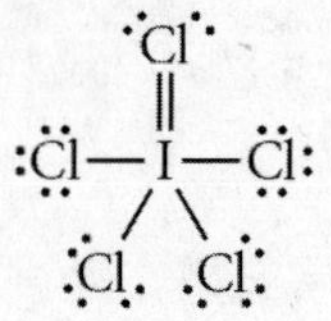

B)

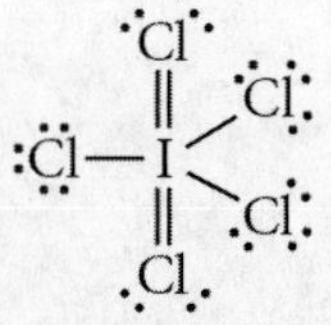

C)

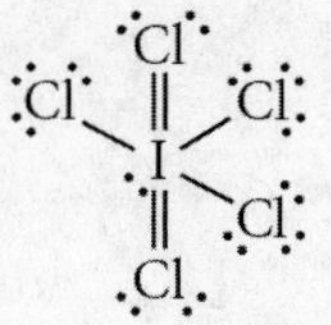

D)

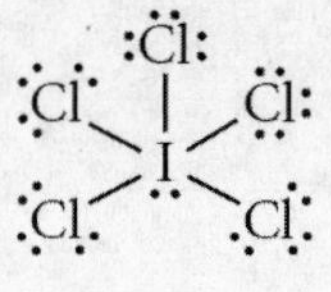

E)

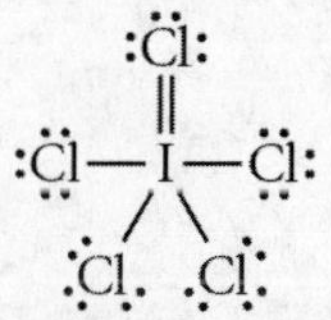

Answer: D

Diff: 2 *Page Ref:* 9.7

46) Choose the best Lewis structure for SF_4.

A)

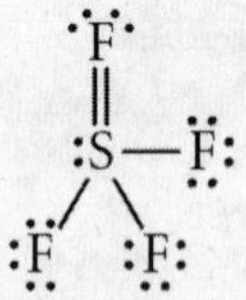

B)

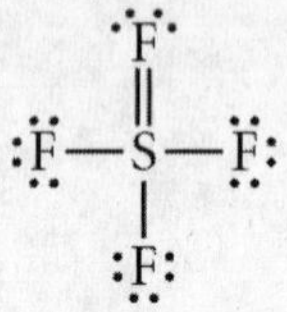

C)

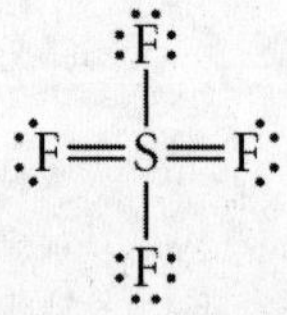

D)

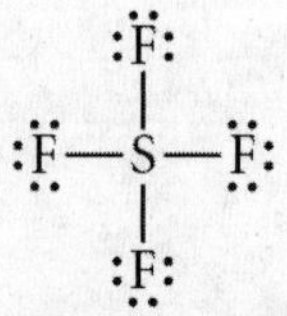

E)

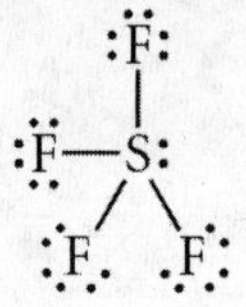

Answer: E
Diff: 2 *Page Ref: 9.7*

47) Choose the best Lewis structure for BF_3.

A)

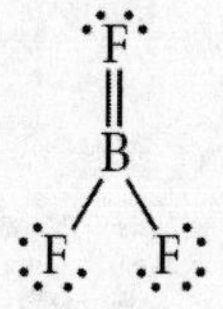

B)

C)

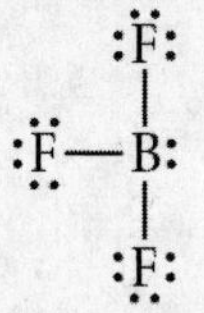

D)

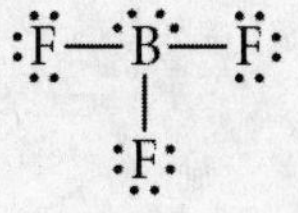

E)

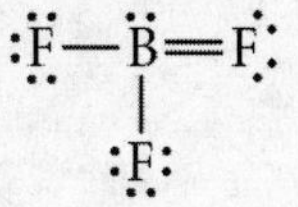

Answer: B

Diff: 2 *Page Ref:* 9.7

48) Choose the best Lewis structure for NO_3^-.

A)

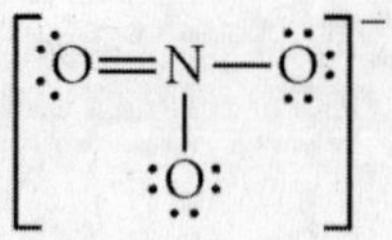

B)

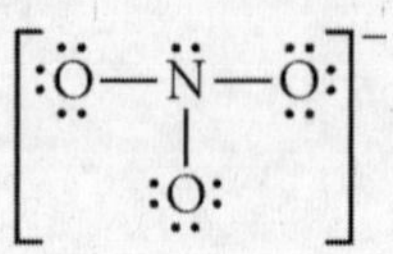

C)

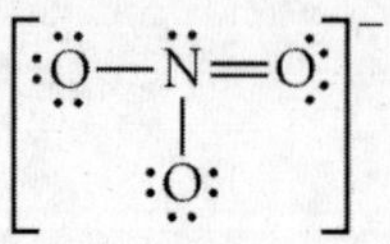

D)

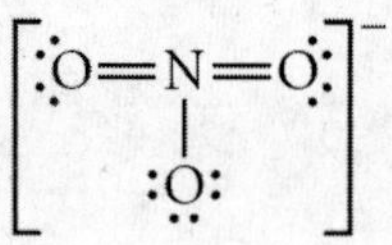

E)

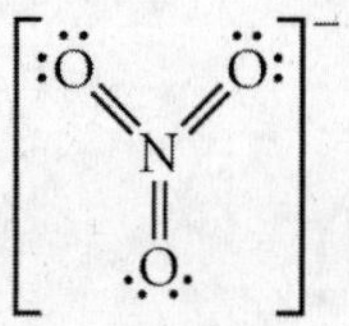

Answer: A

Diff: 2 *Page Ref:* 9.7

49) Give the number of valence electrons for SO_4^{2-}.

A) 32 B) 30 C) 34 D) 28 E) 36

Answer: A

Diff: 2 *Page Ref:* 9.7

50) Choose the best Lewis structure for SO_4^{2-}.

A)

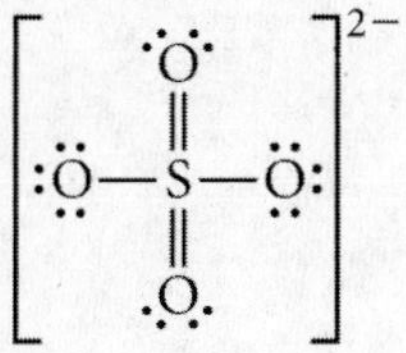

B)

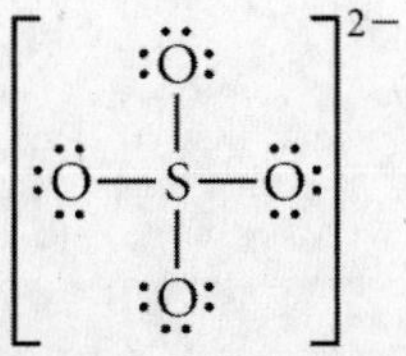

C)

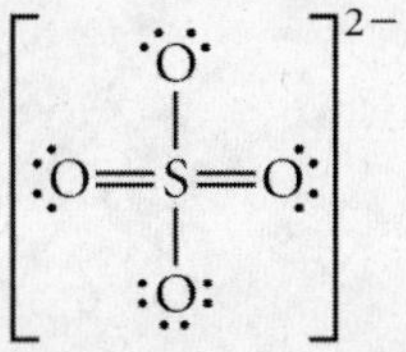

D)

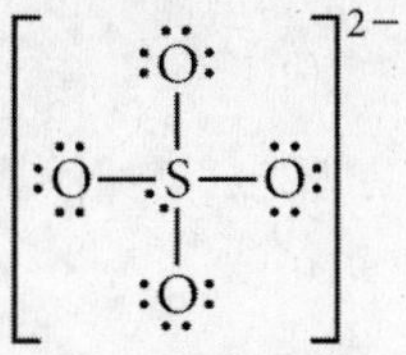

E)

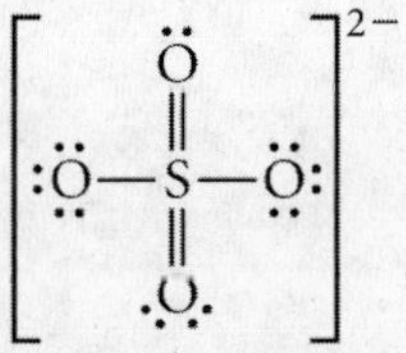

Answer: A
Diff: 2 *Page Ref:* 9.7

51) Choose the best Lewis structure for PO_4^{3-}.

A)

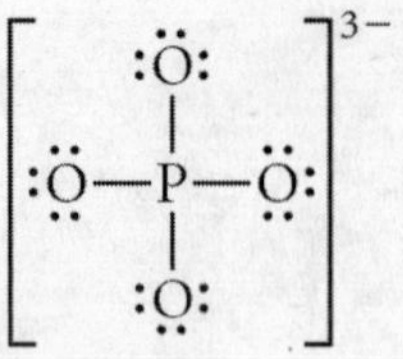

B)

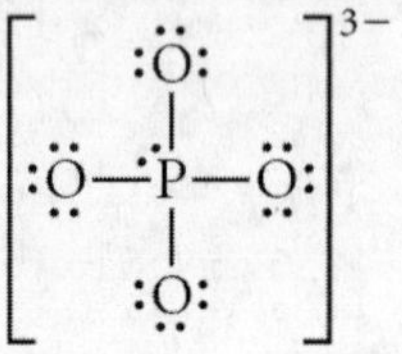

C)

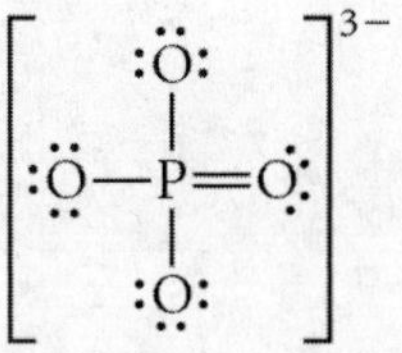

D)

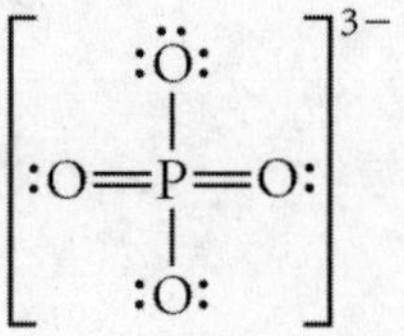

E)

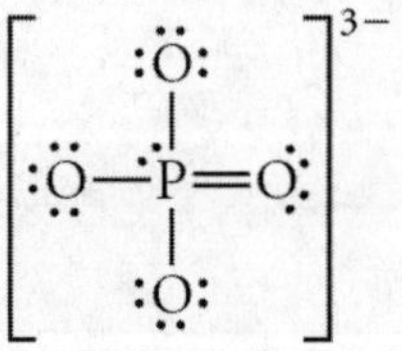

Answer: C

Diff: 2 *Page Ref: 9.7*

52) Choose the best Lewis structure for SeO_4^{2-}.

A)

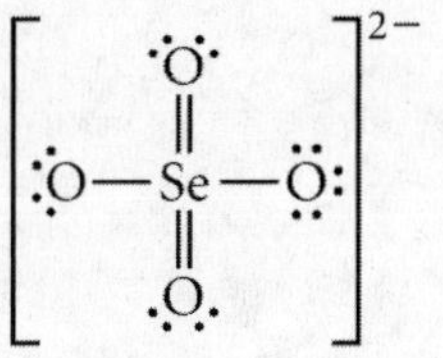

B)

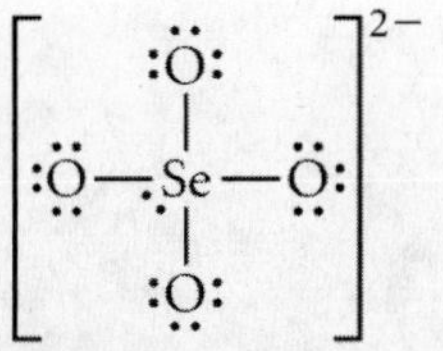

C)

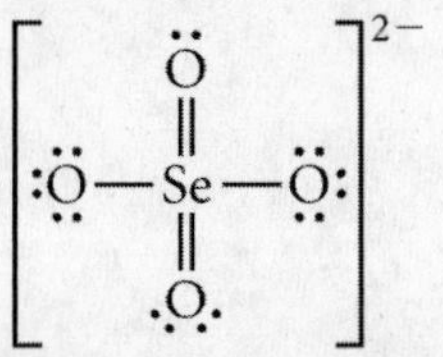

D)

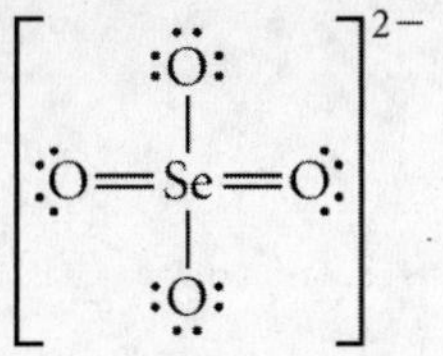

E)

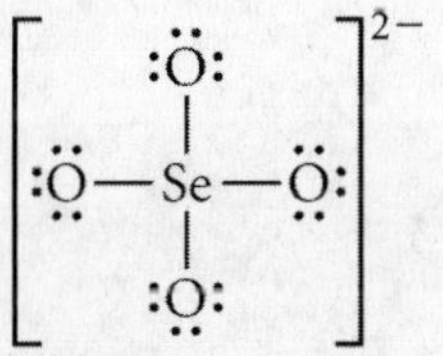

Answer: D

Diff: 2 *Page Ref:* 9.7

53) Choose the best Lewis structure for NH_4^+.

A)

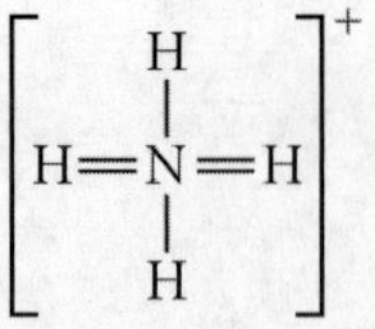

B)

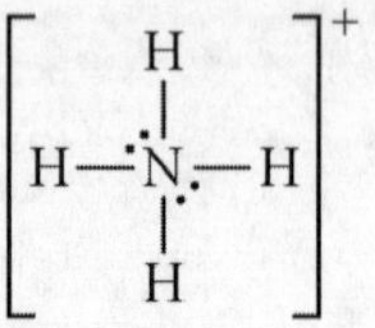

C)

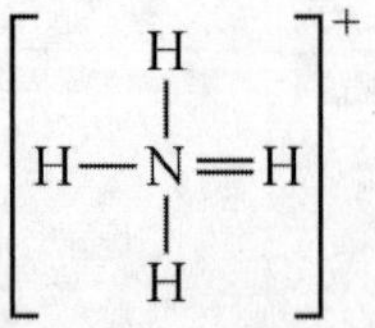

D)

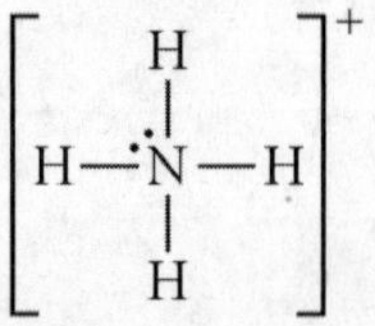

E)

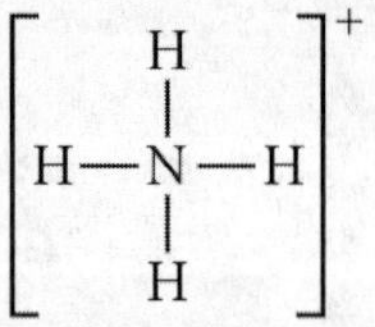

Answer: E

Diff: 2 *Page Ref:* 9.7

54) Draw the Lewis structure for NO_2^- including any valid resonance structures. Which of the following statements is TRUE?

A) The nitrite ion contains one N—O single bond and one N═O double bond.

B) The nitrite ion contains two N—O bonds that are equivalent to $1\frac{1}{2}$ bonds.

C) The nitrite ion contains two N═O double bonds.

D) The nitrite ion contains two N—O single bonds.

E) None of the above are true.

Answer: B

Diff: 2 *Page Ref:* 9.8

55) Draw the Lewis structure for CO_3^{2-} including any valid resonance structures. Which of the following statements is TRUE?

A) The CO_3^{2-} ion contains one C—O single bond and two C=O double bonds.

B) The CO_3^{2-} ion contains two C—O single bonds and one C=O double bond.

C) The CO_3^{2-} ion contains three C—O double bonds.

D) The CO_3^{2-} ion contains two C—O single bonds and one C≡O triple bond.

E) None of the above are true.

Answer: B
Diff: 2 *Page Ref: 9.8*

56) Which of the following resonance structures for OCN^- will contribute most to the correct structure of OCN^-?

A) O(2 lone pairs)=C=N (2 lone pairs)
B) O(1 lone pair)≡C-N(3 lone pairs)
C) O(1 lone pair)=C(2 lp)=N(1 lone pair)
D) O(3 lone pairs)—C≡N(with 1 lone pair)
E) They all contribute equally to the correct structure of OCN^-.

Answer: D
Diff: 2 *Page Ref: 9.8*

57) Using Lewis structures and formal charge, which of the following ions is most stable?

OCN^- ONC^- NOC^-

A) OCN^-
B) ONC^-
C) NOC^-
D) None of these ions are stable according to Lewis theory.
E) All of these compounds are equally stable according to Lewis theory.

Answer: A
Diff: 2 *Page Ref: 9.8*

58) Draw the Lewis structure for SO_4^{2-}. How many equivalent resonance structures can be drawn?

A) 6 B) 2 C) 4 D) 3 E) 8

Answer: A
Diff: 3 *Page Ref: 9.8*

59) Draw the best Lewis structure for Cl_3^-. What is the formal charge on the central Cl atom?

A) -1 B) 0 C) +1 D) +2 E) -2

Answer: A
Diff: 3 *Page Ref: 9.8*

60) Draw the best Lewis structure for the free radical, NO_2. What is the formal charge on the N?

A) 0 B) +1 C) -1 D) +2 E) -2

Answer: B
Diff: 3 *Page Ref: 9.8*

61) Draw the best Lewis structure for CH_3^{-1}. What is the formal charge on the C?
A) 0 B) 1 C) –1 D) 2

Answer: C
Diff: 3 *Page Ref: 9.8*

62) Draw the best Lewis structure for CH_3^{+1}. What is the formal charge on the C?
A) 0 B) 1 C) –1 D) 2

Answer: B
Diff: 3 *Page Ref: 9.8*

63) Draw the best Lewis structure for BrO_4^- and determine the formal charge on bromine.
A) –1 B) +1 C) 0 D) +2 E) +3

Answer: C
Diff: 3 *Page Ref: 9.9*

64) Identify the compound with atoms that have an incomplete octet.
A) ICl_5 B) CO_2 C) BF_3 D) Cl_2 E) CO

Answer: C
Diff: 3 *Page Ref: 9.9*

65) Which compound has the longest carbon–carbon bond length?
A) CH_3CH_3 B) CH_2CH_2
C) HCCH D) all bond lengths are the same

Answer: A
Diff: 1 *Page Ref: 9.10*

66) Which compound has the shortest carbon–carbon bond length?
A) CH_3CH_3 B) CH_2CH_2
C) HCCH D) all bond lengths are the same

Answer: C
Diff: 1 *Page Ref: 9.10*

67) Which compound has the highest carbon–carbon bond strength?
A) CH_3CH_3 B) CH_2CH_2
C) HCCH D) all bond strengths are the same

Answer: C
Diff: 1 *Page Ref: 9.10*

68) Place the following in order of **increasing** bond length.

C–F C–S C–Cl

A) C–S < C–Cl < C–F
B) C–Cl < C–F < C–S
C) C–F < C–S < C–Cl
D) C–F < C–Cl < C–S
E) C–S < C–F < C–Cl

Answer: D
Diff: 1 *Page Ref: 9.10*

69) Place the following in order of **decreasing** bond length.

H–F H–I H–Br

A) H–F > H–Br > H–I
B) H–I > H–F > H–Br
C) H–I > H–Br > H–F
D) H–Br > H–F > H–I
E) H–F > H–I > H–Br

Answer: C
Diff: 1 *Page Ref: 9.10*

70) Place the following in order of **decreasing** XO bond length, where "X" represents the central atom in each of the following compounds or ions.

SiO_3^{2-} CO_2 CO_3^{2-}

A) CO_2 > SiO_3^{2-} > CO_3^{2-}
B) CO_2 > CO_3^{2-} > SiO_3^{2-}
C) CO_3^{2-} > CO_2 > SiO_3^{2-}
D) CO_3^{2-} > SiO_3^{2-} > CO_2
E) SiO_3^{2-} > CO_3^{2-} > CO_2

Answer: E
Diff: 1 *Page Ref: 9.10*

71) Place the following in order of **increasing** bond length.

NO_2^- NO_3^- NO

A) NO < NO_2^- < NO_3^-
B) NO_2^- < NO_3^- < NO
C) NO_3^- < NO < NO_2^-
D) NO < NO_3^- < NO_2^-
E) NO_3^- < NO_2^- < NO

Answer: D
Diff: 1 *Page Ref: 9.10*

72) Rank the following molecules in decreasing bond energy.

Cl_2 Br_2 F_2 I_2

A) I_2 > Br_2 > Cl_2 > F_2
B) Cl_2 > Br_2 > F_2 > I_2
C) I_2 > Cl_2 > Br_2 > F_2
D) Cl_2 > I_2 > F_2 > Br_2

Answer: B
Diff: 2 *Page Ref: 9.10*

73) Identify the bond with the highest bond energy.
A) Si=O
B) N=N
C) C=C
D) C≡N
E) O=O

Answer: D
Diff: 2 *Page Ref: 9.10*

74) Which of the following processes are exothermic?
A) $Cl_2(g) \rightarrow 2Cl(g)$
B) $Br(g) + e^- \rightarrow Br^-(g)$
C) $Li(s) \rightarrow Li(g)$
D) $NaF(s) \rightarrow Na^+(g) + F^-(g)$
E) None of the above are exothermic.

Answer: B
Diff: 2 *Page Ref: 9.10*

75) Which of the following processes are exothermic?
A) the second ionization energy of Mg
B) the sublimation of Li
C) the breaking the bond of I_2
D) the formation of NaBr from its constituent elements in their standard state
E) None of the above are exothermic

Answer: D
Diff: 2 *Page Ref: 9.10*

76) Which of the following processes are endothermic?

A) $K^+(g) + I^-(g) \rightarrow KI(s)$

B) $2\ Br(g) \rightarrow Br_2(g)$

C) $Ca(s) \rightarrow Ca(g)$

D) $2\ Na(s) + \frac{1}{2} O_2(g) \rightarrow Na_2O(s)$

E) None of the above are endothermic.

Answer: C

Diff: 2 *Page Ref: 9.10*

77) Which of the following processes are endothermic?

A) the reaction associated with the lattice energy of LiCl.

B) the reaction associated with the ionization energy of potassium.

C) the reaction associated with the heat of formation of CaS.

D) the formation of F_2 from its elements in their standard states.

E) None of the above are endothermic.

Answer: B

Diff: 2 *Page Ref: 9.10*

78) Use the bond energies provided to estimate $\Delta H°_{rxn}$ for the reaction below.

$$PCl_3(g) + Cl_2(g) \rightarrow PCl_5(l) \qquad \Delta H°_{rxn} = ?$$

Bond	Bond Energy (kJ/mol)
Cl-Cl	243
P-Cl	331

A) -243 kJ B) -419 kJ C) -662 kJ D) -67 kJ E) -905 kJ

Answer: B

Diff: 3 *Page Ref: 9.10*

79) Use the bond energies provided to estimate $\Delta H°_{rxn}$ for the reaction below.

$$2\ Br_2(l) + C_2H_2(g) \rightarrow C_2H_2Br_4(l) \qquad \Delta H°_{rxn} = ?$$

Bond	Bond Energy (kJ/mol)
Br-Br	193
C≡C	837
C-C	347
C-Br	276
C-H	414

A) +407 kJ B) -324 kJ C) -228 kJ D) +573 kJ E) -648 kJ

Answer: C

Diff: 3 *Page Ref: 9.10*

80) Use the bond energies provided to estimate $\Delta H°_{rxn}$ for the reaction below.

$$CH_3OH(l) + 2\,O_2(g) \rightarrow CO_2(g) + 2\,H_2O(g) \qquad \Delta H°_{rxn} = ?$$

Bond	Bond Energy (kJ/mol)
C–H	414
C–O	360
C=O	799
O=O	498
O–H	464

A) +473 kJ B) -91 kJ C) -486 kJ D) -392 kJ E) +206 kJ

Answer: D
Diff: 3 *Page Ref: 9.10*

81) Use the bond energies provided to estimate $\Delta H°_{rxn}$ for the reaction below.

$$XeF_2 + 2\,F_2 \rightarrow XeF_6 \qquad \Delta H°_{rxn} = ?$$

Bond	Bond Energy (kJ/mol)
Xe–F	147
F–F	159

A) -429 kJ B) +159 kJ C) -660 kJ D) +176 kJ E) -270 kJ

Answer: E
Diff: 3 *Page Ref: 9.10*

82) Use the bond energies provided to estimate $\Delta H°_{rxn}$ for the reaction below.

$$C_2H_4(g) + H_2(g) \rightarrow C_2H_6(g) \qquad \Delta H°_{rxn} = ?$$

Bond	Bond Energy (kJ/mol)
C–C	347
C–H	414
C=C	611
C≡C	837
H–H	436

A) -128 kJ B) +98 kJ C) +700 kJ D) -102 kJ E) -166 kJ

Answer: A
Diff: 3 *Page Ref: 9.10*

Algorithmic Questions

1) Identify an ionic bond.
A) Electrons are pooled.
B) Electrons are shared.
C) Electrons are transferred.
D) Protons are gained.
E) Electrons are lost.

Answer: C
Diff: 1 *Page Ref: 9.2*

2) Identify the compound with ionic bonding.

A) NaBr B) Na C) H_2O D) He E) S

Answer: A
Diff: 1 *Page Ref: 9.2*

3) Identify the compound with covalent bonding.

A) NaBr B) Na C) H_2O D) He E) S

Answer: C
Diff: 1 *Page Ref: 9.2*

4) Identify the compound with metallic bonding.

A) NaBr B) Na C) H_2O D) He E) S

Answer: B
Diff: 1 *Page Ref: 9.2*

5) Identify the substance that conducts electricity.

A) NaCl dissolved in water
B) solid NaCl
C) ethanol
D) solid sugar
E) sugar dissolved in water.

Answer: A
Diff: 1 *Page Ref: 9.4*

6) Use Lewis theory to determine the chemical formula for the compound formed between Ba and N.

A) BaN B) Ba_3N_2 C) BaN_2 D) Ba_2N E) Ba_2N_3

Answer: B
Diff: 2 *Page Ref: 9.4*

7) Use Lewis theory to determine the chemical formula for the compound formed between Na and O.

A) NaO B) NaO_2 C) Na_2O D) Na_2O_3 E) Na_3O_2

Answer: C
Diff: 2 *Page Ref: 9.4*

8) Use Lewis theory to determine the chemical formula for the compound formed between Ba and I.

A) BaI B) Ba_2I_3 C) Ba_3I_2 D) BaI_2 E) Ba_2I

Answer: D
Diff: 2 *Page Ref: 9.4*

9) Use Lewis theory to determine the chemical formula for the compound formed between Al and Cl.

A) Al_3Cl_2 B) Al_2Cl_3 C) $AlCl_2$ D) AlCl E) $AlCl_3$

Answer: E
Diff: 2 *Page Ref: 9.4*

10) Use Lewis theory to determine the chemical formula for the compound formed between Li and Br.

A) $LiBr_2$ B) Li_2Br C) Li Br D) Li_2Br_2

Answer: C
Diff: 2 *Page Ref: 9.4*

11) Which of the following ionic compounds would be expected to have the highest lattice energy?

A) LiF B) LiCl C) LiBr D) LiI

Answer: A
Diff: 2 *Page Ref: 9.4*

12) Which of the following ionic compounds would be expected to have the highest lattice energy?

A) Li Cl B) Na Cl C) K Cl D) Rb Cl

Answer: A
Diff: 2 *Page Ref: 9.4*

13) Which ionic compound would be expected to have the highest lattice energy?

A) Rb_2O B) SrO C) In_2O_3 D) CO_2

Answer: C
Diff: 2 *Page Ref: 9.4*

14) Identify the compound with the highest magnitude of lattice energy.

A) $MgCl_2$ B) $BaCl_2$ C) $SrCl_2$ D) $CsCl_2$

Answer: A
Diff: 2 *Page Ref: 9.4*

15) Identify the compound with the lowest magnitude of lattice energy.

A) KCl B) KBr C) SrO D) BaO

Answer: B
Diff: 2 *Page Ref: 9.4*

16) Place the following in order of **increasing** magnitude of lattice energy.

MgO LiI BaS

A) BaS < MgO < LiI
B) LiI < BaS < MgO
C) MgO < BaS < LiI
D) LiI < MgO < BaS
E) MgO < LiI < BaS

Answer: B
Diff: 2 *Page Ref: 9.4*

17) Place the following in order of **decreasing** magnitude of lattice energy.

KF CaS RbI

A) RbI > KF > CaS
B) RbI > CaS > KF
C) CaS > RbI > KF
D) KF > RbI > CaS
E) CaS > KF > RbI

Answer: E
Diff: 2 *Page Ref: 9.4*

18) Choose the compound below that should have the **highest** melting point according to the ionic bonding model.
A) AlN B) MgO C) NaCl D) CaS E) RbI

Answer: A
Diff: 2 *Page Ref: 9.4*

19) Choose the compound below that should have the **highest** melting point according to the ionic bonding model.
A) SrI_2 B) MgF_2 C) $CaCl_2$ D) SrF_2 E) $SrBr_2$

Answer: B
Diff: 2 *Page Ref: 9.4*

20) Choose the compound below that should have the **lowest** melting point according to the ionic bonding model.
A) LiF B) LiCl C) CsI D) KBr E) RbI

Answer: C
Diff: 2 *Page Ref: 9.4*

21) Identify the number of bonding pairs and lone pairs of electrons in N_2.
A) 6 bonding pair and 1 lone pair
B) 4 bonding pair and 2 lone pairs
C) 3 bonding pairs and 2 lone pairs
D) 2 bonding pairs and 1 lone pair
E) 2 bonding pairs and 3 lone pairs

Answer: C
Diff: 1 *Page Ref: 9.5*

22) Of the following elements, which has the highest electronegativity?
A) Si B) P C) Ti D) Ge

Answer: B
Diff: 1 *Page Ref: 9.6*

23) Of the following elements, which has the lowest electronegativity?
A) Sr B) I C) Ba D) At

Answer: C
Diff: 1 *Page Ref: 9.6*

24) Place the following elements in order of **increasing** electronegativity.

Na Rb P

A) P < Na < Rb
B) Na < P < Rb
C) Rb < P < Na
D) Rb < Na < P
E) P < Rb < Na

Answer: D
Diff: 1 *Page Ref: 9.6*

25) Place the following elements in order of **decreasing** electronegativity.

S F Se

A) Se > S > F
B) F > Se > S
C) Se > F > S
D) S > F > Se
E) F > S > Se

Answer: E
Diff: 1 *Page Ref: 9.6*

26) Place the following elements in order of **increasing** electronegativity.

Ba Se Li

A) Ba < Li < Se
B) Li < Se < Ba
C) Ba < Se < Li
D) Se < Ba < Li
E) Se < Li < Ba

Answer: A
Diff: 1 *Page Ref: 9.6*

27) List the following compounds in decreasing electronegativity difference.

F_2 HF KF

A) KF > F_2 > HF
B) F_2 > HF > KF
C) HF > KF > F_2
D) KF > HF > F_2

Answer: D
Diff: 1 *Page Ref: 9.6*

28) Choose the bond below that is **most** polar.
A) C–N
B) C–F
C) N–O
D) C–C
E) Cl–Cl

Answer: B
Diff: 1 *Page Ref: 9.6*

29) Using periodic trends, place the following bonds in order of **increasing** ionic character.

S–Cl   Se–Cl   O–Cl

A) Se–Cl < S–Cl < O–Cl
B) S–Cl < Se–Cl < O–Cl
C) O–Cl < Se–Cl < S–Cl
D) Se–Cl < O–Cl < S–Cl
E) O–Cl < S–Cl < Se–Cl

Answer: E
Diff: 1   *Page Ref: 9.6*

30) Using periodic trends, place the following bonds in order of **decreasing** ionic character.

Sb–F   P–F   As–F

A) Sb–F > As–F > P–F
B) As–F > Sb–F > P–F
C) Sb–F > P–F > As–F
D) P–F > As–F > Sb–F
E) Sb–F > P–F > As–F

Answer: A
Diff: 1   *Page Ref: 9.6*

31) Which molecule or compound below contains a pure covalent bond?
A) Li_2CO_3 B) SCl_6 C) Br_2 D) PCl_3 E) NaCl

Answer: C
Diff: 1   *Page Ref: 9.6*

32) Which molecule or compound below contains a polar covalent bond?
A) C_2H_4 B) ZnS   C) LiI   D) NCl_3 E) AgI

Answer: D
Diff: 1   *Page Ref: 9.6*

33) Which molecule or compound below contains an ionic bond?
A) CO_2
B) C_2Br_4
C) SiF_4
D) OCl_2
E) NH_4NO_3

Answer: E
Diff: 1   *Page Ref: 9.6*

34) The electronegativity is 2.1 for H and 1.9 for Pb. Based on these electronegativities PbH_4 would be expected to

A) be ionic and contain H^- ions.
B) be ionic and contain H^+ ions.
C) have polar covalent bonds with a partial negative charges on the H atoms.
D) have polar covalent bonds with a partial positive charges on the H atoms.

Answer: C
Diff: 1 *Page Ref: 9.6*

35) The compound ClF contains

A) ionic bonds.
B) nonpolar covalent bonds.
C) polar covalent bonds with partial negative charges on the F atoms.
D) polar covalent bonds with partial negative charges on the Cl atoms.

Answer: C
Diff: 1 *Page Ref: 9.6*

36) The phosphorus atom in PCl_3 would be expected to have a

A) partial positive ($\delta+$) charge.
B) partial negative ($\delta-$) charge.
C) 3+ charge.
D) 3- charge.

Answer: A
Diff: 1 *Page Ref: 9.6*

37) The iodine atom in I_2 would be expected to have a

A) charge of 1-.
B) partial charge $\delta-$.
C) partial charge $\delta+$.
D) charge of 0.

Answer: D
Diff: 1 *Page Ref: 9.6*

38) Give the number of valence electrons for SBr_4.

A) 28
B) 30
C) 32
D) 34

Answer: D
Diff: 2 *Page Ref: 9.7*

39) Give the number of pairs of valence electrons for BF_3.

A) 16
B) 8
C) 14
D) 10
E) 12

Answer: E
Diff: 2 *Page Ref: 9.7*

40) In the best Lewis structure for NO^+, what is the formal charge on the N atom?

A) -1
B) 0
C) +1
D) +2

Answer: B
Diff: 2 *Page Ref: 9.8*

41) Which of the following elements can form compounds with an expanded octet?
A) Se
B) C
C) Li
D) F
E) All of the above elements can form compounds with an expanded octet.

Answer: A
Diff: 1 *Page Ref: 9.9*

42) Which of the following elements can form compounds with an expanded octet?
A) O
B) Br
C) F
D) Be
E) None of the above can form compounds with an expanded octet.

Answer: B
Diff: 1 *Page Ref: 9.9*

43) How many of the following elements can form compounds with an expanded octet?

I O Br Xe

A) 2 B) 0 C) 3 D) 1 E) 4

Answer: C
Diff: 2 *Page Ref: 9.9*

44) How many of the following elements can form compounds with an expanded octet?

P Kr Xe B

A) 0 B) 1 C) 2 D) 3 E) 4

Answer: D
Diff: 2 *Page Ref: 9.9*

45) How many lone pairs of electrons are on the As atom in $AsCl_3$?
A) 0 B) 1 C) 2 D) 3

Answer: B
Diff: 2 *Page Ref: 9.9*

46) Which element can expand its valence shell to accommodate more than eight electrons?
A) N B) O C) Br D) He

Answer: C
Diff: 2 *Page Ref: 9.9*

47) Which of the following contains an atom that does **not** obey the octet rule?
A) CsI B) SnO_2 C) ClF_5 D) ClF

Answer: C
Diff: 2 *Page Ref: 9.9*

48) How many lone pairs of electrons are on the S atom in SF_4 ?

A) 0 B) 1 C) 2 D) 3

Answer: B
Diff: 2 *Page Ref: 9.9*

49) How many lone pairs are on the Br atom in $BrCl_2^-$?

A) 0 B) 1 C) 2 D) 3

Answer: D
Diff: 2 *Page Ref: 9.9*

50) Choose the bond below that is the **strongest**.

A) C–F B) C=O C) C–I D) I–I E) C≡N

Answer: E
Diff: 1 *Page Ref: 9.10*

51) Choose the bond below that is the **strongest**.

A) N=O B) N–F C) C–O D) N–C E) N=N

Answer: A
Diff: 1 *Page Ref: 9.10*

52) Choose the bond below that is the **weakest**.

A) Na–Br B) Br–Br C) C=N D) Li–I E) C=N

Answer: B
Diff: 1 *Page Ref: 9.10*

53) Choose the bond below that is the **weakest**.

A) C≡O B) N≡N C) C–I D) C=O E) K–Br

Answer: C
Diff: 1 *Page Ref: 9.10*

Matching Questions

Match the following.

1) Sr–Sr
Diff: 1 *Page Ref: 9.2*

2) Cs–I
Diff: 1 *Page Ref: 9.4*

3) Ca–O
Diff: 1 *Page Ref: 9.4*

4) Se–I
Diff: 1 *Page Ref: 9.5*

5) C=N
Diff: 1 *Page Ref: 9.5*

A) longest covalent bond

B) weakest ionic bond

C) strongest covalent bond

D) metallic bond

E) highest melting point

Answers: 1) D 2) B 3) E 4) A 5) C

Short Answer Questions

1) Describe a covalent bond.

Answer: A bond formed from the sharing of electrons.
Diff: 1 *Page Ref: 9.2*

2) Draw the Lewis Dot structure for Al^{3+}.

Answer: The Al should have no electrons around it, with a 3+ charge.
Diff: 1 *Page Ref: 9.4*

3) Explain why the lattice energy of MgS is approximately 4 times as large as that of NaCl.

Answer: The Mg^{2+} and Na^{+} ion are similar in size as are the $S^{2-}$and Cl^{-} ions. Therefore, the major difference in the lattice energies for these two compounds is due to the difference in the magnitude of the ionic charge on each ion. According to Coulomb's law, Mg and S ions have 4× the attractive force (2+ × 2–) than the Na and Cl ions (1+ × 1–).
Diff: 2 *Page Ref: 9.4*

4) List the most electronegative atom.

Answer: F
Diff: 1 *Page Ref: 9.6*

5) List the least electronegative atom.

Answer: Fr
Diff: 1 *Page Ref: 9.6*

6) Define dipole moment.

Answer: Dipole moment is the measured quantitative value associated with the separation of the partial positive and negative charges found within a molecule.
Diff: 1 *Page Ref: 9.6*

7) How are electron affinity and electronegativity different?

Answer: Electron affinity is the process of a single atom gaining an electron. Electronegativity is the strength of the attraction of a nucleus to a pair of shared (bonding) electrons within a covalent bond. Electronegativity is only important when looking at covalent bonds and electron affinity is only important when considering single atoms gaining electrons to form anions.
Diff: 2 *Page Ref: 9.6*

8) Describe the difference between a pure covalent bond and a polar covalent bond.

Answer: A pure covalent bond occurs when bonding electrons are shared equally (or very close to it) as in the N–N bond. A polar covalent bond is formed between 2 atoms of differing electronegativities. The bonding electrons are unequally shared between the two atoms as in the CO molecule.
Diff: 2 *Page Ref: 9.6*

9) Draw the Lewis structure for the acetate ion, $CH_3CO_2^-$, including any important resonance structures. Label each atom with its formal charge.

Answer: There should be two equivalent resonance structures drawn. All atoms should have a formal charge of "0" except the singly bonded O will have a formal charge of –1.

Diff: 2 *Page Ref: 9.8*

10) Draw the Lewis structure for BrO_3^-. Make sure to include any important resonance structures.

Answer: Three equivalent resonance structures should be drawn, each containing 1 single Br–O bond and 2 double Br–O bonds. Bromine has a lone pair, each double-bonded oxygen has two lone pairs, and the single-bonded oxygen has three lone pairs.

Diff: 2 *Page Ref: 9.8*

11) Define formal charge.

Answer: Formal charge is the charge of an atom that it would have if all bonding electrons were shared equally between the bonded atoms.

Diff: 2 *Page Ref: 9.8*

12) Define bond energy.

Answer: Bond energy is the energy required to break 1 mole of the bond in the gas phase.

Diff: 1 *Page Ref: 9.10*

Chapter 10 Chemical Bonding II: Molecular Shapes, Valence Bond Theory, and Molecular Orbital Theory

Multiple Choice Questions

1) Give the approximate bond angle for a molecule with an octahedral shape.

A) 109.5° B) 180° C) 120° D) 105° E) 90°

Answer: E

Diff: 2 *Page Ref:* 10.2

2) Determine the electron geometry (eg) and molecular geometry(mg) of BCl_3.

A) eg=trigonal planar, mg=trigonal planar
B) eg=tetrahedral, mg=trigonal planar
C) eg=tetrahedral, mg=trigonal pyramidal
D) eg=trigonal planar, mg=bent
E) eg=trigonal bipyramidal, mg= trigonal bipyramidal

Answer: A

Diff: 2 *Page Ref:* 10.2

3) Determine the electron geometry (eg) and molecular geometry (mg) of CO_3^{2-}.

A) eg=tetrahedral, mg=tetrahedral
B) eg=tetrahedral, mg=trigonal pyramidal
C) eg=trigonal planar, mg=bent
D) eg=trigonal planar, mg=trigonal planar
E) eg=tetrahedral, mg=trigonal planar

Answer: D

Diff: 2 *Page Ref:* 10.2

4) Determine the electron geometry (eg) and molecular geometry (mg) of CH_3^{+1}.

A) eg=tetrahedral, mg=tetrahedral
B) eg=tetrahedral, mg=trigonal pyramidal
C) eg=trigonal planar, mg=bent
D) eg=trigonal planar, mg=trigonal planar
E) eg=tetrahedral, mg=trigonal planar

Answer: D

Diff: 2 *Page Ref:* 10.2

5) Determine the electron geometry (eg) and molecular geometry (mg) of SiF_4.

A) eg=tetrahedral, mg=trigonal pyramidal
B) eg=octahedral, mg=square planar
C) eg=trigonal bipyramidal, mg=trigonal pyramidal
D) eg=tetrahedral, mg=bent
E) eg=tetrahedral, mg=tetrahedral

Answer: E

Diff: 2 *Page Ref:* 10.2

6) Determine the electron geometry (eg) and molecular geometry (mg) of PF_5.

A) eg=trigonal bipyramidal, mg=trigonal bipyramidal
B) eg=octahedral, mg=octahedral
C) eg=trigonal bipyramidal, mg=tetrahedral
D) eg=tetrahedral, mg=trigonal pyramidal
E) eg=trigonal planar, mg=octahedral

Answer: A
Diff: 2 *Page Ref: 10.2*

7) Determine the electron geometry (eg) and molecular geometry (mg) of CO_2.

A) eg=tetrahedral, mg=tetrahedral
B) eg=linear, mg=trigonal planar
C) eg=trigonal planar, mg=bent
D) eg=linear, mg=linear
E) eg=trigonal planar, mg=trigonal planar

Answer: D
Diff: 2 *Page Ref: 10.2*

8) The bond angle in NH_3 is

A) 107° B) 104.5° C) 120° D) 109.5° E) 95°

Answer: A
Diff: 2 *Page Ref: 10.3*

9) The bond angle in H_2O is

A) 107° B) 104.5° C) 120° D) 109.5° E) 95°

Answer: B
Diff: 2 *Page Ref: 10.3*

10) Determine the electron geometry (eg) and molecular geometry (mg) of NCl_3.

A) eg=tetrahedral, mg=tetrahedral
B) eg=linear, mg=trigonal planar
C) eg=trigonal planar, mg=bent
D) eg=linear, mg=linear
E) eg=tetrahedral, mg=trigonal pyramidal

Answer: E
Diff: 2 *Page Ref: 10.4*

11) Determine the electron geometry (eg) and molecular geometry (mg) of ICl_3.

A) eg=trigonal planar, mg=trigonal planar
B) eg=trigonal bipyramidal, mg= T-shape
C) eg=trigonal planar, mg=bent
D) eg=trigonal bipyramidal, mg=see-saw
E) eg=tetrahedral, mg=trigonal pyramidal

Answer: B
Diff: 3 *Page Ref: 10.4*

12) Determine the electron geometry (eg) and molecular geometry (mg) of ICl_2^-.

A) eg=tetrahedral, mg=bent
B) eg=tetrahedral, mg=trigonal pyramidal
C) eg=trigonal bipyramidal, mg=linear
D) eg=trigonal bipyramidal, mg=trigonal planar
E) eg=octahedral, mg=linear

Answer: C
Diff: 3 *Page Ref: 10.4*

13) Determine the electron geometry (eg) and molecular geometry (mg) of XeF_2.

A) eg=trigonal bipyramidal, mg=bent
B) eg=linear, mg=linear
C) eg=tetrahedral, mg=linear
D) eg=trigonal bipyramidal, mg=linear
E) eg=tetrahedral, mg=bent

Answer: D
Diff: 3 *Page Ref: 10.4*

14) Determine the electron geometry (eg) and molecular geometry (mg) of XeF_4.

A) eg=tetrahedral, mg=tetrahedral
B) eg=linear, eg=linear
C) eg=tetrahedral, mg=bent
D) eg=trigonal bipyramidal, mg=tetrahedral
E) eg=octahedral, mg=square planar

Answer: E
Diff: 3 *Page Ref: 10.4*

15) Determine the electron geometry (eg) and molecular geometry (mg) of the underlined atom $\underline{C}H_3OCH_3$.

A) eg=tetrahedral, mg=tetrahedral
B) eg=linear, eg=linear
C) eg=tetrahedral, mg=bent
D) eg=trigonal bipyramidal, mg=tetrahedral
E) eg=octahedral, mg=square planar

Answer: A
Diff: 3 *Page Ref: 10.4*

16) Determine the electron geometry (eg) and molecular geometry (mg) of the underlined atom $CH_3\underline{O}CH_3$.

A) eg=tetrahedral, mg=tetrahedral
B) eg=linear, eg=linear
C) eg=tetrahedral, mg=bent
D) eg=trigonal bipyramidal, mg=tetrahedral
E) eg=octahedral, mg=square planar

Answer: C
Diff: 3 *Page Ref: 10.4*

17) Determine the electron geometry (eg) and molecular geometry (mg) of the underlined atom $H_2\underline{C}O$.

A) eg=tetrahedral, mg=tetrahedral
B) eg=trigonal planar, eg=trigonal planar
C) eg=tetrahedral, mg=bent
D) eg=trigonal bipyramidal, mg=tetrahedral
E) eg=octahedral, mg=square planar

Answer: B
Diff: 3 *Page Ref: 10.4*

18) Consider the molecule below. Determine the molecular geometry at each of the 2 labeled carbons.

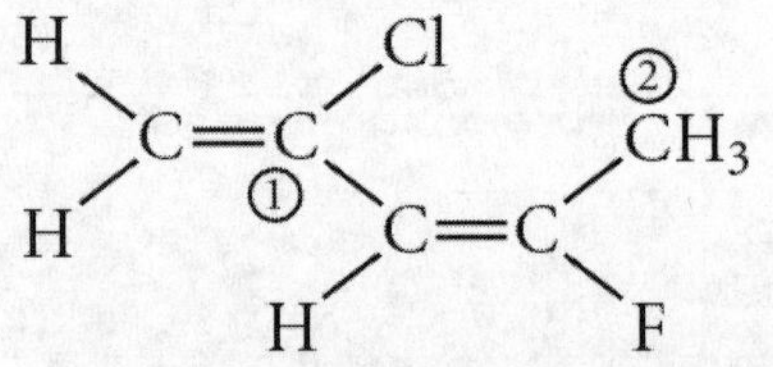

A) C1 = tetrahedral, C2 = linear
B) C1 = trigonal planar, C2= bent
C) C1 = bent, C2 = trigonal planar
D) C1 = trigonal planar, C2 = tetrahedral
E) C1 = trigonal pyramidal, C2 = see-saw

Answer: D
Diff: 3 *Page Ref: 10.4*

19) Consider the molecule below. Determine the molecular geometry at each of the 3 labeled atoms.

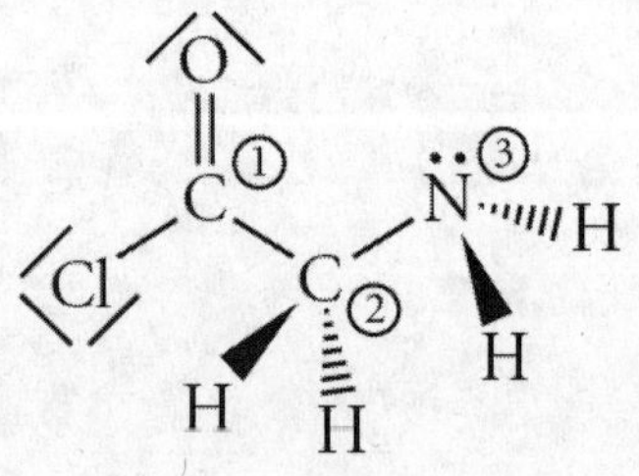

A) 1=trigonal planar, 2=tetrahedral, 3=trigonal pyramidal
B) 1=tetrahedral, 2=tetrahedral, 3=tetrahedral
C) 1=trigonal planar, 2=tetrahedral, 3=tetrahedral
D) 1=tetrahedral, 2=tetrahedral, 3=trigonal planar
E) 1=trigonal planar, 2=trigonal pyramidal, 3=trigonal pyramidal

Answer: A
Diff: 3 *Page Ref: 10.4*

20) Place the following in order of **increasing** X–Se–X bond angle, where X represents the outer atoms in each molecule.

SeO_2 $SeCl_6$ SeF_2

A) $SeCl_6 < SeF_2 < SeO_2$
B) $SeF_2 < SeO_2 < SeCl_6$
C) $SeF_2 < SeCl_6 < SeO_2$
D) $SeO_2 < SeF_2 < SeCl_6$
E) $SeCl_6 < SeO_2 < SeF_2$

Answer: A
Diff: 4 Page Ref: 10.4

21) Place the following in order of **increasing** F–A–F bond angle, where A represents the central atom in each molecule.

PF_3 OF_2 PF_4^+

A) $PF_3 < OF_2 < PF_4^+$
B) $OF_2 < PF_3 < PF_4^+$
C) $OF_2 < PF_4^+ < PF_3$
D) $PF_4^+ < OF_2 < PF_3$
E) $PF_4^+ < PF_3 < OF_2$

Answer: B
Diff: 4 Page Ref: 10.4

22) Place the following in order of **decreasing** X–A–X bond angle, where A represents the central atom and X represents the outer atoms in each molecule.

N_2O NCl_3 NO_2^-

A) $NCl_3 > NO_2^- > N_2O$
B) $NO_2^- > N_2O > NCl_3$
C) $N_2O > NO_2^- > NCl_3$
D) $NCl_3 > N_2O > NO_2^-$
E) $N_2O > NCl_3 > NO_2^-$

Answer: C
Diff: 4 Page Ref: 10.4

23) Place the following in order of **decreasing** X–A–X bond angle, where A represents the central atom and X represents the outer atoms in each molecule.

CS_2 CF_4 SCl_2

A) $CS_2 = SCl_2 > CF_4$
B) $SCl_2 > CF_4 > CS_2$
C) $CF_4 > CS_2 > SCl_2$
D) $CS_2 > CF_4 > SCl_2$
E) $CF_4 > CS_2 > SCl_2$

Answer: D
Diff: 4 *Page Ref: 10.4*

24) Place the following in order of **increasing** X–A–X bond angle, where A represents the central atom and X represents the outer atoms in each molecule.

HCN H_2O H_3O^+

A) $H_2O < H_3O^+ < HCN$
B) $H_3O^+ < H_2O < HCN$
C) $HCN < H_3O^+ < H_2O$
D) $H_2O < HCN < H_3O^+$
E) $H_2O < H_3O^+ < HCN$

Answer: E
Diff: 4 *Page Ref: 10.4*

25) Determine the electron geometry (eg), molecular geometry (mg), and polarity of SO_3.
A) eg=tetrahedral, mg=trigonal pyramidal, polar
B) eg=tetrahedral, mg=tetrahedral, nonpolar
C) eg=trigonal planar, mg=trigonal planar, nonpolar
D) eg= trigonal bipyramidal, mg=trigonal planar, polar
E) eg=trigonal pyramidal, mg=bent, nonpolar

Answer: C
Diff: 2 *Page Ref: 10.5*

26) How many of the following molecules are polar?

$BrCl_3$ CS_2 SiF_4 SO_3

A) 1 B) 2 C) 3 D) 4 E) 0

Answer: A
Diff: 3 *Page Ref: 10.5*

27) How many of the following molecules are polar?

PCl_5 COS XeO_3 $SeBr_2$

A) 2 B) 0 C) 1 D) 3 E) 4

Answer: D
Diff: 3 *Page Ref: 10.5*

28) Determine the electron geometry (eg), molecular geometry (mg), and polarity of SO_2.
A) eg=tetrahedral, mg=bent, polar
B) eg=trigonal planar, mg=bent, polar
C) eg=linear, mg=linear, nonpolar
D) eg=tetrahedral, mg=tetrahedral, nonpolar
E) eg=trigonal pyramidal, mg=trigonal pyramidal, polar

Answer: B
Diff: 3 *Page Ref: 10.5*

29) Determine the electron geometry (eg), molecular geometry (mg), and polarity of PCl_3.
A) eg=tetrahedral, mg=bent, polar
B) eg=trigonal planar, mg=trigonal planar, nonpolar
C) eg=linear, mg=linear, nonpolar
D) eg=tetrahedral, mg=trigonal pyramidal, polar
E) eg=trigonal pyramidal, mg=trigonal pyramidal, polar

Answer: D
Diff: 3 *Page Ref: 10.5*

30) Determine the electron geometry, molecular geometry and polarity of SF_6.
A) eg=trigonal bipyramidal, mg=trigonal bipyramidal, nonpolar
B) eg=tetrahedral, mg=tetrahedral, polar
C) eg=trigonal bipyramidal, mg=see-saw, polar
D) eg=octahedral, mg=trigonal bipyramidal, nonpolar
E) eg=octahedral, mg=octahedral, nonpolar

Answer: E
Diff: 3 *Page Ref: 10.5*

31) Determine the electron geometry (eg), molecular geometry(mg) and polarity of XeO_3.
A) eg=trigonal planar, mg=trigonal planar, nonpolar
B) eg=tetrahedral, mg=trigonal pyramidal, polar
C) eg=trigonal planar, mg=trigonal pyramidal, polar
D) eg=trigonal bipyramidal, mg=trigonal planar, nonpolar
E) eg=octahedral, mg=tetrahedral, nonpolar

Answer: B
Diff: 3 *Page Ref: 10.5*

32) Determine the electron geometry, molecular geometry and polarity of $HBrO_2$.

A) eg=trigonal bipyramidal, mg=trigonal planar, nonpolar
B) eg=octahedral, mg=square planar, nonpolar
C) eg=tetrahedral, mg=trigonal pyramidal, polar
D) eg=tetrahedral, mg=linear, nonpolar
E) eg=linear, mg=linear, polar

Answer: C
Diff: 3 *Page Ref: 10.5*

33) Place the following in order of **increasing** dipole moment.

I. BCl_3 II. BIF_2 III. $BClF_2$

A) I < II = III
B) II < III < I
C) I < II < III
D) II < I < III
E) I < III < II

Answer: E
Diff: 3 *Page Ref: 10.5*

34) Place the following in order of **decreasing** dipole moment.

I. cis–CHCl=CHCl II. trans–CHCl=CHCl III. cis–CHF=CHF

A) III > I > II B) II > I > III C) I > III > II D) II > III > I E) I = III > II

Answer: A
Diff: 3 *Page Ref: 10.5*

35) Identify the number of electron groups around a molecule with sp hybridization.

A) 1 B) 2 C) 3 D) 4 E) 5

Answer: B
Diff: 1 *Page Ref: 10.7*

36) Identify the number of electron groups around a molecule with sp^2 hybridization.

A) 1 B) 2 C) 3 D) 4 E) 5

Answer: C
Diff: 1 *Page Ref: 10.7*

37) Identify the number of electron groups around a molecule with a tetrahedral shape.

A) 1 B) 2 C) 3 D) 4 E) 5

Answer: D
Diff: 1 *Page Ref: 10.7*

38) Identify the number of electron groups around a molecule with a trigonal bipyramidal shape.

A) 1 B) 2 C) 3 D) 4 E) 5

Answer: E
Diff: 1 *Page Ref: 10.7*

39) Identify the number of electron groups around a molecule with an octahedral shape.
A) 6 B) 2 C) 3 D) 4 E) 5

Answer: A
Diff: 1 *Page Ref: 10.7*

40) Give the electron geometry (eg), molecular geometry (mg), and hybridization for NH_3.
A) eg=tetrahedral, mg=trigonal planar, sp^2
B) eg=trigonal pyramidal, mg=trigonal pyramidal, sp^3
C) eg=tetrahedral, mg=trigonal pyramidal, sp^3
D) eg=trigonal pyramidal, mg=tetrahedral, sp^3
E) eg=trigonal planar, mg=trigonal planar, sp^2

Answer: C
Diff: 1 *Page Ref: 10.7*

41) Give the electron geometry (eg), molecular geometry (mg), and hybridization for H_2O.
A) eg=tetrahedral, mg=bent, sp^3
B) eg=trigonal pyramidal, mg=trigonal pyramidal, sp^3
C) eg=tetrahedral, mg=trigonal pyramidal, sp^3
D) eg=bent, mg=bent, sp^2
E) eg=trigonal planar, mg=trigonal planar, sp^2

Answer: A
Diff: 1 *Page Ref: 10.7*

42) Give the electron geometry (eg), molecular geometry (mg), and hybridization for XeF_4.
A) eg=tetrahedral, mg=tetrahedral, sp^3
B) eg=trigonal pyramidal, mg=trigonal pyramidal, sp^3
C) eg=octahedral, mg=square planar, sp^3d^2
D) eg=octahedral, mg=octahedral, sp^3d^2
E) eg=trigonal bipyramidal, mg=seesaw, sp^3d

Answer: C
Diff: 1 *Page Ref: 10.7*

43) Give the hybridization for the O in OF_2.
A) sp B) sp^3 C) sp^2 D) sp^3d E) sp^3d^2

Answer: B
Diff: 3 *Page Ref: 10.7*

44) Give the hybridization for the O in H_3O^+.
A) sp B) sp^3 C) sp^2 D) sp^3d E) sp^3d^2

Answer: B
Diff: 3 *Page Ref: 10.7*

45) Give the hybridization for the S in SO_3.
A) sp B) sp^3 C) sp^2 D) sp^3d E) sp^3d^2

Answer: C
Diff: 3 *Page Ref: 10.7*

46) Give the hybridization for the Br in BrO_4^-.

A) sp B) sp^3d^2 C) sp^3d D) sp^3 E) sp^2

Answer: D
Diff: 3 *Page Ref: 10.7*

47) Give the hybridization for the S in SF_6.

A) sp B) sp^2 C) sp^3 D) sp^3d E) sp^3d^2

Answer: E
Diff: 3 *Page Ref: 10.7*

48) Give the hybridization for the Br in BrF_5.

A) sp^3d^2 B) sp^3d C) sp^3 D) sp^2 E) sp

Answer: A
Diff: 3 *Page Ref: 10.7*

49) Give the hybridization for the Br in $BrCl_3$.

A) sp^3d^2 B) sp^3d C) sp^3 D) sp^2 E) sp

Answer: B
Diff: 3 *Page Ref: 10.7*

50) Give the hybridization for the C in HCN.

A) sp^3d^2 B) sp^3d C) sp^3 D) sp^2 E) sp

Answer: E
Diff: 3 *Page Ref: 10.7*

51) Give the hybridization for the C in HCCH.

A) sp^3d^2 B) sp^3d C) sp^3 D) sp^2 E) sp

Answer: E
Diff: 3 *Page Ref: 10.7*

52) How many of the following molecules have sp^2 hybridization on the central atom?

HCN SO_2 OCl_2 $XeCl_2$

A) 4 B) 3 C) 2 D) 1 E) 0

Answer: D
Diff: 5 *Page Ref: 10.7*

53) Which of the following statements is TRUE?

A) The total number of molecular orbitals formed doesn't always equal the number of atomic orbitals in the set.

B) A bond order of 0 represents a stable chemical bond.

C) When two atomic orbitals come together to form two molecular orbitals, one molecular orbital will be lower in energy than the two separate atomic orbitals and one molecular orbital will be higher in energy than the separate atomic orbitals.

D) Electrons placed in antibonding orbitals stabilize the ion/molecule.

E) All of the above are true.

Answer: C

Diff: 3 *Page Ref: 10.8*

54) Use the molecular orbital diagram shown to determine which of the following is **most** stable.

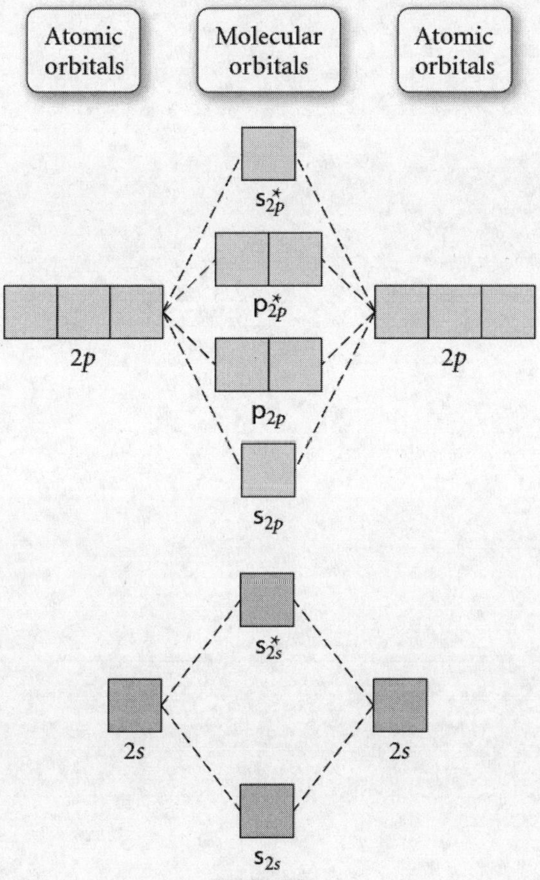

O_2, F_2, Ne_2

A) F_2 B) $F_2{}^{2+}$ C) $Ne_2{}^{2+}$ D) $O_2{}^{2+}$ E) $F_2{}^{2-}$

Answer: D

Diff: 4 *Page Ref: 10.8*

55) Use the molecular orbital diagram shown to determine which of the following are paramagnetic.

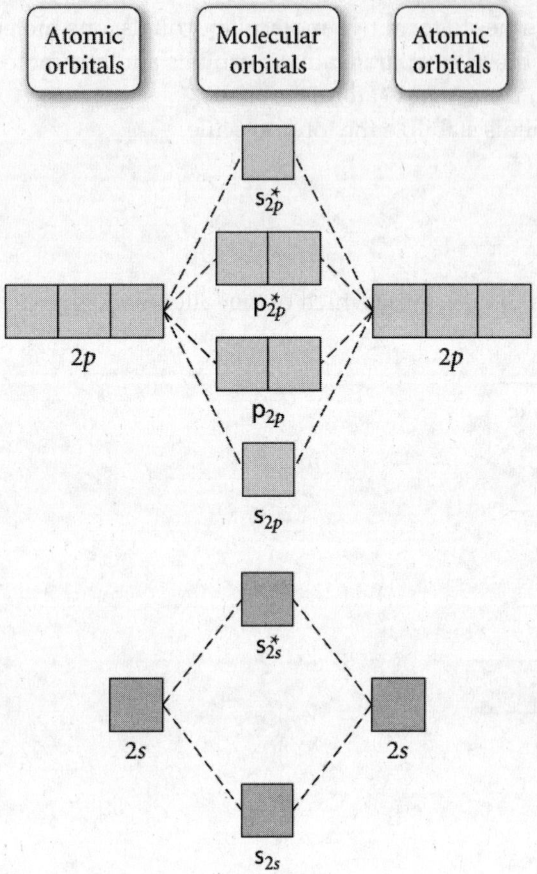

O_2, F_2, Ne_2

A) O_2^{2-}
B) Ne_2^{2+}
C) O_2^{2+}
D) F_2^{2+}
E) None of the above are paramagnetic.

Answer: D
Diff: 4 *Page Ref: 10.8*

56) Use the molecular orbital diagram shown to determine which of the following is **most** stable.

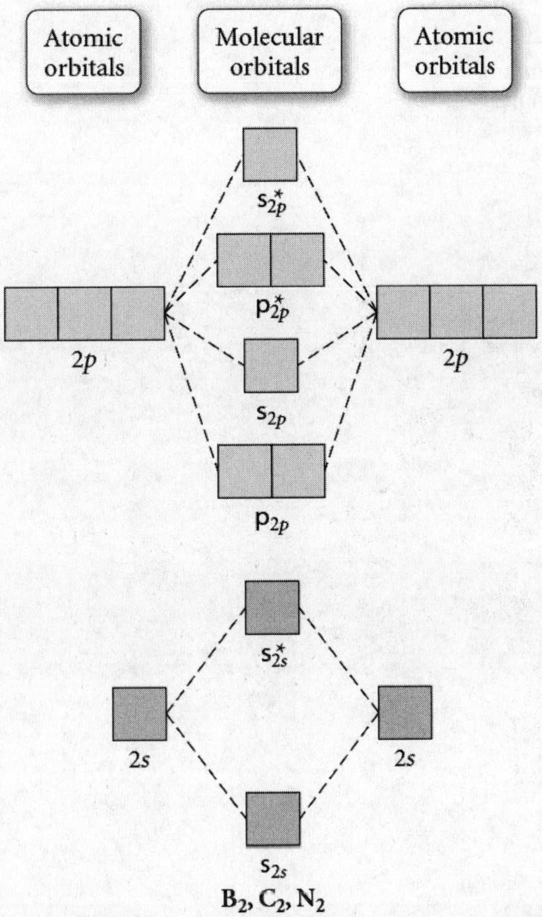

A) C_2^{2+} B) N_2^{2+} C) B_2 D) C_2^{2-} E) B_2^{2+}

Answer: D
Diff: 4 *Page Ref: 10.8*

57) Use the molecular orbital diagram shown to determine which of the following are paramagnetic.

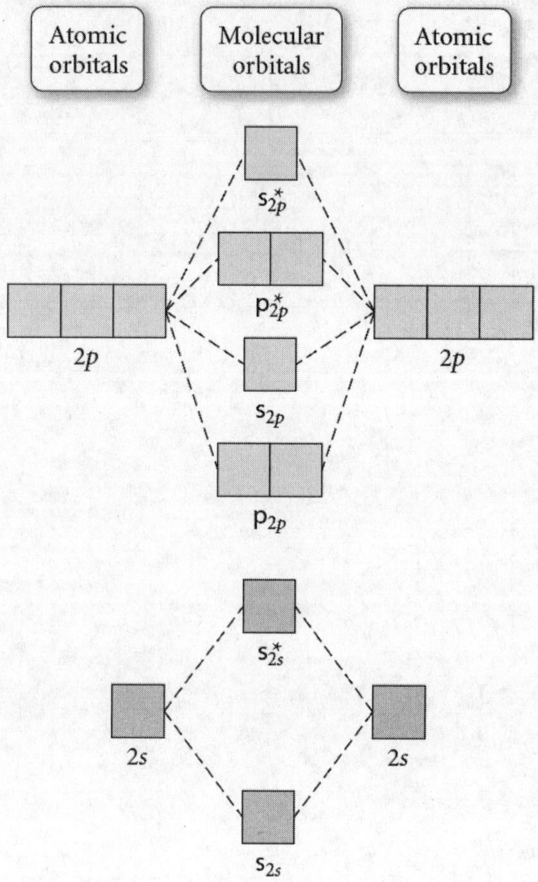

B_2, C_2, N_2

A) B_2^{2+} B) B_2^{2-} C) N_2^{2+} D) C_2^{2-} E) B_2

Answer: E
Diff: 4 *Page Ref: 10.8*

58) Draw the molecular orbital diagram shown to determine which of the following is **most** stable.

A) F_2 B) F_2^{2+} C) Ne_2^{2+} D) O_2^{2+} E) F_2^{2-}

Answer: D
Diff: 5 *Page Ref: 10.8*

59) Draw the molecular orbital diagram shown to determine which of the following is paramagnetic.
A) O_2^{2-}
B) Ne_2^{2+}
C) O_2^{2+}
D) F_2^{2+}
E) None of the above are paramagnetic.

Answer: D
Diff: 5 *Page Ref: 10.8*

60) Draw the molecular orbital diagram shown to determine which of the following is **most** stable.
A) C_2^{2+} B) N_2^{2+} C) B_2 D) C_2^{2-} E) B_2^{2+}

Answer: D
Diff: 5 *Page Ref: 10.8*

61) Draw the molecular orbital diagram shown to determine which of the following is paramagnetic.
A) B_2^{2+} B) B_2^{2-} C) N_2^{2+} D) C_2^{2-} E) B_2

Answer: E
Diff: 5 *Page Ref: 10.8*

Algorithmic Questions

1) What geometric arrangement of charge clouds is expected for an atom that has four charge clouds?
A) trigonal bipyramidal B) octahedral
C) tetrahedral D) square planar

Answer: C
Diff: 1 *Page Ref: 10.2*

2) Give the approximate bond angle for a molecule with a trigonal planar shape.
A) 109.5° B) 180° C) 120° D) 107° E) 90°

Answer: C
Diff: 2 *Page Ref: 10.2*

3) Give the approximate bond angle for a molecule with a tetrahedral shape.
A) 109.5° B) 180° C) 120° D) 105° E) 90°

Answer: A
Diff: 2 *Page Ref: 10.2*

4) Give the approximate bond angle for a molecule with a linear shape.
A) 109.5° B) 180° C) 120° D) 105° E) 90°

Answer: B
Diff: 2 *Page Ref: 10.2*

5) Determine the electron geometry (eg) and molecular geometry (mg) of the underlined carbon in $\underline{C}H_3F$.

A) eg=tetrahedral, mg=tetrahedral
B) eg=linear, mg=trigonal planar
C) eg=trigonal planar, mg=tetrahedral
D) eg=linear, mg=linear
E) eg=trigonal planar, mg=trigonal planar

Answer: A
Diff: 2 *Page Ref: 10.4*

6) What is the molecular geometry of ClF_4^-?

A) seesaw
B) square planar
C) square pyramidal
D) tetrahedral

Answer: B
Diff: 3 *Page Ref: 10.4*

7) What is the molecular geometry of $SbCl_3$?

A) T-shaped
B) tetrahedral
C) trigonal planar
D) trigonal pyramidal

Answer: D
Diff: 3 *Page Ref: 10.4*

8) What is the molecular geometry of SF_5^-?

A) octahedral
B) seesaw
C) square pyramidal
D) trigonal bipyramidal

Answer: C
Diff: 3 *Page Ref: 10.4*

9) Which of the following best describes ClF_2^-? It has a molecular geometry that is

A) linear with no lone pairs on the Cl atom.
B) linear with lone pairs on the Cl atom.
C) nonlinear with no lone pairs on the Cl atom.
D) nonlinear with lone pairs on the Cl atom.

Answer: B
Diff: 3 *Page Ref: 10.4*

10) What is the molecular geometry of SBr_4?

A) seesaw
B) square planar
C) square pyramidal
D) tetrahedral

Answer: A
Diff: 3 *Page Ref: 10.4*

11) Using the VSEPR model, the electron-domain geometry of the central atom in BrF_4^- is ________.

A) linear
B) trigonal planar
C) tetrahedral
D) trigonal bipyramidal
E) octahedral

Answer: E
Diff: 3 *Page Ref: 10.4*

12) Using the VSEPR model, the molecular geometry of the central atom in XeF_2 is ________.

A) linear
B) trigonal planar
C) tetrahedral
D) bent
E) trigonal pyramidal

Answer: A
Diff: 3 *Page Ref: 10.4*

13) Using the VSEPR model, the molecular geometry of the central atom in BF_3 is ________.

A) linear
B) trigonal planar
C) tetrahedral
D) bent
E) trigonal pyramidal

Answer: B
Diff: 3 *Page Ref: 10.4*

14) Using the VSEPR model, the molecular geometry of the central atom in CF_4 is ________.

A) linear
B) trigonal planar
C) tetrahedral
D) bent
E) trigonal pyramidal

Answer: C
Diff: 3 *Page Ref: 10.4*

15) Using the VSEPR model, the molecular geometry of the central atom in SO_2 is ________.

A) linear
B) trigonal planar
C) tetrahedral
D) bent
E) trigonal pyramidal

Answer: D
Diff: 3 *Page Ref: 10.4*

16) Using the VSEPR model, the molecular geometry of the central atom in NCl_3 is ________.

A) linear
B) trigonal planar
C) tetrahedral
D) bent
E) trigonal pyramidal

Answer: E
Diff: 3 *Page Ref: 10.4*

17) Using the VSEPR model, the molecular geometry of the central atom in XeF_4 is ________.

A) tetrahedral
B) seesaw
C) square planar
D) trigonal bipyramidal
E) square pyramidal

Answer: C
Diff: 3 *Page Ref: 10.4*

18) Using the VSEPR model, the electron-domain geometry of the central atom in BF_3 is ________.

A) linear
B) trigonal planar
C) tetrahedral
D) trigonal bipyramidal
E) octahedral

Answer: B
Diff: 3 *Page Ref: 10.4*

19) Using the VSEPR model, the electron-domain geometry of the central atom in SF_2 is ________.

A) linear
B) trigonal planar
C) tetrahedral
D) trigonal bipyramidal
E) octahedral

Answer: C
Diff: 3 *Page Ref: 10.4*

20) Using the VSEPR model, the electron-domain geometry of the central atom in SF_4 is ________.

A) linear
B) trigonal planar
C) tetrahedral
D) trigonal bipyramidal
E) octahedral

Answer: D
Diff: 3 *Page Ref: 10.4*

21) What is the O-B-O bond angle in BO_3^{3-}?

A) less than 109.5°
B) 109.5°
C) 120°
D) greater than 120°

Answer: C
Diff: 3 *Page Ref: 10.4*

22) What are the F-Po-F bond angles in PoF_6 ?

A) 60°
B) 90°
C) 109.5°
D) 120°

Answer: B
Diff: 3 *Page Ref: 10.4*

23) The VSEPR model predicts the H—O—H bond angle in H_3O^+ to be

A) 60°.
B) 90°.
C) less than 109.5° but greater than 90°.
D) 109.5°.

Answer: C
Diff: 3 *Page Ref: 10.4*

24) A pilot checks for water in the gas before flying a small airplane. How does she do it?

A) Drain a little bit of gas from the bottom and look for two layers.
B) Taste it.
C) Shake the wings.
D) Pipet a sample from the top of the tank and look for two layers.
E) Check the oil.

Answer: A
Diff: 1 *Page Ref: 10.5*

25) How many of the following molecules are polar?

$XeCl_2$ COF_2 PCl_4F SF_6

A) 0
B) 3
C) 1
D) 2
E) 4

Answer: B
Diff: 3 *Page Ref: 10.5*

26) How many of the following molecules are polar?

XeO_2 $SiCl_2Br_2$ C_2Cl_2 $SeBr_6$

A) 1
B) 4
C) 2
D) 3
E) 0

Answer: C
Diff: 3 *Page Ref: 10.5*

27) Determine the electron geometry, molecular geometry and polarity of $TeCl_6$.

A) eg=octahedral, mg=octahedral, nonpolar
B) eg=trigonal bipyramidal, mg=trigonal bipyramidal, nonpolar
C) eg=octahedral, mg=tetrahedral, polar
D) eg=trigonal bipyramidal, mg=T-shaped, nonpolar
E) eg=tetrahedral, mg=trigonal pyramidal, polar

Answer: A
Diff: 3 *Page Ref: 10.5*

28) Determine the electron geometry, molecular geometry and polarity of XeF_4.

A) eg=trigonal bipyramidal, mg=trigonal bipyramidal, nonpolar
B) eg=tetrahedral, mg=trigonal bipyramidal, polar
C) eg=trigonal bipyramidal, mg=see-saw, polar
D) eg=octahedral, mg=trigonal pyramidal, nonpolar
E) eg=octahedral, mg=square planar, nonpolar

Answer: E
Diff: 3 *Page Ref: 10.5*

29) Choose the compound below that contains at least one polar covalent bond, but is nonpolar.

A) HCN
B) CCl_4
C) SeF_4
D) ICl_3
E) Both B and C are nonpolar and contain a polar covalent bond.

Answer: B
Diff: 3 *Page Ref: 10.5*

30) Choose the compound below that contains at least one polar covalent bond, but is nonpolar.

A) GeH_2Br_2
B) SBr_2
C) $AsCl_5$
D) CBr_2Cl_2
E) All of the above are nonpolar and contain a polar covalent bond.

Answer: C
Diff: 3 *Page Ref: 10.5*

31) Determine the electron geometry, molecular geometry and polarity of N_2O (N central).

A) eg=linear, mg=linear, nonpolar
B) eg=tetrahedral, mg=linear, polar
C) eg=tetrahedral, mg=bent, polar
D) eg= linear, mg=linear, polar
E) eg=trigonal planar, mg=bent, nonpolar

Answer: D
Diff: 4 *Page Ref: 10.5*

32) The orbital hybridization on the carbon atoms in C_2H_2 is

A) sp. B) sp^2. C) sp^3. D) sp^3d^2.

Answer: A
Diff: 2 *Page Ref: 10.7*

33) The hybrid orbital set used by the central atom in NO_3^- is ________.

A) sp B) sp^2 C) sp^3 D) sp^3d E) sp^3d^2

Answer: B
Diff: 2 *Page Ref: 10.7*

34) The hybrid orbital set used by the central atom in NCl_3 is ________.

A) sp B) sp^2 C) sp^3 D) sp^3d E) sp^3d^2

Answer: C
Diff: 2 *Page Ref: 10.7*

35) The hybrid orbital set used by the central atom in KrF_2 is ________.

A) sp B) sp^2 C) sp^3 D) sp^3d E) sp^3d^2

Answer: D
Diff: 2 *Page Ref: 10.7*

36) Describe a pi bond.
A) side by side overlap of p orbitals
B) end to end overlap of p orbitals
C) s orbital overlapping with the end of a d orbital
D) overlap of two s orbitals
E) p orbital overlapping with a f orbital

Answer: A
Diff: 1 *Page Ref: 10.7*

37) Describe a sigma bond.
A) side by side overlap of f orbitals
B) end to end overlap of p orbitals
C) s orbital overlapping with the side of a p orbital
D) overlap of two f orbitals
E) p orbital overlapping with a f orbital

Answer: B
Diff: 1 *Page Ref: 10.7*

38) A molecule containing a central atom with sp hybridization has a(n) ________ electron geometry.
A) linear
B) trigonal bipyramidal
C) square planar
D) tetrahedral
E) bent

Answer: A
Diff: 1 *Page Ref: 10.7*

39) A molecule containing a central atom with sp^2 hybridization has a(n) ________ electron geometry.

A) linear
B) trigonal bipyramidal
C) trigonal planar
D) octahedral
E) bent

Answer: C
Diff: 1 *Page Ref: 10.7*

40) A molecule containing a central atom with sp^3 hybridization has a(n) ________ electron geometry.

A) linear
B) trigonal planar
C) octahedral
D) tetrahedral
E) bent

Answer: D
Diff: 1 *Page Ref: 10.7*

41) A molecule containing a central atom with sp^3d hybridization has a(n) ________ electron geometry.

A) tetrahedral
B) bent
C) octahedral
D) square planar
E) trigonal bipyramidal

Answer: E
Diff: 1 *Page Ref: 10.7*

42) A molecule containing a central atom with sp^3d^2 hybridization has a(n) ________ electron geometry.

A) octahedral
B) trigonal pyramidal
C) tetrahedral
D) linear
E) tirgonal planar

Answer: A
Diff: 1 *Page Ref: 10.7*

43) Determine the electron geometry (eg) and molecular geometry (mg) of the underlined carbon in $CH_3\underline{C}N$.

A) eg=tetrahedral, mg=linear
B) eg=linear, mg=bent
C) eg=trigonal planar, mg=linear
D) eg=linear, mg=linear
E) eg=bent, mg=tetrahedral

Answer: D
Diff: 2 *Page Ref: 10.7*

44) Determine the electron geometry (eg) and molecular geometry (mg) of $H_2\underline{C}O$.

A) eg=tetrahedral, mg=linear
B) eg=linear, mg=trigonal planar
C) eg=trigonal planar, mg=linear
D) eg=linear, mg=linear
E) eg=trigonal planar, mg=trigonal planar

Answer: E
Diff: 2 *Page Ref: 10.7*

45) Consider the molecule below. Determine the hybridization at each of the 2 labeled carbons.

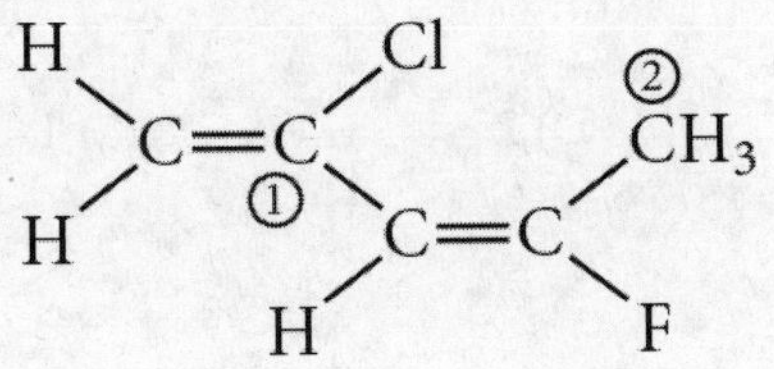

A) C1 = sp^3, C2 = sp^2
B) C1 = sp^2, C2 = sp^2
C) C1 = sp^2, C2 = sp^3d
D) C1 = sp^3d^2, C2 = sp^3d^2
E) C1 = sp^2, C2 = sp^3

Answer: E
Diff: 4 *Page Ref: 10.7*

46) Consider the molecule below. Determine the hybridization at each of the 3 labeled atoms.

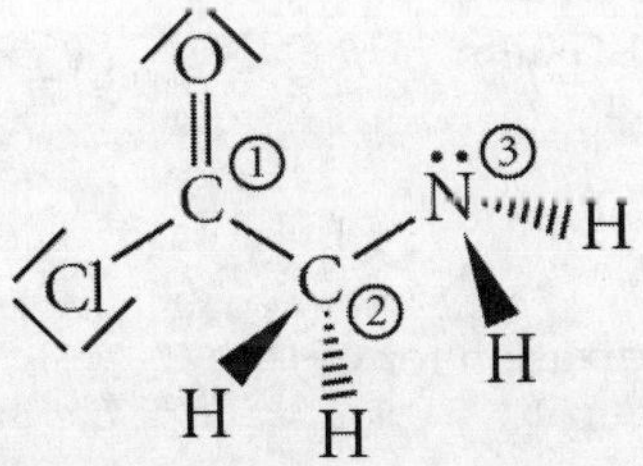

A) 1=sp^2, 2=sp^3, 3=sp^2
B) 1=sp^2, 2=sp^3, 3=sp^3
C) 1=sp^3, 2=sp^3, 3=sp^3
D) 1=sp^3, 2=sp^3, 3=sp^2
E) 1=sp , 2=sp^2, 3=sp^2

Answer: B
Diff: 4 *Page Ref: 10.7*

47) How many of the following molecules have sp hybridization on the central atom?

C_2Br_2 CS_2 O_3 H_2S

A) 0 B) 1 C) 2 D) 3 E) 4

Answer: C
Diff: 4 *Page Ref: 10.7*

48) How many of the following molecules have sp^3 hybridization on the central atom?

$XeCl_4$ CH_4 SF_4 C_2Cl_2

A) 0 B) 4 C) 3 D) 2 E) 1

Answer: E
Diff: 4 *Page Ref: 10.7*

49) How many of the following molecules have sp^3d hybridization on the central atom?

CBr_4 BrF_5 $AsCl_5$ BrF_3

A) 2 B) 0 C) 4 D) 1 E) 3

Answer: A
Diff: 4 *Page Ref: 10.7*

50) How many of the following molecules have sp^3d^2 hybridization on the central atom?

SeF_6 XeF_4 IF_5 $AsCl_5$

A) 1 B) 3 C) 0 D) 2 E) 4

Answer: B
Diff: 4 *Page Ref: 10.7*

51) Give the electron geometry, molecular geometry, and hybridization for CH_3^-.

A) eg = tetrahedral; mg = trigonal pyramidal; sp^3
B) eg = tetrahedral; mg = trigonal planar; sp^3
C) eg = trigonal pyramidal; mg = linear; sp^3
D) eg = trigonal planar; mg = trigonal planar; sp^2

Answer: A
Diff: 5 *Page Ref: 10.7*

52) List the number of sigma bonds and pi bonds in a single bond.

A) 1 sigma, 0 pi B) 0 sigma, 0 pi C) 1 sigma, 1 pi D) 3 sigma, 2 pi

Answer: A
Diff: 5 *Page Ref: 10.7*

53) List the number of sigma bonds and pi bonds in a double bond.
A) 1 sigma, 1 pi
B) 2 sigma, 1 pi
C) 2 sigma, 2 pi
D) 1 sigma, 0 pi

Answer: A
Diff: 5 *Page Ref: 10.7*

54) List the number of sigma bonds and pi bonds in a triple bond.
A) 0 sigma, 1 pi
B) 2 sigma, 1 pi
C) 3 sigma, 0 pi
D) 1 sigma, 2 pi

Answer: D
Diff: 5 *Page Ref: 10.7*

55) Draw the Lewis structure for the molecule CH_2CHCH_3. How many sigma and pi bonds does it contain?
A) 8 sigma, 1 pi
B) 9 sigma, 0 pi
C) 9 sigma, 1 pi
D) 7 sigma, 1 pi
E) 8 sigma, 2 pi

Answer: A
Diff: 5 *Page Ref: 10.7*

56) Draw the Lewis structure for the molecule CH_3CH_2CCH. How many sigma and pi bonds does it contain?
A) 11 sigma, 2 pi
B) 9 sigma, 1 pi
C) 8 sigma, 3 pi
D) 9 sigma, 2 pi
E) 8 sigma, 0 pi

Answer: D
Diff: 5 *Page Ref: 10.7*

57) Consider the following compound. How many sigma and pi bonds does it contain?

$CH_3CHCHCO_2H$

A) 9 sigma, 4 pi
B) 11 sigma, 2 pi
C) 9 sigma, 2 pi
D) 12 sigma, 0 pi
E) 6 sigma, 2 pi

Answer: B
Diff: 5 *Page Ref: 10.7*

58) Draw the Lewis structure for the molecule C_3H_4. How many sigma and pi bonds does it contain?

A) 7 sigma, 0 pi
B) 8 sigma, 0 pi
C) 6 sigma, 2 pi
D) 10 sigma, 2 pi
E) 8 sigma, 2 pi

Answer: C
Diff: 5 *Page Ref: 10.7*

59) How many of the following molecules contain at least one pi bond?

C_2H_6 Cl_2CO C_2Cl_4 HCN

A) 0 B) 1 C) 3 D) 4 E) 2

Answer: C
Diff: 5 *Page Ref: 10.7*

Matching Questions

Match the following.

1) SF_4

Diff: 3 *Page Ref: 10.4*

2) $XeCl_4$

Diff: 3 *Page Ref: 10.4*

3) CH_2F_2

Diff: 2 *Page Ref: 10.5*

4) BCl_3

Diff: 2 *Page Ref: 10.5*

5) sp

Diff: 1 *Page Ref: 10.7*

6) sp^2

Diff: 1 *Page Ref: 10.7*

7) sp^3

Diff: 1 *Page Ref: 10.7*

8) sp^3d

Diff: 1 *Page Ref: 10.7*

9) sp^3d^2

Diff: 1 *Page Ref: 10.7*

10) BeF_2

Diff: 2 *Page Ref: 10.7*

A) octahedral

B) sp^2 hybridized central atom

C) nonpolar, but contains a polar covalent bond

D) tetrahedral

E) octahedral electron geometry

F) polar, but contains no polar bonds

G) trigonal planar

H) linear

I) polar

J) see-saw molecular geometry

K) sp hybridized central atom

L) trigonal bipyramidal

Answers:					
	1) J	2) E	3) I	4) C	5) H
	6) G	7) D	8) L	9) A	10) K

Short Answer Questions

1) Is it possible for a molecule to be nonpolar even though it contains polar bonds? Explain your answer and give an example.

Answer: Yes. The polarity of a molecule depends on the molecular geometry and whether or not all of the dipoles (polar bonds) cancel one another. If the molecular geometry causes all of the dipoles to cancel, the molecule will be nonpolar. An example is CF_4 where there are four polar bonds, but the dipoles sum to 0 making the molecule nonpolar.

Diff: 1 *Page Ref: 10.5*

2) Explain why oil and water do not mix.

Answer: Water molecules are polar and oil molecules are not polar.
Diff: 1 *Page Ref: 10.5*

3) Determine the molecular geometry about each interior atom in the following structure. Sketch the three-dimensional structure and label the interior atoms with the corresponding molecular geometry.

$CH_2CHCCCH_3$

Answer: The sketch should show all of the appropriate multiple bonds, with a double bond between carbons 1 and 2 and a triple bond between carbons 3 and 4. The first two carbons are trigonal planar, the second carbons are linear and the last carbon is tetrahedral.
Diff: 5 *Page Ref: 10.7*

4) Give the electron geometry, molecular geometry, and hybridization for both carbons in CH_3COOH.

Answer: eg = tetrahedral; mg = tetrahedral; sp^3 and
eg = trigonal planar; mg = trigonal planar; sp^2
Diff: 5 *Page Ref: 10.7*

5) Determine the hybridization about each interior atom in the following structure. Sketch the three-dimensional structure and label the interior atoms with their corresponding hybridization.

$CH_2CHCCCH_3$

Answer: The sketch should show all of the appropriate multiple bonds, with a double bond between carbons 1 and 2 and a triple bond between carbons 3 and 4. The first 2 C's are sp^2 hybridized, the next 2 C's are sp hybridized and the last C is sp^3 hybridized.
Diff: 5 *Page Ref: 10.7*

6) According to molecular orbital theory, what is an antibonding orbital?

Answer: An antibonding orbital is formed when 2 orbitals destructively interfere with one another. An antibonding orbital is usually higher in energy than the two separated atoms and has a node in the internuclear region.
Diff: 2 *Page Ref: 10.8*

7) Use molecular orbital theory to determine whether He_2^{2+} or He_2^+ is more stable. Draw the molecular orbital diagram for each and explain your answer.

Answer: The MO diagram should show He_2^{2+} with 2 electrons in the bonding orbital and the MO diagram of He_2^+ should show 2 electrons in the bonding MO and 1 electron in the antibonding MO. The He_2^{2+} ion is more stable since it has a higher bond order (bond order =1) than He_2^+ (bond order = 1/2).
Diff: 3 *Page Ref: 10.8*

Chapter 11 Liquids, Solids, and Intermolecular Forces

Multiple Choice Questions

1) Identify the characteristics of a liquid.
   A) indefinite shape and volume
   B) indefinite shape, but definite volume
   C) definite shape and volume
   D) none of the above
   E) all of the above

Answer: B
Diff: 1 *Page Ref:* 11.2

2) Identify the characteristics of a gas.
   A) indefinite shape and volume
   B) indefinite shape, but definite volume
   C) definite shape and volume
   D) none of the above
   E) all of the above

Answer: A
Diff: 1 *Page Ref:* 11.2

3) Which one of the following has a low density?
   A) gas
   B) liquid
   C) solid
   D) none of the above
   E) all of the above

Answer: A
Diff: 1 *Page Ref:* 11.2

4) Which one of the following has a definite shape and volume?
   A) gas
   B) liquid
   C) solid
   D) none of the above
   E) all of the above

Answer: C
Diff: 1 *Page Ref:* 11.2

5) Give the change in condition to go from a liquid to a gas.
   A) increase heat or reduce pressure
   B) increase heat or increase pressure
   C) cool or reduce pressure
   D) cool or increase pressure
   E) none of the above

Answer: A
Diff: 1 *Page Ref:* 11.2

6) The forces between polar molecules is known as ________.
   A) hydrogen bonding
   B) ion–dipole forces
   C) dipole–dipole forces
   D) dispersion forces
   E) ionic forces

Answer: C
Diff: 1 *Page Ref:* 11.3

7) Which of the following statements is TRUE?
   A) Intermolecular forces are generally stronger than bonding forces.
   B) The potential energy of molecules decrease as they get closer to one another.
   C) Energy is given off when the attraction between two molecules is broken.
   D) Increasing the pressure on a solid usually causes it to become a liquid.
   E) None of the above are true.

Answer: B
Diff: 1 *Page Ref:* 11.3

8) What is the strongest type of intermolecular force present in NH_2CH_3?
   A) dispersion
   B) dipole–dipole
   C) hydrogen bonding
   D) ion–dipole
   E) none of the above

Answer: C
Diff: 1 *Page Ref:* 11.3

9) The two strands in DNA are held together by ________.
   A) dispersion forces
   B) dipole–dipole forces
   C) hydrogen bonding
   D) ion–dipole forces

Answer: C
Diff: 1 *Page Ref:* 11.3

10) How many compounds, of the ones listed below, have hydrogen bonding?

$CH_3CH_2CH_2NH_2$ $CH_3CH_2NHCH_2CH_3$ $(CH_3CH_2)_2NCH_2CH_3$

   A) 2
   B) 3
   C) 1
   D) 0

Answer: A
Diff: 2 *Page Ref:* 11.3

11) Place the following compounds in order of **increasing** strength of intermolecular forces.

CH_4 $CH_3CH_2CH_3$ CH_3CH_3

A) $CH_3CH_2CH_3 < CH_4 < CH_3CH_3$
B) $CH_3CH_2CH_3 < CH_3CH_3 < CH_4$
C) $CH_3CH_3 < CH_4 < CH_3CH_2CH_3$
D) $CH_4 < CH_3CH_2CH_3 < CH_3CH_3$
E) $CH_4 < CH_3CH_3 < CH_3CH_2CH_3$

Answer: E
Diff: 2 *Page Ref: 11.3*

12) Place the following compounds in order of **decreasing** strength of intermolecular forces.

I. $CH_3CH_2CH_2CH_2CH_2CH_3$ II. $(CH_3)_3CCH_3$ III. $(CH_3)_3CCH_2CH_3$

A) III > II > I
B) I > III > II
C) I > II > III
D) II > III > I
E) III > I > II

Answer: B
Diff: 2 *Page Ref: 11.3*

13) Identify the compound that does not have hydrogen bonding.
A) $(CH_3)_3N$ B) H_2O C) CH_3OH D) HF E) CH_3NH_2

Answer: A
Diff: 2 *Page Ref: 11.3*

14) In a liquid, the energy required to increase the surface of the area by a unit amount is called ________.
A) viscosity
B) surface tension
C) dipole-dipole force
D) hydrogen bonding
E) capillary action

Answer: B
Diff: 1 *Page Ref: 11.4*

15) Choose the substance with the lowest viscosity.
A) Cl_3CCCl_3
B) Cl_2CHCH_2Cl
C) $Cl_2CHCHCl_2$
D) $ClCH_2CH_2Cl$
E) Cl_3CCHCl_2

Answer: D
Diff: 2 *Page Ref: 11.4*

16) Choose the substance with the highest viscosity.
A) $SbCl_3$ B) $AsCl_5$ C) ICl_2 D) $BeCl_2$ E) OCl_2

Answer: D
Diff: 2 *Page Ref: 11.4*

17) Identify the term used to describe the ability of a liquid to flow against gravity up a narrow tube.
A) capillary action
B) viscosity
C) surface tension
D) density
E) none of the above

Answer: A
Diff: 2 *Page Ref: 11.4*

18) Describe sweating in humans.
A) It is an endothermic reaction.
B) The sweat evaporates absorbing heat from the body.
C) The skin is cooled.
D) None of the above.
E) All of the above.

Answer: E
Diff: 2 *Page Ref: 11.5*

19) Which of the following statements is TRUE?
A) Vapor pressure increases with temperature.
B) Hydrogen bonds are stronger than covalent bonds.
C) Intermolecular forces hold the atoms in molecules together.
D) Dispersion forces are generally stronger than dipole-dipole forces.
E) None of the above are true.

Answer: A
Diff: 1 *Page Ref: 11.5*

20) Identify the place which has the lowest boiling point of water.
A) Death Valley, 282 feet below sea level
B) A pressurized passenger jet, 35,000 feet
C) New Orleans, sea level
D) Mt. Everest, 29,035 feet
E) Denver, Colorado, 5280 feet

Answer: D
Diff: 1 *Page Ref: 11.5*

21) Identify the place which has the highest boiling point of water.
A) Death Valley, 282 feet below sea level
B) A pressurized passenger jet, 35,000 feet
C) New Orleans, sea level
D) Mt. Everest, 29035 feet
E) Denver, Colorado, 5280 feet

Answer: A
Diff: 1 *Page Ref: 11.5*

22) Which of the following statements is FALSE?
A) The rate of vaporization increases with increasing surface area.
B) The rate of vaporization increases with decreasing strength of intermolecular forces.
C) The rate of vaporization increases with increasing temperature.
D) Molecules with hydrogen bonding are more volatile than compounds with dipole-dipole forces.
E) None of the above are false.

Answer: D
Diff: 1 *Page Ref: 11.5*

23) Which substance below has the strongest intermolecular forces?
A) A_2X, ΔH_{vap}= 39.6 kJ/mol
B) BY_2, ΔH_{vap}= 26.7 kJ/mol
C) C_3X_2, ΔH_{vap}= 36.4 kJ/mol
D) DX_2, ΔH_{vap}= 23.3 kJ/mol
E) EY_3, ΔH_{vap}= 21.5 kJ/mol

Answer: A
Diff: 1 *Page Ref: 11.5*

24) Define boiling.
A) A liquid becomes a gas.
B) A gas becomes a liquid.
C) A gas becomes a solid.
D) A solid becomes a gas.
E) A solid becomes a liquid.

Answer: A
Diff: 1 *Page Ref: 11.5*

25) Give the term for the temperature at which the gas and liquid phases form a supercritical fluid.
A) absolute temperature
B) definite temperature
C) fluid temperature
D) critical temperature
E) solid temperature

Answer: D
Diff: 1 *Page Ref: 11.5*

26) Choose the substance with the highest vapor pressure at a given temperature.
A) SiS_2
B) RbCl
C) CH_3SCH_3
D) BF_3
E) SbH_3

Answer: D
Diff: 2 *Page Ref: 11.5*

27) Which of the following substances would you predict to have the highest ΔH_{vap}?

A) CH_3Cl
B) HCl
C) $HOCH_2CH_2OH$
D) CH_3CH_2OH
E) $CH_3CH_2CH_2CH_3$

Answer: C
Diff: 2 *Page Ref: 11.5*

28) How much energy is required to vaporize 48.7 g of dichloromethane (CH_2Cl_2) at its boiling point, if its ΔH_{vap} is 31.6 kJ/mol?

A) 31.2 kJ B) 6.49 kJ C) 55.1 kJ D) 15.4 kJ E) 18.1 kJ

Answer: E
Diff: 3 *Page Ref: 11.5*

29) How much energy is required to vaporize 98.6 g of ethanol (C_2H_5OH) at its boiling point, if its ΔH_{vap} is 40.5 kJ/mol?

A) 86.7 kJ B) 11.5 kJ C) 18.9 kJ D) 52.8 kJ E) 39.9 kJ

Answer: A
Diff: 3 *Page Ref: 11.5*

30) Place the following substances in order of **increasing** vapor pressure at a given temperature.

SF_6 SiH_4 SF_4

A) $SF_6 < SiH_4 < SF_4$
B) $SiH_4 < SF_4 < SF_6$
C) $SF_6 < SF_4 < SiH_4$
D) $SF_4 < SF_6 < SiH_4$
E) $SiH_4 < SF_6 < SF_4$

Answer: D
Diff: 4 *Page Ref: 11.5*

31) Place the following substances in order of **decreasing** vapor pressure at a given temperature.

PF_5 BrF_3 CF_4

A) $BrF_3 > PF_5 > CF_4$
B) $BrF_3 > CF_4 > PF_5$
C) $PF_5 > BrF_3 > CF_4$
D) $CF_4 > BrF_3 > PF_5$
E) $CF_4 > PF_5 > BrF_3$

Answer: E
Diff: 4 *Page Ref: 11.5*

32) Place the following substances in order of **increasing** vapor pressure at a given temperature.

NF_3 NH_3 BCl_3

A) $NH_3 < NF_3 < BCl_3$
B) $NF_3 < NH_3 < BCl_3$
C) $BCl_3 < NF_3 < NH_3$
D) $NH_3 < BCl_3 < NF_3$
E) $BCl_3 < NH_3 < NF_3$

Answer: A
Diff: 4 *Page Ref: 11.5*

33) Place the following substances in order of **decreasing** vapor pressure at a given temperature.

BeF_2 CH_3OH OF_2

A) $CH_3OH > OF_2 > BeF_2$
B) $BeF_2 > OF_2 > CH_3OH$
C) $OF_2 > CH_3OH > BeF_2$
D) $OF_2 > BeF_2 > CH_3OH$
E) $BeF_2 > CH_3OH > OF_2$

Answer: C
Diff: 4 *Page Ref: 11.5*

34) Place the following substances in order of **decreasing** boiling point.

H_2O N_2 CO

A) $CO > H_2O > N_2$
B) $N_2 > CO > H_2O$
C) $H_2O > CO > N_2$
D) $CO > N_2 > H_2O$
E) $N_2 > H_2O > CO$

Answer: C
Diff: 4 *Page Ref: 11.5*

35) Place the following substances in order of **decreasing** boiling point.

N_2 O_2 H_2

A) $O_2 > H_2 > N_2$
B) $N_2 > H_2 > O_2$
C) $N_2 > O_2 > H_2$
D) $O_2 > N_2 > H_2$
E) $H_2 > N_2 > O_2$

Answer: D
Diff: 4 *Page Ref: 11.5*

36) Determine ΔH_{vap} for a compound that has a measured vapor pressure of 24.3 torr at 273 K and 135 torr at 325 K.

A) 41 kJ/mol B) 79 kJ/mol C) 24 kJ/mol D) 13 kJ/mol E) 34 kJ/mol

Answer: C
Diff: 5 *Page Ref: 11.5*

37) Determine the normal boiling point of a substance whose vapor pressure is 55.1 mm Hg at 35°C and has a ΔH_{vap} of 32.1 kJ/mol.

A) 255 K B) 368 K C) 412 K D) 390. K E) 466 K

Answer: D
Diff: 5 *Page Ref: 11.5*

38) Determine the vapor pressure (in mm Hg) of a substance at 29°C, whose normal boiling point is 76°C and has a ΔH_{vap} of 38.7 kJ/mol.

A) 80 mm Hg
B) 13 mm Hg
C) 21 mm Hg
D) 48 mm Hg
E) 96 mm Hg

Answer: E
Diff: 5 *Page Ref: 11.5*

39) Determine the vapor pressure (in torr) of a substance at 36°C, whose normal boiling point is 84°C and has a ΔH_{vap} of 22.1 kJ/mol.

A) 239 torr B) 31.8 torr C) 41.8 torr D) 147 torr E) 98 torr

Answer: A
Diff: 5 *Page Ref: 11.5*

40) Define deposition.

A) A liquid becomes a gas.
B) A gas becomes a liquid.
C) A gas becomes a solid.
D) A solid becomes a gas.
E) A solid becomes a liquid.

Answer: C
Diff: 1 *Page Ref: 11.6*

41) Define sublimation.
A) the phase transition from solid to gas
B) the phase transition from gas to solid
C) the phase transition from gas to liquid
D) the phase transition from liquid to gas
E) the phase transition from liquid to solid

Answer: A
Diff: 1 *Page Ref: 11.6*

42) Define freezing.
A) the phase transition from solid to gas
B) the phase transition from gas to solid
C) the phase transition from gas to liquid
D) the phase transition from liquid to gas
E) the phase transition from liquid to solid

Answer: E
Diff: 1 *Page Ref: 11.6*

43) Define fusion.
A) the phase transition from solid to liquid
B) the phase transition from gas to solid
C) the phase transition from gas to liquid
D) the phase transition from liquid to gas
E) the phase transition from liquid to solid

Answer: A
Diff: 1 *Page Ref: 11.6*

44) At atmospheric pressure, dry ice ________.
A) freezes B) deposits C) sublimes D) melts E) boils

Answer: C
Diff: 1 *Page Ref: 11.6*

45) At atmospheric pressure, ice ________.
A) freezes B) deposits C) sublimes D) melts E) boils

Answer: D
Diff: 1 *Page Ref: 11.6*

46) How much energy is required to heat 36.0 g H_2O from a liquid at 65°C to a gas at 115°C? The following physical data may be useful.

ΔH_{vap} = 40.7 kJ/mol
C_{liq} = 4.18 J/g °C
C_{gas} = 2.01 J/g °C
C_{sol} = 2.09 J/g °C
$T_{melting}$ = 0 °C
$T_{boiling}$ = 100 °C

A) 63.5 kJ B) 87.7 kJ C) 10.9 kJ D) 52.7 kJ E) 91.7 kJ

Answer: B
Diff: 5 Page Ref: 11.7

47) How much energy must be removed from a 125 g sample of benzene (molar mass = 78.11 g/mol) at 425.0 K to liquify the sample and lower the temperature to 335.0 K? The following physical data may be useful.

ΔH_{vap} = 33.9 kJ/mol
ΔH_{fus} = 9.8 kJ/mol
C_{liq} = 1.73 J/g°C
C_{gas} = 1.06 J/g°C
C_{sol} = 1.51 J/g°C
$T_{melting}$ = 279.0 K
$T_{boiling}$ = 353.0 K

A) 38.9 kJ B) 95.4 kJ C) 67.7 kJ D) 54.3 kJ E) 74.4 kJ

Answer: C
Diff: 5 Page Ref: 11.7

48) How much energy is required to heat 87.1 g acetone (molar mass = 58.08 g/mol) from a solid at −154.0°C to a liquid at −42.0°C? The following physical data may be useful.

ΔH_{fus} = 7.27 kJ/mol
C_{liq} = 2.16 J/g°C
C_{gas} = 1.29 J/g°C
C_{sol} = 1.65 J/g°C
$T_{melting}$ = −95.0°C

A) 8.48 kJ B) 18.5 kJ C) 32.2 kJ D) 29.4 kJ E) 9.97 kJ

Answer: D
Diff: 5 Page Ref: 11.7

49) How much energy must be removed from a 94.4 g sample of benzene (molar mass= 78.11 g/mol) at 322.0 K to solidify the sample and lower the temperature to 205.0 K? The following physical data may be useful.

ΔH_{vap} = 33.9 kJ/mol
ΔH_{fus} = 9.8 kJ/mol
C_{liq} = 1.73 J/g°C
C_{gas} = 1.06 J/g°C
C_{sol} = 1.51 J/g°C
$T_{melting}$ = 279.0 K
$T_{boiling}$ = 353.0 K

A) 17.6 kJ B) 11.8 kJ C) 70.2 kJ D) 10.5 kJ E) 29.4 kJ

Answer: E
Diff: 5 Page Ref: 11.7

50) Define triple point.
A) The temperature, pressure, and density for a gas.
B) The temperature at which the boiling point equals the melting point.
C) The temperature and pressure where liquid, solid, and gas are equally stable and are in equilibrium.
D) The temperature that is unique for a substance.
E) The temperature at which the solid and liquid co-exist.

Answer: C
Diff: 1 Page Ref: 11.8

51) Consider the phase diagram shown. Choose the statement below that is TRUE.

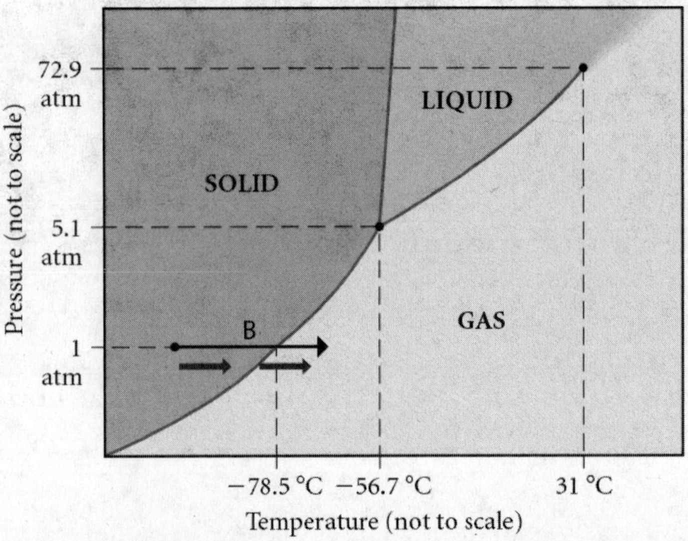

A) The triple point of this substance occurs at a temperature of 31°C.
B) At 10 atm of pressure, there is no temperature where the liquid phase of this substance would exist.
C) The solid phase of this substance is higher in density than the liquid phase.
D) The line separating the solid and liquid phases represents the ΔH_{vap}.
E) None of the above are true.

Answer: C
Diff: 1 *Page Ref: 11.8*

52) Consider the phase diagram below. If the dashed line at 1 atm of pressure is followed from 100 to 500°C, what phase changes will occur (in order of increasing temperature)?

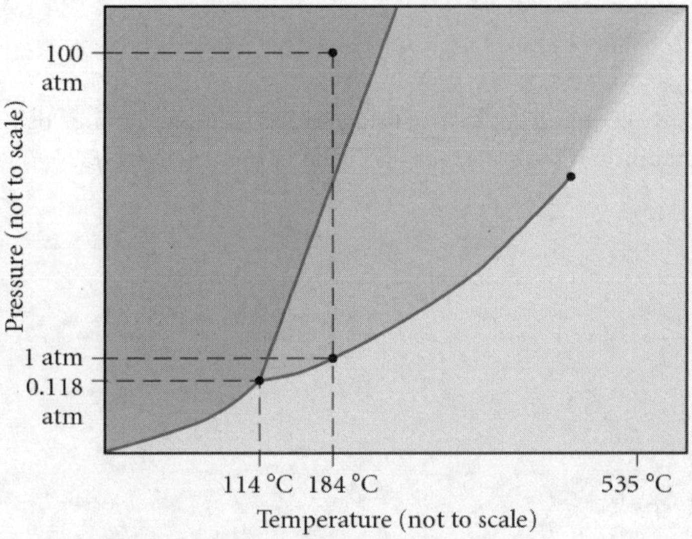

A) condensation, followed by vaporization
B) sublimation, followed by deposition
C) vaporization, followed by deposition
D) fusion, followed by vaporization
E) No phase change will occur under the conditions specified.

Answer: D
Diff: 1 *Page Ref: 11.8*

53) Why is water an extraordinary substance?
A) Water has a low molar mass, yet it is a liquid at room temperature.
B) Water is the main solvent within living organisms.
C) Water has an exceptionally high specific heat capacity.
D) Water has strong hydrogen bonding.
E) All of the above.

Answer: E
Diff: 1 *Page Ref: 11.9*

54) Identify the compound with the highest boiling point.
A) CH_4 B) NH_3 C) HF D) H_2S E) H_2O

Answer: E
Diff: 1 *Page Ref: 11.9*

55) Determine the radius of an Al atom (in pm) if the density of aluminum is 2.71 g/cm^3. Aluminum crystallizes in a face centered cubic structure with an edge length of $2\sqrt{2}$ r.
A) 143 pm B) 227 pm C) 96 pm D) 172 pm E) 193 pm

Answer: A
Diff: 5 *Page Ref: 11.11*

56) A metal crystallizes in a face centered cubic structure and has a density of 11.9 g/cm^3. If the radius of the metal atom is 138 pm, what is the identity of the metal?

A) At B) Pd C) Mn D) Fe E) Cr

Answer: B
Diff: 5 *Page Ref: 11.11*

57) Vanadium crystallizes in a body centered cubic structure and has an atomic radius of 131 pm. Determine the density of vanadium, if the edge length of a bcc structure is $4r/\sqrt{3}$.

A) 3.06 g/cm^3
B) 12.2 g/cm^3
C) 6.11 g/cm^3
D) 2.77 g/cm^3
E) 8.46 g/cm^3

Answer: C
Diff: 5 *Page Ref: 11.11*

58) Identify the type of solid for AgCl.

A) metallic atomic solid
B) ionic solid
C) nonbonding atomic solid
D) molecular solid
E) networking atomic solid

Answer: B
Diff: 1 *Page Ref: 11.12*

59) Identify the type of solid for ice.

A) metallic atomic solid
B) ionic solid
C) nonbonding atomic solid
D) molecular solid
E) networking atomic solid

Answer: D
Diff: 1 *Page Ref: 11.12*

60) Identify the type of solid for gold.

A) metallic atomic solid
B) ionic solid
C) nonbonding atomic solid
D) molecular solid
E) networking atomic solid

Answer: A
Diff: 1 *Page Ref: 11.12*

61) Identify the type of solid for diamond.
A) metallic atomic solid
B) ionic solid
C) nonbonding atomic solid
D) molecular solid
E) networking atomic solid

Answer: E
Diff: 1 *Page Ref: 11.12*

62) Identify the type of solid for argon.
A) metallic atomic solid
B) ionic solid
C) nonbonding atomic solid
D) molecular solid
E) networking atomic solid

Answer: C
Diff: 1 *Page Ref: 11.12*

Algorithmic Questions

1) Which has the smallest dipole–dipole forces?
A) CH_3Cl B) HBr C) O_2 D) NO

Answer: C
Diff: 2 *Page Ref: 11.3*

2) Which is expected to have the largest dispersion forces?
A) C_3H_8 B) $C_{12}H_{26}$ C) F_2 D) Be Cl_2

Answer: B
Diff: 2 *Page Ref: 11.3*

3) Which of the following compounds exhibits hydrogen bonding?
A) CH_3I B) HBr C) CH_3OCH_3 D) CH_3NH_2

Answer: D
Diff: 2 *Page Ref: 11.3*

4) List the compounds in decreasing boiling point order.

$CH_3CH_2CH_3$ Ar CH_3CN

A) $CH_3CH_2CH_3 > CH_3CN > Ar$
B) $Ar > CH_3CH_2CH_3 > CH_3CN$
C) $CH_3CH_2CH_3 > Ar > CH_3CN$
D) $CH_3CN > CH_3CH_2CH_3 > Ar$
E) $Ar > CH_3CN > CH_3CH_2CH_3$

Answer: D
Diff: 2 *Page Ref: 11.3*

5) In liquid propanol,

$CH_3CH_2CH_2OH$

which intermolecular forces are present?

A) Dispersion, hydrogen bonding and dipole-dipole forces are present.
B) Only dipole-dipole and ion-dipole forces are present.
C) Only dispersion and dipole-dipole forces are present.
D) Only hydrogen bonding forces are present.

Answer: A
Diff: 3 *Page Ref: 11.3*

6) Which of the following compounds exhibits only dipole-dipole intermolecular interactions?

A) Br_2
B) CH_3Br
C) CBr_4
D) $BrCH_2CH_2OH$

Answer: B
Diff: 3 *Page Ref: 11.3*

7) The normal boiling point for H_2Se is higher than the normal boiling point for H_2S. This can be explained by

A) larger dipole-dipole forces for H_2Se.
B) larger dispersion forces for H_2Se.
C) larger hydrogen-bond forces for H_2Se.
D) larger dipole-dipole forces, larger dispersion forces, and larger hydrogen-bond forces for H_2Se.

Answer: B
Diff: 3 *Page Ref: 11.3*

8) What is the strongest type of intermolecular force present in Cl_2?

A) ion-dipole
B) dipole-dipole
C) dispersion
D) hydrogen bonding
E) none of the above

Answer: C
Diff: 1 *Page Ref: 11.3*

9) What is the strongest type of intermolecular force present in $CHBr_3$?

A) ion-dipole
B) dispersion
C) hydrogen bonding
D) dipole-dipole
E) none of the above

Answer: D
Diff: 1 *Page Ref: 11.3*

10) What type of intermolecular force causes the dissolution of KF in water?
A) hydrogen bonding
B) dipole–dipole forces
C) ion–dipole force
D) dispersion forces
E) none of the above

Answer: C
Diff: 1 *Page Ref: 11.3*

11) Choose the molecule or compound that exhibits dispersion forces as its strongest intermolecular force.
A) O_2
B) CO
C) HF
D) NaCl
E) All of these have intermolecular forces stronger than dispersion.

Answer: A
Diff: 2 *Page Ref: 11.3*

12) Choose the molecule or compound that exhibits dipole–dipole forces as its strongest intermolecular force.
A) H_2 B) SO_2 C) NH_3 D) CF_4 E) BCl_3

Answer: B
Diff: 2 *Page Ref: 11.3*

13) Choose the compound that exhibits hydrogen bonding as its strongest intermolecular force.
A) SBr_2
B) C_2H_6
C) CH_3OH
D) CH_2Br_2
E) None of the above compounds exhibit hydrogen bonding.

Answer: C
Diff: 2 *Page Ref: 11.3*

14) Place the following compounds in order of **increasing** strength of intermolecular forces.

CS_2 I_2 NH_2CH_3

A) $NH_2CH_3 < CS_2 < I_2$
B) $I_2 < NH_2CH_3 < CS_2$
C) $NH_2CH_3 < I_2 < CS_2$
D) $I_2 < CS_2 < NH_2CH_3$
E) $CS_2 < NH_2CH_3 < I_2$

Answer: D
Diff: 2 *Page Ref: 11.3*

15) Place the following compounds in order of **decreasing** strength of intermolecular forces.

HF H_2 CO_2

A) $HF > CO_2 > H_2$
B) $HF > H_2 > CO_2$
C) $H_2 > CO_2 > HF$
D) $CO_2 > HF > H_2$
E) $CO_2 > H_2 > HF$

Answer: A
Diff: 2 *Page Ref: 11.3*

16) Identify the compound that does not have dipole-dipole forces as its strongest force.
A) CH_2Br_2
B) CH_3OCH_3
C) CH_3Cl
D) $HCBr_3$
E) CO_2

Answer: E
Diff: 2 *Page Ref: 11.3*

17) Identify the compound that has hydrogen bonding.
A) $(CH_3)_3N$ B) N_2 C) CH_3CH_3 D) HI E) H_2O

Answer: E
Diff: 2 *Page Ref: 11.3*

18) Choose the pair of substances that are most likely to form a homogeneous solution.
A) KI and Hg
B) LiCl and C_6H_{14}
C) C_3H_8 and C_2H_5OH
D) F_2 and PF_3
E) NH_3 and CH_3OH

Answer: E
Diff: 2 *Page Ref: 11.3*

19) Choose the pair of substances that are most likely to form a homogeneous solution.
A) C_6H_{14} and $C_{10}H_{20}$
B) KCl and C_5H_{12}
C) N_2O_4 and NH_4I
D) C_6H_{14} and H_2O
E) None of the pairs above will form a homogeneous solution.

Answer: A
Diff: 2 *Page Ref: 11.3*

20) Choose the pair of substances that are most likely to form a homogeneous solution.
A) CBr_4 and SF_2
B) NF_3 and SO_2
C) CO and C_6H_6
D) NH_2CH_3 and CH_4
E) None of the pairs above will form a homogeneous solution.

Answer: B
Diff: 2 *Page Ref:* 11.3

21) Choose the substance with the highest surface tension.
A) $HOCH_2CH_2OH$
B) CH_2Br_2
C) CH_3CH_2Cl
D) CH_3CH_2OH
E) $CH_3CH_2CH_3$

Answer: A
Diff: 2 *Page Ref:* 11.4

22) Choose the substance with the lowest surface tension.
A) CH_3OH
B) $CH_3CH_2CH_2CH_3$
C) C_6H_6
D) H_2O
E) $(CH_3)_2CO$

Answer: B
Diff: 2 *Page Ref:* 11.4

23) Choose the substance with the highest viscosity.
A) $(CH_3CH_2)_2CO$
B) $C_2H_4Cl_2$
C) $HOCH_2CH_2CH_2CH_2OH$
D) CCl_4
E) C_6H_{14}

Answer: C
Diff: 2 *Page Ref:* 11.4

24) Identify the substance with the highest viscosity.
A) gasoline
B) water
C) maple syrup
D) motor oil
E) coffee

Answer: C
Diff: 2 *Page Ref:* 11.4

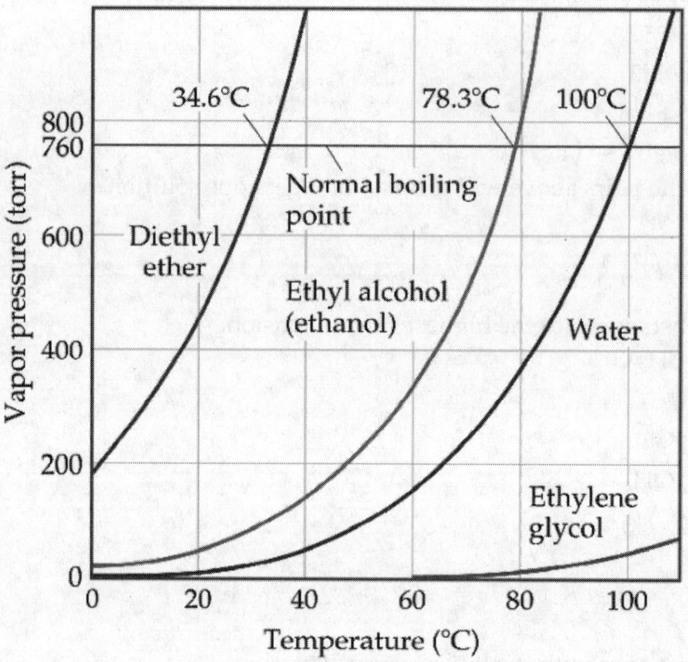

25) Based on the figure above, the boiling point of diethyl ether under an external pressure of 1.32 atm is ________°C.

A) 10 B) 20 C) 30 D) 40 E) 0

Answer: D
Diff: 3 *Page Ref: 11.5*

26) Based on the figure above, the boiling point of ethyl alcohol under an external pressure of 0.0724 atm is ________°C.

A) 80 B) 60 C) 70 D) 40 E) 20

Answer: E
Diff: 3 *Page Ref: 11.5*

27) Based on the figure above, the boiling point of water under an external pressure of 0.316 atm is ________°C.

A) 70 B) 40 C) 60 D) 80 E) 90

Answer: A
Diff: 3 *Page Ref: 11.5*

28) Which of the following compounds has the highest boiling point?

A) H_2O B) H Cl C) H_2S D) N H_3

Answer: A
Diff: 4 *Page Ref: 11.5*

29) Choose the substance with the lowest vapor pressure at a given temperature.

A) CO_2 B) CaI_2 C) BF_3 D) Ar E) PF_5

Answer: B
Diff: 2 *Page Ref: 11.5*

30) Place the following substances in order of **increasing** boiling point.

Ne I_2 N_2

A) Ne < I_2 < N_2
B) I_2 < N_2 < Ne
C) N_2 < I_2 < Ne
D) I_2 < Ne < N_2
E) Ne < N_2 < I_2

Answer: E
Diff: 2 *Page Ref: 11.5*

31) Place the following substances in order of **increasing** boiling point.

CH_3CH_2OH He CH_3OCH_3

A) He < CH_3OCH_3 < CH_3CH_2OH
B) CH_3CH_2OH < He < CH_3OCH_3
C) CH_3CH_2OH < CH_3OCH_3 < He
D) CH_3OCH_3 < He < CH_3CH_2OH
E) He < CH_3CH_2OH < CH_3OCH_3

Answer: A
Diff: 2 *Page Ref: 11.5*

32) Choose the substance with the highest boiling point.
A) CI_4 B) KI C) CS_2 D) HF E) I_2

Answer: B
Diff: 2 *Page Ref: 11.5*

33) Choose the substance with the lowest boiling point.
A) H_2S B) NBr_3 C) O_2 D) Cl_2H_2 E) H_2O_2

Answer: C
Diff: 2 *Page Ref: 11.5*

34) Which of the following substances would you predict to have the highest ΔH_{vap}?
A) Xe B) C_6H_6 C) SiF_4 D) F_2 E) N_2

Answer: B
Diff: 2 *Page Ref: 11.5*

35) How much energy is required to vaporize 158 g of butane (C_4H_{10}) at its boiling point, if its ΔH_{vap} is 24.3 kJ/mol?
A) 15.1 kJ B) 66.1 kJ C) 41.9 kJ D) 2.60 kJ E) 38.4 kJ

Answer: B
Diff: 3 *Page Ref: 11.5*

36) Place the following substances in order of **decreasing** boiling point.

He　　Rn　　H_2

A) He > H_2 > Rn
B) H_2 > He > Rn
C) He > Rn > H_2
D) Ar > Rn > H_2
E) H_2 > Ar > Rn

Answer: D
Diff: 4　*Page Ref: 11.5*

37) The boiling point of water is ________.
A) 0 K　B) 32°C　C) 212 K　D) 100°C　E) 273°C

Answer: D
Diff: 1　*Page Ref: 11.5*

38) The melting point of water is ________.
A) 32°F　B) 32°C　C) 212°C　D) 100°C　E) 273°C

Answer: A
Diff: 1　*Page Ref: 11.6*

39) The freezing point of water is ________.
A) 32°F　B) 32°C　C) 212°C　D) 100°C　E) 273°C

Answer: A
Diff: 1　*Page Ref: 11.6*

40) The heat of vaporization of water at 100°C is 40.66 kJ/mol. Calculate the quantity of heat that is absorbed/released when 9.00 g of steam condenses to liquid water at 100 °C.
A) 20.3 kJ of heat are absorbed.　B) 20.3 kJ of heat are released.
C) 81.3 kJ of heat are absorbed.　D) 81.3 kJ of heat are released.

Answer: B
Diff: 4　*Page Ref: 11.7*

41) Calculate the total quantity of heat required to convert 25.0 g of liquid $CCl_4(l)$ from 35.0°C to gaseous CCl_4 at 76.8°C (the normal boiling point for CCl_4). The specific heat of $CCl_4(l)$ is 0.857 J/(g • °C), its heat of fusion is 3.27 kJ/mol, and its heat of vaporization is 29.82 kJ/mol.
A) 0.896 kJ　B) 1.43 kJ　C) 5.74 kJ　D) 6.28 kJ

Answer: C
Diff: 4　*Page Ref: 11.7*

42) The enthalpy change for converting 1.00 mol of ice at –50.0°C to water at 70.0°C is ________ kJ. The specific heats of ice, water, and steam are 2.09 J/gK, 4.18 J/gK, and 1.84 J/gK, respectively. For H_2O, ΔH_{fus} = 6.01 kJ/mol, and ΔH_{vap} = 40.67 kJ/mol.
A) 12.28　B) 6.41　C) 13.16　D) 7154　E) 9.40

Answer: C
Diff: 4　*Page Ref: 11.7*

43) The enthalpy change for converting 10.0 g of ice at –25.0°C to water at 80.0°C is ________ kJ. The specific heats of ice, water, and steam are 2.09 J/gK, 4.18 J/gK, and 1.84 J/gK, respectively. For H_2O, ΔH_{fus} = 6.01 kJ/mol, and ΔH_{vap} = 40.67 kJ/mol.

A) 12.28   B) 6.16   C) 3870   D) 7.21   E) 9.88

Answer: D
Diff: 4   *Page Ref: 11.7*

44) The fluorocarbon $C_2Cl_3F_3$ has a normal boiling point of 47.6°C. The specific heats of $C_2Cl_3F_3(l)$ and $C_2Cl_3F_3(g)$ are 0.91 J/gK and 0.67 J/gK, respectively. The heat of vaporization of the compound is 27.49 kJ/mol. The heat required to convert 50.0 g of the compound from the liquid at 5.0°C to the gas at 80.0°C is ________ kJ.

A) 8.19   B) 1454   C) 30.51   D) 3031   E) 10.36

Answer: E
Diff: 4   *Page Ref: 11.7*

45) Ethanol (C_2H_5OH) melts at –114°C. The enthalpy of fusion is 5.02 kJ/mol. The specific heats of solid and liquid ethanol are 0.97 J/gK and 2.3 J/gK, respectively. How much heat (kJ) is needed to convert 25.0 g of solid ethanol at –135°C to liquid ethanol at –50°C?

A) 207.3 kJ   B) –12.7 kJ   C) 6.91 kJ   D) 4192 kJ   E) 9.21 kJ

Answer: C
Diff: 4   *Page Ref: 11.7*

46) Ethyl chloride, C_2H_5Cl, is used as a local anesthetic. It works by cooling tissue as it vaporizes; its heat of vaporization is 26.4 kJ/mol. How much heat could be removed by 20.0 g of ethyl chloride?

A) 8.18 kJ   B) 341 kJ   C) 528 kJ   D) 3410 kJ

Answer: A
Diff: 4   *Page Ref: 11.7*

47) How much heat is released when 105 g of steam at 100.0°C is cooled to ice at –15.0°C? The enthalpy of vaporization of water is 40.67 kJ/mol, the enthalpy of fusion for water is 6.01 kJ/mol, the molar heat capacity of liquid water is 75.4 J/(mol • °C), and the molar heat capacity of ice is 36.4 J/(mol • °C).

A) 54.8 kJ   B) 273 kJ   C) 319 kJ   D) 347 kJ

Answer: C
Diff: 4   *Page Ref: 11.7*

48) Assign the appropriate labels to the phase diagram shown below.

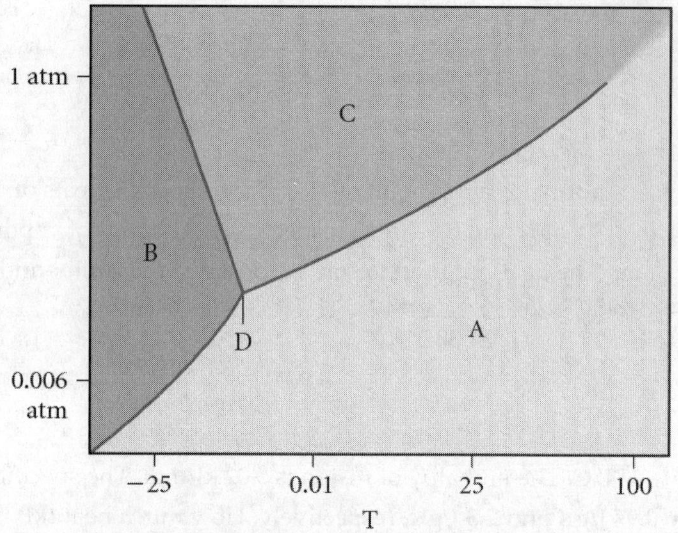

A) A = liquid, B = solid, C = gas, D = critical point
B) A = gas, B = solid, C = liquid, D = triple point
C) A = gas, B = liquid, C = solid, D = critical point
D) A = solid, B = gas, C = liquid, D = triple point
E) A = liquid, B = gas, C = solid, D = critical point

Answer: B
Diff: 1 *Page Ref: 11.8*

49) Give the coordination number for a body-centered cubic cell.
A) 0 B) 6 C) 8 D) 10 E) 12

Answer: C
Diff: 1 *Page Ref: 11.11*

50) What is the edge length of a face-centered cubic unit cell made up of atoms having a radius of 128 pm?
A) 181 pm B) 362 pm C) 512 pm D) 1020 pm

Answer: B
Diff: 5 *Page Ref: 11.11*

51) Nickel has a face-centered cubic structure and has a density of 8.90 g/cm^3. What is its atomic radius?
A) 125 pm B) 249 pm C) 353 pm D) 997 pm

Answer: A
Diff: 5 *Page Ref: 11.11*

52) Gold crystallizes in a face-centered cubic structure. What is the edge length of the unit cell if the atomic radius of gold is 144 pm?
A) 204 pm B) 288 pm C) 333 pm D) 407 pm

Answer: D
Diff: 5 *Page Ref: 11.11*

53) Cesium has a radius of 272 pm and crystallizes in a body-centered cubic structure. What is the edge length of the unit cell?

A) 314 pm B) 385 pm C) 544 pm D) 628 pm

Answer: D
Diff: 5 *Page Ref: 11.11*

54) Lithium crystallizes in a body-centered cubic structure. What is the coordination number of each atom?

A) 4 B) 6 C) 8 D) 12

Answer: C
Diff: 5 *Page Ref: 11.11*

55) NaCl crystallizes in a cubic unit cell with Cl^- ions on each corner and each face. How many Na + and Cl^- ions are in each unit cell of NaCl?

A) 1 Na^+ ion and 1 Cl^- ion
B) 2 Na^+ ions and 2 Cl^- ions
C) 4 Na^+ ions and 4 Cl^- ions
D) 8 Na^+ ions and 8 Cl^- ions

Answer: C
Diff: 5 *Page Ref: 11.11*

56) How many H^- ions are around each Na^+ ion in NaH, which has a cubic unit cell with H^- ions on each corner and each face?

A) 1 B) 4 C) 6 D) 8

Answer: C
Diff: 5 *Page Ref: 11.11*

57) Which of the following forms a molecular solid?

A) CaO B) $C_{10}H_{22}$ C) C, graphite D) gold

Answer: B
Diff: 1 *Page Ref: 11.12*

58) Which of the following forms an ionic solid?

A) Ag B) $C_7H_{15}NH_2$ C) Rb I D) SO_3

Answer: C
Diff: 1 *Page Ref: 11.12*

59) Which type of bonding does Sr form upon solidification?

A) covalent network
B) ionic
C) metallic
D) molecular

Answer: C
Diff: 2 *Page Ref: 11.12*

60) Which of the following is considered a molecular solid?
A) Au
B) NH_4NO_3
C) I_2
D) Rn
E) None of these is a molecular solid.

Answer: C
Diff: 1 *Page Ref: 11.12*

61) Which of the following is considered an ionic solid?
A) $(NH_4)_2CO_3$
B) CBr_4
C) $SeBr_2$
D) XeF_4
E) None of these is an ionic solid.

Answer: A
Diff: 1 *Page Ref: 11.12*

62) Which of the following is considered an atomic solid?
A) F_2
B) CsBr
C) N_2
D) Nb
E) None of these is an atomic solid.

Answer: D
Diff: 1 *Page Ref: 11.12*

63) Which of the following is considered a nonbonding atomic solid?
A) Ne B) Cu C) I_2 D) Ca E) K

Answer: A
Diff: 1 *Page Ref: 11.12*

64) Which of the following substances should have the highest melting point?
A) CO_2 B) SrS C) Kr D) F_2 E) MgO

Answer: E
Diff: 1 *Page Ref: 11.12*

65) Which of the following substances should have the highest melting point?
A) Fe B) Ne C) Xe D) N_2 E) CO

Answer: A
Diff: 1 *Page Ref: 11.12*

Matching Questions

Match the following.

1) LiI
Diff: 1 *Page Ref: 11.3*

2) CH_3OH
Diff: 1 *Page Ref: 11.3*

3) CH_3CH_3
Diff: 1 *Page Ref: 11.3*

4) CH_2F_2
Diff: 1 *Page Ref: 11.3*

5) LiI + H_2O
Diff: 1 *Page Ref: 11.3*

A) ionic bond

B) dipole–dipole forces

C) dispersion forces

D) ion–dipole forces

E) $H_2 + H_2O$

F) hydrogen bonding

Answers: 1) A 2) F 3) C 4) B 5) D

Short Answer Questions

1) Why do O, F and N, when bonded to H, form such strong intermolecular attractions to neighboring molecules? Make sure to be specific.

Answer: Oxygen, fluorine and nitrogen are all very small and highly electronegative elements. When they bond to hydrogen they form extremely polar bonds. These extremely polar bonds formed between small atoms can get very close to interact with other molecules. The closer the molecules get, the more strongly they can interact. In addition, these three elements usually have lone pair electrons which make the attractions to other molecules even stronger.
Diff: 1 *Page Ref: 11.3*

2) Define viscosity.

Answer: Viscosity is the resistance of a liquid to flow.
Diff: 2 *Page Ref: 11.4*

3) Why does the temperature of a substance stay constant during a phase change such as vaporization?

Answer: The energy added to the substance is not used to raise the kinetic energy of the substance. Instead it is "used up" in breaking the intermolecular attractions between particles to take them from a liquid (with intermolecular attractions) to a gas (with no intermolecular attractions).
Diff: 1 *Page Ref: 11.5*

4) Define volatile.

Answer: Liquids that vaporize easily.
Diff: 1 *Page Ref: 11.5*

5) Define boiling point of a liquid.

Answer: The temperature at which the vapor pressure equals the external pressure.
Diff: 1 *Page Ref: 11.5*

6) Define dynamic equilibrium.

Answer: Dynamic equilibrium is when the rate of condensation equals the rate of vaporization.
Diff: 2 *Page Ref: 11.5*

7) Why is the ΔH_{vap} higher than ΔH_{fus} for a given compound?

Answer: Vaporizing a substance requires the complete "breaking" of all intermolecular attractions, whereas the melting of a substance only requires the breaking of a portion of the intermolecular attractions.
Diff: 1 *Page Ref: 11.6*

8) Sketch the phase diagram of benzene. Make sure to label the axes and the different phases of benzene. Use the physical data provided below.

melting point = 279 K
boiling point = 353 K
T_c = 562 K
P_c = 48.4 atm
Triple Point = 0.05 atm, 279 K

Answer: Sketch should include all given information showing fusion and vaporization curves.
Diff: 3 *Page Ref: 11.8*

9) Give the edge length in terms of r for a simple cubic cell.

Answer: 2r
Diff: 1 *Page Ref: 11.9*

10) Describe the difference between the conduction band and the valence band.

Answer: The valence band is the group of highest energy occupied orbitals in the solid. The conduction band is the set of empty orbitals that are higher in energy than the valence band. In metals the difference in energy between the valence band and conduction band is fairly small, making it easy for electrons to move from the valence band to the conduction band.
Diff: 2 *Page Ref: 11.12*

Chapter 12 Solutions

Multiple Choice Questions

1) Choose the statement below that is TRUE.
   A) A solution will form between two substances if the solute-solvent interactions are of comparable strength to the solute-solute and solvent-solvent interactions.
   B) A solution will form between two substances if the solute-solvent interactions are small enough to be overcome by the solute-solute and solvent-solvent interactions.
   C) A solution will form between two substances if the solute-solute interactions are strong enough to overcome the solvent-solvent interactions.
   D) A solution will form between two substances only if the solvent-solvent interactions are weak enough to overcome the solute-solvent interactions.
   E) None of the above are true.

Answer: A
Diff: 1 *Page Ref: 12.2*

2) Give the major force between ethanol and water.
   A) dipole-dipole
   B) dispersion
   C) hydrogen bonding
   D) ion-ion
   E) ion-dipole

Answer: C
Diff: 1 *Page Ref: 12.2*

3) Give the major force in seawater.
   A) dipole-dipole
   B) dispersion
   C) hydrogen bonding
   D) ion-ion
   E) ion-dipole

Answer: E
Diff: 1 *Page Ref: 12.2*

4) Give the major force between acetone and chloroform.
   A) dipole-dipole
   B) dispersion
   C) hydrogen bonding
   D) ion-ion
   E) ion-dipole

Answer: A
Diff: 1 *Page Ref: 12.2*

5) Identify the polar solvent.
   A) acetone
   B) hexane
   C) diethyl ether
   D) toluene
   E) carbon tetrachloride

Answer: A
Diff: 1 *Page Ref: 12.2*

6) Which of the following compounds will be most soluble in ethanol (CH_3CH_2OH)?
   A) trimethylamine ($N(CH_3)_3$)
   B) acetone (CH_3COCH_3)
   C) ethylene glycol ($HOCH_2CH_2OH$)
   D) hexane ($CH_3CH_2CH_2CH_2CH_2CH_3$)
   E) None of these compounds should be soluble in ethanol.

Answer: C
Diff: 1 *Page Ref: 12.2*

7) Which of the following compounds is most soluble in hexane ($CH_3CH_2CH_2CH_2CH_2CH_2CH_3$)?
   A) methanol
   B) ethanol
   C) 1-propanol
   D) 1-butanol
   E) 1-pentanol

Answer: E
Diff: 1 *Page Ref: 12.2*

8) Choose the situation below that would result in a $\Delta H_{solution}$ near 0.
   A) When $|\Delta H_{solvent}| >> |\Delta H_{solute}|$
   B) When $|\Delta H_{solute}| > |\Delta H_{hydration}|$
   C) When $|\Delta H_{solute}| < |\Delta H_{hydration}|$
   D) When $|\Delta H_{solute}|$ is close to $|\Delta H_{hydration}|$
   E) There isn't enough information to determine.

Answer: D
Diff: 1 *Page Ref: 12.3*

9) Choose the situation below that would result in an exothermic $\Delta H_{solution}$.
   A) When $|\Delta H_{solute}| > |\Delta H_{hydration}|$
   B) When $|\Delta H_{solute}|$ is close to $|\Delta H_{hydration}|$
   C) When $|\Delta H_{solute}| < |\Delta H_{hydration}|$
   D) When $|\Delta H_{solvent}| >> |\Delta H_{solute}|$
   E) There isn't enough information to determine.

Answer: C
Diff: 1 *Page Ref: 12.3*

10) Choose the situation below that would result in an endothermic $\Delta H_{solution}$.

A) When $|\Delta H_{solute}| > |\Delta H_{hydration}|$
B) When $|\Delta H_{solute}| < |\Delta H_{hydration}|$
C) When $|\Delta H_{solute}|$ is close to $|\Delta H_{hydration}|$
D) When $|\Delta H_{solvent}| >> |\Delta H_{solute}|$
E) There isn't enough information to determine.

Answer: A
Diff: 1 *Page Ref: 12.3*

11) Determine ΔH_{solute} for KBr if the $\Delta H_{solution}$ (KBr) = +19.9 kJ/mol and the $\Delta H_{hydration}$(KBr)=-670. kJ/mol.

A) +650 kJ/mol
B) −650 kJ/mol
C) +690 kJ/mol
D) −710 kJ/mol
E) −690 kJ/mol

Answer: C
Diff: 2 *Page Ref: 12.3*

12) A solution containing less than the equilibrium amount is called ________.

A) an unsaturated solution
B) a dilute solution
C) a supersaturated solution
D) a concentrated solution
E) a saturated solution

Answer: A
Diff: 2 *Page Ref: 12.4*

13) A solution containing more than the equilibrium amount is called ________.

A) an unsaturated solution
B) a dilute solution
C) a supersaturated solution
D) a concentrated solution
E) a saturated solution

Answer: C
Diff: 2 *Page Ref: 12.4*

14) A solution is formed at room temperature by vigorously dissolving enough of the solid solute so that some solid remains at the bottom of the solution. Which statement below is TRUE?

A) The solution is considered unsaturated.
B) The solution is considered supersaturated.
C) The solution is considered saturated.
D) The solution would be considered unsaturated if it were cooled a bit to increase the solubility of the solid.
E) None of the above are true.

Answer: C
Diff: 1 *Page Ref: 12.4*

15) Which of the following statements is generally TRUE?
A) The solubility of a solid is not dependent on either temperature or pressure.
B) The solubility of a solid is highly dependent on pressure.
C) The solubility of a solid is highly dependent on both pressure and temperature.
D) The solubility of a solid is highly dependent on temperature.
E) None of the above.

Answer: D
Diff: 1 *Page Ref: 12.4*

16) A solid can be purified through what technique?
A) recrystallization
B) dilution
C) dissolution
D) boiling
E) melting

Answer: A
Diff: 1 *Page Ref: 12.4*

17) Which of the following statements is TRUE?
A) In general, the solubility of a solid in water decreases with increasing temperature.
B) In general, the solubility of a gas in water decreases with increasing temperature.
C) The solubility of a gas in water usually increases with decreasing pressure.
D) The solubility of an ionic solid in water decreases with increasing temperature.
E) None of the above statements are true.

Answer: B
Diff: 1 *Page Ref: 12.4*

18) Determine the Henry's law constant for ammonia in water at 25°C if an ammonia pressure of 0.022 atm produces a solution with a concentration of 1.3 M.
A) 59 M/atm
B) 0.017 M/atm
C) 0.029 M/atm
D) 35 M/atm
E) 0.038 M/atm

Answer: A
Diff: 2 *Page Ref: 12.4*

19) Calculate the mass of oxygen (in mg) dissolved in a 5.00 L bucket of water exposed to a pressure of 1.13 atm of air. Assume the mole fraction of oxygen in air to be 0.21 and the Henry's law constant for oxygen in water at this temperature to be 1.3×10^{-3} M/atm.
A) 49.4 mg B) 23.5 mg C) 9.87 mg D) 27.3 mg E) 13.7 mg

Answer: A
Diff: 3 *Page Ref: 12.4*

20) Give the term for the amount of solute in moles per liter of solution.
A) molality
B) molarity
C) mole fraction
D) mole percent
E) mass percent

Answer: B
Diff: 1 *Page Ref: 12.5*

21) Give the term for the amount of solute in moles per kilogram of solvent.
A) molality
B) molarity
C) mole fraction
D) mole percent
E) mass percent

Answer: A
Diff: 1 *Page Ref: 12.5*

22) Which of the following concentration units are temperature dependent?
A) mole fraction
B) molality
C) mass percent
D) molarity
E) none of the above.

Answer: D
Diff: 1 *Page Ref: 12.5*

23) What mass (in g) of NH_3 must be dissolved in 475 g of methanol to make a 0.250 *m* solution?
A) 2.02 g B) 4.94 g C) 1.19 g D) 8.42 g E) 1.90 g

Answer: A
Diff: 2 *Page Ref: 12.5*

24) How many moles of KF are contained in 347 g of water in a 0.175 *m* KF solution?
A) 1.65×10^{-2} mol KF
B) 5.04×10^{-2} mol KF
C) 6.07×10^{-2} mol KF
D) 3.22×10^{-2} mol KF
E) 1.98×10^{-2} mol KF

Answer: C
Diff: 2 *Page Ref: 12.5*

25) How many moles of KF are contained in 244 mL of 0.135 *m* KF solution? The density of the solution is 1.22 g/mL.

A) 4.31×10^{-2} mol KF
B) 4.02×10^{-2} mol KF
C) 3.29×10^{-2} mol KF
D) 2.32×10^{-2} mol KF
E) 1.67×10^{-2} mol KF

Answer: B
Diff: 3 *Page Ref: 12.5*

26) Calculate the molality of a solution that is prepared by mixing 25.5 mL of CH_3OH (d= 0.792 g/mL) and 387 mL of $CH_3CH_2CH_2OH$ (d= 0.811 g/mL).

A) 0.630 *m* B) 0.812 *m* C) 1.57 *m* D) 2.01 *m* E) 4.98 *m*

Answer: D
Diff: 3 *Page Ref: 12.5*

27) Determine the molality of a solution prepared by dissolving 0.500 moles of CaF_2 in 11.5 moles H_2O.

A) 1.88 *m* B) 4.35 *m* C) 5.31 *m* D) 4.14 *m* E) 2.41 *m*

Answer: E
Diff: 3 *Page Ref: 12.5*

28) Determine the molality of an aqueous solution prepared by dissolving 0.112 moles of LiCl in 13.7 moles of H_2O.

A) 0.454 *m* B) 0.220 *m* C) 0.818 *m* D) 0.122 *m* E) 0.153 *m*

Answer: A
Diff: 4 *Page Ref: 12.5*

29) Commercial grade HCl solutions are typically 39.0% (by mass) HCl in water. Determine the molarity of the HCl, if the solution has a density of 1.20 g/mL.

A) 7.79 M B) 10.7 M C) 12.8 M D) 9.35 M E) 13.9 M

Answer: C
Diff: 4 *Page Ref: 12.5*

30) A solution is prepared by dissolving 98.6 g of NaCl in enough water to form 875 mL of solution. Calculate the mass % of the solution if the density of the solution is 1.06 g/mL.

A) 11.3% B) 12.7% C) 9.4% D) 10.6% E) 11.9%

Answer: D
Diff: 3 *Page Ref: 12.5*

31) A solution is prepared by dissolving 49.3 g of KBr in enough water to form 473 mL of solution. Calculate the mass % of KBr in the solution if the density is 1.12 g/mL.

A) 10.4% B) 8.57% C) 10.1% D) 11.7% E) 9.31%

Answer: E
Diff: 3 *Page Ref: 12.5*

32) What mass of NaCl is contained in 24.88 g of a 15.00% by mass solution of NaCl in water?

A) 20.00 g B) 21.15 g C) 3.73 g D) 1.50 g E) 12.44 g

Answer: C
Diff: 3 *Page Ref: 12.5*

33) A solution is prepared by dissolving 38.6 g sucrose ($C_{12}H_{22}O_{11}$) in 495 g of water. Determine the mole fraction of sucrose if the final volume of the solution is 508 mL.

A) 4.09×10^{-3}
B) 7.80×10^{-2}
C) 1.28×10^{-3}
D) 7.23×10^{-2}
E) 2.45×10^{-3}

Answer: A
Diff: 3 *Page Ref: 12.5*

34) A solution is prepared by dissolving 76.3 g NaI in 545 g of water. Determine the mole fraction of NaI if the final volume of the solution is 576 mL.

A) 6.04×10^{-3}
B) 1.65×10^{-2}
C) 1.40×10^{-3}
D) 1.32×10^{-2}
E) 8.84×10^{-2}

Answer: B
Diff: 3 *Page Ref: 12.5*

35) A 4.55 L sample of water contains 0.115 g of sodium ions. Determine the concentration of sodium ions in ppm if the density of the solution is 1.00 g/mL.

A) 52.3 ppm B) 13.2 ppm C) 12.7 ppm D) 25.3 ppm E) 36.5 ppm

Answer: D
Diff: 3 *Page Ref: 12.5*

36) What mass of ethane (CH_3CH_3) is contained in 50.0 mL of a 13.9% by mass solution of ethane in hexane? The density of the solution is 0.611 g/mL.

A) 2.35 g B) 8.49 g C) 11.8 g D) 6.95 g E) 4.25 g

Answer: E
Diff: 3 *Page Ref: 12.5*

37) Commercial grade HCl solutions are typically 39.0% (by mass) HCl in water. Determine the molality of the HCl, if the solution has a density of 1.20 g/mL.

A) 39.0 *m* B) 17.5 *m* C) 6.39 *m* D) 10.7 *m* E) 9.44 *m*

Answer: B
Diff: 4 *Page Ref: 12.5*

38) Identify the colligative property.
A) vapor pressure lowering
B) freezing point depression
C) boiling point elevation
D) osmotic pressure
E) all of the above

Answer: E
Diff: 1 *Page Ref: 12.6*

39) ________ is applied to roads in wintertime so that the ice will melt at a lower temperature.
A) Antifreeze
B) Water
C) Salt
D) Sugar
E) Oil

Answer: C
Diff: 4 *Page Ref: 12.6*

40) Determine the vapor pressure of a solution at 25°C that contains 76.6 g of glucose ($C_6H_{12}O_6$) in 250.0 mL of water. The vapor pressure of pure water at 25°C is 23.8 torr.
A) 70.8 torr B) 72.9 torr C) 23.1 torr D) 22.9 torr E) 7.29 torr

Answer: C
Diff: 3 *Page Ref: 12.6*

41) Determine the freezing point depression of a solution that contains 30.7 g glycerin ($C_3H_8O_3$, molar mass = 92.09 g/mol) in 376 mL of water. Some possibly useful constants for water are $K_f = 1.86°C/m$ and $K_b = 0.512°C/m$.
A) 0.887°C B) 1.65°C C) 3.33°C D) 3.33°C E) 0.654°C

Answer: B
Diff: 3 *Page Ref: 12.6*

42) Determine the freezing point of a solution that contains 78.8 g of naphthalene ($C_{10}H_8$, molar mass = 128.16 g/mol) dissolved in 722 mL of benzene (d = 0.877 g/mL). Pure benzene has a melting point of 5.50°C and a freezing point depression constant of 4.90°C/*m*.
A) 4.76°C B) 4.17°C C) 0.74°C D) 1.33°C E) 1.68°C

Answer: C
Diff: 3 *Page Ref: 12.6*

43) Calculate the freezing point of a solution of 500.0 g of ethylene glycol ($C_2H_6O_2$) dissolved in 500.0 g of water. $K_f = 1.86°C/m$ and $K_b = 0.512°C/m$.
A) 30.0°C B) -30.0°C C) 8.32°C D) -8.32°C E) 70.2°C

Answer: B
Diff: 3 *Page Ref: 12.6*

44) Calculate the boiling point of a solution of 500.0 g of ethylene glycol ($C_2H_6O_2$) dissolved in 500.0 g of water. $K_f = 1.86°C/m$ and $K_b = 0.512°C/m$. Use 100°C as the boiling point of water.
A) 108°C B) 92°C C) 130°C D) 70°C E) 8.3°C

Answer: A
Diff: 3 *Page Ref: 12.6*

45) The boiling point elevation of an aqueous sucrose solution is found to be 0.39°C. What mass of sucrose (molar mass= 342.30 g/mol) would be needed to dissolve in 500.0 g of water? K_b (water) = 0.512°C/*m*.

A) 261 g sucrose
B) 528 g sucrose
C) 762 g sucrose
D) 223 g sucrose
E) 130. g sucrose

Answer: E
Diff: 3 *Page Ref: 12.6*

46) Determine the boiling point of a solution that contains 78.8 g of naphthalene ($C_{10}H_8$, molar mass = 128.16 g/mol) dissolved in 722 mL of benzene (d = 0.877 g/mL). Pure benzene has a boiling point of 80.1°C and a boiling point elevation constant of 2.53°C/*m*.

A) 2.2°C B) 2.5°C C) 82.3°C D) 80.4°C E) 82.6°C

Answer: E
Diff: 3 *Page Ref: 12.6*

47) Give the reason that antifreeze is added to a car radiator.

A) The freezing point is lowered and the boiling point is elevated.
B) The freezing point is elevated and the boiling point is lowered.
C) The freezing point and the boiling point are elevated.
D) The freezing point and the boiling point are lowered.
E) None of the above.

Answer: A
Diff: 1 *Page Ref: 12.6*

48) A compound is found to have a molar mass of 598 g/mol. If 35.8 mg of the compound is dissolved in enough water to make 175 mL of solution at 25°C, what is the osmotic pressure of the resulting solution?

A) 3.42 torr B) 6.36 torr C) 5.01 torr D) 5.99 torr E) 8.36 torr

Answer: B
Diff: 3 *Page Ref: 12.6*

49) A 150.0 mL sample of an aqueous solution at 25°C contains 15.2 mg of an unknown nonelectrolyte compound. If the solution has an osmotic pressure of 8.44 torr, what is the molar mass of the unknown compound?

A) 223 g/mol
B) 294 g/mol
C) 341 g/mol
D) 448 g/mol
E) 195 g/mol

Answer: A
Diff: 4 *Page Ref: 12.6*

50) Determine the vapor pressure of a solution at 55°C that contains 34.2 g NaCl in 375 mL of water. The vapor pressure of pure water at 55°C is 118.1 torr. The van't Hoff factor for NaCl is 1.9

A) 115 torr B) 87.1 torr C) 108 torr D) 112 torr E) 92.8 torr

Answer: D
Diff: 3 *Page Ref: 12.7*

51) The boiling point of an aqueous 1.83 *m* $(NH_4)_2SO_4$ (molar mass = 132.15 g/mol) solution is 102.5°C. Determine the value of the van't Hoff factor for this solute if the K_b for water is 0.512°C/*m*.

A) 3.0 B) 3.6 C) 1.8 D) 2.7 E) 2.3

Answer: D
Diff: 3 *Page Ref: 12.7*

52) Fluids used for an intravenous transfusion must be ________ with bodily fluids.

A) isosmotic
B) hyperosmotic
C) hyposmotic
D) neosmotic
E) magnosmotic

Answer: A
Diff: 1 *Page Ref: 12.7*

53) Solutions having osmotic pressures less than those of body fluids are called ________.

A) isosmotic
B) hyperosmotic
C) hyposmotic
D) neosmotic
E) magnosmotic

Answer: C
Diff: 1 *Page Ref: 12.7*

54) Identify the classification of milk.

A) aerosol
B) solid aerosol
C) foam
D) emulsion
E) solid emulsion

Answer: D
Diff: 1 *Page Ref: 12.8*

55) Identify the classification of opal.

A) aerosol
B) solid aerosol
C) foam
D) emulsion
E) solid emulsion

Answer: E
Diff: 1 *Page Ref: 12.8*

56) Identify the classification of whipped cream.
A) aerosol
B) solid aerosol
C) foam
D) emulsion
E) solid emulsion

Answer: C
Diff: 1 *Page Ref: 12.8*

57) Soap has an ionic and a polar end. It works well to remove oil by
A) surrounding the oil with the nonpolar end, and the water interacts with the polar end.
B) surrounding the oil with the polar end, and the water interacts with the nonpolar end.
C) surrounding the oil and water with the polar end.
D) surrounding the oil and water with the nonpolar end.

Answer: A
Diff: 1 *Page Ref: 12.8*

Algorithmic Questions

1) Which of the following compounds will be most soluble in nonane (C_9H_{20})?
A) pentanol ($CH_3CH_2CH_2CH_2CH_2OH$)
B) benzene (C_6H_6)
C) acetic acid (CH_3CO_2H)
D) ethyl methyl ketone ($CH_3CH_2COH_3$)
E) None of these compounds should be soluble in pentane.

Answer: B
Diff: 1 *Page Ref: 12.2*

2) Identify the compounds that are soluble in both water and hexane.
A) methanol and 1-pentanol
B) ethanol and 1-butanol
C) 1-propanol and 1-butanol
D) 1-butanol and 1-pentanol
E) ethanol and 1-propanol

Answer: E
Diff: 1 *Page Ref: 12.2*

3) Which of the following ions should have the most exothermic $\Delta H_{hydration}$?

A) Na^+ B) Ba^{2+} C) Al^{3+} D) Ca^{2+} E) Sr^{2+}

Answer: C
Diff: 1 *Page Ref: 12.3*

4) Which cation in each set is expected to have the larger (more negative) hydration energy?
I Be^{2+} or Ca^{2+}
II Rb^{+} or Zn^{2+}
A) Be^{2+} in set I and Rb^{+} in set II
B) Be^{2+} in set I and Zn^{2+} in set II
C) Ca^{2+} in set I and Rb^{+} in set II
D) Ca^{2+} in set I and Zn^{2+} in set II

Answer: B
Diff: 2 *Page Ref:* 12.3

5) The Henry's Law constant of methyl bromide, CH_3Br, is $k = 0.159$ mol/(L • atm) at 25°C. What is the solubility of methyl bromide in water at 25°C and at a partial pressure of 270. mm Hg?
A) 0.0565 mol/L
B) 0.355 mol/L
C) 0.448 mol/L
D) 42.9 mol/L

Answer: A
Diff: 3 *Page Ref:* 12.4

6) The Henry's law constant for helium gas in water at 30 °C is 3.70×10^{-4} M/atm. When the partial pressure of helium above a sample of water is 0.650 atm, the concentration of helium in the water is ________ M.
A) 5.69×10^{-4}
B) 1.76×10^{3}
C) 1.30
D) 2.41×10^{-4}
E) 3.70×10^{-4}

Answer: D
Diff: 3 *Page Ref:* 12.4

7) Dynamic equilibrium can be defined as
A) rate of dissolution = rate of deposition
B) rate of dissolution < rate of deposition
C) rate of dissolution > rate of deposition
D) rate of bubbling > rate of dissolving
E) rate of condensing > rate of bubbling

Answer: A
Diff: 2 *Page Ref:* 12.4

8) Dissolving can be defined as
A) rate of dissolution = rate of deposition
B) rate of dissolution < rate of deposition
C) rate of dissolution > rate of deposition
D) rate of bubbling > rate of dissolving
E) rate of condensing > rate of bubbling

Answer: C
Diff: 2 *Page Ref:* 12.4

9) Identify the compound whose solubility is least affected by temperature.

A) K Cl
B) $K_2Cr_2O_7$
C) $NaNO_3$
D) $Pb(NO_3)_3$
E) NaCl

Answer: E
Diff: 1 *Page Ref: 12.4*

10) Identify the compound whose solubility decreases with increasing temperature.

A) Na_2SO_4 B) $K_2Cr_2O_7$ C) $NaNO_3$ D) $KClO_3$ E) $CaCl_2$

Answer: A
Diff: 1 *Page Ref: 12.4*

11) Identify the gas that is dissolved in carbonated sodas.

A) carbon dioxide
B) hydrogen
C) neon
D) carbon monoxide
E) argon

Answer: A
Diff: 1 *Page Ref: 12.4*

12) Which of the following should have the largest Henry's law constant (k_H) in water?

A) Ne B) CO C) Br_2 D) CH_3CH_3 E) CO_2

Answer: B
Diff: 1 *Page Ref: 12.4*

13) Determine the solubility of N_2 in water exposed to air at 25°C if the atmospheric pressure is 2.0 atm. Assume that the mole fraction of nitrogen is 0.78 in air and the Henry's law constant for nitrogen in water at this temperature is 6.1×10^{-4} M/atm.

A) 1.6×10^{-4} M
B) 6.4×10^{-4} M
C) 9.5×10^{-4} M
D) 1.1×10^{-4} M
E) 1.2×10^{-4} M

Answer: C
Diff: 2 *Page Ref: 12.4*

14) Determine the solubility of CO_2 in soda water at 25°C if the pressure of CO_2 is 5.6 atm. The Henry's law constant for carbon dioxide in water at this temperature is 3.4×10^{-2} M/atm.

A) 0.16 M B) 0.53 M C) 0.61 M D) 0.19 M E) 0.78 M

Answer: D
Diff: 2 *Page Ref: 12.4*

15) Determine the partial pressure of oxygen necessary to form an aqueous solution that is 6.5×10^{-4} M O_2 at 25°C. The Henry's law constant for oxygen in water at 25°C is 1.3×10^{-3} M/atm.

A) 1.2 atm B) 0.85 atm C) 0.78 atm D) 2.0 atm E) 0.50 atm

Answer: E
Diff: 2 *Page Ref: 12.4*

16) Calculate the molarity of a solution that contains 0.250 moles of KF in 0.500 L of water.

A) 0.500 M B) 2.00 M C) 0.125 M D) 8.00 M E) 0.750

Answer: A
Diff: 1 *Page Ref: 12.5*

17) Parts per billion requires a multiplication factor of ________.

A) 10^{-9} B) 10^{-3} C) 10^{3} D) 10^{6} E) 10^{9}

Answer: E
Diff: 1 *Page Ref: 12.5*

18) Parts per million requires a multiplication factor of ________.

A) 10^{-9} B) 10^{-3} C) 10^{3} D) 10^{15} E) 10^{6}

Answer: E
Diff: 1 *Page Ref: 12.5*

19) Calculate the molality of a solution formed by dissolving 34.8 g of LiI in 500.0 mL of water.

A) 0.254 *m* B) 0.130 *m* C) 0.696 *m* D) 0.260 *m* E) 0.520 *m*

Answer: E
Diff: 2 *Page Ref: 12.5*

20) What mass of $CuCl_2$ is contained in 90.85 g of a 22.4% by mass solution of $CuCl_2$ in water?

A) 20.4 g B) 4.91 g C) 24.6 g D) 40.6 g E) 11.2 g

Answer: A
Diff: 3 *Page Ref: 12.5*

21) Calculate the mole fraction of the total ions in an aqueous solution prepared by dissolving 0.400 moles of $MgBr_2$ in 850.0 g of water.

A) 0.00841 B) 0.0270 C) 0.00900 D) 0.0252 E) 0.0167

Answer: D
Diff: 3 *Page Ref: 12.5*

22) Calculate the mole fraction of CaI_2 in an aqueous solution prepared by dissolving 0.400 moles of CaI_2 in 850.0 g of water.

A) 0.00841 B) 0.0270 C) 0.00900 D) 0.0252 E) 0.0167

Answer: A
Diff: 3 *Page Ref: 12.5*

23) A 1.00 L sample of water contains 0.0044 g of Cl^- ions. Determine the concentration of chloride ions in ppm if the density of the solution is 1.00 g/mL.

A) 3.3 ppm B) 8.8 ppm C) 4.4 ppm D) 2.2 ppm E) 6.6 ppm

Answer: C
Diff: 3 *Page Ref: 12.5*

24) A solution is 0.0480 *m* LiF. What is the molarity of the solution if the density is 1.10 g/mL?

A) 0.0441 M B) 0.0480 M C) 0.0436 M D) 0.0528 M E) 0.0417 M

Answer: D
Diff: 4 *Page Ref: 12.5*

25) What volume of a 0.716 M KBr solution is needed to provide 30.5 g of KBr?

A) 21.8 mL B) 42.7 mL C) 184 mL D) 357 mL

Answer: D
Diff: 2 *Page Ref: 12.5*

26) What volume of 3.00 M CH_3OH solution is needed to provide 0.220 mol of CH_3OH?

A) 1.52 mL B) 13.6 mL C) 73.3 mL D) 660 mL

Answer: C
Diff: 2 *Page Ref: 12.5*

27) To make a 0.500 M solution, one could take 0.500 moles of solute and add

A) 1.00 L of solvent.
B) 1.00 kg of solvent.
C) enough solvent to make 1.00 L of solution.
D) enough solvent to make 1.00 kg of solution.

Answer: C
Diff: 2 *Page Ref: 12.5*

28) How many grams of KBr are required to make 350. mL of a 0.115 M KBr solution?

A) 0.338 g B) 3.04 g C) 4.79 g D) 40.3 g

Answer: C
Diff: 3 *Page Ref: 12.5*

29) What is the molality of a glucose solution prepared by dissolving 18.0 g of glucose, $C_6H_{12}O_6$, in 125.9 g of water?

A) 7.94×10^{-4} *m* B) 0.143 *m* C) 0.695 *m* D) 0.794 *m*

Answer: D
Diff: 3 *Page Ref: 12.5*

30) To make a 2.00 *m* solution, one could take 2.00 moles of solute and add

A) 1.00 L of solvent.
B) 1.00 kg of solvent.
C) enough solvent to make 1.00 L of solution.
D) enough solvent to make 1.00 kg of solution.

Answer: B
Diff: 3 *Page Ref: 12.5*

31) What molality of pentane is obtained by dissolving 15 g pentane, C_5H_{12}, in 245.0 g hexane, C_6H_{14}?

A) 0.058 *m* B) 0.068 *m* C) 0.85 *m* D) 61 *m*

Answer: C
Diff: 3 *Page Ref: 12.5*

32) A solution of LiCl in water has X_{LiCl} = 0.0900. What is the molality?

A) 4.46 *m* LiCl B) 5.00 *m* LiCl C) 5.49 *m* LiCl D) 9.89 *m* LiCl

Answer: C
Diff: 3 Page Ref: 12.5

33) A solution is prepared by adding 1.43 mol of KCl to 889 g of water. The concentration of KCl is ________ molal.

A) 1.61×10^{-3}
B) 622
C) 0.622
D) 1.27×10^{3}
E) 1.61

Answer: E
Diff: 3 Page Ref: 12.5

34) A solution is prepared by dissolving 16.2 g of benzene (C_6H_6) in 282 g of carbon tetrachloride (CCl_4). The concentration of benzene in this solution is ________ molal. The molar masses of C_6H_6 and CCl_4 are 78.1 g/mol and 154 g/mol, respectively.

A) 7.36×10^{-4}
B) 0.736
C) 0.102
D) 0.0543
E) 5.43

Answer: B
Diff: 3 Page Ref: 12.5

35) What is the weight percent of vitamin C in a solution made by dissolving 6.50 g of vitamin C, $C_6H_8O_6$, in 55.0 g of water?

A) 0.547% B) 1.21% C) 10.6% D) 11.8%

Answer: C
Diff: 3 Page Ref: 12.5

36) What is the weight percent of a caffeine solution made by dissolving 8.35 g of caffeine, $C_8H_{10}N_4O_2$, in 75 g of benzene, C_6H_6?

A) 0.010% B) 0.011% C) 10% D) 11%

Answer: C
Diff: 3 Page Ref: 12.5

37) How much water must be added to 40.0 g of $CaCl_2$ to produce a solution that is 35.0 wt% $CaCl_2$?

A) 54.0 g B) 74.2 g C) 87.5 g D) 114 g

Answer: B
Diff: 3 Page Ref: 12.5

38) A solution of LiCl in water is 20.0 wt% LiCl. What is the mole fraction of LiCl?

A) 0.0960 B) 0.106 C) 0.472 D) 4.44

Answer: A
Diff: 3 Page Ref: 12.5

39) What is the mole fraction of ethanol in a solution made by dissolving 29.2 g of ethanol, C_2H_5OH, in 53.6 g of water?

A) 0.176 B) 0.213 C) 0.352 D) 0.545

Answer: A
Diff: 3 *Page Ref: 12.5*

40) What is the mole fraction of I_2 in a solution made by dissolving 55.6 g of I_2 in 245 g of hexane, C_6H_{14}?

A) 0.0715 B) 0.0770 C) 0.133 D) 0.154

Answer: A
Diff: 3 *Page Ref: 12.5*

41) What is the mole fraction of oxygen in a gas mixture that is 37% oxygen and 63% nitrogen by volume?

A) 0.34 B) 0.37 C) 0.25 D) 0.52

Answer: B
Diff: 3 *Page Ref: 12.5*

42) A solution is 2.25% by weight $NaHCO_3$. How many grams of $NaHCO_3$ are in 150.0 g of solution?

A) 1.50 g B) 3.38 g C) 66.7 g D) 225 g

Answer: B
Diff: 3 *Page Ref: 12.5*

43) At 20°C, an aqueous solution that is 24.0% by mass in ammonium chloride has a density of 1.0674 g/mL. What is the molarity of ammonium chloride in the solution? The formula weight of NH_4Cl is 53.50 g/mol.

A) 5.90 B) 0.479 C) 4.79 D) 0.0445 E) 22.5

Answer: C
Diff: 4 *Page Ref: 12.5*

44) At 20°C, a 2.32 M aqueous solution of ammonium chloride has a density of 1.0344 g/mL. What is the molality of ammonium chloride in the solution? The formula weight of NH_4Cl is 53.50 g/mol.

A) 2.55 B) 0.0449 C) 2.32 D) 0.446 E) 12.00

Answer: A
Diff: 4 *Page Ref: 12.5*

45) At 20°C, a 0.376 M aqueous solution of ammonium chloride has a density of 1.0045 g/mL. What is the mass % of ammonium chloride in the solution? The formula weight of NH_4Cl is 53.50 g/mol.

A) 0.381 B) 0.705 C) 0.374 D) 2.68 E) 2.00

Answer: E
Diff: 4 *Page Ref: 12.5*

46) An aqueous solution is 0.467 M in HCl. What is the molality of the solution if the density is 1.23 g/mL?

A) 0.115 *m* B) 0.268 *m* C) 0.374 *m* D) 0.574 *m* E) 0.318 *m*

Answer: C
Diff: 4 *Page Ref: 12.5*

47) Choose the solvent below that would show the greatest boiling point elevation when used to make a 0.10 *m* nonelectrolyte solution.

A) carbon tetrachloride, K_b = 5.03°C/m
B) benzene, K_b = 2.53°C/m
C) diethyl ether, K_b = 2.02°C/m
D) water, K_b = 0.512°C/m
E) acetic acid, K_b = 3.07°C/m

Answer: A
Diff: 1 *Page Ref: 12.6*

48) A solution is prepared by dissolving 40.0 g of sucrose, $C_{12}H_{22}O_{11}$, in 250. g of water at 25°C. What is the vapor pressure of the solution if the vapor pressure of water at 25 °C is 23.76 mm Hg?

A) 0.198 mm Hg B) 20.5 mm Hg C) 23.6 mm Hg D) 24.0 mm Hg

Answer: C
Diff: 3 *Page Ref: 12.6*

49) At a given temperature the vapor pressures of benzene and toluene are 183 mm Hg and 59.2 mm Hg, respectively. Calculate the total vapor pressure over a solution of benzene and toluene with $X_{benzene}$ = 0.580.

A) 106 mm Hg B) 121 mm Hg C) 131 mm Hg D) 242 mm Hg

Answer: C
Diff: 3 *Page Ref: 12.6*

50) Calculate the freezing point of a solution of 40.0 g methyl salicylate, $C_7H_6O_2$, dissolved in 800. g of benzene, C_6H_6. K_f for benzene is 5.10°C/*m* and the freezing point is 5.50°C for benzene.

A) -2.09°C B) 2.09°C C) 3.41°C D) 7.59°C

Answer: C
Diff: 3 *Page Ref: 12.6*

51) An aqueous solution has a normal boiling point of 103.0°C. What is the freezing point of this solution? For water K_b = 0.51°C/*m* and K_f = 1.86°C/*m*.

A) -0.82°C B) -3.0°C C) -3.6°C D) -11°C

Answer: D
Diff: 3 *Page Ref: 12.6*

52) A solution is prepared by adding 30.00 g of lactose (milk sugar) to 110.0 g of water at 55°C. The partial pressure of water above the solution is ________ torr. The vapor pressure of pure water at 55°C is 118.0 torr. The MW of lactose is 342.3 g/mol.

A) 1.670 B) 94.1 C) 169.4 D) 116.3 E) 92.7

Answer: D
Diff: 4 *Page Ref: 12.6*

53) A solution is prepared by dissolving 7.00 g of glycerin ($C_3H_8O_3$) in 201 g of ethanol (C_2H_5OH). The freezing point of the solution is ________ °C. The freezing point of pure ethanol is −114.6°C at 1 atm. The molal-freezing-point-depression constant (K_f) for ethanol is 1.99°C/*m*. The molar masses of glycerin and of ethanol are 92.1 g/mol and 46.1 g/mol, respectively.

A) −121.3 B) 0.752 C) −107.9 D) −113.8 E) −115.4

Answer: E
Diff: 4 *Page Ref: 12.6*

54) Calculate the freezing point of a solution containing 5.0 grams of KCl and 550.0 grams of water. The molal-freezing-point-depression constant (K_f) for water is 1.86°C/*m*.

A) −0.45 °C B) +0.45 °C C) −0.23 °C D) +0.23 °C E) 1.23 °C

Answer: A
Diff: 4 *Page Ref: 12.6*

55) The osmotic pressure of a solution formed by dissolving 25.0 mg of aspirin ($C_9H_8O_4$) in 0.250 L of water at 25°C is ________ atm.

A) 13.6
B) 1.14×10^{-3}
C) 0.0136
D) 2.45
E) 1.38

Answer: C
Diff: 4 *Page Ref: 12.6*

56) Place the following aqueous solutions of nonvolatile, nonionic compounds in order of **decreasing** osmotic pressure.

I. 0.011 M sucrose II. 0.00095 M galactose III. 0.0060 M glycerin

A) I > III > II
B) I > II > III
C) II > III > I
D) III > I > II
E) II > I > III

Answer: A
Diff: 1 *Page Ref: 12.6*

57) Place the following solutions in order of **increasing** osmotic pressure.

I. 0.15 M $C_2H_6O_2$ II. 0.15 M $BaCl_2$ III. 0.15 M NaI

A) III < I < II
B) II < III < I
C) I < II < III
D) II < I < III
E) I < III < II

Answer: E
Diff: 1 *Page Ref: 12.6*

58) Identify the solute with the highest van't Hoff factor.
A) nonelectrolyte
B) KI
C) $MgSO_4$
D) $CaCl_2$
E) $AlCl_3$

Answer: E
Diff: 1 *Page Ref: 12.7*

59) Identify the solute with the lowest van't Hoff factor.
A) nonelectrolyte
B) KI
C) $MgSO_4$
D) $CaCl_2$
E) $AlCl_3$

Answer: A
Diff: 1 *Page Ref: 12.7*

60) Choose the aqueous solution that has the highest boiling point. These are all solutions of nonvolatile solutes and you should assume ideal van't Hoff factors where applicable.
A) 0.100 *m* $AlCl_3$
B) 0.100 *m* LiBr
C) 0.100 *m* $BaCl_2$
D) 0.100 *m* $C_6H_{12}O_6$
E) They all have the same boiling point.

Answer: A
Diff: 1 *Page Ref: 12.7*

61) Which of the following solutions will have the lowest freezing point?
A) 0.010 *m* K Br
B) 0.010 *m* Na_2CO_3
C) 0.035 *m* CH_3CH_2OH
D) 0.015 *m* $BaCl_2$

Answer: D
Diff: 3 *Page Ref: 12.6*

62) Choose the aqueous solution that has the highest boiling point. These are all solutions of nonvolatile solutes and you should assume ideal van't Hoff factors where applicable.
A) 0.100 *m*KNO_2
B) 0.100 *m* Li_2SO_4
C) 0.200 *m* $C_3H_8O_3$
D) 0.060 *m* Li_3PO_4
E) They all have the same boiling point.

Answer: B
Diff: 1 *Page Ref: 12.7*

63) Choose the aqueous solution with the **lowest** vapor pressure. These are all solutions of nonvolatile solutes and you should assume ideal van't Hoff factors where applicable.

A) 0.120 *m* $C_2H_6O_2$
B) 0.040 *m* $(NH_4)_2SO_4$
C) 0.060 *m* Li_2CO_3
D) 0.030 *m* $RbC_2H_3O_2$
E) They all have the same vapor pressure.

Answer: C
Diff: 1 *Page Ref: 12.7*

64) Choose the aqueous solution with the **highest** vapor pressure. These are all solutions of nonvolatile solutes and you should assume ideal van't Hoff factors where applicable.

A) 0.50 *m* $C_4H_{10}O_4$
B) 0.50 *m* $C_2H_6O_2$
C) 0.50 *m* $C_5H_{12}O_5$
D) 0.50 *m* $C_{11}H_{24}O_{11}$
E) They all have about the same vapor pressure.

Answer: E
Diff: 1 *Page Ref: 12.7*

65) Choose the aqueous solution below with the **lowest** freezing point. These are all solutions of nonvolatile solutes and you should assume ideal van't Hoff factors where applicable.

A) 0.075 *m* LiCl
B) 0.075 *m* $(NH_4)_3PO_4$
C) 0.075 *m* $NaBrO_4$
D) 0.075 *m* KCN
E) 0.075 *m* KNO_2

Answer: B
Diff: 1 *Page Ref: 12.7*

66) Choose the aqueous solution below with the **highest** freezing point. These are all solutions of nonvolatile solutes and you should assume ideal van't Hoff factors where applicable.

A) 0.200 *m* $Mg(ClO_4)_2$
B) 0.200 *m* Na_3PO_3
C) 0.200 *m* $HOCH_2CH_2OH$
D) 0.200 *m* $Ba(NO_3)_2$
E) These all have the same freezing point.

Answer: C
Diff: 1 *Page Ref: 12.7*

67) What is the freezing point of a solution of 7.15 g $MgCl_2$ in 100 g of water? K_f for water is 1.86°C/*m*.

A) −0.140°C B) −1.40°C C) −2.80°C D) −4.18°C

Answer: D
Diff: 3 *Page Ref: 12.7*

68) What is the expected freezing point of a 0.50 *m* solution of Li_2SO_4 in water? K_f for water is 1.86°C/*m*.

A) -0.93°C B) -1.9°C C) -2.8°C D) -6.5°C

Answer: C
Diff: 3 Page Ref: 12.7

69) A KCl solution is prepared by dissolving 25.0 g KCl in 250.0 g of water at 25°C. What is the vapor pressure of the solution if the vapor pressure of water at 25°C is 23.76 mm Hg?

A) 21.6 mm Hg B) 22.7 mm Hg C) 23.2 mm Hg D) 24.9 mm Hg

Answer: B
Diff: 3 Page Ref: 12.7

70) In an ________ solution, red blood cells ________.

A) isomotic, have a normal shape
B) hyperosmotic, have a normal shape
C) hyposmatic, lose water and shrivel
D) isosmotic, lose water and shrivel
E) isosmotic, swell and may burst

Answer: A
Diff: 1 Page Ref: 12.7

71) Define smoke as a colloid.

A) a liquid dispersed in a gas
B) a solid dispersed in a gas
C) a gas dispersed in a liquid
D) a liquid dispersed in a liquid
E) a solid dispersed in a liquid

Answer: B
Diff: 1 Page Ref: 12.8

72) Define fog as a colloid.

A) a liquid dispersed in a gas
B) a solid dispersed in a gas
C) a gas dispersed in a liquid
D) a liquid dispersed in a liquid
E) a solid dispersed in a liquid

Answer: A
Diff: 1 Page Ref: 12.8

73) The scattering of light by a collodial dispersion is known as

A) an emulsion
B) an aerosol.
C) micelle repulsion.
D) Brownian motion.
E) Tyndall effect.

Answer: E
Diff: 1 Page Ref: 12.8

Matching Questions

Match the following.

1) 0.050 *m* NaCl
Diff: 1 *Page Ref: 12.6*

2) 0.050 *m* $C_6H_{12}O_6$ (aqueous)
Diff: 1 *Page Ref: 12.6*

3) 0.0050 *m* CO_2
Diff: 1 *Page Ref: 12.6*

4) 0.020 m NH_4Cl
Diff: 1 *Page Ref: 12.6*

5) 0.050 *m* NaCl
Diff: 1 *Page Ref: 12.6*

6) 0.010 *m* $Al(NO_3)_3$
Diff: 1 *Page Ref: 12.7*

A) solution that is most strongly dependent upon pressure

B) highest boiling point

C) solution with $\Delta T_b = 0.026°C$

D) solution of ionic compound with highest freezing point.

E) largest van't Hoff factor

Answers: 1) B 2) C 3) A 4) D 5) B 6) E

Short Answer Questions

1) Explain why water does not dissolve in gasoline.
Answer: Water is polar and gasoline is nonpolar. Water has hydrogen bonding and gasoline has dispersion forces.
Diff: 1 *Page Ref: 12.2*

2) Why isn't pentanol ($CH_3CH_2CH_2CH_2CH_2OH$) very soluble in water?
Answer: Even though the pentanol molecule contains a hydrogen bonding portion, the bulk of the molecule is nonpolar and only exhibits dispersion forces. The majority of this molecule will not hydrogen bond to water molecules, making it difficult to dissolve.
Diff: 1 *Page Ref: 12.2*

3) What happens to a supersaturated solution of potassium acetate once it is cooled and a small crystal of solid potassium acetate is added?
Answer: The portion of potassium acetate that was dissolved beyond its normal solubility, will precipitate out of the solution. The solution will now be saturated.
Diff: 1 *Page Ref: 12.4*

4) Give the preparation of rock candy.
Answer: A saturated solution of sugar is prepared at an elevated temperature. A string or stick is left to dangle in the solution. The solution is allowed to cool for several days. The sugar crystals grow on the string or stick.
Diff: 1 *Page Ref: 12.4*

5) How does a solution of two volatile components with strong solute-solvent attractions deviate from Raoult's law? Why?

Answer: The vapor pressure of the solution will be lower than predicted by Raoult's law. The strong solute-solvent interactions lower the number of particles that have enough energy to escape the solution (as compared to either pure component), therefore lowering the vapor pressure.

Diff: 1 *Page Ref:* 12.6

6) Define osmosis.

Answer: Osmosis is the flow of solvent from a solution of lower solute concentration to one of higher solute concentration.

Diff: 1 *Page Ref:* 12.6

7) Calculate the osmotic pressure, in torr, of a solution containing 3.00 mg of a sugar (342 g/mole) in 15.0 mL of water at 25°C.

Answer: 10.9 torr

Diff: 3 *Page Ref:* 12.6

8) Explain why the van't Hoff factor for $MgCl_2$ is less than it's predicted value.

Answer: The hydrated Mg^{2+} and Cl^- ions can essentially attract one another in solution and effectively reduce the number of free particles in solution.

Diff: 1 *Page Ref:* 12.7

9) Define Tyndall effect.

Answer: The Tyndall effect is the scattering of light through a colloidal dispersion.

Diff: 1 *Page Ref:* 12.8

10) Define a colloid.

Answer: A colloid is a mixture in which a dispersed substance is finely divided in a dispersing medium.

Diff: 2 *Page Ref:* 12.8

Chapter 13 Chemical Kinetics

Multiple Choice Questions

1) Identify the methods used to monitor a reaction as it occurs in the reaction flask.

A) polarimeter
B) spectrometer
C) pressure measurement
D) none of the above
E) all of the above

Answer: E
Diff: 1 *Page Ref: 13.2*

2) Given the following balanced equation, determine the rate of reaction with respect to [NOCl].

$$2\,NO(g) + Cl_2(g) \rightarrow 2\,NOCl(g)$$

A) Rate $= -\frac{1}{2}\frac{\Delta[NOCl]}{\Delta t}$

B) Rate $= +\frac{1}{2}\frac{\Delta[NOCl]}{\Delta t}$

C) Rate $= -\frac{1}{2}\frac{\Delta[NO]}{\Delta t}$

D) Rate $= -\frac{2\,\Delta[NOCl]}{\Delta t}$

E) It is not possible to determine without more information.

Answer: B
Diff: 2 *Page Ref: 13.2*

3) Given the following balanced equation, determine the rate of reaction with respect to [H_2].

$$N_2(g) + 3\,H_2(g) \rightarrow 2\,NH_3(g)$$

A) Rate $= +\frac{3\,\Delta[H_2]}{\Delta t}$

B) Rate $= -\frac{3\Delta[H_2]}{\Delta t}$

C) Rate $= +\frac{1}{3}\frac{\Delta[H_2]}{\Delta t}$

D) Rate $= -\frac{1}{3}\frac{\Delta[H_2]}{\Delta t}$

E) It is not possible to determine without more information.

Answer: D
Diff: 2 *Page Ref: 13.2*

4) Given the following balanced equation, determine the rate of reaction with respect to $[NH_3]$.

$$N_2(g) + 3\,H_2(g) \rightarrow 2\,NH_3(g)$$

A) Rate $= + \frac{2\,\Delta[NH_3]}{\Delta t}$

B) Rate $= - \frac{2\Delta[NH_3]}{\Delta t}$

C) Rate $= + \frac{1}{2}\frac{\Delta[NH_3]}{\Delta t}$

D) Rate $= - \frac{1}{2}\frac{\Delta[NH_3]}{\Delta t}$

E) It is not possible to determine without more information.

Answer: C
Diff: 2 *Page Ref:* 13.2

5) Given the following balanced equation, determine the rate of reaction with respect to $[N_2]$.

$$N_2(g) + 3\,H_2(g) \rightarrow 2\,NH_3(g)$$

A) Rate $= + \frac{\Delta[N_2]}{\Delta t}$

B) Rate $= - \frac{\Delta[N_2]}{\Delta t}$

C) Rate $= + \frac{1}{2}\frac{\Delta[N_2]}{\Delta t}$

D) Rate $= - \frac{1}{2}\frac{\Delta[N_2]}{\Delta t}$

E) It is not possible to determine without more information.

Answer: B
Diff: 2 *Page Ref:* 13.2

6) Write a balanced reaction for which the following rate relationships are true.

$$\text{Rate} = -\frac{1}{2}\frac{\Delta[N_2O_5]}{\Delta t} = \frac{1}{4}\frac{\Delta[NO_2]}{\Delta t} = \frac{\Delta[O_2]}{\Delta t}$$

A) $2\ N_2O_5 \rightarrow 4\ NO_2 + O_2$
B) $4\ NO_2 + O_2 \rightarrow 2\ N_2O_5$
C) $2\ N_2O_5 \rightarrow NO_2 + 4\ O_2$
D) $\frac{1}{4}\ NO_2 + O_2 \rightarrow \frac{1}{2}\ N_2O_5$
E) $\frac{1}{2}\ N_2O_5 \rightarrow \frac{1}{4}\ NO_2 + O_2$

Answer: A
Diff: 2 *Page Ref: 13.2*

7) Write a balanced reaction for which the following rate relationships are true.

$$\text{Rate} = \frac{1}{2}\frac{\Delta[N_2]}{\Delta t} = \frac{\Delta[O_2]}{\Delta t} = -\frac{1}{2}\frac{\Delta[N_2O]}{\Delta t}$$

A) $\frac{1}{2}\ N_2 + O_2 \rightarrow \frac{1}{2}\ N_2O$
B) $2\ N_2O \rightarrow 2\ N_2 + O_2$
C) $N_2O \rightarrow N_2 + 2\ O_2$
D) $\frac{1}{2}\ N_2O \rightarrow \frac{1}{2}\ N_2 + O_2$
E) $2\ N_2 + O_2 \rightarrow 2\ N_2O$

Answer: B
Diff: 2 *Page Ref: 13.2*

8) Give the characteristic of a first order reaction having only one reactant.
A) The rate of the reaction is not proportional to the concentration of the reactant.
B) The rate of the reaction is proportional to the square of the concentration of the reactant.
C) The rate of the reaction is proportional to the square root of the concentration of the reactant.
D) The rate of the reaction is proportional to the natural logarithm of the concentration of the reactant.
E) The rate of the reaction is directly proportional to the concentration of the reactant.

Answer: E
Diff: 1 *Page Ref: 13.3*

9) Give the characteristic of a zero order reaction having only one reactant.
A) The rate of the reaction is not proportional to the concentration of the reactant.
B) The rate of the reaction is proportional to the square of the concentration of the reactant.
C) The rate of the reaction is proportional to the square root of the concentration of the reactant.
D) The rate of the reaction is proportional to the natural logarithm of the concentration of the reactant.
E) The rate of the reaction is directly proportional to the concentration of the reactant.

Answer: A
Diff: 1 *Page Ref: 13.3*

10) Give the characteristic of a second order reaction having only one reactant.
A) The rate of the reaction is not proportional to the concentration of the reactant.
B) The rate of the reaction is proportional to the square of the concentration of the reactant.
C) The rate of the reaction is proportional to the square root of the concentration of the reactant.
D) The rate of the reaction is proportional to the natural logarithm of the concentration of the reactant.
E) The rate of the reaction is directly proportional to the concentration of the reactant.

Answer: B
Diff: 1 *Page Ref: 13.3*

11) What is the overall order of the following reaction, given the rate law?

$$NO(g) + O_3(g) \rightarrow NO_2(g) + O_2(g) \qquad \text{Rate} = k[NO][O_3]$$

A) 1st order
B) 2nd order
C) 3rd order
D) $1\frac{1}{2}$ order
E) 0th order

Answer: B
Diff: 2 *Page Ref: 13.3*

12) What are the units of k in the following rate law?

$$\text{Rate} = k[X][Y]^2$$

A) $\frac{1}{M\ s^2}$ B) $\frac{1}{M^2\ s}$ C) $M^2\ s$ D) $\frac{M^2}{s}$ E) $\frac{1}{M^3\ s}$

Answer: B
Diff: 2 *Page Ref: 13.3*

13) What are the units of k in the following rate law?

Rate = $k[X]^2[Y]^2$

A) $\frac{1}{M\,s^2}$ B) $\frac{1}{M^2\,s}$ C) $M^2\,s$ D) $\frac{M^2}{s}$ E) $\frac{1}{M^3\,s}$

Answer: E
Diff: 2 *Page Ref:* 13.3

14) What are the units of k in the following rate law?
Rate = k[X][Y]

A) $\frac{M}{s}$ B) Ms C) $\frac{1}{Ms}$ D) $\frac{M^2}{s}$ E) $\frac{s}{M^2}$

Answer: C
Diff: 2 *Page Ref:* 13.3

15) What are the units of k in the following rate law?
Rate = k[X]

A) $\frac{M}{s}$ B) Ms C) $\frac{1}{Ms}$ D) $\frac{1}{s}$ E) $\frac{s}{M}$

Answer: D
Diff: 2 *Page Ref:* 13.3

16) What are the units of k in the following rate law?
Rate = $k[X]^0[Y]^0$

A) $\frac{M}{s}$ B) Ms C) $M^{-1}s^{-1}$ D) $\frac{1}{s}$ E) $\frac{s}{M^2}$

Answer: D
Diff: 2 *Page Ref:* 13.3

17) What are the units of k in the following rate law?
Rate = $k[X][Y]^{1/2}$

A) M/s B) $M^{-1}s^{-1}$ C) $M^{1/2}s^{-1}$ D) $M^{-1/2}s^{-1}$ E) $M^{-1}s^{-1/2}$

Answer: D
Diff: 2 *Page Ref:* 13.3

18) What are the units of k in the following rate law?
Rate = $k[X]^2$

A) Ms^{-1} B) Ms C) $M^{-1}s^{-1}$ D) $M^{-2}s^{-1}$ E) M^2s^{-1}

Answer: C
Diff: 2 *Page Ref:* 13.3

19) Determine the rate law and the value of k for the following reaction using the data provided.

$CO(g) + Cl_2(g) \rightarrow COCl_2(g)$

$[CO]_i$ (M)	$[Cl_2]_i$ (M)	Initial Rate ($M^{-1}s^{-1}$)
0.25	0.40	0.696
0.25	0.80	1.97
0.50	0.80	3.94

A) Rate = 11 $M^{-3/2}s^{-1}$ $[CO][Cl_2]^{3/2}$

B) Rate = 36 $M^{-1.8}s^{-1}$ $[CO][Cl_2]^{2.8}$

C) Rate = 17 $M^{-2}s^{-1}$ $[CO][Cl_2]^2$

D) Rate = 4.4 $M^{-1/2}s^{-1}$ $[CO][Cl_2]^{1/2}$

E) Rate = 18 $M^{-3/2}s^{-1}$ $[CO]^2[Cl_2]^{1/2}$

Answer: A
Diff: 3 *Page Ref: 13.3*

20) Determine the rate law and the value of k for the following reaction using the data provided.

$S_2O_8^{2-}(aq) + 3\ I^-(aq) \rightarrow 2\ SO_4^{2-}(g) + I_3^-(aq)$

$[S_2O_8^{2-}]_i$ (M)	$[I^-]_i$ (M)	Initial Rate ($M^{-1}s^{-1}$)
0.30	0.42	4.54
0.44	0.42	6.65
0.44	0.21	3.33

A) Rate = 120 $M^{-2}s^{-1}$ $[S_2O_8^{2-}]^2[I^-]$

B) Rate = 36 $M^{-1}s^{-1}$ $[S_2O_8^{2-}][I^-]$

C) Rate = 86 $M^{-2}s^{-1}$ $[S_2O_8^{2-}][I^-]^2$

D) Rate = 195 $M^{-3}s^{-1}$ $[S_2O_8^{2-}]^2[I^-]^2$

E) Rate = 23 $M^{-1/2}s^{-1}$ $[S_2O_8^{2-}][I^-]^{1/2}$

Answer: B
Diff: 3 *Page Ref: 13.3*

21) Determine the rate law and the value of k for the following reaction using the data provided.

$NO_2(g) + O_3(g) \rightarrow NO_3(g) + O_2(g)$

$[NO_2]_i$ (M)	$[O_3]_i$ (M)	Initial Rate ($M^{-1}s^{-1}$)
0.10	0.33	1.42
0.10	0.66	2.84
0.25	0.66	7.10

A) Rate = 1360 $M^{-2.5}s^{-1}[NO_2]^{2.5}[O_3]$
B) Rate = 227 $M^{-2.5}s^{-1}[NO_2][O_3]^{2.5}$
C) Rate = 43 $M^{-1}s^{-1}[NO_2][O_3]$
D) Rate = 430 $M^{-2}s^{-1}[NO_2]^2[O_3]$
E) Rate = 130 $M^{-2}s^{-1}[NO_2][O_3]^2$

Answer: C
Diff: 3 *Page Ref: 13.3*

22) Determine the rate law and the value of k for the following reaction using the data provided.

$2\ NO(g) + O_2(g) \rightarrow 2\ NO_2(g)$

$[NO]_i$ (M)	$[O_2]_i$ (M)	Initial Rate ($M^{-1}s^{-1}$)
0.030	0.0055	8.55×10^{-3}
0.030	0.0110	1.71×10^{-2}
0.060	0.0055	3.42×10^{-2}

A) Rate = 57 $M^{-1}s^{-1}[NO][O_2]$
B) Rate = 3.8 $M^{-1/2}s^{-1}[NO][O_2]^{1/2}$
C) Rate = $3.1 \times 10^5\ M^{-3}s^{-1}[NO]^2[O_2]^2$
D) Rate = $1.7 \times 10^3\ M^{-2}s^{-1}[NO]^2[O_2]$
E) Rate = $9.4 \times 10^3\ M^{-2}s^{-1}[NO][O_2]^2$

Answer: D
Diff: 3 *Page Ref: 13.3*

23) Carbon-14 has a half-life of 5720 years and this is a first order reaction. If a piece of wood has converted 75% of the carbon-14, then how old is it?
A) 11440 years
B) 2375 years
C) 4750 years
D) 4290 years
E) 1430 years

Answer: A
Diff: 3 *Page Ref: 13.4*

24) Carbon-14 has a half-life of 5720 years and this is a first order reaction. If a piece of wood has converted 25% of the carbon-14, then how old is it?

A) 11440 years
B) 2375 years
C) 4750 years
D) 4290 years
E) 1430 years

Answer: B
Diff: 3 *Page Ref: 13.4*

25) Which of the following represents the integrated rate law for a first-order reaction?

A) $\ln\frac{[A]_t}{[A]_o} = -kt$

B) $\frac{1}{[A]_t} - \frac{1}{[A]_o} = kt$

C) $[A]_t - [A]_o = -kt$

D) $k = Ae^{(-Ea/RT)}$

E) $\ln\frac{k_2}{k_1} = \frac{E_a}{R}\left(\frac{1}{T}\right) + \ln A$

Answer: A
Diff: 1 *Page Ref: 13.4*

26) Which of the following represents the integrated rate law for a second-order reaction?

A) $\ln\frac{[A]_t}{[A]_o} = -kt$

B) $\frac{1}{[A]_t} - \frac{1}{[A]_o} = kt$

C) $[A]_t - [A]_o = -kt$

D) $k = Ae^{(-Ea/RT)}$

E) $\ln\frac{k_2}{k_1} = \frac{E_a}{R}\left(\frac{1}{T}\right) + \ln A$

Answer: B
Diff: 1 *Page Ref: 13.4*

27) Which of the following represents the integrated rate law for a zeroth-order reaction?

A) $\ln\frac{[A]_t}{[A]_o} = -kt$

B) $\frac{1}{[A]_t} - \frac{1}{[A]_o} = kt$

C) $[A]_t - [A]_o = -kt$

D) $k = Ae^{(-Ea/RT)}$

E) $\ln\frac{k_2}{k_1} = \frac{E_a}{R}\left(\frac{1}{T}\right) + \ln A$

Answer: C
Diff: 1 *Page Ref: 13.4*

28) What data should be plotted to show that experimental concentration data fits a first-order reaction?

A) 1/[reactant] vs. time
B) [reactant] vs. time
C) ln[reactant] vs. time
D) ln(k) vs. 1/T
E) ln(k) vs. E_a

Answer: C
Diff: 1 *Page Ref: 13.4*

29) What data should be plotted to show that experimental concentration data fits a second-order reaction?

A) ln[reactant] vs. time
B) [reactant] vs. time
C) ln(k) vs. 1/T
D) 1/[reactant] vs. time
E) ln(k) vs. E_a

Answer: D
Diff: 1 *Page Ref: 13.4*

30) What data should be plotted to show that experimental concentration data fits a zeroth-order reaction?

A) ln[reactant] vs. time
B) 1/[reactant] vs. time
C) ln(k) vs. 1/T
D) ln(k) vs. E_a
E) [reactant] vs. time

Answer: E
Diff: 1 *Page Ref: 13.4*

31) Which of the following statements is FALSE?
A) The average rate of a reaction decreases during a reaction.
B) It is not possible to determine the rate of a reaction from its balanced equation.
C) The rate of zero order reactions are not dependent on concentration.
D) The half life of a first order reaction is dependent on the initial concentration of reactant.
E) None of the statements are FALSE.

Answer: D
Diff: 1 *Page Ref: 13.4*

32) Which of the following statements is FALSE?
A) The half life of a zero order reaction is dependent on concentration.
B) The half life of a second order reaction is not dependent on concentration.
C) The rate of second order reactions is dependent on concentration.
D) The rate of a first order reaction is dependent on concentraion.
E) None of the statements are FALSE.

Answer: B
Diff: 1 *Page Ref: 13.4*

33) The rate constant for the first–order decomposition of N_2O is 3.40 s^{-1}. What is the half–life of the decomposition?
A) 0.491 s B) 0.204 s C) 0.236 s D) 0.424 s E) 0.294 s

Answer: B
Diff: 2 *Page Ref: 13.4*

34) The half–life for the second–order decomposition of HI is 15.4 s when the initial concentration of HI is 0.67 M. What is the rate constant for this reaction?
A) 1.0×10^{-2} $M^{-1}s^{-1}$
B) 4.5×10^{-2} $M^{-1}s^{-1}$
C) 9.7×10^{-2} $M^{-1}s^{-1}$
D) 2.2×10^{-2} $M^{-1}s^{-1}$
E) 3.8×10^{-2} $M^{-1}s^{-1}$

Answer: C
Diff: 2 *Page Ref: 13.4*

35) The half life for the decay of radium is 1620 years. What is the rate constant for this first–order process?
A) 4.28×10^{-4} yr^{-1}
B) 1.12×10^{-4} yr^{-1}
C) 2.33×10^{-4} yr^{-1}
D) 8.91×10^{-4} yr^{-1}
E) 6.17×10^{-4} yr^{-1}

Answer: A
Diff: 2 *Page Ref: 13.4*

36) The first-order decomposition of cyclopropane has a rate constant of $6.7 \times 10^{-4}\ s^{-1}$. If the initial concentration of cyclopropane is 1.33 M, what is the concentration of cyclopropane after 644 s?

A) 0.43 M B) 0.15 M C) 0.94 M D) 0.86 M E) 0.67 M

Answer: D

Diff: 3 *Page Ref: 13.4*

37) The first-order decomposition of N_2O at 1000 K has a rate constant of $0.76\ s^{-1}$. If the initial concentration of N_2O is 10.9 M, what is the concentration of N_2O after 9.6 s?

A) 7.4×10^{-3} M
B) 1.0×10^{-3} M
C) 1.4×10^{-3} M
D) 3.6×10^{-3} M
E) 8.7×10^{-3} M

Answer: A

Diff: 3 *Page Ref: 13.4*

38) The second-order decomposition of HI has a rate constant of $1.80 \times 10^{-3}\ M^{-1}s^{-1}$. How much HI remains after 27.3 s if the initial concentration of HI is 4.78 M?

A) 4.55 M B) 0.258 M C) 3.87 M D) 2.20 M E) 2.39 M

Answer: C

Diff: 3 *Page Ref: 13.4*

39) Derive an expression for a "1/3-life" for a first-order reaction.

A) $\frac{\ln\frac{[A]_t}{[A]_o}}{-k}$ B) $\frac{0.462}{k}$ C) $\frac{\ln\frac{[A]_t}{[A]_o}}{-3k}$ D) $\frac{1.099}{k}$ E) $\frac{3}{k}$

Answer: D

Diff: 3 *Page Ref: 13.4*

40) Derive an expression for a "1/4-life" for a first-order reaction.

A) $\frac{\ln\frac{[A]_t}{[A]_o}}{-k}$ B) $\frac{0.347}{k}$ C) $\frac{\ln\frac{[A]_t}{[A]_o}}{-4k}$ D) $\frac{4}{k}$ E) $\frac{1.386}{k}$

Answer: E

Diff: 3 *Page Ref: 13.4*

41) The first-order decay of radon has a half-life of 3.823 days. How many grams of radon remain after 7.22 days if the sample initially weighs 250.0 grams?

A) 4.21 g B) 183 g C) 54.8 g D) 76.3 g E) 67.5 g

Answer: E

Diff: 4 *Page Ref: 13.4*

42) The first-order decay of radon has a half-life of 3.823 days. How many grams of radon **decomposes** after 5.55 days if the sample initially weighs 100.0 grams?
A) 83.4 g B) 16.6 g C) 50.0 g D) 36.6 g E) 63.4 g

Answer: E
Diff: 4 Page Ref: 13.4

43) The first-order decomposition of N_2O_5 at 328 K has a rate constant of $1.70 \times 10^{-3}\ s^{-1}$. If the initial concentration of N_2O_5 is 2.88 M, what is the concentration of N_2O_5 after 12.5 minutes?
A) 0.124 M B) 0.805 M C) 2.82 M D) 0.355 M E) 0.174 M

Answer: B
Diff: 4 Page Ref: 13.4

44) The second-order decomposition of NO_2 has a rate constant of $0.255\ M^{-1}s^{-1}$. How much NO_2 **decomposes** in 4.00 s if the initial concentration of NO_2 (1.00 L volume) is 1.33 M?
A) 1.8 mol B) 0.85 mol C) 0.48 mol D) 0.77 mol E) 0.56 mol

Answer: D
Diff: 4 Page Ref: 13.4

45) For a reaction, what generally happens if the temperature is increased?
A) a decrease in k occurs, which results in a faster rate
B) a decrease in k occurs, which results in a slower rate
C) an increase in k occurs, which results in a faster rate
D) an increase in k occurs, which results in a slower rate
E) there is no change with k or the rate

Answer: C
Diff: 1 Page Ref: 13.5

46) Identify the rate-determining step.
A) the slowest step
B) the faster step
C) the fast step
D) always the last step
E) always the second step

Answer: A
Diff: 1 Page Ref: 13.5

47) The first-order rearrangement of CH_3NC is measured to have a rate constant of $3.61 \times 10^{-15}\ s^{-1}$ at 298 K and a rate constant of $8.66 \times 10^{-7}\ s^{-1}$ at 425 K. Determine the activation energy for this reaction.
A) 160. kJ/mol
B) 240. kJ/mol
C) 417 kJ/mol
D) 127 kJ/mol
E) 338 kJ/mol

Answer: A
Diff: 3 Page Ref: 13.5

48) A reaction is followed and found to have a rate constant of $3.36 \times 10^4\ M^{-1}s^{-1}$ at 344 K and a rate constant of $7.69\ M^{-1}s^{-1}$ at 219 K. Determine the activation energy for this reaction.

A) 23.8 kJ/mol
B) 42.0 kJ/mol
C) 11.5 kJ/mol
D) 12.5 kJ/mol
E) 58.2 kJ/mol

Answer: B
Diff: 3 *Page Ref: 13.5*

49) A reaction is found to have an activation energy of 108 kJ/mol. If the rate constant for this reaction is $4.60 \times 10^{-6}\ s^{-1}$ at 275 K, what is the rate constant at 366 K?

A) $12\ s^{-1}$
B) $1.7\ s^{-1}$
C) $0.58\ s^{-1}$
D) $5.4 \times 10^{-5}\ s^{-1}$
E) $1.9 \times 10^{-4}\ s^{-1}$

Answer: C
Diff: 3 *Page Ref: 13.5*

50) A reaction is found to have an activation energy of 38.0 kJ/mol. If the rate constant for this reaction is $1.60 \times 10^2\ M^{-1}s^{-1}$ at 249 K, what is the rate constant at 436 K?

A) $2.38 \times 10^5\ M^{-1}s^{-1}$
B) $1.26 \times 10^3\ M^{-1}s^{-1}$
C) $7.94 \times 10^4\ M^{-1}s^{-1}$
D) $4.20 \times 10^5\ M^{-1}s^{-1}$
E) $3.80 \times 10^4\ M^{-1}s^{-1}$

Answer: D
Diff: 3 *Page Ref: 13.5*

51) If the activation energy for a given compound is found to be 42.0 kJ/mol, with a frequency factor of $8.0 \times 10^{10}\ s^{-1}$, what is the rate constant for this reaction at 298 K?

A) $2.9 \times 10^{-4}\ s^{-1}$
B) $7.4 \times 10^{-4}\ s^{-1}$
C) $1.4 \times 10^9\ s^{-1}$
D) $4.6 \times 10^5\ s^{-1}$
E) $3.5 \times 10^3\ s^{-1}$

Answer: E
Diff: 3 *Page Ref: 13.5*

52) If the activation energy for a given compound is found to be 103 kJ/mol, with a frequency factor of $4.0 \times 10^{13}\ s^{-1}$, what is the rate constant for this reaction at 398 K?

A) $1.2\ s^{-1}$

B) $8.2\ s^{-1}$

C) $3.9 \times 10^{10}\ s^{-1}$

D) $1.7 \times 10^{10}\ s^{-1}$

E) $2.5 \times 10^{7}\ s^{-1}$

Answer: A

Diff: 3 *Page Ref: 13.5*

53) Given the following proposed mechanism, predict the rate law for the overall reaction.

$A_2 + 2B \rightarrow 2AB$ (overall reaction)

<u>Mechanism</u>

$A_2 \rightleftharpoons 2A$ fast

$A + B \rightarrow AB$ slow

A) Rate = k[A][B]

B) Rate = $k[A_2][B]$

C) Rate = $k[A_2][B]^{1/2}$

D) Rate = $k[A_2]$

E) Rate = $k\,[A_2]^{1/2}[B]$

Answer: E

Diff: 3 *Page Ref: 13.6*

54) Which of the following statements is TRUE?

A) The rate constant does not depend on the activation energy for a reaction where the products are lower in energy than the reactants.

B) A catalyst raises the activation energy of a reaction.

C) Rate constants are temperature dependent.

D) The addition of a homogeneous catalyst does not change the activation energy of a given reaction.

E) None of the above are true.

Answer: C

Diff: 1 *Page Ref: 13.7*

55) Identify an homogeneous catalyst.

A) SO_2 over vanadium (V) oxide

B) Pd in H_2 gas

C) Pt with methane

D) H_2SO_4 with concentrated HCl

E) N_2 and H_2 catalyzed by Fe

Answer: D

Diff: 1 *Page Ref: 13.7*

56) Identify an heterogeneous catalyst.

A) CFCs with ozone
B) Pd in H_2 gas
C) KI dissolved in H_2O_2
D) H_2SO_4 with concentrated HCl
E) H_3PO_4 with an alcohol

Answer: B
Diff: 1 *Page Ref: 13.7*

57) In the hydrogenation of double bonds, a catalyst is needed. In the first step, the reactants must come into contact with a metal surface. This step is known as ________.

A) adsorption
B) diffusion
C) reaction
D) desorption
E) none of the above

Answer: A
Diff: 1 *Page Ref: 13.7*

58) In the hydrogenation of double bonds, a catalyst is needed. In the last step, the reactants must escape from the surface into the gas phase. This step is known as ________.

A) adsorption
B) diffusion
C) reaction
D) desorption
E) none of the above

Answer: D
Diff: 1 *Page Ref: 13.7*

59) Biological catalysts that increase the rates of biochemical reactions are known as ________.

A) substrates
B) inhibitors
C) enzymes
D) binders
E) trumanettes

Answer: C
Diff: 1 *Page Ref: 13.7*

Algorithmic Questions

1) Given the following balanced equation, determine the rate of reaction with respect to $[SO_2]$.

$$2\ SO_2(g) + O_2(g) \rightarrow 2\ SO_3(g)$$

A) Rate $= -\frac{1}{2}\frac{\Delta[SO_2]}{\Delta t}$

B) Rate $= +\frac{1}{2}\frac{\Delta[SO_2]}{\Delta t}$

C) Rate $= -\frac{\Delta[SO_2]}{\Delta t}$

D) Rate $= +\frac{2\ \Delta[SO_2]}{\Delta t}$

E) It is not possible to determine without more information.

Answer: A

Diff: 2 *Page Ref:* 13.2

2) Given the following balanced equation, determine the rate of reaction with respect to $[O_2]$.

$$2\ SO_2(g) + O_2(g) \rightarrow 2\ SO_3(g)$$

A) Rate $= -\frac{1}{3}\frac{\Delta[O_2]}{\Delta t}$

B) Rate $= +\frac{1}{3}\frac{\Delta[O_2]}{\Delta t}$

C) Rate $= -\frac{\Delta[O_2]}{\Delta t}$

D) Rate $= +\frac{2\ \Delta[O_2]}{\Delta t}$

E) It is not possible to determine without more information.

Answer: C

Diff: 2 *Page Ref:* 13.2

3) Given the following balanced equation, determine the rate of reaction with respect to $[SO_3]$.

$$2\,SO_2(g) + O_2(g) \rightarrow 2\,SO_3(g)$$

A) $\text{Rate} = -\frac{1}{2}\frac{\Delta[SO_3]}{\Delta t}$

B) $\text{Rate} = +\frac{1}{2}\frac{\Delta[SO_3]}{\Delta t}$

C) $\text{Rate} = -\frac{\Delta[SO_3]}{\Delta t}$

D) $\text{Rate} = -\frac{2\,\Delta[SO_3]}{\Delta t}$

E) It is not possible to determine without more information.

Answer: B

Diff: 2 *Page Ref:* 13.2

4) Given the following balanced equation, determine the rate of reaction with respect to $[O_2]$.

$$2\,O_3(g) \rightarrow 3\,O_2(g)$$

A) $\text{Rate} = +\frac{2\,\Delta[O_2]}{\Delta t}$

B) $\text{Rate} = -\frac{2}{3}\frac{\Delta[O_2]}{\Delta t}$

C) $\text{Rate} = +\frac{1}{3}\frac{\Delta[O_2]}{\Delta t}$

D) $\text{Rate} = -\frac{3\,\Delta[O_2]}{\Delta t}$

E) It is not possible to determine without more information.

Answer: C

Diff: 2 *Page Ref:* 13.2

5) Given the following balanced equation, determine the rate of reaction with respect to $[H_2]$.

$$N_2(g) + 3\ H_2(g) \rightarrow 2\ NH_3(g)$$

A) Rate $= +\dfrac{3\ \Delta[H_2]}{\Delta t}$

B) Rate $= -\dfrac{2}{3}\dfrac{\Delta[H_2]}{\Delta t}$

C) Rate $= +\dfrac{3}{2}\dfrac{\Delta[H_2]}{\Delta t}$

D) Rate $= -\dfrac{1}{3}\dfrac{\Delta[H_2]}{\Delta t}$

E) It is not possible to determine without more information.

Answer: D
Diff: 2 *Page Ref: 13.2*

6) Given the following balanced equation, determine the rate of reaction with respect to $[Cl_2]$. If the rate of Cl_2 loss is 4.24×10^{-2} M/s, what is the rate of formation of NOCl?

$$2\ NO(g) + Cl_2(g) \rightarrow 2\ NOCl(g)$$

A) 4.24×10^{-2} M/s
B) 2.12×10^{-2} M/s
C) 1.06×10^{-1} M/s
D) 8.48×10^{-2} M/s
E) 1.61×10^{-2} M/s

Answer: D
Diff: 3 *Page Ref: 13.3*

7) Given the following balanced equation, determine the rate of reaction with respect to $[Cl_2]$. If the rate of Cl_2 loss is 4.24×10^{-2} M/s, what is the rate of formation of NO?

$$2\ NO(g) + Cl_2(g) \rightarrow 2\ NOCl(g)$$

A) 4.24×10^{-2} M/s
B) 2.12×10^{-2} M/s
C) 1.06×10^{-1} M/s
D) 8.48×10^{-2} M/s
E) 1.61×10^{-2} M/s

Answer: D
Diff: 3 *Page Ref: 13.3*

8) Given the following balanced equation, determine the rate of reaction with respect to $[O_2]$. If the rate of O_2 loss is 2.64 x 10^{-3} M/s, what is the rate of formation of SO_3?

$$2\,SO_2(g) + O_2(g) \rightarrow 2\,SO_3(g)$$

A) 2.64 × 10^{-3} M/s
B) 1.19 × 10^{-3} M/s
C) 1.32 × 10^{-3} M/s
D) 6.60 × 10^{-2} M/s
E) 5.28 × 10^{-3} M/s

Answer: E
Diff: 3 *Page Ref:* 13.3

9) Given the following balanced equation, determine the rate of reaction with respect to $[O_2]$. If the rate of formation of O_2 is 7.78 x 10^{-1} M/s, what is the rate of the loss of O_3?

$$2\,O_3(g) \rightarrow 3\,O_2(g)$$

A) 0.519 M/s B) 1.56 M/s C) 2.34 M/s D) 1.17 M/s E) 7.78 M/s

Answer: A
Diff: 3 *Page Ref:* 13.3

10) What is the overall order of the following reaction, given the rate law?

$X + 2\,Y \rightarrow 4\,Z$ Rate = k[X][Y]

A) 3rd order
B) 5th order
C) 2nd order
D) 1st order
E) 6th order

Answer: C
Diff: 2 *Page Ref:* 13.3

11) What are the units of k in a zero order reaction?

A) $\frac{M}{s}$ B) M C) $M^{-1}s^{-1}$ D) $\frac{M^2}{s}$ E) $\frac{s}{M}$

Answer: A
Diff: 2 *Page Ref:* 13.3

12) What are the units of k in a first order reaction?

A) $\frac{M}{s}$ B) M C) $M^{-1}s^{-1}$ D) $\frac{M^2}{s^2}$ E) $\frac{1}{s}$

Answer: E
Diff: 2 *Page Ref:* 13.3

13) What are the units of k in a second order reaction?

A) $\frac{M}{s}$ B) M C) $M^{-1}s^{-1}$ D) $\frac{M^2}{s}$ E) $\frac{1}{M}$

Answer: C
Diff: 2 *Page Ref: 13.3*

14) The decomposition of dinitrogen pentoxide is described by the chemical equation

$2\ N_2O_5(g) \rightarrow 4\ NO_2(g) + O_2(g)$

If the rate of disappearance of N_2O_5 is equal to 1.60 mol/min at a particular moment, what is the rate of appearance of NO_2 at that moment?

A) 0.800 mol/min B) 1.60 mol/min C) 3.20 mol/min D) 6.40 mol/min

Answer: C
Diff: 3 *Page Ref: 13.3*

15) What is the overall order of the following reaction, given the rate law?

$2\ X + 3\ Y \rightarrow 2\ Z$ Rate = $k[X]^1[Y]^2$

A) 3rd order
B) 5th order
C) 2nd order
D) 1st order
E) 0th order

Answer: A
Diff: 2 *Page Ref: 13.3*

16) What is the overall order of the following reaction, given the rate law?

$2NO(g) + H_2(g) \rightarrow N_2(g) + 2H_2O(g)$ Rate = $k[NO]^2[H_2]$

A) 1st order B) 1st order C) 3rd order D) 4th order E) 5th order

Answer: C
Diff: 2 *Page Ref: 13.3*

17) Given the following rate law, how does the rate of reaction change if the concentration of Y is doubled?

Rate = $k\ [X][Y]^2$

A) The rate of reaction will increase by a factor of 2.
B) The rate of reaction will increase by a factor of 4.
C) The rate of reaction will increase by a factor of 5.
D) The rate of reaction will decrease by a factor of 2.
E) The rate of reaction will remain unchanged.

Answer: B
Diff: 2 *Page Ref: 13.3*

18) Given the following rate law, how does the rate of reaction change if the concentration of X is doubled?

Rate = k [X][Y]^2

A) The rate of reaction will increase by a factor of 2.
B) The rate of reaction will increase by a factor of 3.
C) The rate of reaction will increase by a factor of 5.
D) The rate of reaction will decrease by a factor of 2.
E) The rate of reaction will remain unchanged.

Answer: A
Diff: 2 *Page Ref:* 13.3

19) Given the following rate law, how does the rate of reaction change if the concentration of Y is doubled?

Rate = k [X][Y]

A) The rate of reaction will increase by a factor of 2.
B) The rate of reaction will increase by a factor of 4.
C) The rate of reaction willdecrease by a factor of 3.
D) The rate of reaction will decrease by a factor of 2.
E) The rate of reaction will remain unchanged.

Answer: A
Diff: 2 *Page Ref:* 13.3

20) Given the following rate law, how does the rate of reaction change if the concentration of Y is doubled?

Rate = k [X]^2[Y]^3

A) The rate of reaction will increase by a factor of 5.
B) The rate of reaction will increase by a factor of 2.
C) the rate of reaction will increase by a factor of 8.
D) The rate of reaction will increase by a factor of 3.
E) The rate of reaction will remain unchanged.

Answer: C
Diff: 2 *Page Ref:* 13.3

21) Determine the rate law and the value of k for the following reaction using the data provided.

$2\,N_2O_5(g) \rightarrow 4\,NO_2(g) + O_2(g)$

$[N_2O_5]_i$ (M)	Initial Rate ($M^{-1}s^{-1}$)
0.093	4.84×10^{-4}
0.084	4.37×10^{-4}
0.224	1.16×10^{-3}

A) Rate = $5.6 \times 10^{-2}\ M^{-1}s^{-1}[N_2O_5]^2$

B) Rate = $6.0 \times 10^{-1}\ M^{-2}s^{-1}[N_2O_5]^3$

C) Rate = $1.6 \times 10^{-3}\ M^{1/2}s^{-1}[N_2O_5]^{1/2}$

D) Rate = $1.7 \times 10^{-2}\ M^{-1/2}s^{-1}[N_2O_5]^{3/2}$

E) Rate = $5.2 \times 10^{-3}\ s^{-1}[N_2O_5]$

Answer: E
Diff: 3 *Page Ref:* 13.3

22) The decomposition of dinitrogen pentoxide is described by the chemical equation

$2\,N_2O_5(g) \rightarrow 4\,NO_2(g) + O_2(g)$

If the rate of appearance of NO_2 is equal to 0.560 mol/min at a particular moment, what is the rate of appearance of O_2 at that moment?

A) 0.140 mol/min B) 0.280 mol/min C) 1.12 mol/min D) 2.24 mol/min

Answer: A
Diff: 3 *Page Ref:* 13.3

23) The decomposition of dinitrogen pentoxide is described by the chemical equation

$2\,N_2O_5(g) \rightarrow 4\,NO_2(g) + O_2(g)$

If the rate of appearance of O_2 is equal to 3.00 mol/min at a particular moment, what is the rate of disappearance of N_2O_5 at that moment?

A) 0.750 mol/min B) 1.50 mol/min C) 6.00 mol/min D) 12.0 mol/min

Answer: C
Diff: 3 *Page Ref:* 13.3

24) For a reaction that follows the general rate law, Rate = $k[A][B]^2$, what will happen to the rate of reaction if the concentration of A is increased by a factor of 5.00? The rate will

A) decrease by a factor of 1/25.0. B) decrease by a factor of 1/5.00.
C) increase by a factor of 5.00. D) increase by a factor of 25.0.

Answer: C
Diff: 3 *Page Ref:* 13.3

25) For a reaction that follows the general rate law, Rate = $k[A][B]^2$, what will happen to the rate of reaction if the concentration of B is increased by a factor of 3.00? The rate will

A) decrease by a factor of 1/9.00. B) decrease by a factor of 1/3.00.
C) increase by a factor of 3.00. D) increase by a factor of 9.00.

Answer: D
Diff: 3 *Page Ref:* 13.3

26) What is the overall reaction order for the reaction that has the rate law: Rate = $k[O_2][NO]^2$?

A) zero order B) first order C) second order D) third order

Answer: D
Diff: 3 *Page Ref: 13.3*

27) Given the following rate law, how does the rate of reaction change if the concentration of X is doubled?

$$\text{Rate} = k\,[X]^2[Y]^3$$

A) The rate of reaction will increase by a factor of 9.
B) The rate of reaction will increase by a factor of 5.
C) the rate of reaction will increase by a factor of 6.
D) The rate of reaction will increase by a factor of 4.
E) The rate of reaction will remain unchanged.

Answer: D
Diff: 2 *Page Ref: 13.3*

28) Which of the following represents the equation for a first-order half-life?

A) $t_{1/2} = \dfrac{1}{k[A]_o}$

B) $t_{1/2} = \dfrac{0.693}{k}$

C) $t_{1/2} = \dfrac{2k[A]_o}{1}$

D) $t_{1/2} = \dfrac{2}{k[A]_o}$

E) $t_{1/2} = \dfrac{k}{[A]_o}$

Answer: B
Diff: 1 *Page Ref: 13.4*

29) Which of the following represents the equation for a second-order half-life?

A) $t_{1/2} = \dfrac{1}{k[A]_o}$

B) $t_{1/2} = \dfrac{0.693}{k}$

C) $t_{1/2} = \dfrac{2k[A]_o}{1}$

D) $t_{1/2} = \dfrac{k}{[A]_o}$

E) $t_{1/2} = \dfrac{k}{[A]_o}$

Answer: A
Diff: 1 *Page Ref: 13.4*

30) Which of the following represents the equation for a zero-order half-life?

A) $t_{1/2} = \frac{1}{[A]_o}$

B) $t_{1/2} = \frac{0.693}{k}$

C) $t_{1/2} = \frac{[A]_o}{2k}$

D) $t_{1/2} = \frac{[A]_o}{k}$

E) $t_{1/2} = \frac{2k}{[A]_o}$

Answer: C
Diff: 1 *Page Ref: 13.4*

31) How many half-lives are required for the concentration of reactant to decrease to 25% of its original value?

A) 1 B) 4 C) 1.5 D) 3.5 E) 2

Answer: E
Diff: 1 *Page Ref: 13.4*

32) How many half-lives are required for the concentration of reactant to decrease to 12.5% of its original value?

A) 3 B) 2 C) 2.5 D) 2.75 E) 8

Answer: A
Diff: 1 *Page Ref: 13.4*

33) How many half-lives are required for the concentration of reactant to decrease to 1.56% of its original value?

A) 6 B) 4 C) 24 D) 6.5 E) 7.5

Answer: A
Diff: 1 *Page Ref: 13.4*

34) If the concentration of a reactant is 6.25%, how many half-lives has it gone through?

A) 7 B) 10 C) 3 D) 4 E) 5

Answer: D
Diff: 1 *Page Ref: 13.4*

35) The rate constant for a second-order reaction is 0.54 $M^{-1}s^{-1}$. What is the half-life of this reaction if the initial concentration is 0.33 M?

A) 0.089 s B) 1.8 s C) 0.31 s D) 5.6 s E) 1.3 s

Answer: D
Diff: 2 *Page Ref: 13.4*

36) The rate constant for a first-order reaction is 0.54 $M^{-1}s^{-1}$. What is the half-life of this reaction if the initial concentration is 0.33 M?

A) 0.089 s B) 1.8 s C) 0.31 s D) 5.6 s E) 1.3 s

Answer: E
Diff: 2 *Page Ref: 13.4*

37) The rate constant for a zero-order reaction is 0.54 $M^{-1}s^{-1}$. What is the half-life of this reaction if the initial concentration is 0.33 M?

A) 0.089 s B) 1.8 s C) 0.31 s D) 5.6 s E) 1.3 s

Answer: C

Diff: 2 *Page Ref:* 13.4

38) The rate of disappearance of HBr in the gas phase reaction

$$2HBr(g) \rightarrow H_2(g) + Br_2(g)$$

is 0.301 M s^{-1} at 150°C. The rate of appearance of Br_2 is _______ M s^{-1}.

A) 1.66 B) 0.151 C) 0.0906 D) 0.602 E) 0.549

Answer: B

Diff: 2 *Page Ref:* 13.4

39) The rate of disappearance of HBr in the gas phase reaction

$$2HBr(g) \rightarrow H_2(g) + Br_2(g)$$

is 0.130 M s^{-1} at 150°C. The rate of reaction is _______ M s^{-1}.

A) 3.85 B) 0.0650 C) 0.0169 D) 0.260 E) 0.0860

Answer: B

Diff: 2 *Page Ref:* 13.4

40) The combustion of ethylene proceeds by the reaction

$$C_2H_4(g) + 3O_2(g) \rightarrow 2CO_2(g) + 2H_2O(g)$$

When the rate of disappearance of O_2 is 0.28 M s^{-1}, the rate of appearance of CO_2 is _______ M s^{-1}.

A) 0.19 B) 0.093 C) 0.84 D) 0.42 E) 0.56

Answer: A

Diff: 2 *Page Ref:* 13.4

41) The combustion of ethylene proceeds by the reaction

$$C_2H_4(g) + 3O_2(g) \rightarrow 2CO_2(g) + 2H_2O(g)$$

When the rate of disappearance of O_2 is 0.23 M s^{-1}, the rate of disappearance of C_2H_4 is _______ M s^{-1}.

A) 0.15 B) 0.077 C) 0.69 D) 0.35 E) 0.46

Answer: B

Diff: 2 *Page Ref:* 13.4

42) The isomerization of methylisonitrile to acetonitrile

$$CH_3NC(g) \rightarrow CH_3CN(g)$$

is first order in CH_3NC. The rate constant for the reaction is $9.45 \times 10^{-5}\ s^{-1}$ at 478 K. The half-life of the reaction when the initial $[CH_3NC]$ is 0.030 M is ________ s.

A) 1.06×10^4
B) 5.29×10^3
C) $3.53E \times 10^5$
D) 7.33×10^3
E) 1.36×10^{-4}

Answer: D
Diff: 3 *Page Ref: 13.4*

43) The elementary reaction

$$2NO_2(g) \rightarrow 2NO(g) + O_2(g)$$

is second order in NO_2 and the rate constant at 501 K is $7.93 \times 10^{-3}\ M^{-1}s^{-1}$. The reaction half-life at this temperature when $[NO_2]_0 = 0.45$ M is ________ s.

A) 3.6×10^{-3}
B) 0.011
C) 126
D) 87
E) 280

Answer: E
Diff: 2 *Page Ref: 13.4*

44) The isomerization of methylisonitrile to acetonitrile

$$CH_3NC(g) \rightarrow CH_3CN(g)$$

is first order in CH_3NC. The half life of the reaction is 5.20×10^1 s at 545 K. The rate constant when the initial $[CH_3NC]$ is 0.030 M is ________ s^{-1}.

A) 75.1
B) 0.641
C) 0.0133
D) 1.56
E) 2.84×10^4

Answer: C
Diff: 2 *Page Ref: 13.4*

45) The decomposition of N_2O_5 in solution in carbon tetrachloride proceeds via the reaction

$$2N_2O_5\ (soln) \rightarrow 4NO_2\ (soln) + O_2\ (soln)$$

The reaction is first order and has a rate constant of $4.82 \times 10^{-3}\ s^{-1}$ at 64°C. If the reaction is initiated with 0.058 mol in a 1.00–L vessel, how many moles remain after 151 s?

A) 0.055 M
B) 0.060 M
C) 0.028 M
D) 12 M
E) 2.0×10^3 M

Answer: C
Diff: 2 *Page Ref:* 13.4

46) SO_2Cl_2 decomposes in the gas phase by the reaction

$$SO_2Cl_2(g) \rightarrow SO_2(g) + Cl_2(g)$$

The reaction is first order in SO_2Cl_2 and the rate constant is $3.0 \times 10^{-6}\ s^{-1}$ at 600 K. A vessel is charged with 2.4 atm of SO_2Cl_2 at 600 K. The partial pressure of SO_2Cl_2 at 3.0×10^5 s is ________ atm.

A) 0.76 B) 2.2 C) 0.98 D) 0.29 E) 1.4×10^5

Answer: C
Diff: 2 *Page Ref:* 13.4

47) The reaction that occurs in a Breathalyzer, a device used to determine the alcohol level in a person's bloodstream, is given below. If the rate of appearance of $Cr_2(SO_4)_3$ is 1.64 mol/min at a particular moment, what is the rate of disappearance of C_2H_6O at that moment?

$$2\ K_2Cr_2O_7 + 8\ H_2SO_4 + 3\ C_2H_6O \rightarrow 2\ Cr_2(SO_4)_3 + 2\ K_2SO_4 + 11\ H_2O$$

A) 0.547 mol/min B) 1.09 mol/min C) 2.46 mol/min D) 4.92 mol/min

Answer: C
Diff: 3 *Page Ref:* 13.4

48) For a particular first–order reaction, it takes 24 minutes for the concentration of the reactant to decrease to 25% of its initial value. What is the value for rate constant (in s^{-1}) for the reaction?

A) $2.0 \times 10^{-4}\ s^{-1}$ B) $9.6 \times 10^{-4}\ s^{-1}$ C) $1.2 \times 10^{-2}\ s^{-1}$ D) $5.8 \times 10^{-2}\ s^{-1}$

Answer: B
Diff: 3 *Page Ref:* 13.4

49) The first–order reaction, $SO_2Cl_2 \rightarrow SO_2 + Cl_2$, has a rate constant equal to $2.20 \times 10^{-5}\ s^{-1}$ at 593 K. What percentage of the initial amount of SO_2Cl_2 will remain after 6.00 hours?

A) 1.00% B) 37.8% C) 40.2% D) 62.2%

Answer: D
Diff: 3 *Page Ref:* 13.4

50) The first-order reaction, $2\ N_2O(g) \rightarrow 2\ N_2(g) + O_2(g)$, has a rate constant equal to $0.76\ s^{-1}$ at 1000 K. How long will it take for the concentration of N_2O to decrease to 12% of its initial concentration?

A) 0.62 s B) 2.8 s C) 6.3 s D) 8.4 s

Answer: B
Diff: 3 *Page Ref: 13.4*

51) The isomerization reaction, $CH_3NC \rightarrow CH_3CN$, is first order and the rate constant is equal to $0.46\ s^{-1}$ at 600 K. What is the concentration of CH_3NC after 0.20 minutes if the initial concentration is 0.30 M?

A) 1.2×10^{-3} M B) 2.7×10^{-3} M C) 1.2×10^{-1} M D) 2.7×10^{-1} M

Answer: A
Diff: 3 *Page Ref: 13.4*

52) The following reaction is first order, $C_2H_6 \rightarrow 2\ CH_3$. If the rate constant is equal to $5.5 \times 10^{-4}\ s^{-1}$ at 1000 K, how long will it take for 0.35 mol of C_2H_6 in a 1.00 L container to decrease to 0. 20 mol in the same container?

A) 4.5 min B) 17 min C) 53 min D) 65 min

Answer: B
Diff: 3 *Page Ref: 13.4*

53) The rate constant, *k*, for a first-order reaction is equal to $4.2 \times 10^{-4}\ s^{-1}$. What is the half-life for the reaction?

A) 2.9×10^{-4} s B) 1.2×10^{3} s C) 1.7×10^{3} s D) 2.4×10^{3} s

Answer: C
Diff: 3 *Page Ref: 13.4*

54) The first-order reaction, $SO_2Cl_2 \rightarrow SO_2 + Cl_2$, has a half-life of 8.75 hours at 593 K. How long will it take for the concentration of SO_2Cl_2 to fall to 16.5% of its initial value?

A) 0.143 hr B) 2.28 hr C) 6.99 hr D) 22.7 hr

Answer: D
Diff: 3 *Page Ref: 13.4*

55) For the first-order reaction, $2\ N_2O(g) \rightarrow 2\ N_2(g) + O_2(g)$, what is the concentration of N_2O after 3 half-lives if 0.25 mol of N_2O is initially placed into a 1.00-L reaction vessel?

A) 1.6×10^{-2} M B) 3.1×10^{-2} M C) 6.2×10^{-2} M D) 1.2×10^{-1} M

Answer: B
Diff: 3 *Page Ref: 13.4*

56) Carbon-14, which is present in all living tissue, radioactively decays via a first-order process. A one-gram sample of wood taken from a living tree gives a rate for carbon-14 decay of 13.6 counts per minute. If the half-life for carbon-14 is 5720 years, how old is a wood sample that gives a rate for carbon-14 decay of 11.9 counts per minute?

A) 5.3×10^{2} yr B) 7.6×10^{2} yr C) 1.1×10^{3} yr D) 9.4×10^{3} yr

Answer: C
Diff: 3 *Page Ref: 13.4*

57) In aqueous solution, hypobromite ion, BrO^-, reacts to produce bromate ion, BrO_3^-, and bromide ion, Br^-, according to the following chemical equation.

$3\ BrO^-(aq) \rightarrow BrO_3^-(aq) + 2\ Br^-(aq)$

A plot of $1/[BrO^-]$ vs. time is linear and the slope is equal to $0.056\ M^{-1}s^{-1}$. If the initial concentration of BrO^- is 0.80 M, how long will it take one-half of the BrO^- ion to react?

A) 4.5×10^{-2} s
B) 7.1 s
C) 12 s
D) 22 s

Answer: D
Diff: 3 *Page Ref: 13.4*

58) The second-order reaction $2\ Mn(CO)_5 \rightarrow Mn_2(CO)_{10}$, has a rate constant equal to $3.0 \times 10^9\ M^{-1}\ s^{-1}$ at 25°C. If the initial concentration of $Mn(CO)_5$ is 2.0×10^{-5} M, how long will it take for 90.% of the reactant to disappear?

A) 6.7×10^{-16} s
B) 7.4×10^{-15} s
C) 1.5×10^{-4} s
D) 6.0×10^3 s

Answer: C
Diff: 3 *Page Ref: 13.4*

59) Nitrogen dioxide decomposes at 300°C via a second-order process to produce nitrogen monoxide and oxygen according to the following chemical equation.

$2\ NO_2(g) \rightarrow 2\ NO(g) + O_2(g)$.

A sample of $NO_2(g)$ is initially placed in a 2.50-L reaction vessel at 300°C. If the half-life and the rate constant at 300°C are 11 seconds and $0.54\ M^{-1}\ s^{-1}$, respectively, how many moles of NO_2 were in the original sample?

A) 0.17 mol
B) 0.42 mol
C) 5.9 mol
D) 15 mol

Answer: B
Diff: 3 *Page Ref: 13.4*

60) Hydrogen iodide decomposes at 800 K via a second-order process to produce hydrogen and iodine according to the following chemical equation.

$2HI(g) \rightarrow H_2(g) + I_2(g)$

At 800 K it takes 142 seconds for the initial concentration of HI to decrease from 6.75×10^{-2} M to 3.50×10^{-2} M. What is the rate constant for the reaction at this temperature?

A) $5.12 \times 10^{-4}\ M^{-1}s^{-1}$
B) $9.69 \times 10^{-2}\ M^{-1}s^{-1}$
C) $10.3\ M^{-1}s^{-1}$
D) $1.95 \times 10^3\ M^{-1}s^{-1}$

Answer: B
Diff: 3 *Page Ref: 13.4*

61) Which of the following reactions would you predict to have the smallest orientation factor?

A) $X_2 + Y_2 \rightarrow 2\ XY$
B) $NOCl_2 + NO \rightarrow 2\ NOCl$
C) $N_2 + O_2 \rightarrow 2\ NO$
D) $N + O_2 \rightarrow NO_2$
E) All of these reactions should have nearly identical orientation factors.

Answer: B
Diff: 1 *Page Ref: 13.5*

62) Which of the following reactions would you predict to have the largest orientation factor?

A) $NOCl(g) + NOCl(g) \rightarrow 2NO(g) + Cl_2(g)$

B) $Br_2(g) + H_2C{=}CH_2(g) \rightarrow H_2BrC{-}CBrH_2(g)$

C) $NH_3(g) + BCl_3(g) \rightarrow H_3N{-}BCl_3(g)$

D) $H(g) + F(g) \rightarrow HF(g)$

E) All of these reactions should have nearly identical orientation factors.

Answer: D

Diff: 1 *Page Ref: 13.5*

63) The aquation of tris(1,10–phenanthroline)iron(II) in acid solution takes place according to the equation:

$$Fe(phen)_3^{2+} + 3\,H_3O^+ + 3\,H_2O \rightarrow Fe(H_2O)_6^{2+} + 3\,phenH^+.$$

If the activation energy, E_a, is 126 kJ/mol and the rate constant at 30°C is 9.8×10^{-3} min^{-1}, what is the rate constant at 35°C?

A) 4.4×10^{-3} min^{-1}

B) 2.2×10^{-2} min^{-1}

C) 4.5×10^{1} min^{-1}

D) 2.3×10^{2} min^{-1}

Answer: B

Diff: 3 *Page Ref: 13.5*

64) A particular first–order reaction has a rate constant of 1.35×10^{2} s^{-1} at 25.0°C. What is the magnitude of k at 95.0°C if $E_a = 55.5$ kJ/mol?

A) 9.56×10^{3} s^{-1}

B) 2.85×10^{4} s^{-1}

C) 576 s^{-1}

D) 4.33×10^{87} s^{-1}

E) 1.36×10^{2} s^{-1}

Answer: A

Diff: 4 *Page Ref: 13.5*

65) A particular first–order reaction has a rate constant of 1.35×10^{2} s^{-1} at 25.0°C. What is the magnitude of k at 75.0°C if $E_a = 85.6$ kJ/mol?

A) 3.47×10^{4} s^{-1}

B) 1.92×10^{4} s^{-1}

C) 670 s^{-1}

D) 3.85×10^{6} s^{-1}

E) 1.36×10^{2} s^{-1}

Answer: B

Diff: 4 *Page Ref: 13.5*

66) Which rate law is termolecular?

A) rate = $k[A][B]^2$

B) rate = $k\ [A][B]^3$

C) rate = $k\ [A]^5$

D) rate = $k\ [A][B][C][D]$

E) rate = $k\ [A]^2$

Answer: A

Diff: 1 *Page Ref: 13.6*

67) Which rate law is bimolecular?

A) rate = $k[A][B]^3$

B) rate = $k\ [A][B]$

C) rate = $k\ [A]^3$

D) rate = $k\ [A][B][C][D]$

E) rate = $k\ [A]^2[B]^2$

Answer: B

Diff: 1 *Page Ref: 13.6*

68) Given the following proposed mechanism, predict the rate law for the overall reaction.

$2NO_2 + Cl_2 \rightarrow 2NO_2Cl$ (overall reaction)

<u>Mechanism</u>

$NO_2 + Cl_2 \rightarrow NO_2Cl + Cl$ slow

$NO_2 + Cl \rightarrow NO_2Cl$ fast

A) Rate = $k[NO_2][Cl_2]$

B) Rate = $k[NO_2]^2[Cl_2]^2$

C) Rate − $k[NO_2][Cl]$

D) Rate = $k[NO_2Cl][Cl]^2$

E) Rate = $k[NO_2Cl]^2$

Answer: A

Diff: 1 *Page Ref: 13.6*

Matching Questions

Match the following.

1) k
Diff: 1 *Page Ref: 13.3*

2) *n*, in Rate = $k[A]^n$
Diff: 1 *Page Ref: 13.3*

3) $t_{1/2}$
Diff: 1 *Page Ref: 13.4*

4) E_a
Diff: 1 *Page Ref: 13.5*

5) A
Diff: 1 *Page Ref: 13.5*

A) frequency factor

B) half-life

C) reaction order

D) activation energy

E) rate constant

Answers: 1) E 2) C 3) B 4) D 5) A

Short Answer Questions

1) What happens in the concentration of reactants and products during a chemical reaction?

Answer: The concentration of reactants decrease and the concentration of products increase during a chemical reaction.
Diff: 1 *Page Ref: 13.2*

2) What is the difference between average reaction rate and instantaneous reaction rate?

Answer: An average reaction rate is determined using long periods of time during the reaction. An instantaneous rate is found at a particular moment during reaction. Instantaneous rates usually decrease during the course of a reaction.
Diff: 1 *Page Ref: 13.2*

3) Explain how the order of a reaction can be determined.

Answer: The order of a reaction can only be determined experimentally.
Diff: 1 *Page Ref: 13.3*

4) Define half-life.

Answer: A half-life of a reaction is the time required for the concentration of a reactant to fall to one-half of its initial value.
Diff: 1 *Page Ref: 13.4*

5) Define activation energy.

Answer: The activation energy shows the energy of the molecule as the reaction proceeds.
Diff: 1 *Page Ref: 13.5*

6) Define the frequency factor.

Answer: The frequency factor represents the number of approaches to the activation barrier per unit time.

Diff: 1 *Page Ref: 13.5*

7) Explain what the exponential factor in the Arrhenius equation represents.

Answer: The exponential factor depends on both the activation energy of the reaction and the temperature at which it is done. Using this information, the exponential factor determines what fraction of collisions will have enough energy to overcome the activation energy and result in products.

Diff: 1 *Page Ref: 13.5*

8) Is the activation energy for a forward reaction the same as the activation energy for the reverse of the same reaction? Use a sketch of a reaction energy diagram.

Answer: The relationship between the activation energy of the forward and reverse reactions is related by the energy difference between the reactants and products. If the reactants are higher in energy than the products, the activation energy for the forward reaction will be smaller than the activation energy for the reverse reaction and vice versa. (A sketch showing this relationship should be included).

Diff: 1 *Page Ref: 13.6*

9) What is a catalyst and what function does it serve?

Answer: A catalyst is a substance that is not used up during a reaction. It also does not affect the products obtained. A catalyst merely changes the pathway of the reactants to products and in doing so, reduces the activation energy and speeds up the reaction.

Diff: 1 *Page Ref: 13.7*

10) What function do enzymes serve?

Answer: They are biological catalysts. They lower the activation energy of biological processes by using their three dimensional structure to capture reactants and hold them in a particular orientation that facilitates reaction.

Diff: 1 *Page Ref: 13.7*

Chapter 14 Chemical Equilibrium

Multiple Choice Questions

1) Which of the following statements is FALSE?
   A) When K >> 1, the forward reaction is favored and essentially goes to completion.
   B) When K << 1, the reverse reaction is favored and the forward reaction does not proceed to a great extent.
   C) When K ≈ 1, neither the forward or reverse reaction is strongly favored, and about the same amount of reactants and products exist at equilibrium.
   D) K >> 1 implies that the reaction is very fast at producing products.
   E) None of the above.

Answer: D
Diff: 1 *Page Ref:* 14.3

2) Give the direction of the reaction, if K >> 1.
   A) The forward reaction is favored.
   B) The reverse reaction is favored.
   C) Neither direction is favored.
   D) If the temperature is raised, then the forward reaction is favored.
   E) If the temperature is raised, then the reverse reaction is favored.

Answer: A
Diff: 1 *Page Ref:* 14.3

3) Give the direction of the reaction, if K << 1.
   A) The forward reaction is favored.
   B) The reverse reaction is favored.
   C) Neither direction is favored.
   D) If the temperature is raised, then the forward reaction is favored.
   E) If the temperature is raised, then the reverse reaction is favored.

Answer: B
Diff: 1 *Page Ref:* 14.3

4) Give the direction of the reaction, if K ≈ 1.
   A) The forward reaction is favored.
   B) The reverse reaction is favored.
   C) Neither direction is favored.
   D) If the temperature is raised, then the forward reaction is favored.
   E) If the temperature is raised, then the reverse reaction is favored.

Answer: C
Diff: 1 *Page Ref:* 14.3

5) Express the equilibrium constant for the following reaction.

$$16\ CH_3Cl(g) + 8\ Cl_2(g) \Leftrightarrow 16\ CH_2Cl_2(g) + 8\ H_{2(g)}$$

A) $K = \dfrac{[CH_2Cl_2][H_2]}{[CH_3Cl][Cl_2]}$

B) $K = \dfrac{[CH_2Cl_2]^{16}[H_2\ ]^{8}}{[CH_3Cl]^{16}[Cl_2]^{8}}$

C) $K = \dfrac{[CH_3Cl]^{16}[Cl_2]^{8}}{[CH_2Cl_2]^{16}[H_2]^{8}}$

D) $K = \dfrac{[CH_3Cl][Cl_2]}{[CH_2Cl_2][H_2]}$

E) $K = \dfrac{[CH_3Cl]^{1/2}[Cl_2]}{[CH_2Cl_2]^{1/2}[H_2]}$

Answer: B
Diff: 2 *Page Ref: 14.3*

6) The equilibrium constant is given for one of the reactions below. Determine the value of the missing equilibrium constant.

$H_2(g) + Br_2(g) \rightleftharpoons 2\ HBr(g)$ $K_c = 3.8 \times 10^4$
$2\ HBr(g) \rightleftharpoons H_2(g) + Br_2(g)$ $K_c = ?$

A) 1.9×10^4
B) 5.3×10^{-5}
C) 2.6×10^{-5}
D) 6.4×10^{-4}
E) 1.6×10^3

Answer: C
Diff: 2 *Page Ref: 14.3*

7) The equilibrium constant is given for one of the reactions below. Determine the value of the missing equilibrium constant.

$H_2(g) + Br_2(g) \rightleftharpoons 2\ HBr(g)$ $K_c = 3.8 \times 10^4$
$4\ HBr(g) \rightleftharpoons 2\ H_2(g) + 2\ Br_2(g)$ $K_c = ?$

A) 1.9×10^4
B) 5.1×10^{-3}
C) 2.6×10^{-5}
D) 6.9×10^{-10}
E) 1.6×10^3

Answer: D
Diff: 2 *Page Ref:* 14.3

8) The equilibrium constant is given for one of the reactions below. Determine the value of the missing equilibrium constant.

$2\ SO_2(g) + O_2(g) \rightleftharpoons 2\ SO_3(g)$ $K_c = 1.7 \times 10^6$
$SO_3(g) \rightleftharpoons 1/2\ O_2(g) + SO_2(g)$ $K_c = ?$

A) 3.4×10^2
B) 8.5
C) 1.3×10^3
D) 1.2×10^{-6}
E) 7.7×10^{-4}

Answer: E
Diff: 2 *Page Ref:* 14.3

9) The equilibrium constant is given for two of the reactions below. Determine the value of the missing equilibrium constant.

$A(g) + B(g) \rightleftharpoons AB(g)$ $Kc = 0.24$
$AB(g) + A(g) \rightleftharpoons A_2B(g)$ $K_c = 3.8$
$2\ A(g) + B(g) \rightleftharpoons A_2B(g)$ $K_c = ?$

A) 4.0 B) 0.91 C) 3.6 D) 16 E) 0.063

Answer: B
Diff: 2 *Page Ref:* 14.3

10) The equilibrium constant is given for two of the reactions below. Determine the value of the missing equilibrium constant.

$A(g) + 2B(g) \rightleftharpoons AB_2(g)$	$Kc = 59$
$AB_2(g) + B(g) \rightleftharpoons AB_3(g)$	$K_c = ?$
$A(g) + 3B(g) \rightleftharpoons AB_3(g)$	$K_c = 478$

A) 3.5×10^{-5}
B) 2.8×10^{4}
C) 8.1
D) 0.12
E) 89

Answer: C
Diff: 2 *Page Ref: 14.3*

11) The equilibrium constant is given for two of the reactions below. Determine the value of the missing equilibrium constant.

$2A(g) + B(g) \rightleftharpoons A_2B(g)$	$Kc = ?$
$A_2B(g) + B(g) \rightleftharpoons A_2B_2(g)$	$K_c = 16.4$
$2A(g) + 2B(g) \rightleftharpoons A_2B_2(g)$	$K_c = 28.2$

A) 11.8 B) 0.00216 C) 0.582 D) 462 E) 1.72

Answer: E
Diff: 2 *Page Ref: 14.3*

12) What is Δn for the following equation in relating K_c to K_p?

$$2\,SO_2(g) + O_2(g) \rightleftharpoons 2\,SO_3(g)$$

A) 3 B) -1 C) -2 D) 2 E) 1

Answer: B
Diff: 2 *Page Ref: 14.4*

13) What is Δn for the following equation in relating K_c to K_p?

$$N_2O_4(g) \rightleftharpoons 2NO_2(g)$$

A) 3 B) -1 C) -2 D) 2 E) 1

Answer: E
Diff: 2 *Page Ref: 14.4*

14) What is Δn for the following equation in relating K_c to K_p?

$$C_3H_8(g) + 5\,O_2(g) \rightleftharpoons 3\,CO_2(g) + 4\,H_2O(l)$$

A) 3 B) −1 C) −3 D) 2 E) 1

Answer: C
Diff: 2 *Page Ref: 14.4*

15) What is Δn for the following equation in relating K_c to K_p?

$$4\,NH_3(g) + 3\,O_2(g) \rightleftharpoons 2\,N_2(g) + 6\,H_2O(g)$$

A) 3 B) −1 C) −2 D) 2 E) 1

Answer: E
Diff: 2 *Page Ref: 14.4*

16) What is Δn for the following equation in relating K_c to K_p?

$$SO_3(g) + NO(g) \rightleftharpoons SO_2(g) + NO_2(g)$$

A) 0 B) −1 C) −2 D) 2 E) 1

Answer: A
Diff: 2 *Page Ref: 14.4*

17) In which of the following reactions will $K_c = K_p$?
A) $4\,NH_3(g) + 3\,O_2(g) \rightleftharpoons 2\,N_2(g) + 6\,H_2O(g)$
B) $SO_3(g) + NO(g) \rightleftharpoons SO_2(g) + NO_2(g)$
C) $2\,N_2(g) + O_2(g) \rightleftharpoons 2\,N_2O(g)$
D) $2\,SO_2(g) + O_2(g) \rightleftharpoons 2\,SO_3(g)$
E) None of the above reactions have $K_c = K_p$.

Answer: B
Diff: 2 *Page Ref: 14.4*

18) The reaction below has a K_p value of 3.3×10^{-5}. What is the value of K_c for this reaction at 700 K?

$$2\,SO_3(g) \rightleftharpoons 2\,SO_2(g) + O_2(g)$$

A) 5.7×10^{-7}
B) 1.7×10^{6}
C) 3.3×10^{-5}
D) 3.0×10^{4}
E) 1.9×10^{-3}

Answer: A
Diff: 3 *Page Ref: 14.4*

19) Express the equilibrium constant for the following reaction.

$$P_4(s) + 5\,O_2(g) \rightleftharpoons P_4O_{10}(s)$$

A) $K = \dfrac{[P_4][O_2]^5}{[P_4O_{10}]}$

B) $K = \dfrac{[P_4O_{10}]}{[P_4][O_2]^5}$

C) $K = [O_2]^{-5}$

D) $K = [O_2]^5$

E) $K = \dfrac{[P_4O_{10}]}{[P_4][O_2]^{1/5}}$

Answer: C

Diff: 2 *Page Ref: 14.5*

20) Express the equilibrium constant for the following reaction.

$$P_4O_{10}(s) \rightleftharpoons P_4(s) + 5\,O_2(g)$$

A) $K = \dfrac{[P_4][O_2]^5}{[P_4O_{10}]}$

B) $K = \dfrac{[P_4O_{10}]}{[P_4][O_2]^5}$

C) $K = [O_2]^{-5}$

D) $K = [O_2]^5$

E) $K = \dfrac{[P_4O_{10}]}{[P_4][O_2]^{1/5}}$

Answer: D

Diff: 2 *Page Ref: 14.5*

21) Determine the value of K_c for the following reaction if the equilibrium concentrations are as follows: $[P_4O_{10}]_{eq}$ = 2.000 moles, $[P_4]_{eq}$ = 3.000 moles, $[O_2]_{eq}$ = 4.000 M

$$P_4O_{10}(s) \rightleftharpoons P_4(s) + 5\,O_2(g)$$

A) 20.00 B) 4.000 C) 1.320 D) 1536 E) 1024

Answer: E

Diff: 2 *Page Ref: 14.5*

22) Determine the value of K_c for the following reaction if the equilibrium concentrations are as follows: $[N_2]_{eq}$ = 3.6 M, $[O_2]_{eq}$ = 4.1 M, $[N_2O]_{eq}$ = 3.3×10^{-18} M.

$$2\ N_2(g) + O_2(g) \rightleftharpoons 2\ N_2O(g)$$

A) 2.2×10^{-19}
B) 4.5×10^{18}
C) 2.0×10^{-37}
D) 5.0×10^{36}
E) 4.9×10^{-17}

Answer: C
Diff: 3 *Page Ref: 14.6*

23) Determine the value of K_c for the following reaction if the equilibrium concentrations are as follows: $[N_2]_{eq}$ = 1.5 M, $[H_2]_{eq}$ = 1.1 M, $[NH_3]_{eq}$ = 0.47 M.

$$N_2(g) + 3\ H_2(g) \rightleftharpoons 2\ NH_3(g)$$

A) 3.5 B) 0.28 C) 9.1 D) 0.11 E) 0.78

Answer: D
Diff: 3 *Page Ref: 14.6*

24) Determine the value of K_c for the following reaction if the equilibrium concentrations are as follows: $[HCl]_{eq}$ = 0.13 M, $[HI]_{eq}$ = 5.6×10^{-16} M, $[Cl_2]_{eq}$ = 0.0019 M.

$$2\ HCl(g) + I_2(s) \rightleftharpoons 2\ HI(g) + Cl_2(g)$$

A) 8.2×10^{-18}
B) 2.8×10^{31}
C) 1.2×10^{17}
D) 1.4×10^{-19}
E) 3.5×10^{-32}

Answer: E
Diff: 3 *Page Ref: 14.6*

25) Determine the value of K_c for the following reaction if the equilibrium concentrations are as follows: $[HCl]_{eq} = 0.13$ M, $[HI]_{eq} = 5.6 \times 10^{-16}$ M, $[Cl_2]_{eq} = 0.0019$ M.

$$2\ HI(g) + Cl_2(g) \rightleftharpoons 2\ HCl(g) + I_2(s)$$

A) 8.2×10^{-18}
B) 2.8×10^{31}
C) 1.2×10^{17}
D) 1.4×10^{-19}
E) 3.5×10^{-32}

Answer: B
Diff: 3 *Page Ref: 14.6*

26) Determine the value of K_p for the following reaction if the equilibrium concentrations are as follows: $P(CO)_{eq} = 6.8 \times 10^{-11}$ atm, $P(O_2)_{eq} = 1.3 \times 10^{-3}$ atm, $P(CO_2)_{eq} = 0.041$ atm.

$$2\ CO(g) + O_2(g) \rightleftharpoons 2\ CO_2(g)$$

A) 3.6×10^{-21}
B) 2.8×10^{20}
C) 4.6×10^{11}
D) 2.2×10^{-12}
E) 3.6×10^{-15}

Answer: B
Diff: 3 *Page Ref: 14.6*

27) Calculate the value of $[N_2]_{eq}$ if $[H_2]_{eq} = 2.0$ M, $[NH_3]_{eq} = 0.5$ M, and $K_c = 2$.

$$N_2(g) + 3\ H_2(g) \rightleftharpoons 2\ NH_3(g)$$

A) 0.016 M B) 0.031 M C) 0.062 M D) 0.40 M E) 62.5 M

Answer: A
Diff: 4 *Page Ref: 14.6*

28) Calculate P $[NO]_{eq}$, if P $[NOCl]_{eq} = 0.33$ atm, P $[Cl_2]_{eq} = 0.50$ atm, and $K_p = 1.9 \times 10^{-2}$.

$$2\ NOCl(g) \rightleftharpoons 2\ NO(g) + Cl_2(g)$$

A) 1.7 atm
B) 0.0042 atm
C) 0.30 atm
D) 0.064 atm
E) 0.087 atm

Answer: D
Diff: 4 *Page Ref: 14.6*

29) Which of the following statements is TRUE?
A) Dynamic equilibrium occurs when the rate of the forward reaction equals the rate of the reverse reaction.
B) The equilibrium constant for the forward reaction is equal to the equilibrium constant for the reverse reaction.
C) A reaction quotient (Q) larger than the equilibrium constant (K) means that the reaction will favor the production of more products.
D) Dynamic equilibrium indicates that the amount of reactants and products are equal.
E) All of the above are true.

Answer: A
Diff: 1 *Page Ref: 14.7*

30) In a reaction mixture containing only reactants, what is the value of Q?
A) -1
B) 1
C) ∞
D) 0
E) It cannot be determined without concentrations.

Answer: D
Diff: 1 *Page Ref: 14.7*

31) In a reaction mixture containing only products, what is the value of Q?
A) -1
B) 1
C) ∞
D) 0
E) It cannot be determined without concentrations.

Answer: C
Diff: 1 *Page Ref: 14.7*

32) In a reaction mixture containing reactants and products, each at a concentration of 1M, what is the value of Q?
A) -1
B) 1
C) ∞
D) 0
E) It cannot be determined without concentrations.

Answer: B
Diff: 1 *Page Ref: 14.7*

33) Which of the following statements is TRUE?
A) If $Q < K$, it means the reverse reaction will proceed to form more reactants.
B) If $Q > K$, it means the forward reaction will proceed to form more products.
C) If $Q = K$, it means the reaction is at equilibrium.
D) All of the above are true.
E) None of the above are true.

Answer: C
Diff: 1 *Page Ref: 14.7*

34) Which of the following statements is TRUE?

A) If $Q < K$, it means the reverse reaction will proceed to form more reactants.
B) If $Q > K$, it means the reverse reaction will proceed to form more reactants.
C) If $Q = K$, it means the reaction is not at equilibrium.
D) All of the above are true.
E) None of the above are true.

Answer: B
Diff: 1 *Page Ref: 14.7*

35) Which of the following statements is TRUE?

A) If $Q < K$, it means the forward reaction will proceed to form more products.
B) If $Q > K$, it means the forward reaction will proceed to form more products.
C) If $Q = K$, it means the reaction is not at equilibrium.
D) All of the above are true.
E) None of the above are true.

Answer: A
Diff: 1 *Page Ref: 14.7*

36) Consider the following reaction, equilibrium concentrations, and equilibrium constant at a particular temperature. Determine the equilibrium concentration of $H_2O(g)$.

$C_2H_4(g) + H_2O(g) \rightleftharpoons C_2H_5OH(g)$ $\quad K_c = 9.0 \times 10^3$

$[C_2H_4]_{eq} = 0.015$ M $\quad [C_2H_5OH]_{eq} = 1.69$ M

A) 9.9×10^{-7} M
B) 80. M
C) 1.0 M
D) 1.68 M
E) 0.013 M

Answer: E
Diff: 3 *Page Ref: 14.8*

37) Consider the following reaction, equilibrium concentrations, and equilibrium constant at a particular temperature. Determine the equilibrium concentration of $SO_3(g)$.

$2\ SO_2(g) + O_2(g) \rightleftharpoons 2\ SO_3(g)$ $\quad K_c = 1.7 \times 10^8$

$[SO_2]_{eq} = 0.0034$ M $\quad [O_2]_{eq} = 0.0018$ M

A) 1.9 M
B) 1.0×10^3 M
C) 0.53 M
D) 9.6×10^{-4} M
E) 0.73 M

Answer: A
Diff: 3 *Page Ref: 14.8*

38) Consider the following reaction, equilibrium concentrations, and equilibrium constant at a particular temperature. Determine the equilibrium concentration of $SO_3(g)$.

$$2\ SO_2(g) + O_2(g) \rightleftharpoons 2\ SO_3(g) \qquad K_c = 1.7 \times 10^8$$

$$[SO_3]_{eq} = 0.0034\ M \qquad [O_2]_{eq} = 0.0018\ M$$

A) 2.8×10^{13} M
B) 1.88 M
C) 6.1×10^{-6} M
D) 1.0×10^3 M
E) 1.4 M

Answer: C
Diff: 3 *Page Ref: 14.8*

39) Consider the following reaction, equilibrium concentrations, and equilibrium constant at a particular temperature. Determine the equilibrium pressure of H_2.

$$D_2(g) + H_2(g) \rightleftharpoons 2\ HD(g) \qquad K_p = 1.80$$

$$P(D_2)_{eq} = 1.1 \times 10^{-3}\ atm \qquad P(HD)_{eq} = 2.7 \times 10^{-3}\ atm$$

A) 2.7 atm
B) 1.4 atm
C) 0.73 atm
D) 3.7×10^{-3} atm
E) 8.1×10^{-4} atm

Answer: D
Diff: 3 *Page Ref: 14.8*

40) Consider the following reaction, equilibrium concentrations, and equilibrium constant at a particular temperature. Determine the equilibrium pressure of CO.

$$CO(g) + 2\ H_2(g) \rightleftharpoons CH_3OH(l) \qquad K_p = 2.25 \times 10^4$$

$$P(H_2)_{eq} = 0.52\ atm$$

A) 8.3×10^4 atm
B) 1.2×10^{-5} atm
C) 6.25×10^{-3} atm
D) 8.5×10^{-5} atm
E) 1.6×10^{-4} atm

Answer: E
Diff: 3 *Page Ref: 14.8*

41) Consider the following reaction:

$$Xe(g) + 2\ F_2(g) \rightarrow XeF_4(g)$$

A reaction mixture initially contains 2.24 atm Xe and 4.27 atm F_2. If the equilibrium pressure of Xe is 0.34 atm, find the equilibrium constant (K_p) for the reaction.

A) 25 B) 0.12 C) 0.99 D) 8.3 E) 0.040

Answer: A
Diff: 4 *Page Ref: 14.8*

42) Consider the following reaction:

$$CH_4(g) + 2\ H_2S(g) \rightleftharpoons CS_2(g) + 4\ H_2(g)$$

A reaction mixture initially contains 0.50 M CH_4 and 0.75 M H_2S. If the equilibrium concentration of H_2 is 0.44 M, find the equilibrium constant (K_c) for the reaction.

A) 0.23 B) 0.038 C) 2.9 D) 10. E) 0.34

Answer: B
Diff: 4 *Page Ref: 14.8*

43) Consider the following reaction:

$$COCl_2(g) \rightleftharpoons CO(g) + Cl_2(g)$$

A reaction mixture initially contains 1.6 M $COCl_2$. Determine the equilibrium concentration of CO if K_c for the reaction at this temperature is 8.33×10^{-4}. Calculate this based on the assumption that the answer is negligible compared to 1.6.

A) 4.2×10^{-4} M
B) 1.5×10^{-3} M
C) 3.7×10^{-2} M
D) 2.1×10^{-2} M
E) 1.3×10^{-3} M

Answer: C
Diff: 4 *Page Ref: 14.8*

44) Consider the following reaction:

$$CuS(s) + O_2(g) \rightleftharpoons Cu(s) + SO_2(g)$$

A reaction mixture initially contains 2.9 M O_2. Determine the equilibrium concentration of O_2 if K_c for the reaction at this temperature is 1.5.

A) 1.9 M B) 1.7 M C) 2.2 M D) 1.2 M E) 0.59 M

Answer: D
Diff: 4 *Page Ref: 14.8*

45) Consider the following reaction:

$$2\ H_2O(g) + 2\ SO_2(g) \rightleftharpoons 2\ H_2S(g) + 3\ O_2(g)$$

A reaction mixture initially contains 2.8 M H_2O and 2.6 M SO_2. Determine the equilibrium concentration of H_2S if K_c for the reaction at this temperature is 1.3×10^{-6}.

A) 0.045 M
B) 0.058 M
C) 0.028 M
D) 3.1×10^{-3} M
E) 0.12 M

Answer: E
Diff: 4 *Page Ref: 14.8*

46) Consider the following reaction:

$$CO_2(g) + C(graphite) \rightleftharpoons 2\ CO(g)$$

A reaction mixture initially contains 0.56 atm CO_2 and 0.32 atm CO. Determine the equilibrium pressure of CO if K_p for the reaction at this temperature is 2.25.

A) 0.83 atm B) 0.31 atm C) 0.26 atm D) 0.58 atm E) 0.42 atm

Answer: A
Diff: 5 *Page Ref: 14.8*

47) Consider the following reaction:

$$NO(g) + SO_3(g) \rightleftharpoons NO_2(g) + SO_2(g)$$

A reaction mixture initially contains 0.86 atm NO and 0.86 atm SO_3. Determine the equilibrium pressure of NO_2 if K_p for the reaction at this temperature is 0.0118.

A) 0.78 atm B) 0.084 atm C) 0.012 atm D) 0.85 atm E) 0.048 atm

Answer: B
Diff: 5 *Page Ref: 14.8*

48) The following reaction is exothermic. Which change will shift the equilibrium to the left?

$$2\ SO_2(g) + O_2(g) \rightleftharpoons 2\ SO_3(g)$$

A) raising the temperature
B) adding SO_3
C) removing O_2
D) all of the above
E) none of the above

Answer: D
Diff: 1 *Page Ref: 14.9*

49) The following reaction is exothermic. Which change will shift the equilibrium to the left?

$$2\ SO_2(g) + O_2(g) \rightleftharpoons 2\ SO_3(g)$$

A) raising the temperature
B) decrease pressure
C) increase volume
D) all of the above
E) none of the above

Answer: D
Diff: 1 *Page Ref: 14.9*

50) Consider the following reaction at equilibrium. What effect will adding more SO_3 have on the system?

$$SO_2(g) + NO_2(g) \rightleftharpoons SO_3(g) + NO(g)$$

A) The reaction will shift in the direction of products.
B) The reaction will shift to decrease the pressure.
C) No change will occur since SO_3 is not included in the equilibrium expression.
D) The reaction will shift in the direction of reactants.
E) The equilibrium constant will decrease.

Answer: D
Diff: 2 *Page Ref: 14.9*

51) Consider the following reaction at equilibrium. What effect will removing NO_2 have on the system?

$$SO_2(g) + NO_2(g) \rightleftharpoons SO_3(g) + NO(g)$$

A) The reaction will shift in the direction of products.
B) The reaction will shift to decrease the pressure.
C) No change will occur since SO_3 is not included in the equilibrium expression.
D) The reaction will shift in the direction of reactants.
E) The equilibrium constant will decrease.

Answer: D
Diff: 2 *Page Ref: 14.9*

52) Consider the following reaction at equilibrium. What effect will adding more H_2S have on the system?

$$2\ H_2S(g) + 3\ O_2(g) \rightleftharpoons 2\ H_2O(g) + 2\ SO_2(g)$$

A) The reaction will shift to the left.
B) No change will be observed.
C) The equilibrium constant will decrease.
D) The equilibrium constant will increase.
E) The reaction will shift in the direction of products.

Answer: E
Diff: 2 *Page Ref: 14.9*

53) Consider the following reaction at equilibrium. What effect will removing H_2O have on the system?

$$2\ H_2S(g) + 3\ O_2(g) \rightleftharpoons 2\ H_2O(g) + 2\ SO_2(g)$$

A) The reaction will shift to the left.
B) No change will be observed.
C) The equilibrium constant will decrease.
D) The equilibrium constant will increase.
E) The reaction will shift in the direction of products.

Answer: E
Diff: 2 *Page Ref: 14.9*

54) Consider the following reaction at equilibrium. What effect will increasing the volume of the reaction mixture have on the system?

$$2\ H_2S(g) + 3\ O_2(g) \rightleftharpoons 2\ H_2O(g) + 2\ SO_2(g)$$

A) The reaction will shift to the right in the direction of products.
B) No effect will be observed.
C) The reaction will shift to the left in the direction of reactants.
D) The equilibrium constant will decrease.
E) The equilibrium constant will increase.

Answer: C
Diff: 2 *Page Ref: 14.9*

55) Consider the following reaction at equilibrium. What effect will increasing the pressure of the reaction mixture have on the system?

$$2\,H_2S(g) + 3\,O_2(g) \rightleftharpoons 2\,H_2O(g) + 2\,SO_2(g)$$

A) The reaction will shift to the right in the direction of products.
B) No effect will be observed.
C) The reaction will shift to the left in the direction of reactants.
D) The equilibrium constant will decrease.
E) The equilibrium constant will increase.

Answer: A
Diff: 2 *Page Ref: 14.9*

56) Consider the following reaction at equilibrium. What effect will adding 1 mole of Ar to the reaction mixture have on the system?

$$2\,H_2S(g) + 3\,O_2(g) \rightleftharpoons 2\,H_2O(g) + 2\,SO_2(g)$$

A) The reaction will shift to the right in the direction of products.
B) No effect will be observed.
C) The reaction will shift to the left in the direction of reactants.
D) The equilibrium constant will decrease.
E) The equilibrium constant will increase.

Answer: B
Diff: 2 *Page Ref: 14.9*

57) Consider the following reaction at equilibrium. What effect will adding some C have on the system?

$$CO_2(g) + C(graphite) \rightleftharpoons 2\,CO(g)$$

A) No effect will be observed since C is not included in the equilibrium expression.
B) The equilibrium constant will decrease.
C) The reaction will shift to the left in the direction of reactants.
D) The equilibrium constant will increase.
E) The reaction will shift to the right in the direction of products.

Answer: A
Diff: 2 *Page Ref: 14.9*

58) Consider the following reaction at equilibrium. What effect will adding CO_2 have on the system?

$$CO_2(g) + C(graphite) \rightleftharpoons 2\ CO(g)$$

A) No effect will be observed since CO_2 is not included in the equilibrium expression.
B) The equilibrium constant will decrease.
C) The reaction will shift to the left in the direction of reactants.
D) The equilibrium constant will increase.
E) The reaction will shift to the right in the direction of products.

Answer: E
Diff: 2 *Page Ref: 14.9*

59) Consider the following reaction at equilibrium. What effect will reducing the volume of the reaction mixture have on the system?

$$CuS(s) + O_2(g) \rightleftharpoons Cu(s) + SO_2(g)$$

A) The equilibrium constant will decrease.
B) No effect will be observed.
C) The reaction will shift to the right in the direction of products.
D) The equilibrium constant will increase.
E) The reaction will shift to the left in the direction of reactants.

Answer: B
Diff: 2 *Page Ref: 14.9*

60) Consider the following reaction at equilibrium. What effect will increasing the pressure of the reaction mixture have on the system?

$$CuS(s) + O_2(g) \rightleftharpoons Cu(s) + SO_2(g)$$

A) The equilibrium constant will decrease.
B) No effect will be observed.
C) The reaction will shift to the right in the direction of products.
D) The equilibrium constant will increase.
E) The reaction will shift to the left in the direction of reactants.

Answer: B
Diff: 2 *Page Ref: 14.9*

61) Consider the following reaction at equilibrium. What effect will reducing the volume of the reaction mixture have on the system?

$$Xe(g) + 2\ F_2(g) \rightarrow XeF_4(g)$$

A) The reaction will shift to the left in the direction of reactants.
B) The equilibrium constant will increase.
C) No effect will be observed.
D) The reaction will shift to the right in the direction of products.
E) The equilibrium constant will decrease.

Answer: D
Diff: 2 *Page Ref:* 14.9

62) Consider the following reaction at equilibrium. What effect will reducing the pressure of the reaction mixture have on the system?

$$Xe(g) + 2\ F_2(g) \rightarrow XeF_4(g)$$

A) The reaction will shift to the left in the direction of reactants.
B) The equilibrium constant will increase.
C) No effect will be observed.
D) The reaction will shift to the right in the direction of products.
E) The equilibrium constant will decrease.

Answer: A
Diff: 2 *Page Ref:* 14.9

63) Consider the following reaction at equilibrium. What effect will increasing the temperature have on the system?

$$Fe_3O_4(s) + CO(g) \rightleftharpoons 3\ FeO(s) + CO_2(g) \qquad \Delta H° = +35.9\ kJ$$

A) The reaction will shift to the left in the direction of reactants.
B) The equilibrium constant will increase.
C) The equilibrium constant will decrease.
D) No effect will be observed.
E) The reaction will shift to the right in the direction of products.

Answer: E
Diff: 2 *Page Ref:* 14.9

64) Consider the following reaction at equilibrium. What effect will decreasing the temperature have on the system?

$$CO_2(g) + 2\ H_2O(l) \rightleftharpoons CH_4(g) + 2\ O_2(g) \qquad \Delta H° = +890\ kJ$$

A) The reaction will shift to the left in the direction of reactants.
B) The equilibrium constant will decrease.
C) The equilibrium constant will increase.
D) The reaction will shift to the right in the direction of products.
E) No effect will be observed.

Answer: A
Diff: 2 *Page Ref:* 14.9

65) Consider the following reaction at equilibrium. What effect will increasing the temperature have on the system?

$$C_3H_8(g) + 5\ O_2(g) \rightleftharpoons 3\ CO_2(g) + 4\ H_2O(l) \qquad \Delta H° = -2220\ kJ$$

A) The reaction will shift to the right in the direction of products.
B) The reaction will shift to the left in the direction of reactants.
C) The equilibrium constant will increase.
D) The equilibrium constant will decrease.
E) No effect will be observed.

Answer: B
Diff: 2 *Page Ref:* 14.9

66) Consider the following reaction at equilibrium. What effect will reducing the volume have on the system?

$$C_3H_8(g) + 5\ O_2(g) \rightleftharpoons 3\ CO_2(g) + 4\ H_2O(l) \qquad \Delta H° = -2220\ kJ$$

A) The reaction will shift to the right in the direction of products.
B) The reaction will shift to the left in the direction of reactants.
C) The equilibrium constant will increase.
D) The equilibrium constant will decrease.
E) No effect will be observed.

Answer: A
Diff: 2 *Page Ref:* 14.9

67) Consider the following reaction at equilibrium. What effect will reducing the pressure have on the system?

$$C_3H_8(g) + 5\ O_2(g) \rightleftharpoons 3\ CO_2(g) + 4\ H_2O(l) \qquad \Delta H° = -2220\ kJ$$

A) The reaction will shift to the right in the direction of products.
B) The reaction will shift to the left in the direction of reactants.
C) The equilibrium constant will increase.
D) The equilibrium constant will decrease.
E) No effect will be observed.

Answer: B
Diff: 2 *Page Ref: 14.9*

Algorithmic Questions

1) Express the equilibrium constant for the following reaction.

$$4\ CH_3Cl(g) + 2\ Cl_2(g) \Leftrightarrow 4\ CH_2Cl_2(g) + 2\ H_{2(g)}$$

A) $K = \dfrac{[CH_2Cl_2]^4[H_2]^2}{[CH_3Cl]^4[Cl_2]^2}$

B) $K = \dfrac{[CH_2Cl_2]^2[H_2]}{[CH_3Cl]^2[Cl_2]}$

C) $K = \dfrac{[CH_3Cl]^4[Cl_2]^2}{[CH_2Cl_2]^4[H_2]^2}$

D) $K = \dfrac{[CH_3Cl][Cl_2]}{[CH_2Cl_2][H_2]}$

E) $K = \dfrac{[CH_3Cl][Cl_2]^{1/2}}{[CH_2Cl_2][H_2]^{1/2}}$

Answer: A
Diff: 2 *Page Ref: 14.3*

2) Express the equilibrium constant for the following reaction.

$$6\ N_2(g) + 18\ H_2(g) \Leftrightarrow 12\ NH_3(g)$$

A) $K = \dfrac{[N_2][H_2]^{1/3}}{[NH_3]^{1/2}}$

B) $K = \dfrac{[NH_3]}{[N_2]^2[H_2]^3}$

C) $K = \dfrac{[NH_3]^{12}}{[N_2]^6[H_2]^{18}}$

D) $K = \dfrac{[N_2]^6[H_2]^{18}}{[NH_3]^{12}}$

E) $K = \dfrac{[NH_3]^{1/2}}{[N_2][H_2]^{1/3}}$

Answer: C
Diff: 2 *Page Ref: 14.3*

3) Express the equilibrium constant for the following reaction.

$$12\ NH_3(g) \Leftrightarrow 6\ N_2(g) + 18\ H_2(g)$$

A) $K = \dfrac{[N_2][H_2]^{1/3}}{[NH_3]^{1/2}}$

B) $K = \dfrac{[NH_3]}{[N_2]^2[H_2]^3}$

C) $K = \dfrac{[NH_3]^{12}}{[N_2]^6[H_2]^{18}}$

D) $K = \dfrac{[N_2]^6[H_2]^{18}}{[NH_3]^{12}}$

E) $K = \dfrac{[NH_3]^{1/2}}{[N_2][H_2]^{1/3}}$

Answer: D
Diff: 2 *Page Ref: 14.3*

4) Express the equilibrium constant for the following reaction.

$$4PCl_5(g) \rightleftharpoons 4PCl_3(g) + 4Cl_2(g)$$

A) $K = \frac{[PCl_5]^4}{[PCl_3]^4[Cl_2]^4}$

B) $K = \frac{[PCl_3]^2[Cl_2]^2}{[PCl_5]^2}$

C) $K = \frac{[PCl_3][Cl]^2}{[PCl_5]}$

D) $K = \frac{[PCl_3]^4[Cl_2]^4}{[PCl_5]^4}$

E) $K = \frac{[PCl_5]^{1/2}}{[PCl_3]^{1/2}[Cl_2]^{1/2}}$

Answer: D
Diff: 2 *Page Ref: 14.3*

5) Express the equilibrium constant for the following reaction.

$$5H_2(g) + 5Br_2(g) \rightleftharpoons 10\ HBr(g)$$

A) $K = \frac{[HBr]}{[H_2]^{1/2}[Br_2]^{1/2}}$

B) $K = \frac{[H_2]^5[Br_2]^5}{[HBr]^{10}}$

C) $K = \frac{[H_2][Br_2]}{[HBr]}$

D) $K = \frac{[H_2]^{1/2}[Br_2]^{1/2}}{[HBr]}$

E) $K = \frac{[HBr]^{10}}{[H_2]^5[Br_2]^5}$

Answer: E
Diff: 2 *Page Ref: 14.3*

6) Express the equilibrium constant for the following reaction.

$$8\ P(g) + 12\ Cl_2(g) \rightleftharpoons 8\ PCl_3(g)$$

A) $K = \frac{[PCl_3]^8}{[P]^8[Cl_2]^{12}}$

B) $K = \frac{[PCl_3]^{1/2}}{[P]^{1/2}[Cl_2]^{1/3}}$

C) $K = \frac{[PCl_3]}{[P][Cl_2]^{3/2}}$

D) $K = \frac{[P]^8[Cl_2]^{12}}{[PCl_3]^8}$

E) $K = \frac{[P][Cl_2]^{3/2}}{[PCl_3]}$

Answer: A
Diff: 2 *Page Ref: 14.3*

7) Express the equilibrium constant for the following reaction.

$$P(g) + 3/2\ Cl_2(g) \rightleftharpoons PCl_3(g)$$

A) $K = \frac{[PCl_3]^2}{[P]^2[Cl_2]^3}$

B) $K = \frac{[PCl_3]^{1/2}}{[P]^{1/2}[Cl_2]^{1/3}}$

C) $K = \frac{[PCl_3]}{[P][Cl_2]^{3/2}}$

D) $K = \frac{[P]^{1/2}[Cl_2]^{1/3}}{[PCl_3]^{1/2}}$

E) $K = \frac{[P][Cl_2]^{3/2}}{[PCl_3]}$

Answer: C
Diff: 2 *Page Ref: 14.3*

8) The equilibrium constant is given for one of the reactions below. Determine the value of the missing equilibrium constant.

$2\ HD(g) \rightleftharpoons H_2(g) + D_2(g)$ $K_c = 0.28$
$6\ H_2(g) + 6\ D_2(g) \rightleftharpoons 12\ HD(g)$ $K_c = ?$

A) 0.00048 B) 1.2 C) 1.62 D) 2075 E) 0.81

Answer: D
Diff: 2 *Page Ref: 14.3*

9) The equilibrium constant is given for one of the reactions below. Determine the value of the missing equilibrium constant.

$N_2O_4(g) \rightleftharpoons 2\ NO_2(g)$ $K_c = 1.46$
$5\ N_2O_4(g) \rightleftharpoons 10\ NO_2(g)$ $K_c = ?$

A) 1.46 B) 0.292 C) 6.63 D) 1.08 E) 7.30

Answer: C
Diff: 2 *Page Ref: 14.3*

10) For the reaction: $N_2(g) + 2\ O_2(g) \rightleftharpoons 2\ NO_2(g)$, $K_c = 8.3 \times 10^{-10}$ at 25°C. What is the concentration of N_2 gas at equilibrium when the concentration of NO_2 is five times the concentration of O_2 gas?

A) 3.3×10^{-11} M B) 1.7×10^{-10} M C) 6.0×10^{9} M D) 3.0×10^{10} M

Answer: D
Diff: 3 *Page Ref: 14.4*

11) The K_p for the reaction below is 1.49×10^{8} at 100.0°C:

$CO(g) + Cl_2(g) \rightarrow COCl_2(g)$

In an equilibrium mixture of the three gases, $P_{CO} = P_{Cl_2} = 2.22 \times 10^{-4}$ atm. The partial pressure of the product, phosgene ($COCl_2$), is ________ atm.

A) 7.34
B) 3.02×10^{15}
C) 3.31×10^{-16}
D) 3.31×10^{4}
E) 6.67×10^{11}

Answer: A
Diff: 3 *Page Ref: 14.4*

12) At 900.0 K, the equilibrium constant (K_p) for the following reaction is 0.345.

$$2SO_2 + O_2(g) \rightarrow 2SO_3(g)$$

At equilibrium, the partial pressure of SO_2 is 35.0 atm and that of O_2 is 15.9 atm. The partial pressure of SO_3 is ________ atm.

A) 82.0
B) 4.21×10^{-3}
C) 192
D) 6.20×10^{-4}
E) 40.2

Answer: A
Diff: 3 *Page Ref: 14.4*

13) The decomposition of ammonia is: $2\ NH_3(g) \rightarrow N_2(g) + 3\ H_2(g)$. If K_p is 1.5×10^3 at 400°C, what is the partial pressure of ammonia at equilibrium when N_2 is 0.10 atm and H_2 is 0.15 atm?

A) 2.2×10^{-7} atm
B) 4.7×10^{-4} atm
C) 2.1×10^3 atm
D) 4.4×10^6 atm

Answer: B
Diff: 3 *Page Ref: 14.4*

14) Phosphorous trichloride and phosphorous pentachloride equilibrate in the presence of molecular chlorine according to the reaction:

$$PCl_3(g) + Cl_2(g) \rightarrow PCl_5(g)$$

An equilibrium mixture at 450 K contains
$P_{PCl_3} = 0.124$ atm,
$P_{Cl_2} = 0.157$ atm, and
$P_{PCl_5} = 1.30$ atm. What is the value of K_p at this temperature?

A) 66.7
B) 1.50×10^{-2}
C) 2.53×10^{-2}
D) 1.02
E) 4.63

Answer: A
Diff: 3 *Page Ref: 14.4*

15) What is Δn for the following equation in relating K_c to K_p?

$$N_2(g) + 3\ Cl_2(g) \rightleftharpoons 2\ NCl_3(g)$$

A) 4
B) -4
C) -2
D) 2
E) 1

Answer: C
Diff: 2 *Page Ref: 14.4*

16) What is Δn for the following equation in relating K_c to K_p?

$$CH_4(g) + H_2O(g) \rightleftharpoons CO(g) + 3\, H_2(g)$$

A) -3 B) -1 C) -2 D) 2 E) 0

Answer: D
Diff: 2 *Page Ref: 14.4*

17) In which of the following reactions will $K_c = K_p$?

A) $H_2(g) + I_2(g) \rightleftharpoons 2\, HI(g)$
B) $CH_4(g) + H_2O(g) \rightleftharpoons CO(g) + 3\, H_2(g)$
C) $N_2O_4(g) \rightleftharpoons 2NO_2(g)$
D) $CO(g) + 2\, H_2(g) \rightleftharpoons CH_3OH(g)$
E) $N_2(g) + 3\, Cl_2(g) \rightleftharpoons 2\, NCl_3(g)$

Answer: A
Diff: 2 *Page Ref: 14.4*

18) The reaction below has a K_c value of 61. What is the value of K_p for this reaction at 400 K?

$$N_2(g) + 3\, H_2(g) \rightleftharpoons 2\, NH_3(g)$$

A) 15
B) 61
C) 1.8
D) 5.7×10^{-2}
E) 1.9×10^{-2}

Answer: D
Diff: 3 *Page Ref: 14.4*

19) The reaction below has a K_p value of 41. What is the value of K_c for this reaction at 400 K?

$$N_2(g) + 3\, H_2(g) \rightleftharpoons 2\, NH_3(g)$$

A) 3.8×10^{-2} B) 4.4×10^4 C) 26 D) 2.3×10^{-5} E) 41

Answer: B
Diff: 3 *Page Ref: 14.4*

20) The reaction below has a K_c value of 1.0×10^{12}. What is the value of K_p for this reaction at 200 K?

$$2\,SO_2(g) + O_2(g) \rightleftharpoons 2\,SO_3(g)$$

A) 6.1×10^{-14}
B) 1.0×10^{12}
C) 1.6×10^{-11}
D) 1.6×10^{13}
E) 6.1×10^{10}

Answer: E
Diff: 3 *Page Ref: 14.4*

21) What is Δn for the following equation in relating K_c to K_p?

$$P_4(s) + 5\,O_2(g) \rightleftharpoons P_4O_{10}(s)$$

A) 3 B) 4 C) -2 D) -6 E) -5

Answer: E
Diff: 2 *Page Ref: 14.4*

22) What is Δn for the following equation in relating K_c to K_p?

$$2\,Rb(s) + 2\,H_2O(l) \rightleftharpoons 2\,RbOH(aq) + H_2(g)$$

A) 3 B) -1 C) -2 D) 2 E) 1

Answer: A
Diff: 2 *Page Ref: 14.4*

23) What is Δn for the following equation in relating K_c to K_p?

$$KBrO_3(s) \rightleftharpoons KBrO(s) + O_2(g)$$

A) 3 B) -1 C) -2 D) 2 E) 1

Answer: E
Diff: 2 *Page Ref: 14.4*

24) For which reaction will $K_p = K_c$?
A) $C(s) + O_2(g) \rightleftharpoons CO_2(g)$
B) $2\,HgO(s) \rightleftharpoons Hg(l) + O_2(g)$
C) $CaCO_3(s) \rightleftharpoons CaO(s) + CO_2(g)$
D) $H_2CO_3(s) \rightleftharpoons H_2O(l) + CO_2(g)$
E) $2\,H_2O(l) \rightleftharpoons 2\,H_2(g) + O_2(g)$

Answer: A
Diff: 2 *Page Ref: 14.4*

25) Express the equilibrium constant for the following reaction.

$$2\ K(s) + 2\ H_2O(l) \rightleftharpoons 2\ KOH(aq) + H_2(g)$$

A) $K = \dfrac{[\ KOH]^2[H_2]}{[K]^2[H_2O]^2}$

B) $K = [H_2][KOH]^{-2}$

C) $K = \dfrac{[K]^2[H_2O]^2}{[\ KOH]^2[H_2]}$

D) $K = [H_2][KOH]^2$

E) $K = \dfrac{[\ KOH]^{1/2}[H_2]}{[K]^{1/2}[H_2O]^{1/2}}$

Answer: D
Diff: 2 *Page Ref: 14.5*

26) Express the equilibrium constant for the following reaction.

$$NaClO_3(s) \rightleftharpoons NaClO(s) + O_2(g)$$

A) $K = \dfrac{[NaClO][O_2]}{[\ NaClO_3]}$

B) $K = [O_2]^{-1}$

C) $K = \dfrac{[NaClO]^2[O_2]^2}{[\ NaClO_3]^2}$

D) $K = \dfrac{[\ NaClO_3]}{[NaClO][O_2]}$

E) $K = [O_2]$

Answer: E
Diff: 2 *Page Ref: 14.5*

27) An equilibrium mixture of CO, O_2 and CO_2 at a certain temperature contains 0.0010 M CO_2 and 0.0100 M O_2. At this temperature, K_c equals 1.4×10^2 for the reaction:

$$2\ CO(g) + O_2(g) \rightleftharpoons 2\ CO_2(g).$$

What is the equilibrium concentration of CO?

A) 7.1×10^{-7} M B) 8.4×10^{-4} M C) 1.4×10^{-2} M D) 1.2×10^{-1} M

Answer: B
Diff: 4 *Page Ref: 14.6*

28) At a certain temperature, K_c equals 1.4×10^2 for the reaction:

$$2\,CO(g) + O_2(g) \rightleftharpoons 2\,CO_2(g).$$

If a 3.00–L flask contains 0.400 mol of CO_2 and 0.100 mol of O_2 at equilibrium, how many moles of CO are also present in the flask?

A) 0.555 mol B) 0.185 mol C) 0.107 mol D) 0.0114 mol

Answer: B
Diff: 4 *Page Ref: 14.6*

29) Determine the value of K_c for the following reaction if the equilibrium concentrations are as follows: $[PBr_5]_{eq} = 0.56$ M, $[PBr_3]_{eq} = 0.23$ M, $[Br_2]_{eq} = 3.3$ M.

$$PBr_5(g) \rightleftharpoons PCl_3(g) + Br_2(g)$$

A) 1.4 B) 0.93 C) 0.74 D) 1.1 E) 0.76

Answer: A
Diff: 3 *Page Ref: 14.6*

30) Determine the value of K_c for the following reaction if the equilibrium concentrations are as follows: $[H_2]_{eq} = 0.14$ M, $[Br_2]_{eq} = 0.39$ M, $[HBr]_{eq} = 1.6$ M.

$$H_2(g) + Br_2(g) \rightleftharpoons 2\,HBr(g)$$

A) 2.1×10^{-2}
B) 29
C) 47
D) 3.4×10^{-2}
E) 8.7×10^{-2}

Answer: C
Diff: 3 *Page Ref: 14.6*

31) Determine the value of K_p for the following reaction if the equilibrium concentrations are as follows: $P(NOCl)_{eq} = 0.22$ atm, $P(NO)_{eq} = 0.10$ atm, $P(Cl_2)_{eq} = 0.11$ atm.

$$2\,NOCl(g) \rightleftharpoons 2\,NO(g) + Cl_2(g)$$

A) 1.8 B) 44 C) 2.3×10^2 D) 2.3×10^{-2} E) 0.53

Answer: D
Diff: 3 *Page Ref: 14.6*

32) At a certain temperature, nitrogen and hydrogen react to form ammonia:

$$N_2(g) + 3\ H_2(g) \rightleftharpoons 2\ NH_3(g).$$

When initial amounts of N_2, H_2, and NH_3 are mixed, the concentration of NH_3 increases. Which statement below is TRUE?

A) $K_c < Q$
B) $K_c > Q$
C) $K_c = Q$
D) More information is needed to make a statement about K_c.

Answer: B
Diff: 3 *Page Ref: 14.7*

33) Consider the following reaction and its equilibrium constant:

$$SO_2(g) + NO_2(g) \rightleftharpoons SO_3(g) + NO(g) \qquad K_c = 0.33$$

A reaction mixture contains 0.41 M SO_2, 0.14 M NO_2, 0.12 M SO_3 and 0.14 M NO. Which of the following statements is TRUE concerning this system?

A) The reaction will shift in the direction of reactants.
B) The equilibrium constant will decrease.
C) The reaction will shift in the direction of products.
D) The reaction quotient will decrease.
E) The system is at equilibrium.

Answer: C
Diff: 3 *Page Ref: 14.7*

34) Consider the following reaction and its equilibrium constant:

$$4\ CuO(s) + CH_4(g) \rightleftharpoons CO_2(g) + 4\ Cu(s) + 2\ H_2O(g) \qquad K_c = 1.10$$

A reaction mixture contains 0.22 M CH_4, 0.67 M CO_2 and 1.3 M H_2O. Which of the following statements is TRUE concerning this system?

A) The reaction will shift in the direction of products.
B) The equilibrium constant will increase.
C) The reaction quotient will increase.
D) The reaction will shift in the direction of reactants.
E) The system is at equilibrium.

Answer: D
Diff: 3 *Page Ref: 14.7*

35) Consider the following reaction and its equilibrium constant:

$I_2(g) + Br_2(g) \rightleftharpoons 2\ IBr(g)$ $\quad K_c = 1.1 \times 10^2$

A reaction mixture contains 0.35 M I_2, 0.31 M Br_2 and 3.5 M IBr. Which of the following statements is TRUE concerning this system?

A) The reaction will shift in the direction of products.
B) The reaction will shift in the direction of reactants.
C) The reaction quotient will decrease.
D) The equilibrium constant will increase.
E) The system is at equilibrium.

Answer: E
Diff: 3 *Page Ref:* 14.7

36) Consider the following reaction and its equilibrium constant:

$I_2(g) \rightleftharpoons 2I(g)$ $\quad K_p = 0.209$

A reaction mixture contains 0.89atm I_2 and1.601.60 atm I. Which of the following statements is TRUE concerning this system?

A) The reaction will shift in the direction of reactants.
B) The reaction quotient will increase.
C) The reaction will shift in the direction of products.
D) The equilibrium constant will decrease.
E) The system is at equilibrium.

Answer: A
Diff: 3 *Page Ref:* 14.7

37) Consider the following reaction, equilibrium concentrations, and equilibrium constant at a particular temperature. Determine the equilibrium concentration of $NO_2(g)$.

$N_2O_4(g) \rightleftharpoons 2\ NO_2(g)$ $\quad K_c = 0.21$

$[N_2O_4]_{eq} = 0.033$ M

A) 1.44 M
B) 8.3×10^{-2} M
C) 12×10^{-2} M
D) 6.9×10^{-3} M
E) 144 M

Answer: B
Diff: 3 *Page Ref:* 14.8

38) Consider the following reaction, equilibrium concentrations, and equilibrium constant at a particular temperature. Determine the equilibrium concentration of $CO_2(g)$.

$NH_2COONH_4(s) \rightleftharpoons 2\ NH_3(g) + CO_2(g)$ $K_c = 1.58 \times 10^{-8}$

$[NH_3]_{eq} = 2.3 \times 10^{-3}$ M

A) 0.053 M
B) 3.6×10^{-11} M
C) 3.0×10^{-3} M
D) 6.9×10^{-6} M
E) 0.022 M

Answer: C
Diff: 3 *Page Ref: 14.8*

39) For the isomerization reaction:

butane $\rightleftharpoons$ isobutane

K_p equals 25 at 500°C. If the initial pressures of butane and isobutane are 20. atm and 0.0 atm, respectively, what are the pressures of the two gases at equilibrium?
A) *P*(butane) = 0.77 atm and *P*(isobutane) = 19.2 atm
B) *P*(butane) = 0.80 atm and *P*(isobutane) = 20. atm
C) *P*(butane) = 19.2 atm and *P*(isobutane) = 0.77 atm
D) *P*(butane) = 20 atm and *P*(isobutane) = 0.80 atm

Answer: A
Diff: 5 *Page Ref: 14.8*

40) Cyclohexane, C_6H_{12}, undergoes a molecular rearrangement in the presence of $AlCl_3$ to form methylcyclopentane, $CH_3C_5H_9$, according to the equation:

$C_6H_{12} \rightleftharpoons CH_3C_5H_9$

If $K_c = 0.143$ at 25°C for this reaction, find the equilibrium concentrations of C_6H_{12} and $CH_3C_5H_9$ if the initial concentrations are 0.200 M and 0.075 M, respectively.
A) $[C_6H_{12}] = 0.041$ M, $[CH_3C_5H_9] = 0.041$ M
B) $[C_6H_{12}] = 0.159$ M, $[CH_3C_5H_9] = 0.116$ M
C) $[C_6H_{12}] = 0.241$ M, $[CH_3C_5H_9] = 0.034$ M
D) $[C_6H_{12}] = 0.253$ M, $[CH_3C_5H_9] = 0.022$ M

Answer: C
Diff: 5 *Page Ref: 14.8*

41) At a certain temperature the equilibrium constant, K_c, equals 0.11 for the reaction:

$$2\ ICl(g) \rightleftharpoons I_2(g) + Cl_2(g).$$

What is the equilibrium concentration of ICl if 0.45 mol of I_2 and 0.45 mol of Cl_2 are initially mixed in a 2.0-L flask?

A) 0.14 M B) 0.17 M C) 0.27 M D) 0.34 M

Answer: A
Diff: 5 Page Ref: 14.8

42) The equilibrium constant, K_p, equals 3.40 for the isomerization reaction:

cis-2-butene $\rightleftharpoons$ *trans*-2-butene.

If a flask initially contains 0.250 atm of *cis*-2-butene and 0.145 atm of *trans*-2-butene, what is the equilibrium pressure of each gas?

A) *P*(*cis*-2-butene) = 0.0426 atm and *P*(*trans*-2-butene) = 0.145 atm
B) *P*(*cis*-2-butene) = 0.0471 atm and *P*(*trans*-2-butene) = 0.160 atm
C) *P*(*cis*-2-butene) = 0.0735 atm and *P*(*trans*-2-butene) = 0.250 atm
D) *P*(*cis*-2-butene) = 0.0898 atm and *P*(*trans*-2-butene) = 0.305 atm

Answer: D
Diff: 5 Page Ref: 14.8

43) Phosphorus pentachloride decomposes to phosphorus trichloride at high temperatures according to the equation:

$$PCl_5(g) \rightleftharpoons PCl_3(g) + Cl_2(g)$$

At 250° 0.125 M PCl_5 is added to the flask. If $K_c = 1.80$, what are the equilibrium concentrations of each gas?

A) $[PCl_5] = 0.00765$ M, $[PCl_3] = 0.117$ M, and $[Cl_2] = 0.117$ M
B) $[PCl_5] = 0.0625$ M, $[PCl_3] = 0.335$ M, and $[Cl_2] = 0.335$ M
C) $[PCl_5] = 1.80$ M, $[PCl_3] = 1.80$ M, and $[Cl_2] = 1.80$ M
D) $[PCl_5] = 3.96$ M, $[PCl_3] = 3.83$ M, and $[Cl_2] = 3.83$ M

Answer: A
Diff: 5 Page Ref: 14.8

44) K_c is 1.67×10^{20} at 25°C for the formation of iron(III) oxalate complex ion:

$$Fe^{3+}(aq) + 3\ C_2O_4^{2-}(aq) \rightleftharpoons [Fe(C_2O_4)_3]^{3-}(aq).$$

If 0.0200 M Fe^{3+} is initially mixed with 1.00 M oxalate ion, what is the concentration of Fe^{3+} ion at equilibrium?

A) 1.44×10^{-22} M B) 0.0100 M C) 8.35×10^{19} M D) 6.94×10^{21} M

Answer: A
Diff: 5 Page Ref: 14.8

45) Dinitrogen tetroxide partially decomposes according to the following equilibrium:

$N_2O_4(g) \rightarrow 2NO_2(g)$

A 1.000-L flask is charged with 3.00×10^{-2} mol of N_2O_4. At equilibrium, 2.36×10^{-2} mol of N_2O_4 remains. K_{eq} for this reaction is ________.

A) 0.723
B) 0.391
C) 0.212
D) 6.94×10^{-3}
E) 1.92×10^{-4}

Answer: D
Diff: 5 *Page Ref: 14.8*

46) Identify the change that will always shift the equilibrium to the right.

A) remove reactant
B) increase product
C) decrease volume
D) increase pressure
E) increase volume

Answer: C
Diff: 2 *Page Ref: 14.9*

47) The following reaction is exothermic. Which change will shift the equilibrium to the left?

$2\ SO_2(g) + O_2(g) \rightleftharpoons 2\ SO_3(g)$

A) raising the temperature
B) adding argon
C) decrease volume
D) all of the above
E) none of the above

Answer: A
Diff: 1 *Page Ref: 14.9*

48) Cyclohexane (C_6H_{12}) undergoes a molecular rearrangement in the presence of $AlCl_3$ to form methylcyclopentane (MCP) according to the equation:

$C_6H_{12} \rightleftharpoons MCP$

If $K_c = 0.143$ at 25°C for this reaction, predict the direction in which the system will shift if the initial concentrations of C_6H_{12} and MCP are 0.0400 M and 0.0200 M, respectively. The system

A) will shift left.
B) will shift right.
C) is already at equilibrium.
D) is not at equilibrium and will remain in an unequilibrated state.

Answer: A
Diff: 2 *Page Ref: 14.9*

49) The equilibrium constant is equal to 5.00 at 1300 K for the reaction:

$$2\ SO_2(g) + O_2(g) \rightleftharpoons 2\ SO_3(g).$$

If initial concentrations are $[SO_2] = 6.00$ M, $[O_2] = 0.45$ M, and $[SO_3] = 9.00$ M, the system is
A) at equilibrium.
B) not at equilibrium and will remain in an unequilibrated state.
C) not at equilibrium and will shift to the left to achieve an equilibrium state.
D) not at equilibrium and will shift to the right to achieve an equilibrium state.

Answer: A
Diff: 2 *Page Ref: 14.9*

50) The equilibrium constant, K_p, equals 3.40 at 25°C for the isomerization reaction:

cis–2–butene ⇌ *trans*–2–butene.

If a flask initially contains 5.00 atm of each gas, in what direction will the system shift to reach equilibrium?
A) It will shift left.
B) It will shift right.
C) The system is already at equilibrium.
D) The system is not at equilibrium and will remain in an unequilibrated state.

Answer: B
Diff: 3 *Page Ref: 14.9*

Matching Questions

Match the following.

1) $K << 1$
Diff: 1 *Page Ref: 14.3*

2) $K \approx 1$
Diff: 1 *Page Ref: 14.3*

3) $Q < K$
Diff: 1 *Page Ref: 14.7*

4) $Q > K$
Diff: 1 *Page Ref: 14.7*

5) $Q = K$
Diff: 1 *Page Ref: 14.7*

6) $Q >> 1$
Diff: 1 *Page Ref: 14.7*

A) reaction is at equilibrium

B) reaction favors formation of more products

C) reverse reaction is favored

D) reaction has a larger amount of products than reactants

E) reaction does not strongly favor reactants or products

F) reaction will favor formation of reactants

Answers: 1) C 2) E 3) B 4) F 5) A
6) D

Short Answer Questions

1) Explain dynamic equilibrium. Use the generic reaction A(g) $\rightleftharpoons$ B(g) to explain.

Answer: In this reaction, dynamic equilibrium means that the ratio of [B]/[A] is constant. Even though the ratio of their concentrations is constant, the forward and reverse reactions continue to occur. The rates of the forward and reverse reactions are equal in order to keep the ratio of [B]/[A] constant.

Diff: 1 *Page Ref: 14.2*

2) Why is an equilibrium constant unitless?

Answer: Each concentration or pressure for a reactant or product, when used to calculate the equilibrium expression is actually a ratio to a reference concentration or pressure. Each ratio is therefore unitless before it is used to calculate into the equilibrium expression.

Diff: 1 *Page Ref: 14.4*

3) Can the K_p and K_c for a reaction ever have the same value? Why or why not?

Answer: Yes, they can have the same value. Since K_p depends on pressure and K_c depends on concentration, the two are related through a factor of RT (gas constant times temperature). The two constants can only have the same value in reactions where the number of moles of gaseous reactants is equal to the number of moles of gaseous products. When this is true, then factors of RT will all cancel making $K_p = K_c$.

Diff: 2 *Page Ref: 14.4*

4) Why aren't solids or liquids included in an equilibrium expression?

Answer: The equilibrium constant relates different concentrations of reactants and products to one another. Since the concentrations of liquids and solids are constant, their concentration(s) becomes part of the constant value of the equilibrium constant.

Diff: 1 *Page Ref: 14.5*

5) How is the reaction quotient different from an equilibrium constant for a given reaction?

Answer: The expression for both are identical. However, the equilibrium constant only holds true for systems AT equilibrium. The reaction quotient can be calculated at concentrations/pressures away from equilibrium and is used to determine which direction a reaction must go to achieve equilibrium.

Diff: 1 *Page Ref: 14.7*

6) Define Le Chatelier's Principle.

Answer: Le Chatelier's Principle states that when a chemical system at equilibrium is disturbed, the system shifts in a direction that minimizes the disturbance.

Diff: 1 *Page Ref: 14.9*

7) How will equilibrium be affected, with the same number of moles of gas on both sides, if the pressure is increased?

Answer: There will be no effect on equilibrium.

Diff: 2 *Page Ref: 14.9*

Chapter 15 Acids and Bases

Multiple Choice Questions

1) ________ is found in carbonated beverages due to the reaction of carbon dioxide with water.

A) CH_3COOH
B) H_2CO_3
C) $HCOOH$
D) C_6H_5COOH
E) CH_3CH_2COOH

Answer: B
Diff: 1 *Page Ref: 15.2*

2) Which of the following species is amphoteric?

A) CO_3^{2-}
B) HF
C) NH_4^+
D) HPO_4^{2-}
E) None of the above are amphoteric.

Answer: D
Diff: 1 *Page Ref: 15.3*

3) What is the conjugate acid of HCO_3^- ?

A) H_3O^+ B) H_2O C) CO_3^{2-} D) OH^- E) H_2CO_3

Answer: E
Diff: 1 *Page Ref: 15.3*

4) What is the conjugate base of $H_2PO_4^-$?

A) HPO_4^{2-} B) PO_4^{3-} C) H_3PO_4 D) H_3O^+ E) OH^-

Answer: A
Diff: 1 *Page Ref: 15.3*

5) Which of the following is NOT a conjugate acid-base pair?

A) NH_4^+/NH_3
B) H_3O^+/OH^-
C) H_2SO_3/HSO_3^-
D) $C_2H_3O_2^-/HC_2H_3O_2$
E) All of the above are conjugate acid-base pairs.

Answer: B
Diff: 1 *Page Ref: 15.3*

6) Identify the weak diprotic acid.
A) CH_3COOH
B) HCOOH
C) H_3PO_4
D) H_2SO_4
E) H_2CO_3
Answer: E
Diff: 1 *Page Ref:* 15.4

7) Identify the weak diprotic acid.
A) HNO_3 B) H_3PO_4 C) H_2SO_3 D) $HClO_4$ E) H_2SO_4
Answer: C
Diff: 1 *Page Ref:* 15.4

8) Give the characteristics of a strong acid.
A) ionizes completely in aqueous solutions
B) has equilibrium far to the right
C) has a polar bond
D) has a weaker bond to hydrogen
E) all of the above
Answer: E
Diff: 1 *Page Ref:* 15.4

9) The stronger the acid, then which of the following is TRUE?
A) The stronger the conjugate acid.
B) The stronger the conjugate base.
C) The weaker the conjugate base.
D) The weaker the conjugate acid.
E) None of the above.
Answer: C
Diff: 1 *Page Ref:* 15.4

10) Place the following in order of increasing acid strength.

$HBrO_2$ $HBrO_3$ HBrO $HBrO_4$

A) $HBrO_2 < HBrO_4 < HBrO < HBrO_3$
B) $HBrO < HBrO_2 < HBrO_3 < HBrO_4$
C) $HBrO_2 < HBrO_3 < HBrO_4 < HBrO$
D) $HBrO_4 < HBrO_2 < HBrO_3 < HBrO$
E) $HBrO < HBrO_4 < HBrO_3 < HBrO_2$
Answer: B
Diff: 1 *Page Ref:* 15.4

11) Which of the following statements is TRUE?

A) A strong acid is composed of a proton and an anion that have a very strong attraction for one another.

B) A weak base is composed of a cation and an anion with a very weak attraction between them.

C) A strong acid has a strong conjugate base.

D) The conjugate base of a very weak acid is stronger than the conjugate base of a strong acid.

E) None of the above statements are true.

Answer: D

Diff: 1 *Page Ref: 15.4*

12) Which of the following acids will have the strongest conjugate base?

A) HCl B) $HClO_4$ C) HNO_3 D) HCN E) HI

Answer: D

Diff: 1 *Page Ref: 15.4*

13) Which of the following solutions would have the highest pH? Assume that they are all 0.10 M in acid at 25 °C. The acid is followed by its K_a value.

A) HF, 3.5×10^{-4}

B) HCN, 4.9×10^{-10}

C) HNO_2, 4.6×10^{-4}

D) $HCHO_2$, 1.8×10^{-4}

E) $HClO_2$, 1.1×10^{-2}

Answer: B

Diff: 1 *Page Ref: 15.4*

14) Which of the following is TRUE?

A) A neutral solution contains $[H_2O] = [H_3O^+]$

B) An neutral solution does not contain any H_3O^+ or OH^-

C) An acidic solution has $[H_3O^+] > [OH^-]$

D) A basic solution does not contain H_3O^+

E) None of the above are true.

Answer: C

Diff: 1 *Page Ref: 15.5*

15) What is the concentration of hydroxide ions in pure water at 30.0 °C, if K_w at this temperature is 1.47×10^{-14}?

A) 1.00×10^{-7} M

B) 1.30×10^{-7} M

C) 1.47×10^{-7} M

D) 8.93×10^{-8} M

E) 1.21×10^{-7} M

Answer: E

Diff: 2 *Page Ref: 15.5*

16) What is the pH of pure water at 40.0°C if the K_w at this temperature is 2.92×10^{-14}?

A) 6.767 B) 0.465 C) 7.000 D) 7.233 E) 8.446

Answer: A
Diff: 2 *Page Ref:* 15.5

17) What is the K_w of pure water at 50.0°C, if the pH is 6.630?

A) 2.34×10^{-7}
B) 5.50×10^{-14}
C) 2.13×10^{-14}
D) 1.00×10^{-14}
E) There is not enough information to calculate the K_w.

Answer: B
Diff: 2 *Page Ref:* 15.5

18) Calculate the concentration of H_3O^+ in a solution that contains 5.5×10^{-5} M OH^- at 25°C. Identify the solution as acidic, basic or neutral.

A) 1.8×10^{-10} M, basic
B) 1.8×10^{-10} M, acidic
C) 5.5×10^{-10} M, neutral
D) 9.2×10^{-1} M, acidic
E) 9.2×10^{-1} M, basic

Answer: A
Diff: 2 *Page Ref:* 15.5

19) Calculate the concentration of OH^- in a solution that contains 3.9×10^{-4} M H_3O^+ at 25°C. Identify the solution as acidic, basic or neutral.

A) 2.6×10^{-11} M, acidic
B) 2.6×10^{-11} M, basic
C) 3.9×10^{-4} M, neutral
D) 2.7×10^{-2} M, basic
E) 2.7×10^{-2} M, acidic

Answer: A
Diff: 2 *Page Ref:* 15.5

20) Calculate the hydronium ion concentration in an aqueous solution with a pH of 9.85 at 25°C.

A) 7.1×10^{-5} M
B) 4.2×10^{-10} M
C) 8.7×10^{-10} M
D) 6.5×10^{-5} M
E) 1.4×10^{-10} M

Answer: E
Diff: 2 *Page Ref:* 15.5

21) Calculate the hydroxide ion concentration in an aqueous solution with a pH of 9.85 at 25°C.

A) 7.1×10^{-5} M

B) 4.2×10^{-10} M

C) 8.7×10^{-10} M

D) 6.5×10^{-5} M

E) 1.4×10^{-10} M

Answer: A

Diff: 2 *Page Ref:* 15.5

22) Calculate the hydronium ion concentration in an aqueous solution with a pOH of 9.85 at 25°C.

A) 7.1×10^{-5} M

B) 4.2×10^{-10} M

C) 8.7×10^{-10} M

D) 6.5×10^{-5} M

E) 1.4×10^{-10} M

Answer: A

Diff: 2 *Page Ref:* 15.5

23) Calculate the hydroxide ion concentration in an aqueous solution with a pOH of 9.85 at 25°C.

A) 7.1×10^{-5} M

B) 4.2×10^{-10} M

C) 8.7×10^{-10} M

D) 6.5×10^{-5} M

E) 1.4×10^{-10} M

Answer: E

Diff: 2 *Page Ref:* 15.5

24) Determine the pH of a 0.023 M HNO_3 solution.

A) 12.36 B) 3.68 C) 1.64 D) 2.30 E) 2.49

Answer: C

Diff: 2 *Page Ref:* 15.5

25) Determine the pH of a 0.00598 M $HClO_4$ solution.

A) 11.777 B) 6.434 C) 7.566 D) 2.223 E) 3.558

Answer: D

Diff: 2 *Page Ref:* 15.5

26) Determine the pOH of a 0.00598 M $HClO_4$ solution.

A) 11.777 B) 6.434 C) 7.566 D) 2.223 E) 3.558

Answer: A

Diff: 2 *Page Ref:* 15.5

27) Calculate the pH of a solution that contains 7.8×10^{-6} M OH^- at 25°C.

A) 1.28 B) 5.11 C) 12.72 D) 8.89 E) 9.64

Answer: D

Diff: 3 *Page Ref:* 15.5

28) Calculate the pOH of a solution that contains 7.8×10^{-6} M OH^- at 25°C.

A) 1.28 B) 5.11 C) 12.72 D) 8.89 E) 9.64

Answer: B

Diff: 3 Page Ref: 15.5

29) Calculate the hydroxide ion concentration in an aqueous solution with a pH of 4.33 at 25°C.

A) 2.1×10^{-10} M

B) 9.7×10^{-10} M

C) 4.7×10^{-5} M

D) 3.8×10^{-5} M

E) 6.3×10^{-6} M

Answer: A

Diff: 3 Page Ref: 15.5

30) Calculate the hydronium ion concentration in an aqueous solution with a pH of 4.33 at 25°C.

A) 2.1×10^{-10} M

B) 9.7×10^{-10} M

C) 4.7×10^{-5} M

D) 3.8×10^{-5} M

E) 6.3×10^{-6} M

Answer: C

Diff: 3 Page Ref: 15.5

31) Calculate the hydronium ion concentration in an aqueous solution with a pOH of 4.33 at 25°C.

A) 2.1×10^{-10} M

B) 9.7×10^{-10} M

C) 4.7×10^{-5} M

D) 3.8×10^{-5} M

E) 6.3×10^{-6} M

Answer: A

Diff: 3 Page Ref: 15.5

32) Determine the $[H_3O^+]$ in a 0.265 M HClO solution. The K_a of HClO is 2.9×10^{-8}.

A) 1.1×10^{-10} M

B) 7.7×10^{-9} M

C) 1.3×10^{-6} M

D) 4.9×10^{-4} M

E) 8.8×10^{-5} M

Answer: E

Diff: 3 Page Ref: 15.6

33) Find the percent ionization of a 0.337 M HF solution. The K_a for HF is 3.5×10^{-4}.

A) 1.1 %

B) 1.2×10^{-2} %

C) 3.2 %

D) 3.5×10^{-2} %

E) 4.7 %

Answer: C

Diff: 3 *Page Ref: 15.6*

34) Determine the pH of a 0.461 M $C_6H_5CO_2H$ M solution if the K_a of $C_6H_5CO_2H$ is 6.5×10^{-5}.

A) 2.26 B) 4.52 C) 11.74 D) 9.48 E) 5.48

Answer: A

Diff: 4 *Page Ref: 15.6*

35) Determine the K_a of an acid whose 0.294 M solution has a pH of 2.80.

A) 1.2×10^{-5}

B) 8.5×10^{-6}

C) 2.7

D) 4.9×10^{-7}

E) 5.4×10^{-3}

Answer: B

Diff: 4 *Page Ref: 15.6*

36) Which of the following bases is the STRONGEST? The base is followed by its K_b.

A) $(CH_3CH_2)_2NH$, 8.6×10^{-4}

B) CH_3NH_2, 4.4×10^{-4}

C) $C_6H_5NH_2$, 4.0×10^{-10}

D) NH_3, 1.76×10^{-5}

E) C_5H_5N, 1.7×10^{-9}

Answer: A

Diff: 1 *Page Ref: 15.7*

37) Which of the following bases is the WEAKEST? The base is followed by its K_b value.

A) $HOCH_2CH_2NH_2$, 3.2×10^{-5}

B) NH_3, 1.76×10^{-5}

C) C_5H_5N, 1.7×10^{-9}

D) $(CH_3CH_2)_3N$, 5.2×10^{-4}

E) Since these are all weak bases, they have the same strength.

Answer: C

Diff: 1 *Page Ref: 15.7*

38) Determine the pH in a 0.235 M NaOH solution.

A) 13.76 B) 0.24 C) 13.37 D) 0.63 E) 12

Answer: C

Diff: 2 *Page Ref: 15.7*

39) Determine the pOH in a 0.235 M NaOH solution.

A) 13.76 B) 0.24 C) 13.37 D) 0.63 E) 12

Answer: D
Diff: 2 *Page Ref: 15.7*

40) Determine the $[OH^-]$ concentration in a 0.169 M $Ca(OH)_2$ solution.

A) 0.338 M
B) 0.169 M
C) 5.92×10^{-14} M
D) 2.96×10^{-14} M
E) 0.298 M

Answer: A
Diff: 2 *Page Ref: 15.7*

41) Determine the $[OH^-]$ concentration of a 0.741 M KOH solution at 25°C.

A) 1.34×10^{-13} M
B) 1.34×10^{-14} M
C) 0.741 M
D) 7.41 M
E) none of the above

Answer: C
Diff: 2 *Page Ref: 15.7*

42) Determine the pH of a 0.188 M NH_3 solution at 25°C. The K_b of NH_3 is 1.76×10^{-5}.

A) 5.480 B) 2.740 C) 8.520 D) 11.260 E) 12.656

Answer: D
Diff: 3 *Page Ref: 15.7*

43) Determine the pOH of a 0.188 M NH_3 solution at 25°C. The K_b of NH_3 is 1.76×10^{-5}.

A) 5.480 B) 2.740 C) 8.520 D) 11.260 E) 12.656

Answer: B
Diff: 3 *Page Ref: 15.7*

44) Determine the pH of a 0.227 M C_5H_5N solution at 25°C. The K_b of C_5H_5N is 1.7×10^{-9}.

A) 4.59 B) 9.41 C) 4.71 D) 10.14 E) 9.29

Answer: E
Diff: 3 *Page Ref: 15.7*

45) Determine the pOH of a 0.227 M C_5H_5N solution at 25°C. The K_b of C_5H_5N is 1.7×10^{-9}.

A) 4.59 B) 9.41 C) 4.71 D) 10.14 E) 9.29

Answer: C
Diff: 3 *Page Ref: 15.7*

46) List an ingredient that is in antacid.
A) $Al(OH)_3$
B) $Mg(OH)_2$
C) $CaCO_3$
D) all of the above
E) none of the above

Answer: D
Diff: 1 Page Ref: 15.8

47) Which one of the following will form a basic solution in water?
A) $NaC_2H_3O_2$
B) LiCN
C) $KClO_2$
D) LiBrO
E) All of the above will form basic solutions.

Answer: B
Diff: 1 Page Ref: 15.8

48) Determine the K_b for CN^- at 25°C. The K_a for HCN is 4.9×10^{-10}.
A) 4.9×10^{-14}
B) 2.3×10^{-9}
C) 1.4×10^{-5}
D) 2.0×10^{-5}
E) 3.7×10^{-7}

Answer: D
Diff: 2 Page Ref: 15.8

49) Determine the K_a for $CH_3NH_3^+$ at 25°C. The K_b for CH_3NH_2 is 4.4×10^{-4}.
A) 3.1×10^{-10}
B) 6.8×10^{-11}
C) 5.6×10^{-10}
D) 2.3×10^{-3}
E) 2.3×10^{-11}

Answer: E
Diff: 2 Page Ref: 15.8

50) Determine the pH of a 0.62 M NH_4NO_3 solution at 25°C. The K_b for NH_3 is 1.76×10^{-5}.
A) 2.48 B) 9.27 C) 11.52 D) 4.73 E) 9.45

Answer: D
Diff: 3 Page Ref: 15.8

51) Determine the pH of a 0.22 M NaF solution at 25°C. The K_a of HF is 3.5×10^{-5}.
A) 10.20 B) 5.10 C) 8.90 D) 11.44 E) 2.56

Answer: C
Diff: 3 Page Ref: 15.8

52) In a triprotic acid, which K_a has the highest value?

A) K_{a1} B) K_{a2} C) K_{a3} D) K_{b1} E) K_{b2}

Answer: A
Diff: 1 Page Ref: 15.9

53) Determine the pH of a 0.18 M H_2CO_3 solution. Carbonic acid is a diprotic acid whose $K_{a1} = 4.3 \times 10^{-7}$ and $K_{a2} = 5.6 \times 10^{-11}$.

A) 11.00 B) 10.44 C) 5.50 D) 4.31 E) 3.56

Answer: E
Diff: 5 Page Ref: 15.9

54) Determine the pH of a solution that is 0.15 M $HClO_2$ ($K_a = 1.1 \times 10^{-2}$) and 0.15 M HClO ($K_a = 2.9 \times 10^{-8}$).

A) 4.18 B) 9.82 C) 12.55 D) 1.39 E) 3.55

Answer: D
Diff: 5 Page Ref: 15.9

55) Determine the concentration of CO_3^{2-} ions in a 0.18 M H_2CO_3 solution. Carbonic acid is a diprotic acid whose $K_{a1} = 4.3 \times 10^{-7}$ and $K_{a2} = 5.6 \times 10^{-11}$.

A) 2.8×10^{-4} M
B) 3.2×10^{-6} M
C) 5.6×10^{-11} M
D) 4.3×10^{-7} M
E) 6.9×10^{-8} M

Answer: C
Diff: 5 Page Ref: 15.9

56) Identify the strongest acid.

A) H_2O
B) H_2S
C) H_2Se
D) H_2Te
E) not enough information is available

Answer: D
Diff: 1 Page Ref: 15.9

57) Identify the weakest acid.

A) HF
B) HCl
C) HBr
D) HI
E) not enough information is available

Answer: A
Diff: 1 Page Ref: 15.9

58) Acid rain consists primarily of ________.
A) nitric and sulfuric acids
B) sulfuric acid
C) nitric acid
D) acetic acid
E) benzoic acid

Answer: A
Diff: 1 *Page Ref: 15.12*

Algorithmic Questions

1) The stomach excretes ________ to kill microorganisms and to activate enzymes that break down food.
A) CH_3COOH
B) HF
C) $LiHCO_3$
D) HCl
E) HNO_3

Answer: D
Diff: 1 *Page Ref: 15.2*

2) Identify the acid that is in car batteries.
A) H_2SO_4
B) HNO_3
C) H_2CO_3
D) CH_3COOH
E) HCl

Answer: A
Diff: 1 *Page Ref: 15.2*

3) Identify the acid that is in vinegar.
A) H_2SO_4
B) HNO_3
C) Na_2CO_3
D) CH_3COOH
E) HF

Answer: D
Diff: 1 *Page Ref: 15.2*

4) Identify the base that is in baking soda.
A) RbOH B) NaOH C) $NaHCO_3$ D) Li_2CO_3 E) NH_3

Answer: C
Diff: 1 *Page Ref: 15.2*

5) Identify the base that is in Drano.
A) KOH B) NaOH C) $KHCO_3$ D) Na_2CO_3 E) NH_3

Answer: B
Diff: 1 *Page Ref: 15.2*

6) When dissolved in water, which compound is generally considered to be an Arrhenius acid?
A) CH_3CO_2H B) NaOH C) Na_2CO_3 D) CH_3CH_2OH

Answer: A
Diff: 1 *Page Ref: 15.3*

7) Which of the following is an Arrhenius acid?
A) H_2SO_4
B) LiOH
C) NH_2CH_3
D) CH_3CCl_3
E) More than one of these is an Arrhenius acid.

Answer: A
Diff: 1 *Page Ref: 15.3*

8) Which of the following is an Arrhenius base?
A) CH_3CO_2H
B) LiOH
C) CH_3OH
D) NaBr
E) More than one of these compounds is an Arrhenius base.

Answer: B
Diff: 1 *Page Ref: 15.3*

9) Which BrØnsted-Lowry acid is not considered to be a strong acid in water?
A) HI B) HBr C) H_2SO_3 D) H NO_3

Answer: C
Diff: 2 *Page Ref: 15.3*

10) Which of the following is a BrØnsted-Lowry acid?
A) NH_4^+ B) CCl_4 C) NH_2^- D) NH_3 E) Br_2

Answer: A
Diff: 1 *Page Ref: 15.3*

11) Which of the following is a BrØnsted-Lowry base?
A) CBr_4
B) HCN
C) NH_3
D) Cl_2
E) None of the above are BrØnsted-Lowry bases.

Answer: C
Diff: 1 *Page Ref: 15.3*

12) Identify the diprotic acid.
A) HNO_3
B) HI
C) CH_3COOH
D) H_2SO_4
E) $HBrO_4$

Answer: D
Diff: 1 *Page Ref: 15.4*

13) Identify the triprotic acid.
A) HNO_3 B) H_3PO_4 C) H_2CO_3 D) HFO_4 E) H_2SO_4

Answer: B
Diff: 1 *Page Ref: 15.4*

14) Which of the following is a STRONG acid?
A) $C_6H_5CO_2H$
B) HCN
C) $HClO_4$
D) NH_4^+
E) H_2O

Answer: C
Diff: 1 *Page Ref: 15.4*

15) Which of the following is a WEAK acid?
A) $HClO_4$ B) H_2SO_4 C) HBr D) HCN E) HNO_3

Answer: D
Diff: 1 *Page Ref: 15.4*

16) How many of the following are WEAK acids?

HNO_2 HF HNO_3 $H_2PO_4^-$

A) 0 B) 1 C) 3 D) 4 E) 2

Answer: C
Diff: 1 *Page Ref: 15.4*

17) Identify the food that is NOT acidic.
A) beer
B) apple
C) egg white
D) soda
E) wine

Answer: C
Diff: 1 *Page Ref: 15.5*

18) Calculate the pOH of a solution that contains 3.9×10^{-5} M H_3O^+ at 25°C.

A) 5.41 B) 4.41 C) 9.59 D) 8.59 E) 0.59

Answer: C
Diff: 2 *Page Ref: 15.5*

19) Calculate the pH of a solution that contains 3.9×10^{-5} M H_3O^+ at 25°C.

A) 5.41 B) 4.41 C) 9.59 D) 8.59 E) 0.59

Answer: B
Diff: 2 *Page Ref: 15.5*

20) Calculate the pH of a solution that contains 2.4×10^{-5} M H_3O^+ at 25°C.

A) 2.40 B) 9.38 C) 4.62 D) 8.38 E) 5.62

Answer: C
Diff: 2 *Page Ref: 15.5*

21) Calculate the pOH of a solution that contains 2.4×10^{-5} M H_3O^+ at 25°C.

A) 2.40 B) 9.38 C) 4.62 D) 8.38 E) 5.62

Answer: B
Diff: 2 *Page Ref: 15.5*

22) Calculate the pOH in an aqueous solution with a pH of 7.85 at 25°C.

A) 4.15 B) 5.15 C) 6.15 D) 7.15 E) 8.15

Answer: C
Diff: 2 *Page Ref: 15.5*

23) Calculate the hydronium ion concentration in an aqueous solution that contains 2.50×10^{-6} M in hydroxide ion.

A) 4.00×10^{-7} M B) 4.00×10^{-8} M C) 4.00×10^{-9} M D) 5.00×10^{-9} M

Answer: C
Diff: 3 *Page Ref: 15.5*

24) Calculate the hydroxide ion concentration in an aqueous solution that contains 3.50×10^{-4} M in hydronium ion.

A) 2.86×10^{-3} M B) 2.86×10^{-10} M C) 2.86×10^{-11} M D) 3.50×10^{-11} M

Answer: C
Diff: 3 *Page Ref: 15.5*

25) A solution with a hydrogen ion concentration of 3.25×10^{-6} M is ________ and has a hydroxide ion concentration of ________.

A) acidic, 3.08×10^{-8} M
B) acidic, 3.08×10^{-9} M
C) basic, 3.08×10^{-8} M
D) basic, 3.08×10^{-9} M

Answer: B
Diff: 3 *Page Ref: 15.5*

26) A solution with a hydroxide ion concentration of 4.15 × 10^{-6} M is ________ and has a hydrogen ion concentration of ________.

A) acidic, 2.41 × 10^{-8} M
B) acidic, 2.41 × 10^{-9} M
C) basic, 2.41 × 10^{-8} M
D) basic, 2.41 × 10^{-9} M

Answer: D
Diff: 3 *Page Ref: 15.5*

27) Calculate the pH for an aqueous solution of acetic acid that contains 2.15 × 10^{-3} M hydronium ion.

A) 4.65 × 10^{-12} M
B) 2.15 × 10^{-3} M
C) 2.67
D) 11.33

Answer: C
Diff: 3 *Page Ref: 15.5*

28) Calculate the pH for an aqueous solution of pyridine that contains 2.15 × 10^{-4} M hydroxide ion.

A) 4.65 × 10^{-11}
B) 2.15 × 10^{-4}
C) 3.67
D) 10.33

Answer: D
Diff: 3 *Page Ref: 15.5*

29) What is the hydroxide ion concentration and the pH for a hydrochloric acid solution that has a hydronium ion concentration of 1.50 × 10^{-2} M?

A) 6.67 × 10^{-12} M, 2.82
B) 6.67 × 10^{-12} M, 11.18
C) 6.67 × 10^{-13} M, 1.82
D) 6.67 × 10^{-13} M, 12.17

Answer: C
Diff: 3 *Page Ref: 15.5*

30) What is the hydronium ion concentration and the pH for an aqueous solution of NH_3 that has a hydroxide ion concentration of 2.25 × 10^{-2} M?

A) 4.44 × 10^{-12} M and 2.65
B) 4.44 × 10^{-12} M and 11.35
C) 4.44 × 10^{-13} M and 1.65
D) 4.44 × 10^{-13} M and 12.35

Answer: D
Diff: 3 *Page Ref: 15.5*

31) What is the hydronium ion concentration of an acid rain sample that has a pH of 3.45?

A) 2.82 × 10^{-11} M
B) 3.55 × 10^{-4} M
C) 3.45 M
D) 10.55 M

Answer: B
Diff: 3 *Page Ref: 15.5*

32) What is the hydroxide ion concentration of a lye solution that has a pH of 9.20?

A) 6.31 × 10^{-10} M
B) 1.58 × 10^{-5} M
C) 4.80 M
D) 9.20 M

Answer: B
Diff: 3 *Page Ref: 15.5*

33) What is the pH of a 0.020 M $HClO_4$ solution?

A) 0.020
B) 0.040
C) 1.70
D) 12.30

Answer: C
Diff: 3 *Page Ref: 15.5*

34) An aqueous solution at 25.0°C contains $[H^+] = 0.099$ M. What is the pH of the solution?

A) 1.00
B) -1.00
C) 13.0
D) 0.0990
E) 1.00×10^{-13}

Answer: A
Diff: 3 *Page Ref: 15.5*

35) The pH of an aqueous solution at 25.0°C is 10.66. What is the molarity of H^+ in this solution?

A) 2.2×10^{-11}
B) 4.6×10^{-4}
C) 3.3
D) 1.1×10^{-13}
E) 4.6×10^{10}

Answer: A
Diff: 3 *Page Ref: 15.5*

36) Calculate the molarity of hydroxide ion in an aqueous solution that has a pOH of 5.00.

A) 1.0×10^{-5}
B) 9.00
C) 1.0×10^{-9}
D) 5.0×10^{-14}
E) 9.0×10^{-14}

Answer: A
Diff: 3 *Page Ref: 15.5*

37) Which of the following acids is the STRONGEST? The acid is followed by its K_a value.

A) HF, 3.5×10^{-4}
B) HCN, 4.9×10^{-10}
C) HNO_2, 4.6×10^{-4}
D) H COOH, 1.8×10^{-4}
E) $HClO_2$, 1.1×10^{-2}

Answer: E
Diff: 1 *Page Ref: 15.6*

38) Which of the following acids is the WEAKEST? The acid is followed by its K_a value.

A) HF, 3.5×10^{-4}
B) HCN, 4.9×10^{-10}
C) HNO_2, 4.6×10^{-4}
D) H COOH, 1.8×10^{-4}
E) $HClO_2$, 1.1×10^{-2}

Answer: B
Diff: 1 *Page Ref: 15.6*

39) What is the hydronium ion concentration of a 0.500 M acetic acid solution with $K_a = 1.8 \times 10^{-5}$? The equation for the dissociation of acetic acid is:

$$CH_3CO_2H(aq) + H_2O(l) \rightleftharpoons H_3O^+(aq) + CH_3CO_2^-(aq).$$

A) 3.0×10^{-2} M B) 4.2×10^{-2} M C) 3.0×10^{-3} M D) 4.2×10^{-3} M

Answer: C
Diff: 4 Page Ref: 15.6

40) What is the hydronium ion concentration of a 0.150 M hypochlorous acid solution with $K_a = 3.5 \times 10^{-8}$? The equation for the dissociation of hypochlorous acid is:

$$HOCl(aq) + H_2O(l) \rightleftharpoons H_3O^+(aq) + OCl^-(aq).$$

A) 1.9×10^{-4} M B) 7.2×10^{-4} M C) 2.8×10^{-5} M D) 7.2×10^{-5} M

Answer: D
Diff: 4 Page Ref: 15.6

41) The acid-dissociation constant at 25.0°C for hypochlorous acid (HClO) is 3.0×10^{-8}. At equilibrium, the molarity of H_3O^+ in a 0.010 M solution of HClO is ________.

A) 1.7×10^{-5}
B) 0.010
C) 5.8×10^{-10}
D) 4.76
E) 2.00

Answer: A
Diff: 4 Page Ref: 15.6

42) Calculate the pH of a 0.800 M $NaCH_3CO_2$ solution. K_a for acetic acid, CH_3CO_2H, is 1.8×10^{-5}.

A) 2.42 B) 4.68 C) 9.32 D) 11.58

Answer: C
Diff: 4 Page Ref: 15.6

43) Calculate the pH of a 1.60 M KBrO solution. K_a for hypobromous acid, HBrO, is 2.0×10^{-9}.

A) 2.55 B) 4.25 C) 9.75 D) 11.45

Answer: D
Diff: 4 Page Ref: 15.6

44) Which of the following is a STRONG base?

A) I^- B) NH_3 C) CH_3OH D) NO_3^- E) LiOH

Answer: E
Diff: 1 Page Ref: 15.7

45) Which of the following is a WEAK base?

A) $NH(CH_3)_2$

B) N_2

C) NaOH

D) CH_2Cl_2

E) None of the above are weak bases.

Answer: A

Diff: 1 *Page Ref: 15.7*

46) Determine the $[OH^-]$ concentration in a 0.235 M NaOH solution.

A) 4.25×10^{-14} M

B) 0.470 M

C) 2.13×10^{-14} M

D) 0.118 M

E) 0.235 M

Answer: E

Diff: 2 *Page Ref: 15.7*

47) Determine the pH of a 0.741 M LiOH solution at 25°C.

A) 0.130 B) 13.87 C) 0.741 D) 13.26 E) 1.48

Answer: B

Diff: 2 *Page Ref: 15.7*

48) Determine the pOH of a 0.741 M LiOH solution at 25°C.

A) 0.130 B) 13.87 C) 0.741 D) 13.26 E) 1.48

Answer: A

Diff: 2 *Page Ref: 15.7*

49) Determine the pH of a 0.232 M $Mg(OH)_2$ solution at 25°C.

A) 12.73 B) 13.37 C) 13.67 D) 0.333 E) 0.634

Answer: C

Diff: 2 *Page Ref: 15.7*

50) Determine the pOH of a 0.348 M $Ba(OH)_2$ solution at 25°C.

A) 13.08 B) 13.54 C) 13.84 D) 0.157 E) 0.458

Answer: D

Diff: 2 *Page Ref: 15.7*

51) Determine the $[OH^-]$ concentration of a 0.123 M $Sr(OH)_2$ solution at 25°C.

A) 4.06×10^{-14} M

B) 0.123 M

C) 0.246 M

D) 8.13×10^{-14} M

E) 0.0615 M

Answer: C

Diff: 2 *Page Ref: 15.7*

52) A 7.0×10^{-3} M aqueous solution of $Ca(OH)_2$ at 25.0°C has a pH of ________.

A) 12.15
B) 1.85
C) 1.4×10^{-2}
D) 7.1×10^{-13}
E) 11.85

Answer: A
Diff: 3 *Page Ref: 15.7*

53) What is the pH of a 0.040 M $Ba(OH)_2$ solution?

A) 1.10 B) 1.40 C) 12.60 D) 12.90

Answer: D
Diff: 3 *Page Ref: 15.7*

54) What is the pH of a 0.100 M NH_3 solution that has $K_b = 1.8 \times 10^{-5}$? The equation for the dissociation of NH_3 is

$$NH_3(aq) + H_2O(l) \rightleftharpoons NH_4^+(aq) + OH^-(aq).$$

A) 1.87 B) 2.87 C) 11.13 D) 10.13

Answer: C
Diff: 3 *Page Ref: 15.7*

55) What is the pH of a 0.30 M pyridine solution that has $K_b = 1.9 \times 10^{-9}$? The equation for the dissociation of pyridine is

$$C_5H_5N(aq) + H_2O(l) \rightleftharpoons C_5H_5NH^+(aq) + OH^-(aq).$$

A) 4.62 B) 8.72 C) 9.38 D) 10.38

Answer: C
Diff: 3 *Page Ref: 15.7*

56) Determine the ammonia concentration of an aqueous solution that has a pH of 11.00. The equation for the dissociation of NH_3 ($K_b = 1.8 \times 10^{-5}$) is

$$NH_3(aq) + H_2O(l) \rightleftharpoons NH_4^+(aq) + OH^-(aq).$$

A) 3.0 M B) 0.056 M C) 1.8×10^{-2} M D) 1.0×10^{-3} M

Answer: B
Diff: 3 *Page Ref: 15.7*

57) The acid-dissociation constant of hydrocyanic acid (HCN) at 25.0°C is 4.9×10^{-10}. What is the pH of an aqueous solution of 0.080 M sodium cyanide (NaCN)?

A) 11.11
B) 2.89
C) 1.3×10^{-3}
D) 7.8×10^{-12}
E) 3.9×10^{-11}

Answer: A
Diff: 4 *Page Ref: 15.7*

58) Calculate the pH of a 1.60 M CH_3NH_3Cl solution. K_b for methylamine, CH_3NH_2, is 3.7×10^{-4}.

A) 1.61 B) 5.18 C) 8.82 D) 12.39

Answer: A
Diff: 5 *Page Ref: 15.7*

59) The base-dissociation constant of ethylamine ($C_2H_5NH_2$) is 6.4×10^{-4} at 25.0°C. The $[H^+]$ in a 1.6×10^{-2} M solution of ethylamine is _______ M.

A) 3.5×10^{-12}
B) 2.9×10^{-3}
C) 3.1×10^{-12}
D) 3.2×10^{-3}
E) 11.46

Answer: C
Diff: 5 *Page Ref: 15.7*

60) Which one of the following salts, when dissolved in water, produces the solution with the *highest* pH?

A) $KHCO_3$ B) $CsClO_4$
C) RaO D) $CH_3CH_3NH_3Cl$

Answer: C
Diff: 1 *Page Ref: 15.8*

61) If an equal number of moles of the weak acid HCN and the strong base KOH are added to water, is the resulting solution acidic, basic, or neutral?

A) acidic
B) basic
C) neutral
D) There is insufficient information provided to answer this question.

Answer: B
Diff: 1 *Page Ref: 15.8*

62) Which one of the following salts, when dissolved in water, produces the solution with the *highest* pH?

A) RbI B) RbBr C) RbCl D) RbF

Answer: D
Diff: 1 *Page Ref: 15.8*

63) Which one of the following salts, when dissolved in water, produces the solution with a pH *closest* to 7.00?

A) NH_4Br B) Ca O C) K HSO_4 D) CsI

Answer: D
Diff: 1 *Page Ref: 15.8*

64) Which one of the following salts, when dissolved in water, produces the solution with the *lowest* pH?

A) NaCl B) KCl C) $MgCl_2$ D) $AlCl_3$

Answer: D
Diff: 1 *Page Ref: 15.8*

65) Which one of the following will form an acidic solution in water?

A) NH_4Cl
B) KF
C) KI
D) KNO_3
E) None of the above solutions will be acidic.

Answer: A
Diff: 1 *Page Ref: 15.8*

66) Which of the following is a polyprotic acid?

A) HI
B) H_2SO_4
C) HCN
D) CCl_4
E) $HC_2H_3O_2$

Answer: B
Diff: 1 *Page Ref: 15.9*

67) Which of the following is a triprotic acid?

A) HI
B) H_2SO_4
C) H_3PO_4
D) CCl_4
E) $HC_2H_3O_2$

Answer: C
Diff: 1 *Page Ref: 15.9*

68) Calculate the pH of a 0.080 M carbonic acid solution, $H_2CO_3(aq)$, that has the stepwise dissociation constants $K_{a1} = 4.3 \times 10^{-7}$ and $K_{a2} = 5.6 \times 10^{-11}$.

A) 1.10 B) 3.73 C) 6.37 D) 10.25

Answer: B
Diff: 5 *Page Ref: 15.9*

69) Calculate the pH of a 0.60 M H_2SO_3, solution that has the stepwise dissociation constants $K_{a1} = 1.5 \times 10^{-2}$ and $K_{a2} = 6.3 \times 10^{-8}$.

A) 1.02 B) 1.06 C) 1.82 D) 2.04

Answer: A
Diff: 5 *Page Ref: 15.9*

70) Calculate the concentration of bicarbonate ion, HCO_3^-, in a 0.010 M H_2CO_3 solution that has the stepwise dissociation constants $K_{a1} = 4.3 \times 10^{-7}$ and $K_{a2} = 5.6 \times 10^{-11}$.

A) 6.6×10^{-5} M
B) 4.3×10^{-7} M
C) 4.3×10^{-9} M
D) 5.6×10^{-11} M

Answer: A
Diff: 5 *Page Ref: 15.9*

71) What is the pH of a 0.40 M H_2Se solution that has the stepwise dissociation constants $K_{a1} = 1.3 \times 10^{-4}$ and $K_{a2} = 1.0 \times 10^{-11}$?

A) 2.14
B) 3.89
C) 4.28
D) 5.57

Answer: A
Diff: 5 *Page Ref: 15.9*

72) What is the selenide ion concentration $[Se^{2-}]$ for a 0.100 M H_2Se solution that has the stepwise dissociation constants of $K_{a1} = 1.3 \times 10^{-4}$ and $K_{a2} = 1.0 \times 10^{-11}$?

A) 3.6×10^{-3} M
B) 1.3×10^{-4} M
C) 1.3×10^{-5} M
D) 1.0×10^{-11} M

Answer: D
Diff: 5 *Page Ref: 15.9*

73) Identify the strongest acid.

A) HFO_4
B) HFO_3
C) HFO_2
D) HFO
E) Not enough information is given.

Answer: A
Diff: 1 *Page Ref: 15.10*

74) Which of the following is a Lewis acid?

A) BBr_3
B) CCl_4
C) NH_3
D) $CHBr_3$
E) None of the above are Lewis acids.

Answer: A
Diff: 1 *Page Ref: 15.11*

75) Which of the following is a Lewis base?

A) AlF_3
B) H_2O
C) SiF_4
D) C_5H_{12}
E) None of the above are Lewis bases.

Answer: B
Diff: 1 *Page Ref: 15.11*

76) List the compound in marble that is destroyed by acid rain.

A) NH_4OH B) $NaHCO_3$ C) $CaCO_3$ D) KOH E) LiOH

Answer: C

Diff: 1 *Page Ref: 15.12*

Matching Questions

Match the following.

1) Arrhenius acid
Diff: 1 *Page Ref: 15.3*

2) Arrhenius base
Diff: 1 *Page Ref: 15.3*

3) Brønsted–Lowry acid
Diff: 1 *Page Ref: 15.3*

4) Brønsted–Lowry base
Diff: 1 *Page Ref: 15.3*

5) Lewis acid
Diff: 1 *Page Ref: 15.11*

6) Lewis base
Diff: 1 *Page Ref: 15.11*

A) proton donor

B) produces protons in aqueous solution

C) proton acceptor

D) produces hydroxide ions in aqueous solution

E) electron pair acceptor

F) electron pair donor

Answers: 1) B 2) D 3) A 4) C 5) E 6) F

Short Answer Questions

1) What does the term *amphoteric* mean?

Answer: A substance is amphoteric if it can behave as an acid or a base, depending on the environment.

Diff: 1 *Page Ref: 15.3*

2) What is the difference between a strong and weak acid?

Answer: A strong acid ionizes/dissociates completely into H^+ and an anion, whereas a weak acid only ionizes/dissociates to a small extent to form H^+ and an anion.

Diff: 1 *Page Ref: 15.4*

3) What is the *autoionization* of water?

Answer: Autoionization of water is the equilibrium process where one water molecule will transfer its proton to another. This forms hydroxide ions and hydronium ions in equal numbers in solution. This autoionization process of water is what allows us to relate hydronium and hydroxide ion concentrations in aqueous solutions of acids and bases.

Diff: 1 *Page Ref: 15.5*

4) Do both protons ionize instantaneously from a diprotic acid such as H_2CO_3? Explain your answer.

Answer: No. Polyprotic acids ionize in stages. A small portion of the first protons ionize (according to K_{a1}), followed by the second proton ionization (according to K_{a2}). The second proton can only be lost from the small number of acid molecules that have already lost their first ionizable hydrogen.

Diff: 1 *Page Ref: 15.9*

5) Describe the relationship between molecular structure and acid strength.

Answer: 1) The stronger the bond between a hydrogen atom and its neighbor, the weaker the acid (the less ionization that will occur)
2) The more electron withdrawing atoms near the hydrogen atom, the weaker the bond to the hydrogen and the stronger the acid will be (more ionization will occur).

Diff: 1 *Page Ref: 15.10*

6) Describe a molecule that can be a Lewis acid.

Answer: A molecule with an incomplete octet can serve as a Lewis acid.

Diff: 1 *Page Ref: 15.11*

7) List the major source of energy in the United States.

Answer: fossil fuel

Diff: 1 *Page Ref: 15.12*

8) List the compound that is formed from the reaction of carbon dioxide with water.

Answer: carbonic acid, H_2CO_3

Diff: 1 *Page Ref: 15.12*

Chapter 16 Aqueous Ionic Equilibrium

Multiple Choice Questions

1) Animals will lick up ethylene glycol (antifreeze) due to its sweet taste. The antidote for ethylene glycol poisioning is the administration of ________.
   A) ethyl alcohol (alcoholic drinks)
   B) isopropyl alcohol (rubbing alcohol)
   C) mineral oil (laxative)
   D) vinegar
   E) sodium bicarbonate (baking soda)

Answer: A
Diff: 1 Page Ref: 16.1

2) An important buffer in the blood is a mixture of ________.
   A) sodium chloride and hydrochloric acid
   B) hydrochloric acid and sodium hydroxide
   C) carbonic acid and bicarbonate ion
   D) acetic acid and bicarbonate ion
   E) acetic acid and carbonate ion

Answer: C
Diff: 1 Page Ref: 16.2

3) Identify a good buffer.
   A) small amounts of both a weak acid and its conjugate base
   B) significant amounts of both a strong acid and a strong base
   C) small amounts of both a strong acid and a strong base
   D) significant amounts of both a weak acid and a strong acid
   E) significant amounts of both a weak acid and its conjugate base

Answer: E
Diff: 1 Page Ref: 16.2

4) If the pKa of $HCHO_2$ is 3.74 and the pH of an $HCHO_2/NaCHO_2$ solution is 3.11, which of the following is TRUE?
   A) $[HCHO_2] < [NaCHO_2]$
   B) $[HCHO_2] = [NaCHO_2]$
   C) $[HCHO_2] << [NaCHO_2]$
   D) $[HCHO_2] > [NaCHO_2]$
   E) It is not possible to make a buffer of this pH from $HCHO_2$ and $NaCHO_2$.

Answer: D
Diff: 2 Page Ref: 16.2

5) If the pKa of $HCHO_2$ is 3.74 and the pH of an $HCHO_2/NaCHO_2$ solution is 3.89, which of the following is TRUE?

A) $[HCHO_2] < [NaCHO_2]$
B) $[HCHO_2] = [NaCHO_2]$
C) $[HCHO_2] > [NaCHO_2]$
D) $[HCHO_2] >> [NaCHO_2]$
E) It is not possible to make a buffer of this pH from $HCHO_2$ and $NaCHO_2$.

Answer: A
Diff: 2 *Page Ref: 16.2*

6) If the pKa of $HCHO_2$ is 3.74 and the pH of an $HCHO_2/NaCHO_2$ solution is 3.74, which of the following is TRUE?

A) $[HCHO_2] > [NaCHO_2]$
B) $[HCHO_2] = [NaCHO_2]$
C) $[HCHO_2] < [NaCHO_2]$
D) $[HCHO_2] << [NaCHO_2]$
E) It is not possible to make a buffer of this pH from $HCHO_2$ and $NaCHO_2$.

Answer: B
Diff: 2 *Page Ref: 16.2*

7) Calculate the pH of a buffer that is 0.225 M $HC_2H_3O_2$ and 0.162 M $KC_2H_3O_2$. The K_a for $HC_2H_3O_2$ is 1.8×10^{-5}.

A) 4.89 B) 9.11 C) 4.74 D) 9.26 E) 4.60

Answer: E
Diff: 3 *Page Ref: 16.2*

8) Calculate the pH of a solution formed by mixing 250.0 mL of 0.15 M NH_4Cl with 100.0 mL of 0.20 M NH_3. The K_b for NH_3 is 1.8×10^{-5}.

A) 9.13 B) 9.25 C) 9.53 D) 4.74 E) 8.98

Answer: E
Diff: 3 *Page Ref: 16.2*

9) Calculate the pH of a solution formed by mixing 250.0 mL of 0.15 M NH_4Cl with 200.0 mL of 0.12 M NH_3. The K_b for NH_3 is 1.8×10^{-5}.

A) 9.06 B) 9.45 C) 4.55 D) 4.74 E) 9.26

Answer: A
Diff: 3 *Page Ref: 16.2*

10) Calculate the pH of a solution formed by mixing 100.0 mL of 0.20 M HClO with 200.0 mL of 0.30 M KClO. The K_a for HClO is 2.9×10^{-8}.

A) 5.99 B) 8.01 C) 7.54 D) 7.06 E) 6.46

Answer: B
Diff: 4 *Page Ref: 16.2*

11) Calculate the pH of a solution formed by mixing 200.0 mL of 0.30 M HClO with 100.0 mL of 0.20 M KClO. The K_a for HClO is 2.9×10^{-8}.

A) 5.99 B) 8.01 C) 7.54 D) 7.06 E) 6.46

Answer: D
Diff: 4 *Page Ref: 16.2*

12) Calculate the pH of a solution formed by mixing 200.0 mL of 0.30 M HClO with 300.0 mL of 0.20 M KClO. The K_a for HClO is 2.9×10^{-8}.

A) 5.99 B) 8.01 C) 7.54 D) 7.06 E) 6.46

Answer: C
Diff: 4 *Page Ref: 16.2*

13) Calculate the pH of a solution formed by mixing 150.0 mL of 0.10 M $HC_7H_5O_2$ with 100.0 mL of 0.30 M $NaC_7H_5O_2$. The K_a for $HC_7H_5O_2$ is 6.5×10^{-5}.

A) 4.19 B) 9.69 C) 4.49 D) 4.31 E) 10.51

Answer: C
Diff: 4 *Page Ref: 16.2*

14) Calculate the pH of a solution formed by mixing 250.0 mL of 0.15 M $HCHO_2$ with 100.0 mL of 0.20 M $LiCHO_2$. The K_a for $HCHO_2$ is 1.8×10^{-4}.

A) 3.87 B) 3.74 C) 10.53 D) 3.47 E) 10.13

Answer: D
Diff: 4 *Page Ref: 16.2*

15) A 1.00 L buffer solution is 0.150 M in $HC_7H_5O_2$ and 0.250 M in $LiC_7H_5O_2$. Calculate the pH of the solution after the addition of 100.0 mL of 1.00 M HCl. The K_a for $HC_7H_5O_2$ is 6.5×10^{-5}.

A) 4.19 B) 5.03 C) 4.41 D) 3.34 E) 3.97

Answer: E
Diff: 4 *Page Ref: 16.2*

16) A 1.00 L buffer solution is 0.250 M in HF and 0.250 M in NaF. Calculate the pH of the solution after the addition of 100.0 mL of 1.00 M HCl. The K_a for HF is 3.5×10^{-4}.

A) 3.09 B) 4.11 C) 3.82 D) 3.46 E) 2.78

Answer: A
Diff: 4 *Page Ref: 16.2*

17) A 1.00 L buffer solution is 0.250 M in HF and 0.250 M in LiF. Calculate the pH of the solution after the addition of 0.150 moles of solid LiOH. Assume no volume change upon the addition of base. The K_a for HF is 3.5×10^{-4}.

A) 3.46 B) 4.06 C) 2.85 D) 3.63 E) 4.24

Answer: B
Diff: 4 *Page Ref: 16.2*

18) A 1.50 L buffer solution is 0.250 M in HF and 0.250 M in NaF. Calculate the pH of the solution after the addition of 0.0500 moles of solid NaOH. Assume no volume change upon the addition of base. The K_a for HF is 3.5×10^{-4}.

A) 3.34 B) 3.46 C) 3.57 D) 3.63 E) 2.89

Answer: C
Diff: 5 *Page Ref: 16.2*

19) A 1.50 L buffer solution is 0.250 M in HF and 0.250 M in NaF. Calculate the pH of the solution after the addition of 0.100 moles of solid NaOH. Assume no volume change upon the addition of base. The K_a for HF is 3.5×10^{-4}.

A) 3.22 B) 3.82 C) 3.69 D) 3.09 E) 4.46

Answer: C
Diff: 5 *Page Ref: 16.2*

20) Which of the following is TRUE?
A) An effective buffer has a [base]/[acid] ratio in the range of 10 - 100.
B) A buffer is most resistant to pH change when [acid] = [conjugate base]
C) An effective buffer has very small absolute concentrations of acid and conjugate base.
D) A buffer can not be destroyed by adding too much strong base. It can only be destroyed by adding too much strong acid.
E) None of the above are true.

Answer: B
Diff: 1 *Page Ref: 16.3*

21) Define buffer capacity.
A) Buffer capacity is the amount of acid or base that can be added to a buffer without destroying its effectiveness.
B) Buffer capacity is the amount of acid that can be added until all of the base is used up.
C) Buffer capacity is the amount of base that can be added until all of the acid is used up.
D) Buffer capacity is the amount of acid that can be added until all of the acid is used up.
E) Buffer capacity is the amount of base that can be added until all of the base is used up.

Answer: A
Diff: 1 *Page Ref: 16.3*

22) Which of the following is TRUE?
A) The equivalence point is where the amount of acid equals the amount of base during any acid–base titration.
B) At the equivalence point, the pH is always 7.
C) An indicator is not pH sensitive.
D) A titration curve is a plot of pH vs. the [base]/[acid] ratio.
E) None of the above are true.

Answer: A
Diff: 1 *Page Ref: 16.4*

23) When titrating a strong monoprotic acid and KOH at 25°C, the
A) pH will be less than 7 at the equivalence point.
B) pH will be greater than 7 at the equivalence point.
C) titration will require more moles of base than acid to reach the equivalence point.
D) pH will be equal to 7 at the equivalence point.
E) titration will require more moles of acid than base to reach the equivalence point.

Answer: D
Diff: 1 *Page Ref: 16.4*

24) When titrating a weak monoprotic acid with NaOH at 25°C, the
A) pH will be less than 7 at the equivalence point.
B) pH will be equal to 7 at the equivalence point.
C) pH will be greater than 7 at the equivalence point.
D) titration will require more moles of base than acid to reach the equivalence point.
E) titration will require more moles of acid than base to reach the equivalence point.

Answer: C
Diff: 1 *Page Ref: 16.4*

25) When titrating a monoprotic strong acid with a weak base at 25°C, the
A) pH will be 7 at the equivalence point.
B) pH will be greater than 7 at the equivalence point.
C) titration will require more moles of the base than acid to reach the equivalence point.
D) titration will require more moles of acid than base to reach the equivalence point.
E) pH will be less than 7 at the equivalence point.

Answer: E
Diff: 1 *Page Ref: 16.4*

26) A 100.0 mL sample of 0.18 M $HClO_4$ is titrated with 0.27 M LiOH. Determine the pH of the solution before the addition of any LiOH.
A) 1.74 B) 1.05 C) 0.74 D) 0.57 E) 1.57

Answer: C
Diff: 2 *Page Ref: 16.4*

27) A 100.0 mL sample of 0.18 M $HClO_4$ is titrated with 0.27 M LiOH. Determine the pH of the solution after the addition of 30.0 mL of LiOH.
A) 0.86 B) 1.21 C) 2.00 D) 1.12 E) 2.86

Answer: D
Diff: 3 *Page Ref: 16.4*

28) A 100.0 mL sample of 0.18 M $HClO_4$ is titrated with 0.27 M LiOH. Determine the pH of the solution after the addition of 50.0 mL of LiOH.
A) 12.48 B) 0.68 C) 2.35 D) 1.52 E) 3.22

Answer: D
Diff: 3 *Page Ref: 16.4*

29) A 100.0 mL sample of 0.18 M $HClO_4$ is titrated with 0.27 M LiOH. Determine the pH of the solution after the addition of 66.67 mL of LiOH (this is the equivalence point).
A) 0.97 B) 13.03 C) 2.76 D) 11.24 E) 7.00

Answer: E
Diff: 3 *Page Ref: 16.4*

30) A 100.0 mL sample of 0.180 M $HClO_4$ is titrated with 0.270 M LiOH. Determine the pH of the solution after the addition of 75.0 mL of LiOH.
A) 12.1 B) 2.65 C) 11.35 D) 1.89 E) 13.06

Answer: A
Diff: 3 *Page Ref: 16.4*

31) A 100.0 mL sample of 0.18 M $HClO_4$ is titrated with 0.27 M LiOH. Determine the pH of the solution after the addition of 100.0 mL of LiOH.
A) 13.13 B) 12.65 C) 1.35 D) 0.87 E) 12.95

Answer: B
Diff: 3 *Page Ref: 16.4*

32) A 100.0 mL sample of 0.20 M HF is titrated with 0.10 M KOH. Determine the pH of the solution before the addition of any KOH. The K_a of HF is 3.5×10^{-4}.
A) 4.15 B) 0.70 C) 2.08 D) 3.46 E) 1.00

Answer: C
Diff: 3 *Page Ref: 16.4*

33) A 100.0 mL sample of 0.20 M HF is titrated with 0.10 M KOH. Determine the pH of the solution after the addition of 100.0 mL of KOH. The K_a of HF is 3.5×10^{-4}.
A) 2.08 B) 3.15 C) 4.33 D) 3.46 E) 4.15

Answer: D
Diff: 3 *Page Ref: 16.4*

34) A 100.0 mL sample of 0.10 M NH_3 is titrated with 0.10 M HNO_3. Determine the pH of the solution before the addition of any HNO_3. The K_b of NH_3 is 1.8×10^{-5}.
A) 4.74 B) 9.26 C) 11.13 D) 13.00 E) 12.55

Answer: C
Diff: 3 *Page Ref: 16.4*

35) A 100.0 mL sample of 0.10 M NH_3 is titrated with 0.10 M HNO_3. Determine the pH of the solution after the addition of 50.0 mL of HNO_3. The K_b of NH_3 is 1.8×10^{-5}.
A) 4.74 B) 7.78 C) 7.05 D) 9.26 E) 10.34

Answer: D
Diff: 3 *Page Ref: 16.4*

36) A 100.0 mL sample of 0.10 M NH_3 is titrated with 0.10 M HNO_3. Determine the pH of the solution after the addition of 150.0 mL of HNO_3. The K_b of NH_3 is 1.8×10^{-5}.
A) 1.70 B) 6.44 C) 7.56 D) 12.30 E) 2.30

Answer: A
Diff: 3 *Page Ref: 16.4*

37) A 100.0 mL sample of 0.10 M NH_3 is titrated with 0.10 M HNO_3. Determine the pH of the solution after the addition of 200.0 mL of HNO_3. The K_b of NH_3 is 1.8×10^{-5}.

A) 6.44 B) 1.48 C) 2.00 D) 12.52 E) 12.00

Answer: B
Diff: 3 Page Ref: 16.4

38) A 100.0 mL sample of 0.10 M $Ca(OH)_2$ is titrated with 0.10 M HBr. Determine the pH of the solution before the addition of any HBr.

A) 12.86 B) 13.00 C) 13.30 D) 0.70 E) 1.00

Answer: C
Diff: 3 Page Ref: 16.4

39) A 100.0 mL sample of 0.10 M $Ca(OH)_2$ is titrated with 0.10 M HBr. Determine the pH of the solution after the addition of 100.0 mL HBr.

A) 2.00 B) 12.00 C) 1.30 D) 12.70 E) 7.00

Answer: D
Diff: 3 Page Ref: 16.4

40) A 100.0 mL sample of 0.10 M $Ca(OH)_2$ is titrated with 0.10 M HBr. Determine the pH of the solution after the addition of 200.0 mL HBr.

A) 2.62 B) 2.00 C) 1.48 D) 12.52 E) 7.00

Answer: E
Diff: 3 Page Ref: 16.4

41) A 100.0 mL sample of 0.10 M $Ca(OH)_2$ is titrated with 0.10 M HBr. Determine the pH of the solution after the addition of 300.0 mL HBr.

A) 1.60 B) 1.30 C) 1.00 D) 12.40 E) 1.12

Answer: A
Diff: 3 Page Ref: 16.4

42) A 100.0 mL sample of 0.10 M $Ca(OH)_2$ is titrated with 0.10 M HBr. Determine the pH of the solution after the addition of 400.0 mL HBr.

A) 1.00 B) 1.40 C) 1.22 D) 1.30 E) 2.11

Answer: B
Diff: 3 Page Ref: 16.4

43) A 100.0 mL sample of 0.20 M HF is titrated with 0.10 M KOH. Determine the pH of the solution after the addition of 300.0 mL of KOH. The K_a of HF is 3.5×10^{-4}.

A) 12.40 B) 9.33 C) 5.06 D) 8.94 E) 12.00

Answer: A
Diff: 4 Page Ref: 16.4

44) A 100.0 mL sample of 0.20 M HF is titrated with 0.10 M KOH. Determine the pH of the solution after the addition of 400.0 mL of KOH. The K_a of HF is 3.5×10^{-4}.

A) 13.08 B) 12.60 C) 13.85 D) 12.30 E) 12.78

Answer: B
Diff: 4 Page Ref: 16.4

45) A 100.0 mL sample of 0.10 M NH_3 is titrated with 0.10 M HNO_3. Determine the pH of the solution after the addition of 100.0 mL of HNO_3. The K_b of NH_3 is 1.8×10^{-5}.

A) 6.58 B) 10.56 C) 8.72 D) 3.44 E) 5.28

Answer: E
Diff: 4 *Page Ref: 16.4*

46) A 100.0 mL sample of 0.20 M HF is titrated with 0.10 M KOH. Determine the pH of the solution after the addition of 200.0 mL of KOH. The K_a of HF is 3.5×10^{-4}.

A) 9.62 B) 7.00 C) 3.46 D) 10.54 E) 8.14

Answer: E
Diff: 5 *Page Ref: 16.4*

47) Determine the molar solubility of AgBr in a solution containing 0.150 M NaBr. K_{sp} (AgBr) = 7.7×10^{-13}.

A) 8.8×10^{-7} M
B) 3.9×10^{-13} M
C) 5.8×10^{-5} M
D) 5.1×10^{-12} M
E) 0.150 M

Answer: D
Diff: 2 *Page Ref: 16.5*

48) Determine the molar solubility of AgBr in a solution containing 0.200 M NaBr. K_{sp} (AgBr) = 7.7×10^{-13}.

A) 8.8×10^{-7} M
B) 1.54×10^{-13} M
C) 5.8×10^{-5} M
D) 3.8×10^{-12} M
E) 0.200 M

Answer: D
Diff: 2 *Page Ref: 16.5*

49) Determine the molar solubility of $PbSO_4$ in pure water. K_{sp} ($PbSO_4$) = 1.82×10^{-8}.

A) 1.82×10^{-8} M
B) 1.35×10^{-4} M
C) 9.1×10^{-9} M
D) 3.31×10^{-16} M
E) 4.48×10^{-4} M

Answer: B
Diff: 2 *Page Ref: 16.5*

50) Determine the molar solubility of $MgCO_3$ in pure water. K_{sp} $(MgCO_3) = 6.82 \times 10^{-6}$.

A) 6.82×10^{-6} M
B) 3.41×10^{-6} M
C) 4.65×10^{-3} M
D) 2.61×10^{-3} M
E) 3.25×10^{-4} M

Answer: D
Diff: 2 *Page Ref: 16.5*

51) Determine the molar solubility of AgI in pure water. K_{sp} (AgI) $= 8.51 \times 10^{-17}$.

A) 9.22×10^{-9} M
B) 4.26×10^{-17} M
C) 8.51×10^{-17} M
D) 2.77×10^{-6} M
E) 4.40×10^{-6} M

Answer: A
Diff: 2 *Page Ref: 16.5*

52) The molar solubility of ZnS is 1.6×10^{-12} M in pure water. Calculate the K_{sp} for ZnS.

A) 8.0×10^{-13}
B) 3.2×10^{-12}
C) 1.6×10^{-35}
D) 2.6×10^{-24}
E) 6.80×10^{-5}

Answer: D
Diff: 2 *Page Ref: 16.5*

53) The molar solubility of CuI is 2.26×10^{-6} M in pure water. Calculate the K_{sp} for CuI.

A) 5.11×10^{-12}
B) 4.52×10^{-6}
C) 1.50×10^{-3}
D) 4.62×10^{-17}
E) 1.02×10^{-11}

Answer: A
Diff: 2 *Page Ref: 16.5*

54) Determine the molar solubility of $Fe(OH)_2$ in pure water. K_{sp} for $Fe(OH)_2 = 4.87 \times 10^{-17}$.

A) 2.44×10^{-17} M
B) 1.62×10^{-17} M
C) 4.03×10^{-9} M
D) 3.65×10^{-6} M
E) 2.30×10^{-6} M

Answer: E
Diff: 3 *Page Ref: 16.5*

55) Determine the molar solubility of BaF_2 in pure water. K_{sp} for $BaF_2 = 2.45 \times 10^{-5}$.

A) 1.83×10^{-2} M
B) 1.23×10^{-5} M
C) 2.90×10^{-2} M
D) 4.95×10^{-3} M
E) 6.13×10^{-6} M

Answer: A
Diff: 3 *Page Ref: 16.5*

56) Determine the molar solubility for $Al(OH)_3$ in pure water. K_{sp} for $Al(OH)_3 = 1.3 \times 10^{-33}$.

A) 3.6×10^{-12} M
B) 2.2×10^{-10} M
C) 4.8×10^{-35} M
D) 2.6×10^{-9} M
E) 6.0×10^{-19} M

Answer: D
Diff: 3 *Page Ref: 16.5*

57) Determine the molar solubility for $Pb_3(PO_4)_2$ in pure water. Ksp for $Pb_3(PO_4)_2$ is 1.0×10^{-54}.

A) 4.1×10^{-28} M
B) 5.8×10^{-10} M
C) 1.1×10^{-11} M
D) 6.2×10^{-12} M
E) 1.0×10^{-54} M

Answer: D
Diff: 3 *Page Ref: 16.5*

58) Determine the molar solubility for $Zn(OH)_2$ in pure water. Ksp for $Zn(OH)_2$ is 3.00×10^{-17}.

A) 3.0×10^{-17} M
B) 7.5×10^{-18} M
C) 3.11×10^{-6} M
D) 1.96×10^{-6} M
E) 5.5×10^{-19} M

Answer: D
Diff: 3 *Page Ref: 16.5*

59) The molar solubility of $Ba_3(PO_4)_2$ is 8.89×10^{-9} M in pure water. Calculate the K_{sp} for $Ba_3(PO_4)_2$.

A) 4.94×10^{-49}
B) 5.55×10^{-41}
C) 5.33×10^{-37}
D) 8.16×10^{-31}
E) 6.00×10^{-39}

Answer: E
Diff: 3 *Page Ref: 16.5*

60) The molar solubility of Ag_2S is 1.26×10^{-16} M in pure water. Calculate the K_{sp} for Ag_2S.

A) 1.59×10^{-32}
B) 1.12×10^{-8}
C) 6.81×10^{-63}
D) 3.78×10^{-12}
E) 8.00×10^{-48}

Answer: E
Diff: 3 *Page Ref: 16.5*

61) Determine the molar solubility of BaF_2 in a solution containing 0.0750 M LiF. K_{sp} $(BaF_2) = 1.7 \times 10^{-6}$.

A) 2.3×10^{-5} M
B) 8.5×10^{-7} M
C) 1.2×10^{-2} M
D) 0.0750 M
E) 3.0×10^{-4} M

Answer: E
Diff: 3 *Page Ref: 16.5*

62) Determine the molar solubility of $Al(OH)_3$ in a solution containing 0.0500 M $AlCl_3$. K_{sp} $(Al(OH)_3) = 1.3 \times 10^{-33}$.

A) 2.6 x 10–9M
B) 5.2×10^{-31} M
C) 2.6×10^{-32} M
D) 1.04×10^{-29} M
E) 6.5×10^{-35} M

Answer: D
Diff: 3 *Page Ref: 16.5*

63) Stalactites and stalagmites form as ________ precipitates out of the water evaporating in underground caves.

A) hydrochloric acid
B) sodium hydroxide
C) sodium chloride
D) calcium carbonate
E) sodium bicarbonate

Answer: D
Diff: 3 *Page Ref: 16.5*

64) Give the equation for an unsaturated solution in comparing Q with K_{sp}.

A) $Q > K_{sp}$
B) $Q < K_{sp}$
C) $Q = K_{sp}$
D) $Q \neq K_{sp}$
E) none of the above

Answer: B
Diff: 1 *Page Ref: 16.6*

65) Give the equation for a saturated solution in comparing Q with K_{sp}.

A) $Q > K_{sp}$
B) $Q < K_{sp}$
C) $Q = K_{sp}$
D) $Q \neq K_{sp}$
E) none of the above

Answer: C
Diff: 1 *Page Ref: 16.6*

66) Give the equation for a supersaturated solution in comparing Q with K_{sp}.

A) $Q > K_{sp}$
B) $Q < K_{sp}$
C) $Q = K_{sp}$
D) $Q \neq K_{sp}$
E) none of the above

Answer: A
Diff: 1 *Page Ref: 16.6*

67) A solution containing $AgNO_3$ is mixed with a solution of NaCl to form a solution that is 0.10 M in $AgNO_3$ and 0.075 M in NaCl. What will happen once these solutions are mixed? K_{sp} (AgCl) = 1.77×10^{-10}.

A) Nothing will happen since the molar solubility of AgCl is higher than the solution concentrations.
B) Silver chloride will precipitate out of solution, leaving an unsaturated solution of AgCl.
C) Silver chloride will precipitate out of solution, leaving a saturated AgCl solution.
D) Nothing will happen since NaCl and $AgNO_3$ are both soluble compounds.
E) There is not enough information to say anything about this solution.

Answer: C
Diff: 2 *Page Ref: 16.6*

68) A solution containing $CaCl_2$ is mixed with a solution of $Li_2C_2O_4$ to form a solution that is 2.1×10^{-5} M in calcium ion and 4.75×10^{-5} M in oxalate ion. What will happen once these solutions are mixed? K_{sp} (CaC_2O_4) = 2.3×10^{-9}.

A) A precipitate will form since Q > K_{sp} for calcium oxalate.
B) Nothing will happen since both calcium chloride and lithium oxalate are soluble compounds.
C) Nothing will happen since calcium oxalate is extremely soluble.
D) Nothing will happen since K_{sp} > Q for all possible precipitants.
E) There is not enough information to determine.

Answer: D
Diff: 2 *Page Ref: 16.6*

69) A solution containing $CaCl_2$ is mixed with a solution of $Li_2C_2O_4$ to form a solution that is 3.5×10^{-4} M in calcium ion and 2.33×10^{-4} M in oxalate ion. What will happen once these solutions are mixed? K_{sp} (CaC_2O_4) = 2.3×10^{-9}.

A) Nothing will happen since both calcium chloride and lithium oxalate are soluble compounds.
B) Nothing will happen K_{sp} > Q for all possible precipitants.
C) A precipitate will form as calcium oxalate is not soluble to any extent.
D) A precipitate will form since Q > K_{sp} for calcium oxalate.
E) There is not enough information to determine.

Answer: D
Diff: 2 *Page Ref: 16.6*

70) A solution contains 0.021 M Cl^- and 0.017 M I^-. A solution containing copper (I) ions is added to selectively precipitate one of the ions. At what concentration of copper (I) ion will a precipitate begin to form? What is the identity of the precipitate? $K_{sp}(CuCl) = 1.0 \times 10^{-6}$, $K_{sp}(CuI) = 5.1 \times 10^{-12}$.

A) 3.0×10^{-10} M, CuI
B) 3.0×10^{-10} M, CuCl
C) 4.8×10^{-5} M, CuCl
D) 4.8×10^{-5} M, CuI
E) No precipitate will form at any concentration of copper (I).

Answer: A
Diff: 3 *Page Ref: 16.6*

71) A solution contains 0.036 M Cu^{2+} and 0.044 M Fe^{2+}. A solution containing sulfide ions is added to selectively precipitate one of the metal ions from solution. At what concentration of sulfide ion will a precipitate begin to form? What is the identity of the precipitate? $K_{sp}(CuS) = 1.3 \times 10^{-36}$, $K_{sp}(FeS) = 6.3 \times 10^{-18}$.

A) 1.4×10^{-16} M, FeS
B) 3.6×10^{-35} M, CuS
C) 3.6×10^{-35} M, FeS
D) 1.4×10^{-16} M, CuS
E) No precipitate will form at any concentration of sulfide ion.

Answer: B
Diff: 3 *Page Ref: 16.6*

72) A solution contains 2.2×10^{-3} M in Cu^{2+} and 0.33 M in LiCN. If the K_f for $Cu(CN)_4{}^{2-}$ is 1.0×10^{25}, how much copper ion remains at equilibrium?

A) 3.8×10^{-24} M
B) 1.9×10^{-26} M
C) 6.7×10^{-28} M
D) 2.9×10^{-27} M
E) 4.6×10^{-25} M

Answer: B
Diff: 4 *Page Ref: 16.8*

73) A solution contains 3.8×10^{-2} M in Al^{3+} and 0.29 M in NaF. If the K_f for $AlF_6{}^{3-}$ is 7×10^{19}, how much aluminum ion remains at equilibrium?

A) 1.1×10^{-19} M
B) 3.1×10^{-22} M
C) 9.1×10^{-19} M
D) 1.9×10^{-21} M
E) 4.4×10^{-20} M

Answer: C
Diff: 4 *Page Ref: 16.8*

Algorithmic Questions

1) Which of the following solutions is a good buffer system?
   A) A solution that is 0.10 M NaCl and 0.10 M HCl
   B) A solution that is 0.10 M HCN and 0.10 M LiCN
   C) A solution that is 0.10 M NaOH and 0.10 M HNO_3
   D) A solution that is 0.10 M HNO_3 and 0.10 M $NaNO_3$
   E) A solution that is 0.10 M HCN and 0.10 M KI

Answer: B
Diff: 1 *Page Ref: 16.2*

2) Which of the following solutions is a good buffer system?
   A) A solution that is 0.10 M $HC_2H_3O_2$ and 0.10 M $LiC_2H_3O_2$
   B) A solution that is 0.10 M HBr and 0.10 M $KC_2H_3O_2$
   C) A solution that is 0.10 M HI and 0.10 M NH_4^+
   D) A solution that is 0.10 M NaOH and 0.10 M KOH
   E) None of the above are buffer systems.

Answer: A
Diff: 1 *Page Ref: 16.2*

3) Which one of the following statements is TRUE?
   A) A buffer is an aqueous solution composed of two weak acids.
   B) A buffer can absorb an unlimited amount of base.
   C) A buffer resists pH change by neutralizing added acids and bases.
   D) A buffer does not change pH when strong acid is added.
   E) None of the above are true.

Answer: C
Diff: 1 *Page Ref: 16.2*

4) A solution is prepared by dissolving 0.23 mol of hydrazoic acid and 0.27 mol of sodium azide in water sufficient to yield 1.00 L of solution.The addition of 0.05 mol of NaOH to this buffer solution causes the pH to increase slightly. The pH does not increase drastically because the NaOH reacts with the ________ present in the buffer solution. The K_a of hydrazoic acid is 1.9×10^{-5}.
   A) H_2O
   B) H_3O^+
   C) azide
   D) hydrazoic acid
   E) This is a buffer solution: the pH does not change upon addition of acid or base.

Answer: D
Diff: 2 *Page Ref: 16.2*

5) A solution is prepared by dissolving 0.23 mol of chloroacetic acid and 0.27 mol of sodium chloroacetate in water sufficient to yield 1.00 L of solution. The addition of 0.05 mol of HCl to this buffer solution causes the pH to drop slightly. The pH does not decrease drastically because the HCl reacts with the __________ present in the buffer solution. The K_a of chloroacetic acid is 1.36×10^{-3}.

A) H_2O
B) H_3O^+
C) chloroacetate ion
D) chloroacetic acid
E) This is a buffer solution: the pH does not change upon addition of acid or base.

Answer: C
Diff: 2 *Page Ref: 16.2*

6) Calculate the pH of a solution that is 0.295 M in sodium formate ($NaHCO_2$) and 0.205 M in formic acid (HCO_2H). The K_a of formic acid is 1.77×10^{-4}.

A) 3.910 B) 3.587 C) 13.84 D) 10.10 E) 4.963

Answer: A
Diff: 3 *Page Ref: 16.2*

7) Calculate the pH of a solution that is 0.210 M in nitrous acid (HNO_2) and 0.290 M in potassium nitrite (KNO_2). The acid dissociation constant of nitrous acid is 4.50×10^{-4}.

A) 3.487 B) 3.210 C) 13.86 D) 10.51 E) 4.562

Answer: A
Diff: 4 *Page Ref: 16.2*

8) Calculate the percent ionization of nitrous acid in a solution that is 0.222 M in nitrous acid (HNO_2) and 0.278 M in potassium nitrite (KNO_2). The acid dissociation constant of nitrous acid is 4.50×10^{-4}.

A) 55.6
B) 0.162
C) 15.5
D) 2.78×10^{-3}
E) 3.448

Answer: B
Diff: 4 *Page Ref: 16.2*

9) Calculate the percent ionization of nitrous acid in a solution that is 0.249 M in nitrous acid. The acid dissociation constant of nitrous acid is 4.50×10^{-4}.

A) 1.12×10^{-4}
B) 0.0450
C) 4.25
D) 0.342
E) 5.53

Answer: C
Diff: 4 *Page Ref: 16.2*

10) What is the pH of a buffer solution that is 0.211 M in lactic acid and 0.111 M in sodium lactate? The K_a of lactic acid is 1.4×10^{-4}.

A) 14.28 B) 10.43 C) 5.48 D) 3.57 E) 4.13

Answer: D

Diff: 4 *Page Ref: 16.2*

11) What is the pH of a buffer solution that is 0.255 M in hypochlorous acid (HClO) and 0.333 M in sodium hypochlorite? The K_a of hypochlorous acid is 3.8×10^{-8}.

A) 13.88 B) 6.46 C) 8.49 D) 7.30 E) 7.54

Answer: E

Diff: 4 *Page Ref: 16.2*

12) What is the pH of a solution prepared by mixing 25.00 mL of 0.10 M CH_3CO_2H with 25.00 mL of 0.010 M CH_3CO_2Na? Assume that the volume of the solutions are additive and that $K_a = 1.8 \times 10^{-5}$ for CH_3CO_2H.

A) 2.87 B) 3.74 C) 4.75 D) 5.74

Answer: B

Diff: 4 *Page Ref: 16.2*

13) What is the pH of a solution prepared by mixing 50.00 mL of 0.10 M NH_3 with 5.00 mL of 0.10 M NH_4Cl? Assume that the volume of the solutions are additive and that $K_b = 1.8 \times 10^{-5}$ for NH_3.

A) 8.25 B) 9.28 C) 10.26 D) 11.13

Answer: C

Diff: 4 *Page Ref: 16.2*

14) What is the pH of a solution prepared by mixing 50.00 mL of 0.10 M methylamine, CH_3NH_2, with 20.00 mL of 0.10 M methylammonium chloride, CH_3NH_3Cl? Assume that the volume of the solutions are additive and that $K_b = 3.70 \times 10^{-4}$ for methylamine.

A) 10.17 B) 10.57 C) 10.97 D) 11.78

Answer: C

Diff: 4 *Page Ref: 16.2*

15) What is the pH of a buffer system prepared by dissolving 10.70 grams of NH_4Cl and 25.00 mL of 12 M NH_3 in enough water to make 1.000 L of solution? $K_b = 1.80 \times 10^{-5}$ for NH_3.

A) 9.08 B) 9.26 C) 9.43 D) 11.32

Answer: C

Diff: 4 *Page Ref: 16.2*

16) What is the pH of a solution prepared by mixing 100.00 mL of 0.020 M $Ca(OH)_2$ with 50.00 mL of 0.300 M NaOH? Assume that the volumes are additive.

A) 13.05 B) 13.10 C) 13.28 D) 13.58

Answer: B

Diff: 4 *Page Ref: 16.2*

17) What is the hydronium ion concentration in a solution prepared by mixing 50.00 mL of 0.10 M HCN with 50.00 mL of 0.030 M NaCN? Assume that the volumes of the solutions are additive and that $K_a = 4.9 \times 10^{-10}$ for HCN.

A) 1.5×10^{-10} M
B) 4.9×10^{-10} M
C) 1.6×10^{-9} M
D) 7.0×10^{-6} M

Answer: C
Diff: 4 *Page Ref: 16.2*

18) What is the $[CH_3CO_2^-]/[CH_3CO_2H]$ ratio necessary to make a buffer solution with a pH of 4. 34? $K_a = 1.8 \times 10^{-5}$ for CH_3CO_2H.

A) 0.39:1
B) 0.91:1
C) 1.09:1
D) 2.5:1

Answer: A
Diff: 4 *Page Ref: 16.2*

19) You wish to prepare an $HC_2H_3O_2$ buffer with a pH of 4.34. If the pK_a of is 4.74, what ratio of $C_2H_3O_2^-/HC_2H_3O_2$ must you use?

A) 0.10
B) 0.40
C) 0.40
D) 2.51
E) 4.0

Answer: C
Diff: 2 *Page Ref: 16.2*

20) You wish to prepare an $HC_2H_3O_2$ buffer with a pH of 5.14. If the pK_a of is 4.74, what ratio of $C_2H_3O_2^-/HC_2H_3O_2$ must you use?

A) 0.10
B) 0.40
C) 0.40
D) 2.51
E) 6.0

Answer: D
Diff: 2 *Page Ref: 16.2*

21) A buffer solution is 0.100 M in both $HC_7H_5O_2$ and $KC_7H_5O_2$ and has a pH of 4.19. Which of the following pH values would you expect from the addition of a small amount of a dilute solution of a strong base?

A) 3.89
B) 3.69
C) 5.79
D) 4.49
E) There is not enough information to determine.

Answer: D
Diff: 2 *Page Ref: 16.2*

22) Calculate the pH of a buffer that is 0.032 M HF and 0.032 M NaF. The K_a for HF is 3.5×10^{-4}.

A) 2.86
B) 9.31
C) 10.54
D) 3.46
E) 4.79

Answer: D
Diff: 3 *Page Ref: 16.2*

23) Calculate the pH of a buffer that is 0.020 M HF and 0.040 M NaF. The K_a for HF is 3.5×10^{-4}.

A) 2.06
B) 4.86
C) 3.16
D) 3.36
E) 3.76

Answer: E
Diff: 3 *Page Ref: 16.2*

24) Calculate the pH of a buffer that is 0.060 M HF and 0.030 M KF. The K_a for HF is 3.5×10^{-4}.
A) 2.06 B) 4.86 C) 3.16 D) 3.56 E) 3.76

Answer: C
Diff: 3 *Page Ref: 16.2*

25) Calculate the pH of a buffer that is 0.94 M HClO and 0.059 M NaClO. The K_a for HClO is 2.9×10^{-8}.
A) 7.54 B) 6.67 C) 7.74 D) 6.36 E) 7.34

Answer: E
Diff: 3 *Page Ref: 16.2*

26) Calculate the pH of a buffer that is 0.145 M $HC_2H_3O_2$ and 0.202 M $KC_2H_3O_2$. The K_a for $HC_2H_3O_2$ is 1.8×10^{-5}.
A) 4.89 B) 9.01 C) 4.74 D) 5.05 E) 4.60

Answer: A
Diff: 3 *Page Ref: 16.2*

27) The highest pH for an effective buffer occurs when the base is how many times as concentrated as the acid?
A) 8 B) 5 C) 10 D) 25 E) 40

Answer: C
Diff: 1 *Page Ref: 16.3*

28) Which of the following acids (listed with pK_a values) and their conjugate base would form a buffer with a pH of 8.10?
A) $HC_2H_3O_2$, $pK_a = 4.74$
B) H_2SO_3, $pK_a = 1.77$
C) HClO, $pK_a = 7.54$
D) HIO, $pK_a = 10.64$
E) HNO_2, $pK_a = 3.34$

Answer: C
Diff: 1 *Page Ref: 16.3*

29) Which of the following acids (listed with K_a values) and their conjugate base would form a buffer with a pH of 2.34?
A) C_6H_5COOH, $K_a = 6.5 \times 10^{-5}$
B) HIO_3, $K_a = 1.7 \times 10^{-1}$
C) HClO, $K_a = 2.9 \times 10^{-8}$
D) HF, $K_a = 3.5 \times 10^{-4}$
E) $HClO_2$, $K_a = 1.1 \times 10^{-2}$

Answer: E
Diff: 1 *Page Ref: 16.3*

30) A 1.0 L buffer solution is 0.250 M $HC_2H_3O_2$ and 0.050 M $LiC_2H_3O_2$. Which of the following actions will destroy the buffer?
A) adding 0.050 moles of $LiC_2H_3O_2$
B) adding 0.050 moles of $HC_2H_3O_2$
C) adding 0.050 moles of HCl
D) adding 0.050 moles of NaOH
E) None of the above will destroy the buffer.

Answer: C
Diff: 1 *Page Ref: 16.3*

31) A 1.0 L buffer solution is 0.050 M $HC_2H_3O_2$ and 0.250 M $NaC_2H_3O_2$. Which of the following actions will destroy the buffer?
A) adding 0.050 moles of KOH
B) adding 0.050 moles of HCl
C) adding 0.050 moles of $HC_2H_3O_2$
D) adding 0.050 moles of $NaC_2H_3O_2$
E) All of the above will destroy the buffer.

Answer: A
Diff: 1 *Page Ref: 16.3*

32) Identify the pH of normal blood.

A) 6.9 B) 7.2 C) 7.4 D) 7.6 E) 7.8

Answer: C
Diff: 1 *Page Ref: 16.3*

33) Identify the most common indicator.
A) alizarin
B) thymol blue
C) crystal violet
D) phenolphthalein
E) thymolphthalein

Answer: D
Diff: 1 *Page Ref: 16.4*

34) Identify the indicator that can be used at the lowest pH.
A) phenol red
B) thymol blue
C) crystal violet
D) phenolphthalein
E) m-nitrophenol

Answer: C
Diff: 1 *Page Ref: 16.4*

35) Identify the indicator that has two endpoints.
A) phenol red
B) alizarin
C) crystal violet
D) phenolphthalein
E) alizarin yellow R

Answer: B
Diff: 1 *Page Ref: 16.4*

36) Identify the indicator that can be used at the highest pH.
A) alizarin
B) thymol blue
C) crystal violet
D) phenolphthalein
E) alizarin yellow R

Answer: E
Diff: 1 *Page Ref: 16.4*

37) What is the pH of a solution made by mixing 15.00 mL of 0.100 M HCl with 50.00 mL of 0.100 M KOH? Assume that the volumes of the solutions are additive.
A) 11.73 B) 12.73 C) 1.27 D) 2.27 E) 7.00

Answer: B
Diff: 3 *Page Ref: 16.4*

38) What volume of 5.00×10^{-3} M HNO_3 is needed to titrate 80.00 mL of 5.00×10^{-3} M $Ca(OH)_2$ to the equivalence point?
A) 10.0 mL B) 40.0 mL C) 80.0 mL D) 160. mL

Answer: D
Diff: 3 *Page Ref: 16.4*

39) What is the pH of a solution made by mixing 25.00 mL of 0.100 M HCl with 40.00 mL of 0.100 M KOH? Assume that the volumes of the solutions are additive.
A) 0.64 B) 1.64 C) 12.36 D) 13.36

Answer: C
Diff: 3 *Page Ref: 16.4*

40) How many milliliters of 0.0850 M NaOH are required to titrate 25.0 mL of 0.0720 M HBr to the equivalence point?
A) 21.2 B) 0.245 C) 3.92 D) 0.153 E) 29.5

Answer: A
Diff: 3 *Page Ref: 16.4*

41) How many milliliters of 0.120 M NaOH are required to titrate 50.0 mL of 0.0998 M butanoic acid to the equivalence point? The K_a of butanoic acid is 1.5×10^{-5}.
A) 4.90 B) 50.0 C) 41.6 D) 60.1 E) 4.65

Answer: C
Diff: 3 *Page Ref: 16.4*

42) A 25.0 mL sample of 0.150 M hydrofluoric acid is titrated with a 0.150 M NaOH solution. What is the pH at the equivalence point? The K_a of hydrofluoric acid is 3.5×10^{-4}.
A) 10.17 B) 10.83 C) 3.17 D) 7.00 E) 8.17

Answer: E
Diff: 4 *Page Ref: 16.4*

43) A 25.0-mL sample of 0.150 M butanoic acid is titrated with a 0.150 M NaOH solution. What is the pH before any base is added? The K_a of butanoic acid is 1.5×10^{-5}.
A) 2.83
B) 1.5×10^{-3}
C) 4.82
D) 4.00
E) 1.0×10^{4}

Answer: A
Diff: 4 *Page Ref: 16.4*

44) A 25.0 mL sample of 0.150 M hypochlorous acid is titrated with a 0.150 M NaOH solution. What is the pH after 26.0 mL of base is added? The K_a of hypochlorous acid is 3.0×10^{-8}.
A) 2.54 B) 11.47 C) 7.00 D) 7.51 E) 7.54

Answer: B
Diff: 4 *Page Ref: 16.4*

45) What is the approximate pH at the equivalence point of a weak acid-strong base titration if 25 mL of aqueous formic acid requires 29.80 mL of 0.3567 M NaOH? $K_a = 1.8 \times 10^{-4}$ for formic acid.
A) 2.06 B) 5.48 C) 8.52 D) 11.94

Answer: C
Diff: 4 *Page Ref: 16.4*

46) What is the approximate pH at the equivalence point of a weak acid-strong base titration if 25 mL of aqueous hydrofluoric acid requires 30.00 mL of 0.400 M NaOH? $K_a = 6.76 \times 10^{-4}$ for HF.
A) 1.74 B) 5.75 C) 8.25 D) 12.26

Answer: C
Diff: 4 *Page Ref: 16.4*

47) Formic acid (HCO_2H, $K_a = 1.8 \times 10^{-4}$) is the principal component in the venom of stinging ants. What is the molarity of a formic acid solution if 25.00 mL of the formic acid solution requires 29.80 mL of 0.0567 M NaOH to reach the equivalence point?
A) 0.0134 M B) 0.0476 M C) 0.0567 M D) 0.0676 M

Answer: D
Diff: 4 *Page Ref: 16.4*

48) Sodium hypochlorite, NaOCl, is the active ingredient in household bleach. What is the concentration of hypochlorite ion if 20.00 mL of bleach requires 32.00 mL of 0.500 M HCl to reach the equivalence point?

A) 0.300 M B) 0.312 M C) 0.800 M D) 1.30 M

Answer: C
Diff: 4 *Page Ref: 16.4*

49) What is the pH of the resulting solution if 25.00 mL of 0.10 M acetic acid is added to 10.00 mL of 0.10 M NaOH? Assume that the volumes of the solutions are additive. $K_a = 1.8 \times 10^{-5}$ for CH_3CO_2H.

A) 9.43 B) 9.08 C) 4.92 D) 4.57

Answer: D
Diff: 5 *Page Ref: 16.4*

50) What is the pH of a solution made by mixing 10.00 mL of 0.10 M acetic acid with 10.00 mL of 0.10 M KOH? Assume that the volumes of the solutions are additive. $K_a = 1.8 \times 10^{-5}$ for CH_3CO_2H.

A) 5.28 B) 7.00 C) 8.72 D) 10.02

Answer: C
Diff: 5 *Page Ref: 16.4*

51) What is the pH of a solution made by mixing 30.00 mL of 0.10 M acetic acid with 50.00 mL of 0.100 M KOH? Assume that the volumes of the solutions are additive. $K_a = 1.8 \times 10^{-5}$ for CH_3CO_2H.

A) 8.26 B) 9.26 C) 11.13 D) 12.40

Answer: D
Diff: 5 *Page Ref: 16.4*

52) What is the pH at the equivalence point of a weak base-strong acid titration if 20.00 mL of NaOCl requires 28.30 mL of 0.50 M HCl? $K_a = 3.0 \times 10^{-8}$ for HOCl.

A) 0.30 B) 3.18 C) 3.76 D) 4.03

Answer: D
Diff: 5 *Page Ref: 16.4*

53) What is the pH of the resulting solution if 45 mL of 0.432 M methylamine, CH_3NH_2, is added to 15 mL of 0.234 M HCl? Assume that the volumes of the solutions are additive. $K_a = 2.70 \times 10^{-11}$ for $CH_3NH_3^+$.

A) 2.77 B) 4.09 C) 9.91 D) 11.23

Answer: D
Diff: 5 *Page Ref: 16.4*

54) A 25.0-mL sample of 0.150 M hydrazoic acid is titrated with a 0.150 M NaOH solution. What is the pH after 13.3 mL of base is added? The K_a of hydrazoic acid is 1.9×10^{-5}.

A) 4.45 B) 1.34 C) 3.03 D) 4.78 E) 4.66

Answer: D
Diff: 5 *Page Ref: 16.4*

55) Calculate the solubility (in g/L) of calcium fluoride in water at 25°C if the K_{sp} for CaF_2 is 1.5×10^{-10}.

A) 9.6×10^{-4} g/L B) 2.6×10^{-2} g/L C) 3.3×10^{-2} g/L D) 4.1×10^{-2} g/L

Answer: B
Diff: 3 *Page Ref: 16.5*

56) Calculate the K_{sp} for zinc hydroxide if the solubility of $Zn(OH)_2$ in pure water is 2.1×10^{-4} g/L.

A) 9.9×10^{-18} B) 3.8×10^{-17} C) 1.5×10^{-3} D) 2.9×10^{-3}

Answer: B
Diff: 3 *Page Ref: 16.5*

57) What is the molar solubility of $Mg(OH)_2$ in a basic solution with a pH of 12.50? K_{sp} for $Mg(OH)_2$ is 5.6×10^{-12}.

A) 1.8×10^{-10} M B) 5.6×10^{-9} M C) 2.4×10^{-6} M D) 1.1×10^{-4} M

Answer: B
Diff: 3 *Page Ref: 16.5*

58) Calculate the molar solubility of thallium chloride in 0.40 M NaCl at 25°C. K_{sp} for TlCl is 1.7×10^{-4}.

A) 6.8×10^{-5} M B) 4.2×10^{-4} M C) 8.2×10^{-3} M D) 1.3×10^{-2} M

Answer: B
Diff: 3 *Page Ref: 16.5*

59) In which of the following solutions would solid $PbBr_2$ be expected to be the *least* soluble at 25°C?

A) 0.1 M HBr B) 0.1 M NaBr C) 0.1 M $CaBr_2$ D) 0.1 M KNO_3

Answer: C
Diff: 3 *Page Ref: 16.5*

60) What is the molar solubility of magnesium carbonate ($MgCO_3$) in water? The solubility-product constant for $MgCO_3$ is 3.5×10^{-8} at 25°C.

A) 1.8×10^{-8}
B) 7.0×10^{-8}
C) 7.46
D) 2.6×10^{-4}
E) 1.9×10^{-4}

Answer: E
Diff: 4 *Page Ref: 16.5*

61) What is the molar solubility of barium fluoride (BaF_2) in water? The solubility-product constant for BaF_2 is 1.7×10^{-6} at 25°C.

A) 6.5×10^{-4}
B) 1.2×10^{-2}
C) 1.8×10^{-3}
D) 7.5×10^{-3}
E) 5.7×10^{-7}

Answer: D
Diff: 4 *Page Ref: 16.5*

62) What is the molar solubility of AgCl in 0.30 M NH_3? K_{sp} for AgCl is 1.8×10^{-10} and K_f for $Ag(NH_3)_2^+$ is 1.7×10^7.

A) 1.3×10^{-5} M B) 1.6×10^{-2} M C) 1.7×10^{-2} M D) 5.5×10^{-2} M

Answer: B
Diff: 4 *Page Ref: 16.5*

63) What is the molar solubility of AgCl in 0.10 M NaCN if the colorless complex ion $Ag(CN)_2^-$ forms? K_{sp} for AgCl is 1.8×10^{-10} and K_f for $Ag(CN)_2^-$ is 1.0×10^{21}.

A) 0.050 M B) 0.10 M C) 0.20 M D) 0.40 M

Answer: A
Diff: 4 *Page Ref: 16.5*

64) Which of the following compounds will have the highest molar solubility in pure water?

A) $PbSO_4$, $K_{sp} = 1.82 \times 10^{-8}$
B) $MgCO_3$, $K_{sp} = 6.82 \times 10^{-6}$
C) AgCN, $K_{sp} = 5.97 \times 10^{-17}$
D) PbS, $K_{sp} = 9.04 \times 10^{-29}$
E) NiS, $K_{sp} = 3.00 \times 10^{-20}$

Answer: B
Diff: 1 *Page Ref: 16.5*

65) Which of the following compounds will be more soluble in acidic solution than in pure water?

A) $PbBr_2$
B) FeS
C) $Ca(ClO_4)_2$
D) CuI
E) None of the above will be more soluble in acidic solution.

Answer: B
Diff: 1 *Page Ref: 16.5*

66) Identify the salts that are in hard water.

A) $CaCO_3$ and $MgCO_3$
B) $MgSO_4$ and $CaSO_4$
C) NaCl and KBr
D) $KHSO_4$ and Na_2SO_4
E) NaOCl and HOCl

Answer: A
Diff: 1 *Page Ref: 16.5*

67) Which of the following compounds solubility will not be affected by a low pH in solution?

A) AgCl B) $Sr(OH)_2$ C) CaF_2 D) CuS E) $SrCO_3$

Answer: A
Diff: 1 *Page Ref: 16.5*

68) Determine the molar solubility of $CaSO_4$ in a solution containing 0.12 M Na_2SO_4. K_{sp} ($CaSO_4$) = 2.4×10^{-5}.

A) 2.9×10^{-6} M
B) 2.0×10^{-4} M
C) 5.8×10^{-10} M
D) 4.9×10^{-3} M
E) 0.10 M

Answer: B
Diff: 2 *Page Ref: 16.5*

69) Determine the molar solubility of CuCl in a solution containing 0.060 M KCl. K_{sp} (CuCl) = 1.0×10^{-6}.

A) 1.0×10^{-12} M
B) 6.0×10^{-8} M
C) 1.7×10^{-5} M
D) 1.0×10^{-3} M
E) 0.050 M

Answer: C
Diff: 2 *Page Ref: 16.5*

70) Give the expression for the solubility product constant for $PbCl_2$.

A) $\frac{[Pb^{2+}][Cl^-]^2}{[PbCl_2]}$

B) $\frac{[PbCl_2]}{[Pb^{2+}][2Cl^-]^2}$

C) $[Pb^{2+}][Cl^-]^2$

D) $\frac{[Pb^{2+}]^2[Cl^-]}{[PbCl_2]}$

E) $[Pb^{2+}]^2[2Cl^-]$

Answer: C
Diff: 2 *Page Ref: 16.5*

71) Give the expression for the solubility product constant for BaF_2.

A) $\frac{[BaF_2]}{[Ba^{2+}][F^-]^2}$

B) $\frac{[Ba^{2+}][F^-]^2}{[BaF_2]}$

C) $[Ba^{2+}]^2[F^-]$

D) $[Ba^{2+}][F^-]^2$

E) $[Ba^{2+}][2F^-]$

Answer: D

Diff: 2 *Page Ref:* 16.5

72) Give the expression for the solubility product constant for $Ca_3(PO_4)_2$.

A) $\frac{[2Ca^{2+}]^3[3PO_4^{3-}]^2}{Ca_3(PO_4)_2}$

B) $\frac{Ca_3(PO_4)_2}{[2Ca^{2+}]^3[3PO_4^{3-}]^2}$

C) $\frac{[3Ca^{2+}]^2[2PO_4^{3-}]^3}{Ca_3(PO_4)_2}$

D) $[Ca^{2+}]^2[PO_4^{3-}]^3$

E) $[Ca^{2+}]^3[PO_4^{3-}]^2$

Answer: E

Diff: 2 *Page Ref:* 16.5

73) Give the expression for the solubility product constant for $Cr_2(CO_3)_3$.

A) $[Cr^{3+}]^2[CO_3^{2-}]^3$

B) $\frac{[2Cr^{3+}]^2[3CO_3^{2-}]^3}{Cr_2(CO_3)_3}$

C) $\frac{Cr_2(CO_3)_3}{[Cr^{3+}]^2[CO_3^{2-}]^3}$

D) $\frac{[2Cr^{3+}]^3[3CO_3^{2-}]^2}{Cr_2(CO_3)_3}$

E) $[3Cr^{3+}]^3[2CO_3^{2-}]^2$

Answer: A

Diff: 2 *Page Ref:* 16.5

74) Which of the following compounds will have the highest molar solubility in pure water?

A) HgS, $K_{sp} = 1.6 \times 10^{-54}$

B) CuS, $K_{sp} = 1.27 \times 10^{-36}$

C) $Fe(OH)_3$, $K_{sp} = 2.79 \times 10^{-39}$

D) ZnS, $K_{sp} = 2.0 \times 10^{-25}$

E) Ag_2S, $K_{sp} = 6.0 \times 10^{-51}$

Answer: C
Diff: 2 *Page Ref: 16.5*

75) 0.10 M potassium chromate is slowly added to a solution containing 0.50 M $AgNO_3$ and 0.50 M $Ba(NO_3)_2$. What is the Ag^+ concentration when $BaCrO_4$ just starts to precipitate? The K_{sp} for Ag_2CrO_4 and $BaCrO_4$ are 1.1×10^{-12} and 1.2×10^{-10}, respectively.

A) 6.5×10^{-5} M B) 1.3×10^{-4} M C) 3.2×10^{-4} M D) 6.8×10^{-2} M

Answer: D
Diff: 3 *Page Ref: 16.6*

76) A solution of NaF is added dropwise to a solution that is 0.0144 M in Ba^{2+}. When the concentration of F^- exceeds ________ M, BaF_2 will precipitate. Neglect volume changes. For BaF_2, $K_{sp} = 1.7 \times 10^{-6}$.

A) 5.9×10^{-5}

B) 1.1×10^{-2}

C) 2.4×10^{-8}

D) 2.7×10^{-3}

E) 1.2×10^{4}

Answer: B
Diff: 5 *Page Ref: 16.6*

77) Identify the compound that is acid-insoluble.

A) $PbCl_2$

B) As_2S_3

C) NiS

D) $Ca_3(PO_4)_2$

E) LiCl

Answer: B
Diff: 1 *Page Ref: 16.7*

78) Identify the compound that is base-insoluble.

A) $PbCl_2$

B) As_2S_3

C) ZnS

D) $Ca_3(PO_4)_2$

E) KCl

Answer: C
Diff: 1 *Page Ref: 16.7*

79) A sample contains $Ba_3(PO_4)_2$, CdS, AgCl, NH_4Cl, and ZnS. Identify the precipitate after the addition of 6 M HCl.

A) $Ca_3(PO_4)_2$
B) HgS
C) AgCl
D) NH_4Cl
E) NiS

Answer: C
Diff: 5 *Page Ref: 16.7*

80) A sample contains $Ba_3(PO_4)_2$, CdS, AgCl, NH_4Cl, and ZnS. Identify the precipitate after the addition of 6 M HCl, then H_2S and 0.2 M HCl.

A) $Ba_3(PO_4)_2$
B) CuS
C) AgCl
D) NH_4Cl
E) MnS

Answer: B
Diff: 5 *Page Ref: 16.7*

81) A sample contains $Ba_3(PO_4)_2$, CdS, AgCl, NH_4Cl, and ZnS. Identify the precipitate after the addition of 6 M HCl; H_2S and 0.2 M HCl; and OH^- to a pH of 8.

A) $Ba_3(PO_4)_2$
B) HgS
C) AgCl
D) NH_4Cl
E) MnS

Answer: E
Diff: 5 *Page Ref: 16.7*

82) A sample contains $Ba_3(PO_4)_2$, CdS, AgCl, NH_4Cl, and ZnS. Identify the precipitate after the addition of 6 M HCl; H_2S and 0.2 M HCl; OH^- to a pH of 8; and $(NH_4)_2HPO_4$ with NH_3.

A) $Ca_3(PO_4)_2$
B) HgS
C) AgCl
D) NH_4Cl
E) ZnS

Answer: A
Diff: 5 *Page Ref: 16.7*

83) A sample contains $Ba_3(PO_4)_2$, CdS, AgCl, NH_4Cl, and ZnS. Identify the soluble ions after the addition of 6 M HCl; H_2S and 0.2 M HCl; OH^- to a pH of 8; and $(NH_4)_2HPO_4$ with NH_3.

A) $Ba_3(PO_4)_2$
B) HgS
C) AgCl
D) NH_4Cl
E) FeS

Answer: D
Diff: 5 *Page Ref: 16.7*

84) A ligand is a molecule or ion that acts as a

A) Lewis acid
B) Brønsted-Lowry base
C) Arrhenius base
D) Lewis base
E) conjugate base

Answer: D
Diff: 1 *Page Ref: 16.8*

85) Describe the solubility of $Al(OH)_3$ with respect to pH.

A) soluble at low pH, insoluble in pH-neutral solution, and soluble at high pH
B) soluble at low pH, insoluble in pH-neutral solution, and insoluble at high pH
C) insoluble at low pH,insoluble in pH-neutral solution, and soluble at high pH
D) insolubleinsoluble at low pH, in pH-neutral solution, and at high pH
E) pH has no effect on the solubility

Answer: A
Diff: 1 *Page Ref: 16.8*

86) Gives what happens at low pH for aluminum hydroxide.

A) $Al(H_2O)_6{}^{3+}$ dissolves
B) $Al(H_2O)_4(OH)_2{}^-$ precipitates
C) $Al(OH)_3$ precipitates
D) Al precipitates
E) Al dissolves

Answer: A
Diff: 1 *Page Ref: 16.8*

87) Gives what happens at neutral pH for aluminum hydroxide.

A) $Al(H_2O)_6{}^{3+}$ precipitates
B) $Al(H_2O)_2(OH)_4{}^-$ dissolves
C) $Al(OH)_3$ precipitates
D) Al precipitates
E) Al dissolves

Answer: C
Diff: 1 *Page Ref: 16.8*

88) Gives what happens at high pH for aluminum hydroxide.
A) $Al(H_2O)_3^{3+}$ precipitates
B) $Al(H_2O)_2(OH)_4^-$ dissolves
C) $Al(OH)_5$ dissolves
D) Al precipitates
E) Al dissolves

Answer: B
Diff: 1 *Page Ref: 16.8*

Matching Questions

Match the following.

1) pH = 7
Diff: 1 *Page Ref: 16.4*

2) pH = pK_a
Diff: 1 *Page Ref: 16.4*

3) pH > 7
Diff: 1 *Page Ref: 16.4*

4) pH < 7
Diff: 1 *Page Ref: 16.4*

5) pH = pK_{a2}
Diff: 1 *Page Ref: 16.4*

A) equivalence point of a strong acid/strong base titration

B) half-way to equivalence point of a weak acid/strong base titration

C) equivalence point of a weak base/strong acid titration

D) 3/4 of the way to second equivalence point of a diprotic acid/strong base titration

E) equivalence point of a weak acid/strong base titration

Answers: 1) A 2) B 3) E 4) C 5) D

Short Answer Questions

1) Give the name of the compound that is in antifreeze and is toxic to pets.

Answer: ethylene glycol
Diff: 1 *Page Ref: 16.1*

2) Define a buffer.

Answer: Buffers contain significant amounts of both a weak acid and its conjugate base or a weak base and its conjugate acid, enabling the buffer to neutralize added acid or added base.
Diff: 1 *Page Ref: 16.2*

3) Sketch the titration curve for a strong acid titrated with a strong base. Make sure to indicate the equivalence point (and whether it is acidic, basic or neutral) and the buffer region.

Answer: See figure in middle of page 732.
Diff: 1 *Page Ref: 16.4*

4) Describe the information necessary (and why) to chose a good indicator for a weak acid/strong base titration.

Answer: In order to chose an appropriate indicator, you must know what the pH should be at the equivalence point of your titration. For a weak acid/strong base, the pH would be basic. The pK_a of the indicator chosen should be very close to the pH of the titration's equivalence point.

Diff: 1 *Page Ref: 16.4*

5) Sketch the titration curve for a monoprotic weak acid titrated with a strong base. Make sure to indicate the equivalence point (and whether it is acidic, basic or neutral) and the buffer region.

Answer: See figure in middle of page 621.

Diff: 2 *Page Ref: 16.4*

6) Sketch the titration curve for a diprotic acid. Make sure to label the equivalence point(s) and the relative amounts of base that were added to reach them.

Answer: See Figure 16.9 in text.

Diff: 3 *Page Ref: 16.4*

7) Identify the compound that is in stalactites and stalagmites.

Answer: calcium carbonate

Diff: 1 *Page Ref: 16.5*

8) Consider the K_{sp} values for two compounds: MZ, $K_{sp} = 1.5 \times 10^{-20}$ and MZ_2, $K_{sp} = 1.5 \times 10^{-20}$. Why don't these compounds have the same molar solubility?

Answer: These two compounds have the same solubility product constant, but that does not mean they have the same molar solubilities. Since the first compound contains 2 ions and the second compound contains 3 ions, their solubility product expressions differ and therefore their molar solubilities must differ. Compound MZ_2 will have a higher molar solubility.

Diff: 1 *Page Ref: 16.6*

9) Explain the common ion effect with respect to molar solubility.

Answer: The common ion effect occurs when a solubility equilibrium is shifted due to a second compound that contains an ion common with the first. For example, the molar solubility of MA ($MA \rightleftharpoons M^+ + A^-$) would be higher in pure water than if MA were dissolved in an aqueous solution that already contained another compound with A^-.

Diff: 1 *Page Ref: 16.6*

Chapter 17 Free Energy and Thermodynamics

Multiple Choice Questions

1) Which of the following statements is TRUE?
   A) There is a "heat tax" for every energy transaction.
   B) A spontaneous reaction is always a fast reaction.
   C) The entropy of a system always decreases for a spontaneous process.
   D) Perpetual motion machines are a possibility in the near future.
   E) None of the above are true.

Answer: A
Diff: 1 *Page Ref: 17.1*

2) The ________ Law of Thermodynamics states the energy is conserved in chemical processes.
   A) Zero B) First C) Second D) Third E) Fourth

Answer: B
Diff: 3 *Page Ref: 17.1*

3) Identify the process that is spontaneous
   A) rusting of iron
   B) electrolysis
   C) photosynthesis
   D) bread rising
   E) burning gasoline

Answer: A
Diff: 1 *Page Ref: 17.2*

4) ________ is a thermodynamic function that increases with the number of energetically equivalent ways to arrange components of a system to achieve a particular state.
   A) Heat of reaction
   B) Free energy
   C) Entropy
   D) Enthalpy
   E) Molar equivalence

Answer: C
Diff: 1 *Page Ref: 17.3*

5) The ________ Law of Thermodynamics states that for any spontaneous reaction, the entropy of the universe increases.
   A) Zero B) First C) Second D) Third E) Fourth

Answer: C
Diff: 3 *Page Ref: 17.3*

6) Which of the following processes have a $\Delta S > 0$?
A) $CH_3OH(l) \rightarrow CH_3OH(s)$
B) $N_2(g) + 3\ H_2(g) \rightarrow 2\ NH_3(g)$
C) $CH_4(g) + H_2O(g) \rightarrow CO(g) + 3\ H_2(g)$
D) $Na_2CO_3(s) + H_2O(g) + CO_2(g) \rightarrow 2\ NaHCO_3(s)$
E) All of the above processes have a $\Delta S > 0$.

Answer: C
Diff: 2 *Page Ref:* 17.3

7) Which of the following statements is TRUE?
A) Entropy is not a state function.
B) Endothermic processes decrease the entropy of the surroundings, at constant T and P.
C) Endothermic processes are never spontaneous.
D) Exothermic processes are always spontaneous.
E) None of the above are true.

Answer: B
Diff: 1 *Page Ref:* 17.4

8) What is the sign of ΔS_{univ} for a biological system?
A) positive
B) negative
C) zero
D) It depends on the biological system.

Answer: A
Diff: 1 *Page Ref:* 17.4

9) Which of the following relationships is correct at constant T and P?
A) ΔG is proportional to $-\Delta S_{univ}$
B) $\Delta G > 0$ represents a spontaneous process
C) $\Delta G > 0$ represents an increase in kinetic energy
D) $\Delta G < 0$ represents a nonspontaneous process
E) All of the above are correct

Answer: A
Diff: 2 *Page Ref:* 17.5

10) Consider a reaction that has a positive ΔH and a positive ΔS. Which of the following statements is TRUE?
A) This reaction will be spontaneous only at high temperatures.
B) This reaction will be spontaneous at all temperatures.
C) This reaction will be nonspontaneous at all temperatures.
D) This reaction will be nonspontaneous only at high temperatures.
E) It is not possible to determine without more information.

Answer: A
Diff: 2 *Page Ref:* 17.5

11) Consider a reaction that has a positive ΔH and a positive ΔS. Which of the following statements is TRUE?
A) This reaction will be spontaneous only at low temperatures.
B) This reaction will be spontaneous at all temperatures.
C) This reaction will be nonspontaneous at all temperatures.
D) This reaction will be nonspontaneous only at low temperatures.
E) It is not possible to determine without more information.

Answer: D
Diff: 2 *Page Ref:* 17.5

12) Consider a reaction that has a negative ΔH and a positive ΔS. Which of the following statements is TRUE?
A) This reaction will be spontaneous only at high temperatures.
B) This reaction will be spontaneous at all temperatures.
C) This reaction will be nonspontaneous at all temperatures.
D) This reaction will be nonspontaneous only at high temperatures.
E) It is not possible to determine without more information.

Answer: B
Diff: 2 *Page Ref:* 17.5

13) Consider a reaction that has a positive ΔH and a negative ΔS. Which of the following statements is TRUE?
A) This reaction will be spontaneous only at high temperatures.
B) This reaction will be spontaneous at all temperatures.
C) This reaction will be nonspontaneous at all temperatures.
D) This reaction will be nonspontaneous only at high temperatures.
E) It is not possible to determine without more information.

Answer: C
Diff: 2 *Page Ref:* 17.5

14) Consider a reaction that has a negative ΔH and a negative ΔS. Which of the following statements is TRUE?
A) This reaction will be spontaneous only at high temperatures.
B) This reaction will be spontaneous at all temperatures.
C) This reaction will be nonspontaneous at all temperatures.
D) This reaction will be nonspontaneous only at high temperatures.
E) It is not possible to determine without more information.

Answer: D
Diff: 2 *Page Ref:* 17.5

15) Consider a reaction that has a negative ΔH and a negative ΔS. Which of the following statements is TRUE?
A) This reaction will be spontaneous only at low temperatures.
B) This reaction will be spontaneous at all temperatures.
C) This reaction will be nonspontaneous at all temperatures.
D) This reaction will be nonspontaneous only at low temperatures.
E) It is not possible to determine without more information.

Answer: A
Diff: 2 *Page Ref:* 17.5

16) For the following example, identify the following.

$$2\,N_2O(g) \rightarrow 2\,N_2(g) + O_2(g)$$

A) a negative ΔH and a negative ΔS
B) a positive ΔH and a negative ΔS
C) a negative ΔH and a positive ΔS
D) a positive ΔH and a positive ΔS
E) It is not possible to determine without more information.

Answer: C
Diff: 2 *Page Ref: 17.5*

17) For the following example, identify the following.

$$3O_2(g) \rightarrow 2O_3(g)$$

A) a negative ΔH and a negative ΔS
B) a positive ΔH and a negative ΔS
C) a negative ΔH and a positive ΔS
D) a positive ΔH and a positive ΔS
E) It is not possible to determine without more information.

Answer: B
Diff: 2 *Page Ref: 17.5*

18) For the following example, identify the following.

$$H_2O(l) \rightarrow H_2O(g)$$

A) a negative ΔH and a negative ΔS
B) a positive ΔH and a negative ΔS
C) a negative ΔH and a positive ΔS
D) a positive ΔH and a positive ΔS
E) It is not possible to determine without more information.

Answer: D
Diff: 2 *Page Ref: 17.5*

19) For the following example, identify the following.

$$H_2O(l) \rightarrow H_2O(s)$$

A) a negative ΔH and a negative ΔS
B) a positive ΔH and a negative ΔS
C) a negative ΔH and a positive ΔS
D) a positive ΔH and a positive ΔS
E) It is not possible to determine without more information.

Answer: A
Diff: 2 *Page Ref: 17.5*

20) Above what temperature does the following reaction become nonspontaneous?

$$2\ H_2S(g) + 3\ O_2(g) \rightarrow 2\ SO_2(g) + 2\ H_2O(g) \qquad \Delta H = -1036\ kJ;\ \Delta S = -153.2\ J/K$$

A) 6.762×10^3 K
B) 158.7 K
C) 298 K
D) This reaction is nonspontaneous at all temperatures.
E) This reaction is spontaneous at all temperatures.

Answer: A
Diff: 3 *Page Ref: 17.5*

21) Above what temperature does the following reaction become nonspontaneous?

$$FeO(s) + CO(g) \rightarrow CO_2(g) + Fe(s) \qquad \Delta H = -11.0\ kJ;\ \Delta S = -17.4\ J/K$$

A) 632 K
B) 298 K
C) 191 K
D) This reaction is nonspontaneous at all temperatures.
E) This reaction is spontaneous at all temperatures.

Answer: A
Diff: 3 *Page Ref: 17.5*

22) Below what temperature does the following reaction become nonspontaneous?

$$2\ HNO_3(aq) + NO(g) \rightarrow 3\ NO_2(g) + H_2O(l) \qquad \Delta H = +136.5\ kJ;\ \Delta S = +287.5\ J/K$$

A) 39.2 K
B) 151 K
C) 475 K
D) This reaction is nonspontaneous at all temperatures.
E) This reaction is spontaneous at all temperatures.

Answer: C
Diff: 3 *Page Ref: 17.5*

23) The ________ Law of Thermodynamics states the entropy of a perfect crystal at absolute zero is zero.

A) Zero B) First C) Second D) Third E) Fourth

Answer: D
Diff: 3 *Page Ref: 17.6*

24) For a given compound, list the decreasing order of entropy for a liquid, solid, and gas

A) solid > gas > liquid
B) liquid > solid > gas
C) gas > liquid > solid
D) gas > solid > liquid
E) solid > liquid > gas

Answer: C
Diff: 3 *Page Ref: 17.6*

25) Which of the following statements is TRUE?

A) Entropy is an extensive property.
B) Entropy is not temperature dependent.
C) Exothermic processes decrease the entropy of the surroundings.
D) $\Delta S_{universe}$ is always greater than zero for a nonspontaneous process.
E) None of the above are true.

Answer: A
Diff: 1 *Page Ref: 17.6*

26) Identify the statement that is FALSE.

A) The entropy of a gas is greater than the entropy of a liquid.
B) Entropy generally increases with increasing molecular complexity.
C) Free atoms have greater entropy than molecules.
D) Entropy increases with dissolution.
E) For noble gasses, entropy increases with size.

Answer: C
Diff: 1 *Page Ref: 17.6*

27) Place the following in order of increasing entropy at 298 K.

Ne Xe He Ar Kr

A) He < Kr < Ne < Ar < Xe
B) Xe < Kr < Ar < Ne < He
C) Ar < He < Ar < Ne < Kr
D) Ar < Ne < Xe < Kr < He
E) He < Ne < Ar < Kr < Xe

Answer: E
Diff: 2 *Page Ref: 17.6*

28) Place the following in order of increasing molar entropy at 298 K.

NO CO SO

A) NO < CO < SO
B) SO < CO < NO
C) SO < NO < CO
D) CO < SO < NO
E) CO < NO < SO

Answer: E
Diff: 2 *Page Ref: 17.6*

29) Place the following in order of decreasing molar entropy at 298 K.

H_2 Cl_2 F_2

A) $H_2 > Cl_2 > F_2$
B) $Cl_2 > H_2 > F_2$
C) $F_2 > Cl_2 > H_2$
D) $H_2 > F_2 > Cl_2$
E) $Cl_2 > F_2 > H_2$

Answer: E
Diff: 2 *Page Ref:* 17.6

30) Place the following in order of increasing standard molar entropy.

$H_2O(l)$ $H_2O(g)$ $H_2O(s)$

A) $H_2O(g) < H_2O(l) < H_2O(s)$
B) $H_2O(s) < H_2O(l) < H_2O(g)$
C) $H_2O(g) < H_2O(s) < H_2O(l)$
D) $H_2O(l) < H_2O(s) < H_2O(g)$
E) $H_2O(s) < H_2O(g) < H_2O(l)$

Answer: B
Diff: 2 *Page Ref:* 17.6

31) Place the following in order of decreasing standard molar entropy.

$N_2O_4(g)$ $NO(g)$ $NO_2(g)$

A) $N_2O_4 > NO_2 > NO$
B) $NO > NO_2 > N_2O_4$
C) $N_2O_4 > NO > NO_2$
D) $NO > N_2O_4 > NO_2$
E) $NO_2 > NO > N_2O_4$

Answer: A
Diff: 2 *Page Ref:* 17.6

32) Which one of the following has the highest standard molar entropy, $S°$, at 25°C?
A) $H_2(g)$ B) $F_2(g)$ C) $O_2(g)$ D) $N_2(g)$ E) $Cl_2(g)$

Answer: E
Diff: 2 *Page Ref:* 17.6

33) Calculate $\Delta S°_{rxn}$ for the following reaction. The S° for each species is shown below the reaction.

	$C_2H_2(g)$ +	$H_2(g)$ →	$C_2H_4(g)$
S°(J/mol•K)	200.9	130.7	219.3

A) +112.3 J/K
B) +550.9 J/K
C) −112.3 J/K
D) +337.1 J/K
E) −550.9 J/K

Answer: C
Diff: 3 Page Ref: 17.6

34) Calculate $\Delta S°_{rxn}$ for the following reaction. The S ° for each species is shown below the reaction.

	$C_2H_2(g)$ +	$2\ H_2(g)$ →	$C_2H_6(g)$
S°(J/mol•K)	200.9	130.7	229.2

A) +303.3 J/K
B) +560.8 J/K
C) −102.4 J/K
D) −233.1 J/K
E) 229.2 J/K

Answer: D
Diff: 3 Page Ref: 17.6

35) Calculate $\Delta S°_{rxn}$ for the following reaction. The S° for each species is shown below the reaction.

	$4\ NH_3(g)$ +	$5\ O_2(g)$ →	$4\ NO(g)$ +	$6\ H_2O(g)$
S°(J/mol•K)	192.8	205.2	210.8	188.8

A) +287.4 J/K
B) −401.2 J/K
C) +160.0 J/K
D) −336.6 J/K
E) +178.8 J/K

Answer: E
Diff: 3 Page Ref: 17.6

36) Calculate ΔS°_{rxn} for the following reaction. The S° for each species is shown below the reaction.

	$N_2H_4(l) + H_2(g) \rightarrow 2\ NH_3(g)$		
S° (J/mol•K)	121.2	130.7	192.8

A) +133.7 J/K
B) −59.1 J/K
C) +118.2 J/K
D) −202.3 J/K
E) +178.9 J/K

Answer: A
Diff: 3 *Page Ref: 17.6*

37) Calculate ΔS°_{rxn} for the following reaction. The S° for each species is shown below the reaction.

	$P_4(g) + 10\ Cl_2(g) \rightarrow 4\ PCl_5(g)$		
S°(J/mol•K)	280.0	223.1	364.6

A) −138.5 J/K
B) −1052.6 J/K
C) +171.3 J/K
D) −583.6 J/K
E) +2334.6 J/K

Answer: B
Diff: 3 *Page Ref: 17.6*

38) What can change the direction of a reversible reaction?
A) change in temperature
B) change in pressure
C) change in volume
D) none of the above
E) all of the above

Answer: E
Diff: 1 *Page Ref: 17.7*

39) Give the name of the reaction that achieves the theoretical limits with respect to free energy in thermodynamics.
A) reversible reaction
B) forward reaction
C) reverse reaction
D) equilibrium reaction
E) irreversible reaction

Answer: A
Diff: 1 *Page Ref: 17.7*

40) Give the name of the reaction that does not achieve the theoretical limits with respect to free energy in thermodynamics.
A) reversible reaction
B) forward reaction
C) reverse reaction
D) equilibrium reaction
E) irreversible reaction

Answer: E
Diff: 1 *Page Ref: 17.7*

41) Given the following equation,

$$N_2O(g) + NO_2(g) \rightarrow 3\ NO(g) \qquad \Delta G°_{rxn} = -23.0\ kJ$$

Calculate $\Delta G°_{rxn}$ for the following reaction.

$$3\ NO(g) \rightarrow N_2O(g) + NO_2(g)$$

A) −23.0 kJ B) 69.0 kJ C) −69.0 kJ D) −7.67 kJ E) 23.0 kJ

Answer: E
Diff: 2 *Page Ref: 17.7*

42) Determine $\Delta G°_{rxn}$ using the following information.

$$H_2(g) + CO(g) \rightarrow CH_2O(g) \qquad \Delta H° = +1.9\ kJ;\ \Delta S° = -109.6\ J/K$$

A) +57.7 kJ B) −30.8 kJ C) +34.6 kJ D) −41.5 kJ E) +17.3 kJ

Answer: C
Diff: 3 *Page Ref: 17.7*

43) Calculate the $\Delta G°_{rxn}$ using the following information.

	$4\ HNO_3(g)$ +	$5\ N_2H_4(l)$ →	$7\ N_2(g)$ +	$12\ H_2O(l)$	$\Delta G°_{rxn}$ = ?
$\Delta G°_f$ (kJ/mol)	−73.5	149.3		−237.1	

A) -3.298×10^3 kJ
B) −312.9 kJ
C) $+2.845 \times 10^3$ kJ
D) +110.7 kJ
E) −954.7 kJ

Answer: A
Diff: 3 *Page Ref: 17.7*

44) Calculate the ΔG°_{rxn} using the following information.

	$2\ HNO_3(aq)$ +	$NO(g) \rightarrow$	$3\ NO_2(g)$ +	$H_2O(l)$	ΔG°_{rxn} = ?
ΔG°_f (kJ/mol)	-110.9	87.6	51.3	-237.1	

A) -162.5 kJ B) +51.0 kJ C) -54.5 kJ D) +171.1 kJ E) -87.6 kJ

Answer: B
Diff: 3 *Page Ref: 17.7*

45) Use Hess's law to calculate ΔG°_{rxn} using the following information.

$CO(g) \rightarrow C(s) + 1/2\ O_2(g)$ ΔG°_{rxn} = ?

$CO_2(g) \rightarrow C(s) + O_2(g)$ ΔG°_{rxn} = +394.4 kJ
$CO(g) + 1/2\ O_2(g) \rightarrow CO_2(g)$ ΔG°_{rxn} = -257.2 kJ

A) -60.0 kJ B) +651.6 kJ C) -265.8 kJ D) +137.2 kJ E) +523.0 kJ

Answer: D
Diff: 3 *Page Ref: 17.7*

46) Use Hess's law to calculate ΔG°_{rxn} using the following information.

$ClO(g) + O_3(g) \rightarrow Cl(g) + 2\ O_2(g)$ ΔG°_{rxn} = ?

$2\ O_3(g) \rightarrow 3\ O_2(g)$ ΔG°_{rxn} = +489.6 kJ
$Cl(g) + O_3(g) \rightarrow ClO(g) + O_2(g)$ ΔG°_{rxn} = -34.5 kJ

A) -472.4 kJ B) -210.3 kJ C) +455.1 kJ D) +262.1 kJ E) +524.1 kJ

Answer: E
Diff: 3 *Page Ref: 17.7*

47) Calculate the ΔG°_{rxn} using the following information.

	$2\ HNO_3(aq)$ +	$NO(g) \rightarrow$	$3\ NO_2(g)$ +	$H_2O(l)$	ΔG°_{rxn} = ?
ΔH°_f (kJ/mol)	-207.0	91.3	33.2	-285.8	
S°(J/mol•K	146.0	210.8	240.1	70.0	

A) -151 kJ B) -85.5 kJ C) +50.8 kJ D) +222 kJ E) -186 kJ

Answer: C
Diff: 4 *Page Ref: 17.7*

48) Calculate the ΔG°_{rxn} using the following information.

	$4 HNO_3(g)$ +	$5 N_2H_4(l)$ →	$7 N_2(g)$ +	$12 H_2O(l)$	ΔG°_{rxn} = ?
ΔH°_f (kJ/mol)	−133.9	50.6		−285.8	
S°(J/mol·K)	266.9	121.2	191.6	70.0	

A) $+4.90 \times 10^3$ kJ
B) $+3.90 \times 10^3$ kJ
C) -2.04×10^3 kJ
D) -3.15×10^3 kJ
E) -3.30×10^3 kJ

Answer: E
Diff: 4 *Page Ref: 17.7*

49) Calculate the ΔG°_{rxn} using the following information.

	$2 H_2S(g)$ +	$3 O_2(g)$ →	$2 SO_2(g)$ +	$2 H_2O(g)$	ΔG°_{rxn} = ?
ΔH°_f (kJ/mol)	−20.6		−296.8	−241.8	
S°(J/mol·K)	205.8	205.2	248.2	188.8	

A) −990.3 kJ B) +108.2 kJ C) −466.1 kJ D) +676.2 kJ E) −147.1 kJ

Answer: A
Diff: 4 *Page Ref: 17.7*

50) Use Hess's law to calculate ΔG°_{rxn} using the following information.

$NO(g) + O(g) \rightarrow NO_2(g)$ ΔG°_{rxn} = ?

$2 O_3(g) \rightarrow 3 O_2(g)$ ΔG°_{rxn} = +489.6 kJ
$O_2(g) \rightarrow 2 O(g)$ ΔG°_{rxn} = +463.4 kJ
$NO(g) + O_3(g) \rightarrow NO_2(g) + O_2(g)$ ΔG°_{rxn} = − 199.5 kJ

A) +753.5 kJ
B) +277.0 kJ
C) −676.0 kJ
D) −1152.5 kJ
E) −225.7 kJ

Answer: C
Diff: 4 *Page Ref: 17.7*

51) Calculate ΔG_{rxn} at 298 K under the conditions shown below for the following reaction.

$3\ O_2(g) \rightarrow 2\ O_3(g)$ $\Delta G° = +326$ kJ

$P(O_2) = 0.41$ atm, $P(O_3) = 5.2$ atm

A) +341 kJ B) +17.8 kJ C) +332 kJ D) −47.4 kJ E) −109 kJ

Answer: A
Diff: 4 *Page Ref: 17.8*

52) Calculate ΔG_{rxn} at 298 K under the conditions shown below for the following reaction.

$2\ Hg(g) + O_2(g) \rightarrow 2\ HgO(s)$ $\Delta G° = -180.8$ kJ

$P(Hg) = 0.025$ atm, $P(O_2) = 0.037$ atm

A) +207 kJ B) −154.4 kJ C) −26.5 kJ D) −164 kJ E) +60.7 kJ

Answer: B
Diff: 4 *Page Ref: 17.8*

53) Calculate ΔG_{rxn} at 298 K under the conditions shown below for the following reaction.

$CaCO_3(s) \rightarrow CaO(s) + CO_2(g)$ $\Delta G° = +131.1$ kJ

$P(CO_2) = 0.033$ atm

A) −49.3 kJ B) −8.32 kJ C) +122.6 kJ D) +39.7 kJ E) +43.3 kJ

Answer: C
Diff: 4 *Page Ref: 17.8*

54) Which of the following is NOT true for ΔG_{rxn}?
A) If $\Delta G°_{rxn} > 0$, the reaction is spontaneous in the forward direction.
B) If $Q = 1$, then $\Delta G_{rxn} = \Delta G°_{rxn}$.
C) If $\Delta G°_{rxn} = 0$, the reaction is spontaneous in the reverse direction.
D) If $\Delta G°_{rxn} > 0$, the reaction is spontaneous in the reverse direction.
E) Under equilibrium conditions, $\Delta G_{rxn} = 0$.

Answer: A
Diff: 4 *Page Ref: 17.8*

55) Choose the statement below that is TRUE.
A) $K > 1$, $\Delta G°_{rxn}$ is positive.
B) $K < 1$, $\Delta G°_{rxn}$ is negative.
C) $\Delta G°_{rxn} = 0$ at equilibrium.
D) $\Delta G_{rxn} = 0$ at equilibrium.
E) None of the above statements are true.

Answer: D
Diff: 1 *Page Ref: 17.9*

56) What is true if ln K is negative?

A) ΔG^{o}_{rxn} is positive and the reaction is spontaneous in the forward direction.

B) ΔG^{o}_{rxn} is negative and the reaction is spontaneous in the forward direction.

C) ΔG^{o}_{rxn} is negative and the reaction is spontaneous in the reverse direction.

D) ΔG^{o}_{rxn} is positive and the reaction is spontaneous in the reverse direction.

E) ΔG^{o}_{rxn} is zero and the reaction is at equilibrium.

Answer: D
Diff: 1 *Page Ref: 17.9*

57) What is true if ln K is positive?

A) ΔG^{o}_{rxn} is positive and the reaction is spontaneous in the forward direction.

B) ΔG^{o}_{rxn} is negative and the reaction is spontaneous in the forward direction.

C) ΔG^{o}_{rxn} is negative and the reaction is spontaneous in the reverse direction.

D) ΔG^{o}_{rxn} is positive and the reaction is spontaneous in the reverse direction.

E) ΔG^{o}_{rxn} is zero and the reaction is at equilibrium.

Answer: B
Diff: 1 *Page Ref: 17.9*

58) What is true if ln K is zero?

A) ΔG^{o}_{rxn} is positive and the reaction is spontaneous in the forward direction.

B) ΔG^{o}_{rxn} is negative and the reaction is spontaneous in the forward direction.

C) ΔG^{o}_{rxn} is negative and the reaction is spontaneous in the reverse direction.

D) ΔG^{o}_{rxn} is positive and the reaction is spontaneous in the reverse direction.

E) ΔG^{o}_{rxn} is zero and the reaction is at equilibrium.

Answer: E
Diff: 1 *Page Ref: 17.9*

59) Which of the following reactions will have the largest equilibrium constant (K) at 298 K?

A) $CaCO_3(s) \rightarrow CaO(s) + CO_2(g)$ $\Delta G° = +131.1$ kJ

B) $2\ Hg(g) + O_2(g) \rightarrow 2\ HgO(s)$ $\Delta G° = -180.8$ kJ

C) $3\ O_2(g) \rightarrow 2\ O_3(g)$ $\Delta G° = +326$ kJ

D) $Fe_2O_3(s) + 3\ CO(g) \rightarrow 2\ Fe(s) + 3\ CO_2(g)$ $\Delta G° = -28.0$ kJ

E) It is not possible to determine without more information.

Answer: B
Diff: 2 *Page Ref: 17.9*

60) Use the free energies of formation given below to calculate the equilibrium constant (K) for the following reaction at 298 K.

	$2\ HNO_3(aq)$ +	$NO(g)$ →	$3\ NO_2(g)$ +	$H_2O(l)$	K = ?
$\Delta G°_f$ (kJ/mol)	-110.9	87.6	51.3	-237.1	

A) 8.71×10^8
B) 0.980
C) 1.15×10^{-9}
D) 1.02
E) 5.11×10^{-4}

Answer: C
Diff: 3 *Page Ref: 17.9*

61) Determine the equilibrium constant for the following reaction at 298 K.

$SO_3(g) + H_2O(g) \rightarrow H_2SO_4(l)$ $\Delta G° = -90.5$ kJ

A) 1.37×10^{-16}
B) 4.78×10^{11}
C) 9.11×10^{-8}
D) 7.31×10^{15}
E) 0.964

Answer: D
Diff: 3 *Page Ref: 17.9*

62) Determine the equilibrium constant for the following reaction at 298 K.

$Cl(g) + O_3(g) \rightarrow ClO(g) + O_2(g)$ $\Delta G° = -34.5$ kJ

A) 5.66×10^5
B) 0.986
C) 8.96×10^{-7}
D) 4.98×10^{-4}
E) 1.12×10^6

Answer: E
Diff: 3 *Page Ref: 17.9*

63) Determine the equilibrium constant for the following reaction at 498 K.

$2\ Hg(g) + O_2(g) \rightarrow 2\ HgO(s)$ $\Delta H° = -304.2$ kJ; $\Delta S° = -414.2$ J/K

A) 1.87×10^{10}
B) 8.10×10^{31}
C) 2.31×10^{-22}
D) 5.34×10^{-11}
E) 4.33×10^{21}

Answer: A
Diff: 4 *Page Ref: 17.9*

64) Determine the equilibrium constant for the following reaction at 655 K.

$HCN(g) + 2\ H_2(g) \rightarrow CH_3NH_2(g)$ $\Delta H° = -158$ kJ; $\Delta S° = -219.9$ J/K

A) 3.99×10^{12}
B) 13.0
C) 2.51×10^{-13}
D) 3.26×10^{-12}
E) 3.07×10^{11}

Answer: B
Diff: 4 *Page Ref: 17.9*

65) Determine the equilibrium constant for the following reaction at 549 K.

$CH_2O(g) + 2\ H_2(g) \rightarrow CH_4(g) + H_2O(g)$ $\Delta H° = -94.9$ kJ; $\Delta S° = -224.2$ J/K

A) 481
B) 1.07×10^{9}
C) 2.08×10^{-3}
D) 9.35×10^{-10}
E) 1.94×10^{-12}

Answer: C
Diff: 4 *Page Ref: 17.9*

Algorithmic Questions

1) In which of the following processes does the molecules become more orderly?
A) water freezing
B) ice melting
C) water evaporating
D) salt dissolving in water
E) dry ice subliming

Answer: A
Diff: 1 *Page Ref: 17.3*

2) Identify the change in state that does not have an increase in entropy.
   A) water freezing
   B) water boiling
   C) ice melting
   D) dry ice subliming
   E) rubbing alcohol evaporating

Answer: A
Diff: 1 *Page Ref: 17.3*

3) Which of the following processes have a $\Delta S > 0$?
   A) $2\ NH_3(g) + CO_2(g) \rightarrow NH_2CONH_2(aq) + H_2O(l)$
   B) lithium fluoride forms from its elements
   C) $2\ HF(g) \rightarrow H_2(g) + F_2(l)$
   D) potassium iodide dissolves in pure water
   E) All of the above processes have a $\Delta S > 0$.

Answer: D
Diff: 2 *Page Ref: 17.3*

4) Which of the following processes have a $\Delta S < 0$?
   A) water freezes
   B) methyl alcohol condenses
   C) propanol (g, at 555 K) → propanol (g, at 400 K)
   D) carbon dioxide(g) → carbon dioxide(s)
   E) All of the above processes have a $\Delta S < 0$.

Answer: E
Diff: 2 *Page Ref: 17.3*

5) Which of the following processes shows a decrease in entropy of the system?
   A) $2\ NO(g) + O_2(g) \rightarrow 2\ NO_2(g)$
   B) $COBr_2(g) \rightarrow CO(g) + Br_2(g)$
   C) $CH_3OH(l) \rightarrow CO(g) + 2H_2(g)$
   D) $KBrO_3(s) \rightarrow K^+(aq) + BrO_3^-(aq)$
   E) None of the above will show a decrease in entropy.

Answer: A
Diff: 2 *Page Ref: 17.3*

6) Consider the following reaction at constant P. Use the information here to determine the value of ΔS_{surr} at 298 K. Predict whether or not this reaction will be spontaneous at this temperature.

$N_2(g) + 2\ O_2(g) \rightarrow 2\ NO_2(g)$ $\qquad$ $\Delta H = +66.4$ kJ

A) $\Delta S_{surr} = +223$ J/K, reaction is spontaneous spontaneous
B) $\Delta S_{surr} = -223$ J/K, reaction is not spontaneous
C) $\Delta S_{surr} = -66.4$ J/K, reaction is not spontaneous
D) $\Delta S_{surr} = +2656$ kJ/K, reaction is not spontaneous
E) $\Delta S_{surr} = -66.4$ J/K, it is not possible to predict the spontaneity of this reaction without more information.

Answer: B
Diff: 2 *Page Ref: 17.4*

7) Consider the following reaction at constant P. Use the information here to determine the value of ΔS_{surr} at 355 K. Predict whether or not this reaction will be spontaneous at this temperature.

$2\ NO(g) + O_2(g) \rightarrow 2\ NO_2(g)$ $\qquad$ $\Delta H = -114$ kJ

A) $\Delta S_{surr} = +114$ kJ/K, reaction is spontaneous
B) $\Delta S_{surr} = +114$ kJ/K, reaction is not spontaneous
C) $\Delta S_{surr} = +321$ J/K, reaction is spontaneous
D) $\Delta S_{surr} = -321$ J/K, reaction is spontaneous
E) $\Delta S_{surr} = +321$ J/K, it is not possible to predict the spontaneity of this reaction without more information.

Answer: C
Diff: 2 *Page Ref: 17.4*

8) Consider the following reaction at constant P. Use the information here to determine the value of ΔS_{surr} at 398 K. Predict whether or not this reaction will be spontaneous at this temperature.

$4\ NH_3(g) + 3\ O_2(g) \rightarrow 2\ N_2(g) + 6\ H_2O(g)$ $\qquad$ $\Delta H = -1267$ kJ

A) $\Delta S_{surr} = +12.67$ kJ/K, reaction is spontaneous
B) $\Delta S_{surr} = -12.67$ kJ/K, reaction is spontaneous
C) $\Delta S_{surr} = +50.4$ kJ/K, reaction is not spontaneous
D) $\Delta S_{surr} = +3.18$ kJ/K, reaction is spontaneous
E) $\Delta S_{surr} = -3.18$ kJ/K, it is not possible to predict the spontaneity of this reaction without more information.

Answer: D
Diff: 2 *Page Ref: 17.4*

9) Place the following in order of increasing molar entropy at 298 K.

CO_2 C_9H_{20} CO

A) CO_2 < C_9H_{20} < CO
B) C_9H_{20} < CO_2 < CO
C) CO < CO_2 < C_9H_{20}
D) C_9H_{20} < CO< CO_2
E) CO_2 < CO< C_9H_{20}

Answer: C
Diff: 2 *Page Ref:* 17.6

10) Place the following in order of decreasing molar entropy at 298 K.

HF N_2H_4 Ar

A) Ar > N_2H_4 > HF
B) Ar > HF > N_2H_4
C) N_2H_4 > Ar> HF
D) N_2H_4 > HF > Ar
E) HF > N_2H_4 > Ar

Answer: D
Diff: 2 *Page Ref:* 17.6

11) Place the following in order of decreasing standard molar entropy.

NaF(s) Li_3PO_4(aq) NaF(aq)

A) NaF(s) > NaF(aq) > Li_3PO_4(aq)
B) NaF(aq) > NaF(s) > Li_3PO_4(aq)
C) Li_3PO_4(aq) > NaF(aq) > NaF(s)
D) NaF(s) > Li_3PO_4(aq) > NaF(aq)
E) NaF(aq) > Li_3PO_4(aq) > NaF(s)

Answer: C
Diff: 2 *Page Ref:* 17.6

12) Which one of the following has the highest standard molar entropy, *S*°, at 25°C?
A) I_2(g) B) F_2(g) C) Br_2(l) D) O_2(g) E) Cl_2(g)

Answer: A
Diff: 2 *Page Ref:* 17.6

13) Identify the compound that is NOT an allotrope of carbon.
A) diamond
B) ethanol
C) graphite
D) charcoal
E) buckyball

Answer: B
Diff: 2 *Page Ref: 17.6*

14) Which one of the following would be expected to have the lowest standard molar entropy, $S°$, at 25°C?
A) $C_{10}H_{22}(s)$ B) $C_{10}H_{22}(l)$ C) $C_{14}H_{30}(s)$ D) $C_{14}H_{30}OH(l)$

Answer: A
Diff: 2 *Page Ref: 17.6*

15) Identify the compound with the lowest standard free energy of formation.
A) NaCl(s)
B) $O_2(g)$
C) NO(g)
D) $O_3(g)$
E) It is hard to determine.

Answer: A
Diff: 1 *Page Ref: 17.7*

16) Identify the compound with the highest standard free energy of formation.
A) NaCl(s)
B) $H_2(g)$
C) NO(g)
D) $O_3(g)$
E) It is hard to determine.

Answer: D
Diff: 1 *Page Ref: 17.7*

17) Identify the compound with the standard free energy of formation equal to zero.
A) KBr(s)
B) $H_2(g)$
C) NO(g)
D) $O_3(g)$
E) It is hard to determine.

Answer: B
Diff: 1 *Page Ref: 17.7*

18) Given the following equation,

$$H_2O(g) + CO(g) \rightarrow H_2(g) + CO_2(g) \qquad \Delta G^\circ_{rxn} = -28.6 \text{ kJ}$$

Calculate ΔG°_{rxn} for the following reaction.

$$5\ H_2O(g) + 5\ CO(g) \rightarrow 5\ H_2(g) + 5\ CO_2(g)$$

A) –143.0 kJ B) +143.0 kJ C) –5.72 kJ D) +5.72 kJ E) –28.6 kJ

Answer: A
Diff: 2 *Page Ref:* 17.7

19) Given the following equation,

$$H_2O(g) + CO(g) \rightarrow H_2(g) + CO_2(g) \qquad \Delta G^\circ_{rxn} = -28.6 \text{ kJ}$$

Calculate ΔG°_{rxn} for the following reaction.

$$5\ H_2(g) + 5\ CO_2(g) \rightarrow 5\ H_2O(g) + 5\ CO(g)$$

A) –143.0 kJ B) +143.0 kJ C) –5.72 kJ D) +5.72 kJ E) –28.6 kJ

Answer: B
Diff: 2 *Page Ref:* 17.7

20) Given the following equation,

$$C_3H_8(g) + 5\ O_2(g) \rightarrow 3\ CO_2(g) + 4\ H_2O(g) \qquad \Delta G^\circ_{rxn} = -2074 \text{ kJ}$$

Calculate ΔG°_{rxn} for the following reaction.

$$7\ C_3H_8(g) + 35\ O_2(g) \rightarrow 21\ CO_2(g) + 28\ H_2O(g)$$

A) –2074 kJ
B) +14518 kJ
C) –14518 kJ
D) +2074 kJ
E) –296 kJ

Answer: C
Diff: 2 *Page Ref:* 17.7

21) Given the following equation,

$$C_3H_8(g) + 5\,O_2(g) \rightarrow 3\,CO_2(g) + 4\,H_2O(g) \qquad \Delta G^\circ_{rxn} = -2074\text{ kJ}$$

Calculate ΔG°_{rxn} for the following reaction.

$$21\,CO_2(g) + 28\,H_2O(g) \rightarrow 7\,C_3H_8(g) + 35\,O_2\,g)$$

A) -2074 kJ
B) +14518 kJ
C) -14518 kJ
D) +2074 kJ
E) -296 kJ

Answer: B
Diff: 2 *Page Ref:* 17.7

22) Given the following equation,

$$N_2O(g) + NO_2(g) \rightarrow 3\,NO(g) \qquad \Delta G^\circ_{rxn} = -23.0\text{ kJ}$$

Calculate ΔG°_{rxn} for the following reaction.

$$6\,N_2O(g) + 6\,NO_2(g) \rightarrow 18\,NO(g)$$

A) -23.0 kJ B) 138 kJ C) -138 kJ D) -3.83 kJ E) 23.0 kJ

Answer: C
Diff: 2 *Page Ref:* 17.7

23) Given the following equation,

$$N_2O(g) + NO_2(g) \rightarrow 3\,NO(g) \qquad \Delta G^\circ_{rxn} = -23.0\text{ kJ}$$

Calculate ΔG°_{rxn} for the following reaction.

$$18\,NO(g) \rightarrow 6\,N_2O(g) + 6\,NO_2(g)$$

A) -23.0 kJ B) 138 kJ C) -138 kJ D) -3.83 kJ E) 23.0 kJ

Answer: B
Diff: 2 *Page Ref:* 17.7

24) Estimate ΔG°_{rxn} for the following reaction at 449.0 K.

$$CH_2O(g) + 2\,H_2(g) \rightarrow CH_4(g) + H_2O(g) \qquad \Delta H^\circ = -94.9\text{ kJ};\ \Delta S^\circ = -224.2\text{ J/K}$$

A) +5.8 kJ B) +12.9 kJ C) +101 kJ D) +2.4 kJ E) -4.2 kJ

Answer: A
Diff: 2 *Page Ref:* 17.7

25) Estimate $\Delta G°_{rxn}$ for the following reaction at 417 K.

$HCN(g) + 2\ H_2(g) \rightarrow CH_3NH_2(g)$ $\Delta H° = -158.0$ kJ; $\Delta S° = -219.9$ J/K

A) +61.9 kJ B) −66.3 kJ C) +66.3 kJ D) −250 kJ E) +250 kJ

Answer: B

Diff: 2 *Page Ref:* 17.7

26) Estimate $\Delta G°_{rxn}$ for the following reaction at 825 K.

$2\ Hg(g) + O_2(g) \rightarrow 2\ HgO(s)$ $\Delta H° = -304.2$ kJ; $\Delta S° = -414.2$ J/K

A) −2.63 kJ B) −37 kJ C) +37 kJ D) +645 kJ E) −645 kJ

Answer: C

Diff: 2 *Page Ref:* 17.7

27) Determine $\Delta G°_{rxn}$ for the following reaction at 338 K.

$FeO(s) + CO(g) \rightarrow Fe(s) + CO_2(g)$ $\Delta H° = -11.0$ kJ; $\Delta S° = -17.4$ J/K

A) +191.0 kJ B) −5.1 kJ C) +5.1 kJ D) −16.1 kJ E) +16.1 kJ

Answer: B

Diff: 3 *Page Ref:* 17.7

28) Determine $\Delta G°_{rxn}$ for the following reaction at 358 K.

$CaCO_3(s) \rightarrow CaO(s) + CO_2(g)$ $\Delta H° = +179.2$ kJ; $\Delta S° = +160.2$ J/K

A) +179.2 kJ B) +236.6 kJ C) −236.6 kJ D) +121.8 kJ E) −121.8 kJ

Answer: D

Diff: 3 *Page Ref:* 17.7

29) Calculate the $\Delta G°_{rxn}$ using the following information.

	$2\ H_2S(g)$	$+\ 3\ O_2(g) \rightarrow$	$2\ SO_2(g)$	$+\ 2\ H_2O(g)$	$\Delta G°_{rxn} = ?$
$\Delta G°_f$ (kJ/mol)	−33.4		−300.1	−228.6	

A) +990.6 kJ B) −495.3 kJ C) +528.7 kJ D) +66.8 kJ E) −990.6 kJ

Answer: E

Diff: 3 *Page Ref:* 17.7

30) The value of $\Delta G°$ at 141.0°C for the formation of phosphorous trichloride from its constituent elements,

$$P_2(g) + 3Cl_2(g) \rightarrow 2PCl_3(g)$$

is ________ kJ/mol. At 25.0°C for this reaction, $\Delta H°$ is -720.5 kJ/mol, $\Delta G°$ is -642.9 kJ/mol, and $\Delta S°$ is -263.7 J/K.

A) -612.3
B) 3.65×10^4
C) 1.08×10^5
D) -683.3
E) -829.7

Answer: A
Diff: 4 Page Ref: 17.7

31) The value of $\Delta G°$ at 100.0°C for the formation of calcium chloride from its constituent elements:

$$Ca(s) + Cl_2(g) \rightarrow CaCl_2(s)$$

is ________ kJ/mol. At 25.0°C for this reaction, $\Delta H°$ is -795.8 kJ/mol, $\Delta G°$ is -748.1 kJ/mol, and $\Delta S°$ is -159.8 J/K.

A) -855.4 B) -736.1 C) 5.88×10^4 D) -779.8 E) 1.52×10^4

Answer: B
Diff: 4 Page Ref: 17.7

32) For a given reaction, $\Delta H = -19.9$ kJ/mol and $\Delta S = -55.5$ J/Kmol. The reaction will have $\Delta G = 0$ at ________ K. Assume that ΔH and ΔS do not vary with temperature.

A) 359 B) 2789 C) 298 D) 2.79 E) 0.359

Answer: A
Diff: 4 Page Ref: 17.7

33) For a given reaction, $\Delta H = +35.5$ kJ/mol and $\Delta S = +83.6$ J/Kmol. The reaction is spontaneous ________. Assume that ΔH and ΔS do not vary with temperature.

A) at $T < 425$ K
B) at $T > 425$ K
C) at all temperatures
D) at $T > 298$ K
E) at $T < 298$ K

Answer: B
Diff: 4 Page Ref: 17.7

34) Under which of the following conditions would one mole of He have the highest entropy, S?

A) 17°C and 15 L B) 127°C and 15 L
C) 17°C and 25 L D) 127°C and 25 L

Answer: D
Diff: 4 Page Ref: 17.8

35) Calculate ΔG_{rxn} at 298 K under the conditions shown below for the following reaction.

$SO_3(g) + H_2O(g) \rightarrow H_2SO_4(l)$ $\Delta G° = -90.5$ kJ

$P(SO_3) = 0.20$ atm, $P(H_2O) = 0.88$ atm

A) -90.5 kJ B) -91.9 kJ C) +91.9 kJ D) -89.1 kJ E) +89.1 kJ

Answer: D
Diff: 4 *Page Ref: 17.8*

36) Calculate ΔG_{rxn} at 298 K under the conditions shown below for the following reaction.

$Fe_2O_3(s) + 3\ CO(g) \rightarrow 2\ Fe(s) + 3\ CO_2(g)$ $\Delta G° = -28.0$ kJ

$P(CO) = 1.4$ atm, $P(CO_2) = 2.1$ atm

A) +31.0 kJ B) -31.0 kJ C) -28.0 kJ D) +25.0 kJ E) -25.0 kJ

Answer: E
Diff: 4 *Page Ref: 17.8*

37) In the Haber process, ammonia is synthesized from nitrogen and hydrogen:

$N_2(g) + 3H_2(g) \rightarrow 2NH_3(g)$

$\Delta G°$ at 298 K for this reaction is -33.3 kJ/mol. The value of ΔG at 298 K for a reaction mixture that consists of 1.9 atm N_2, 1.6 atm H_2, and 0.65 atm NH_3 is ________.

A) -1.8
B) -3.86×10^3
C) -7.25×10^3
D) -104.5
E) -40.5

Answer: E
Diff: 5 *Page Ref: 17.9*

38) Phosphorous and chlorine gases combine to produce phosphorous trichloride:

$P_2(g) + 3Cl_2(g) \rightarrow 2PCl_3(g)$

$\Delta G°$ at 298 K for this reaction is -642.9 kJ/mol. The value of ΔG at 298 K for a reaction mixture that consists of 1.5 atm P_2, 1.6 atm Cl_2, and 0.65 atm PCl_3 is ________.

A) -44.2
B) -3.88×10^3
C) -7.28×10^3
D) -708.4
E) -649.5

Answer: E
Diff: 5 *Page Ref: 17.9*

Matching Questions

Match the following.

1) $Q = K$
Diff: 1 *Page Ref: 17.9*

2) $Q < K$
Diff: 1 *Page Ref: 17.9*

3) $Q > K$
Diff: 1 *Page Ref: 17.9*

4) $Q = 1$
Diff: 1 *Page Ref: 17.9*

5) $Q > 1$
Diff: 1 *Page Ref: 17.9*

6) $Q < 1$
Diff: 1 *Page Ref: 17.9*

A) equilibrium

B) $\Delta G < \Delta G°$

C) $K = 0$

D) $\Delta G > \Delta G°$

E) standard state

F) $\Delta G > 0$

G) $\Delta G < 0$

Answers: 1) A 2) G 3) F 4) E 5) D 6) B

Short Answer Questions

1) Why is heating your home with gas more efficient than heating it with electricity?

Answer: Heating with gas only requires a one step energy transfer process. Heating with electricity takes multiple steps to make the electricity and transfer it to your home before it is used to heat your home. During each step of energy transfer, some energy is lost to the surroundings (nature's heat tax). Therefore, generally, the more energy transfer steps in the process, the less efficient it is.
Diff: 1 *Page Ref: 17.1*

2) Why can't we say that a spontaneous reaction is a fast reaction?

Answer: Spontaneity is a thermodynamic quantity that determines if a reaction will occur and to what extent. The speed of a reaction is not related to the spontaneity. A spontaneous reaction can be very slow.
Diff: 1 *Page Ref: 17.2*

3) Why can endothermic reactions be spontaneous?

Answer: Even though an endothermic process requires energy from the surroundings, it can be spontaneous if the disorder (entropy) increase of the process overcomes that energy cost. Overall the entropy of the universe must increase during the process.
Diff: 1 *Page Ref: 17.3*

4) Define entropy.

Answer: Entropy is a thermodynamic function that increases with a number of energetically equivalent ways to arrange the components of a system to achieve a particular state.
Diff: 1 *Page Ref: 17.3*

5) How many microstates are possible in a collection of four particles that are present, with two particles each in two connected flasks? Sketch them below.

Answer: There are six microstates. The diagrams should be similar to the diagrams on page 777.
Diff: 1 *Page Ref: 17.3*

6) Define the second law of thermodynamics.

Answer: For any spontaneous process, the entropy of the universe increases ($\Delta S_{universe} > 0$).
Diff: 1 *Page Ref: 17.3*

7) What is "free" energy? Give a fictitious example.

Answer: Consider a reaction that is exothermic ($\Delta H = -100$ kJ), but has a negative entropy ($\Delta S = -20$ J/K, -6 kJ at 298 K). The amount of free energy in this example is -94 kJ. In this example, free energy is the maximum amount of energy available to do work after taking into account the amount of energy used (from the reaction) to raise the entropy of the surroundings to compensate for the decrease in entropy of the system.
Diff: 1 *Page Ref: 17.5*

8) Give the standard states for a gas, liquid, solid, and solution.

Answer: For a gas -- 1 atm; for a liquid -- 1 atm and 25°C; for a solid -- 1 atm and 25°C; for a solution -- 1 M and 1 atm.
Diff: 1 *Page Ref: 17.6*

9) Define the third law of thermodynamics.

Answer: The entropy of a perfect crystal at absolute zero (0 K) is zero.
Diff: 1 *Page Ref: 17.6*

10) Define allotrope.

Answer: An allotrope is two or more forms of an element existing in the same state.
Diff: 1 *Page Ref: 17.6*

11) How is a nonspontaneous process made spontaneous?

Answer: The nonspontaneous process is coupled with a process that is highly spontaneous.
Diff: 1 *Page Ref: 17.7*

12) Why is the quantity of energy required to recharge a battery greater than the quantity of work done?

Answer: Some of the energy is lost as heat.
Diff: 1 *Page Ref: 17.7*

Chapter 18 Electrochemistry

Multiple Choice Questions

1) What element is being reduced in the following redox reaction?

$$MnO_4^-(aq) + H_2C_2O_4(aq) \rightarrow Mn^{2+}(aq) + CO_2(g)$$

A) C B) O C) Mn D) H

Answer: C
Diff: 2 *Page Ref: 18.2*

2) What element is being oxidized in the following redox reaction?

$$MnO_4^-(aq) + H_2C_2O_4(aq) \rightarrow Mn^{2+}(aq) + CO_2(g)$$

A) C B) O C) Mn D) H

Answer: A
Diff: 2 *Page Ref: 18.2*

3) What element is being reduced in the following redox reaction?

$$H_2O_2(l) + ClO_2(aq) \rightarrow ClO_2^-(aq) + O_2(g)$$

A) H B) O C) Cl D) N E) C

Answer: C
Diff: 2 *Page Ref: 18.2*

4) What element is being oxidized in the following redox reaction?

$$H_2O_2(l) + ClO_2(aq) \rightarrow ClO_2^-(aq) + O_2(g)$$

A) H B) O C) Cl D) N E) C

Answer: B
Diff: 2 *Page Ref: 18.2*

5) What element is being oxidized in the following redox reaction?

$$Cr(OH)_4^-(aq) + ClO^-(aq) \rightarrow CrO_4^{2-}(aq) + Cl^-(aq)$$

A) Cr B) O C) H D) Cl

Answer: A
Diff: 2 *Page Ref: 18.2*

6) What element is being reduced in the following redox reaction?

$$Cr(OH)_4^-(aq) + ClO^-(aq) \rightarrow CrO_4^{2-}(aq) + Cl^-(aq)$$

A) Cr
B) O
C) H
D) Cl

Answer: D
Diff: 2 *Page Ref: 18.2*

7) What element is being oxidized in the following redox reaction?

$$C_3H_8O_2(aq) + KMnO_4(aq) \rightarrow C_3H_2O_4K_2(aq) + MnO_2(aq)$$

A) C
B) H
C) O
D) K
E) Mn

Answer: A
Diff: 2 *Page Ref: 18.2*

8) What element is being reduced in the following redox reaction?

$$C_3H_8O_2(aq) + KMnO_4(aq) \rightarrow C_3H_2O_4K_2(aq) + MnO_2(aq)$$

A) C
B) H
C) O
D) K
E) Mn

Answer: E
Diff: 2 *Page Ref: 18.2*

9) What element is being oxidized in the following redox reaction?

$$C_3H_8O_2(aq) + K_2Cr_2O_7(aq) \rightarrow C_3H_4O_4(aq) + Cr^{+3}(aq)$$

A) C
B) H
C) O
D) K
E) Cr

Answer: A
Diff: 2 *Page Ref: 18.2*

10) What element is being reduced in the following redox reaction?

$$C_3H_8O_2(aq) + K_2Cr_2O_7(aq) \rightarrow C_3H_4O_4(aq) + Cr^{+3}(aq)$$

A) C
B) H
C) O
D) K
E) Cr

Answer: E
Diff: 2 *Page Ref: 18.2*

11) Balance the following redox reaction if it occurs in H_2SO_4. What are the coefficients in front of $C_3H_8O_2$ and H_2SO_4 in the balanced reaction?

$$C_3H_8O_2(aq) + K_2Cr_2O_7(aq) \rightarrow C_3H_4O_4(aq) + Cr_2(SO_4)_3(aq)$$

A) $C_3H_8O_2 = 3$, $H_2SO_4 = 8$
B) $C_3H_8O_2 = 2$, $H_2SO_4 = 8$
C) $C_3H_8O_2 = 1$, $H_2SO_4 = 16$
D) $C_3H_8O_2 = 3$, $H_2SO_4 = 16$
E) $C_3H_8O_2 = 2$, $H_2SO_4 = 4$

Answer: D
Diff: 4 *Page Ref: 18.2*

12) Balance the following redox reaction if it occurs in H_2SO_4. What are the coefficients in front of H_2O and $Cr_2(SO_4)_3$ in the balanced reaction?

$$C_3H_8O_2(aq) + K_2Cr_2O_7(aq) \rightarrow C_3H_4O_4(aq) + Cr_2(SO_4)_3(aq)$$

A) $H_2O = 1$, $Cr_2(SO_4)_3 = 5$
B) $H_2O = 5$, $Cr_2(SO_4)_3 = 2$
C) $H_2O = 11$, $Cr_2(SO_4)_3 = 4$
D) $H_2O = 22$, $Cr_2(SO_4)_3 = 4$
E) $H_2O = 8$, $Cr_2(SO_4)_3 = 2$

Answer: D
Diff: 4 *Page Ref: 18.2*

13) Balance the following redox reaction if it occurs in acidic solution. What are the coefficients in front of $H_2C_2O_4$ and H_2O in the balanced reaction?

$$MnO_4^-(aq) + H_2C_2O_4(aq) \rightarrow Mn^{2+}(aq) + CO_2(g)$$

A) $H_2C_2O_4 = 5$, $H_2O = 8$
B) $H_2C_2O_4 = 1$, $H_2O = 1$
C) $H_2C_2O_4 = 5$, $H_2O = 1$
D) $H_2C_2O_4 = 1$, $H_2O = 4$
E) $H_2C_2O_4 = 3$, $H_2O = 2$

Answer: A
Diff: 4 *Page Ref: 18.2*

14) Balance the following reaction in KOH. What are the coefficients in front of $C_3H_8O_2$ and $KMnO_4$ in the balanced reaction?

$$C_3H_8O_2(aq) + KMnO_4(aq) \rightarrow C_3H_2O_4K_2(aq) + MnO_2(aq)$$

A) $C_3H_8O_2$ = 3, $KMnO_4$ = 8
B) $C_3H_8O_2$ = 3, $KMnO_4$ = 4
C) $C_3H_8O_2$ = 8, $KMnO_4$ = 3
D) $C_3H_8O_2$ = 2, $KMnO_4$ = 4
E) $C_3H_8O_2$ = 3, $KMnO_4$ = 2

Answer: A
Diff: 4 *Page Ref: 18.2*

15) Balance the following reaction in KOH. What are the coefficients in front of H_2O and KOH in the balanced reaction and list which side of the equation that H_2O and KOH appear?

$$C_3H_8O_2(aq) + KMnO_4(aq) \rightarrow C_3H_2O_4K_2(aq) + MnO_2(aq)$$

A) H_2O = 1, KOH = 4, right
B) H_2O = 4, KOH = 1, left
C) H_2O = 8, KOH = 2 right
D) H_2O = 3, KOH = 2, right
E) H_2O = 4, KOH = 2, left

Answer: C
Diff: 4 *Page Ref: 18.2*

16) Balance the following redox reaction if it occurs in acidic solution. What are the coefficients in front of H^+ and Fe^{3+} in the balanced reaction?

$$Fe^{2+}(aq) + MnO_4^-(aq) \rightarrow Fe^{3+}(aq) + Mn^{2+}(aq)$$

A) H^+ = 2, Fe^{3+} = 3
B) H^+ = 8, Fe^{3+} = 5
C) H^+ = 3, Fe^{3+} = 2
D) H^+ = 5, Fe^{3+} = 1
E) H^+ = 8, Fe^{3+} = 1

Answer: B
Diff: 4 *Page Ref: 18.2*

17) Balance the following redox reaction if it occurs in basic solution. What are the coefficients in front of Br_2 and OH^- in the balanced reaction?

$Br_2(l) \rightarrow BrO_3^-(aq) + Br^-(aq)$

A) $Br_2 = 1, OH^- = 2$
B) $Br_2 = 2, OH^- = 5$
C) $Br_2 = 3, OH^- = 3$
D) $Br_2 = 3, OH^- = 6$
E) $Br_2 = 1, OH^- = 6$

Answer: D
Diff: 4 *Page Ref: 18.2*

18) Balance the following redox reaction if it occurs in basic solution. What are the coefficients in front of ClO_2 and H_2O in the balanced reaction?

$H_2O_2(l) + ClO_2(aq) \rightarrow ClO_2^-(aq) + O_2(g)$

A) $ClO_2 = 1, H_2O = 1$
B) $ClO_2 = 1, H_2O = 2$
C) $ClO_2 = 4, H_2O = 3$
D) $ClO_2 = 4, H_2O = 2$
E) $ClO_2 = 2, H_2O = 2$

Answer: E
Diff: 4 *Page Ref: 18.2*

19) Balance the following redox reaction if it occurs in basic solution. What are the coefficients in front of $Cr(OH)_4^-$ and ClO^- in the balanced reaction?

$Cr(OH)_4^-(aq) + ClO^-(aq) \rightarrow CrO_4^{2-}(aq) + Cl^-(aq)$

A) $Cr(OH)_4^- = 2, ClO^- = 3$
B) $Cr(OH)_4^- = 1, ClO^- = 1$
C) $Cr(OH)_4^- = 1, ClO^- = 2$
D) $Cr(OH)_4^- = 2, ClO^- = 6$
E) $Cr(OH)_4^- = 6, ClO^- = 5$

Answer: A
Diff: 4 *Page Ref: 18.2*

20) Identify the location of oxidation in an electrochemical cell.
A) the anode
B) the cathode
C) the electrode
D) the salt bridge
E) the socket

Answer: A
Diff: 1 *Page Ref: 18.3*

21) Identify the location of reduction in an electrochemical cell.
A) the anode
B) the cathode
C) the electrode
D) the salt bridge
E) the socket

Answer: B
Diff: 1 *Page Ref: 18.3*

22) Define a salt bridge.
A) A pathway, composed of salt water, that ions pass through.
B) A pathway in which no ions flow.
C) A pathway between the cathode and anode in which ions are reduced.
D) A pathway between the cathode and anode in which ions are oxidized.
E) A pathway by which counterions can flow between the half-cells without the solutions in the half-cell totally mixing.

Answer: E
Diff: 1 *Page Ref: 18.3*

23) Determine the cell notation for the redox reaction given below.

$$3\ Cl_2(g) + 2\ Fe(s) \rightarrow 6\ Cl^-(aq) + 2\ Fe^{3+}(aq)$$

A) $Cl_2(g) \mid Cl^-(aq) \mid Pt \parallel Fe(s) \mid Fe^{3+}(aq)$
B) $Cl^-(aq) \mid Cl_2(g) \mid Pt \parallel Fe^{3+}(aq) \mid Fe(s)$
C) $Fe^{3+}(aq) \mid Fe(s) \parallel Cl^-(aq) \mid Cl_2(g) \mid Pt$
D) $Fe(s) \mid Cl_2(g) \parallel Fe^{3+}(aq) \mid Cl^-(aq) \mid Pt$
E) $Fe(s) \mid Fe^{3+}(aq) \parallel Cl_2(g) \mid Cl^-(aq) \mid Pt$

Answer: E
Diff: 3 *Page Ref: 18.3*

24) What statement is NOT true about standard electrode potentials?

A) $E°_{cell}$ is positive for spontaneous reactions.

B) Electrons will flow from more negative electrode to more positive electrode.

C) The electrode potential of the standard hydrogen electrode is exactly zero.

D) $E°_{cell}$ is the difference in voltage between the anode and the cathode.

E) The electrode in any half-cell with a greater tendency to undergo reduction is positively charged relative to the standard hydrogen electrode and therefore has a positive E°.

Answer: D

Diff: 1 *Page Ref: 18.4*

25) Which of the following is the weakest oxidizing agent?

A) $H_2O_2(aq)$

B) $Fe^{3+}(aq)$

C) $ClO_2(g)$

D) $F^-(s)$

E) Fe(s)

Answer: D

Diff: 1 *Page Ref: 18.4*

26) Use the standard half-cell potentials listed below to calculate the standard cell potential for the following reaction occurring in an electrochemical cell at 25°C. (The equation is balanced.)

$$3\ Cl_2(g) + 2\ Fe(s) \rightarrow 6\ Cl^-(aq) + 2\ Fe^{3+}(aq)$$

$Cl_2(g) + 2\ e^- \rightarrow 2\ Cl^-(aq)$ E° = +1.36 V

$Fe^{3+}(aq) + 3\ e^- \rightarrow Fe(s)$ E° = -0.04 V

A) +4.16 V B) -1.40 V C) -1.32 V D) +1.32 V E) +1.40 V

Answer: E

Diff: 2 *Page Ref: 18.4*

27) Use the standard half-cell potentials listed below to calculate the standard cell potential for the following reaction occurring in an electrochemical cell at 25°C. (The equation is balanced.)

$$Pb(s) + Br_2(l) \rightarrow Pb^{2+}(aq) + 2\ Br^-(aq)$$

$Pb^{2+}(aq) + 2\ e^- \rightarrow Pb(s)$ E° = -0.13 V

$Br_2(l) + 2\ e^- \rightarrow 2\ Br^-(aq)$ E° = +1.07 V

A) +1.20 V B) +0.94 V C) -0.94 V D) -1.20 V E) -0.60 V

Answer: A

Diff: 2 *Page Ref: 18.4*

28) Use the standard half-cell potentials listed below to calculate the standard cell potential for the following reaction occurring in an electrochemical cell at 25°C. (The equation is balanced.)

$$Sn(s) + 2\ Ag^{+}(aq) \rightarrow Sn^{2+}(aq) + 2\ Ag(s)$$

$Sn^{2+}(aq) + 2\ e^{-} \rightarrow Sn(s)$ E° = −0.14 V
$Ag^{+}(aq) + e^{-} \rightarrow Ag(s)$ E° = +0.80 V

A) +1.74 V B) +0.94 V C) +1.08 V D) −1.08 V E) −1.74 V

Answer: B
Diff: 2 *Page Ref: 18.4*

29) Use the standard half-cell potentials listed below to calculate the standard cell potential for the following reaction occurring in an electrochemical cell at 25°C. (The equation is balanced.)

$$Mg(s) + Cu^{2+}(aq) \rightarrow Cu(s) + Mg^{2+}(aq)$$

$Mg^{2+}(aq) + 2\ e^{-} \rightarrow Mg(s)$ E° = −2.38 V
$Cu^{2+}(aq) + 2\ e^{-} \rightarrow Cu(s)$ E° = +0.34 V

A) +2.04 V B) −2.04 V C) +2.72 V D) −1.36 V E) +1.36 V

Answer: C
Diff: 2 *Page Ref: 18.4*

30) Use the standard half-cell potentials listed below to calculate the standard cell potential for the following reaction occurring in an electrochemical cell at 25°C. (The equation is balanced.)

$$2\ K(s) + I_2(s) \rightarrow 2\ K^{+}(aq) + 2\ I^{-}(aq)$$

$K^{+}(aq) + e^{-} \rightarrow K(s)$ E° = −2.93 V
$I_2(s) + 2\ e^{-} \rightarrow 2\ I^{-}(aq)$ E° = +0.54 V

A) +6.40 V B) +1.85 V C) −5.32 V D) +3.47 V E) +5.32 V

Answer: D
Diff: 2 *Page Ref: 18.4*

31) Identify the characteristics of a spontaneous reaction.

A) $\Delta G° < 0$
B) $\Delta E°_{cell} > 0$
C) $K > 1$
D) all of the above
E) none of the above

Answer: D
Diff: 2 *Page Ref: 18.4*

32) How many electrons are transferred in the following reaction? (The reaction is unbalanced.)

$I_2(s) + Fe(s) \rightarrow Fe^{3+}(aq) + I^-(aq)$

A) 1 B) 2 C) 6 D) 3 E) 4

Answer: C
Diff: 2 *Page Ref: 18.5*

33) Which of the following reactions would have the smallest value of K at 298 K?
A) $A + B \rightarrow C$; $E°_{cell} = +1.22$ V
B) $A + 2 B \rightarrow C$; $E°_{cell} = +0.98$ V
C) $A + B \rightarrow 2 C$; $E°_{cell} = -0.030$ V
D) $A + B \rightarrow 3 C$; $E°_{cell} = +0.15$ V
E) More information is needed to determine.

Answer: C
Diff: 2 *Page Ref: 18.5*

34) Use the tabulated half-cell potentials to calculate ΔG° for the following balanced redox reaction.

$Pb^{2+}(aq) + Cu(s) \rightarrow Pb(s) + Cu^{2+}(aq)$

A) -41 kJ B) -0.47 kJ C) +46 kJ D) +91 kJ E) -21 kJ

Answer: D
Diff: 3 *Page Ref: 18.5*

35) Use the tabulated half-cell potentials to calculate ΔG° for the following balanced redox reaction.

$3 I_2(s) + 2 Fe(s) \rightarrow 2 Fe^{3+}(aq) + 6 I^-(aq)$

A) -1.1×10^2 kJ
B) $+4.9 \times 10^1$ kJ
C) -9.7×10^1 kJ
D) $+2.3 \times 10^2$ kJ
E) -3.3×10^2 kJ

Answer: E
Diff: 3 *Page Ref: 18.5*

36) Use the tabulated half-cell potentials to calculate ΔG° for the following redox reaction.

$$2\ Al(s) + 3\ Mg^{2+}(aq) \rightarrow 2\ Al^{3+}(aq) + 3\ Mg(s)$$

A) $+4.1 \times 10^2$ kJ
B) $+1.4 \times 10^2$ kJ
C) -2.3×10^2 kJ
D) -7.8×10^2 kJ
E) $+6.8 \times 10^2$ kJ

Answer: A
Diff: 3 *Page Ref: 18.5*

37) Which of the following reactions would be the most spontaneous at 298 K?
A) $A + 2\ B \rightarrow C$; $E°_{cell} = +0.98$ V
B) $A + B \rightarrow 2\ C$; $E°_{cell} = -0.030$ V
C) $A + B \rightarrow 3\ C$; $E°_{cell} = +0.15$ V
D) $A + B \rightarrow C$; $E°_{cell} = +1.22$ V
E) More information is needed to determine.

Answer: D
Diff: 3 *Page Ref: 18.5*

38) Use the tabulated half-cell potentials to calculate the equilibrium constant (K) for the following balanced redox reaction at 25°C.

$$2\ Al(s) + 3\ Mg^{2+}(aq) \rightarrow 2\ Al^{3+}(aq) + 3\ Mg(s)$$

A) 1.1×10^{72}
B) 8.9×10^{-73}
C) 1.1×10^{-72}
D) 1.0×10^{24}
E) 4.6×10^{31}

Answer: C
Diff: 4 *Page Ref: 18.5*

39) Use the tabulated half-cell potentials to calculate the equilibrium constant (K) for the following balanced redox reaction at 25°C.

$$3\ I_2(s) + 2\ Fe(s) \rightarrow 2\ Fe^{3+}(aq) + 6\ I^-(aq)$$

A) 3.5×10^{-59}
B) 1.1×10^{17}
C) 2.4×10^{58}
D) 8.9×10^{-18}
E) 1.7×10^{29}

Answer: C
Diff: 4 *Page Ref: 18.5*

40) Use the tabulated half-cell potentials to calculate the equilibrium constant (K) for the following balanced redox reaction at 25°C.

$$Pb^{2+}(aq) + Cu(s) \rightarrow Pb(s) + Cu^{2+}(aq)$$

A) 7.9×10^{-8}
B) 8.9×10^{7}
C) 7.9×10^{15}
D) 1.3×10^{-16}
E) 1.1×10^{-8}

Answer: D
Diff: 4 *Page Ref: 18.5*

41) Calculate the cell potential for the following reaction that takes place in an electrochemical cell at 25°C.

$Sn(s) \mid Sn^{2+}(aq, 1.8\ M) \parallel Ag^{+}(aq, 0.055\ M) \mid Ag(s)$

A) –0.94 V B) –0.85 V C) +1.02 V D) +0.98 V E) +0.86 V

Answer: E
Diff: 3 *Page Ref: 18.6*

42) Calculate the cell potential for the following reaction that takes place in an electrochemical cell at 25°C.

$Sn(s) \mid Sn^{2+}(aq, 0.022\ M) \parallel Ag^{+}(aq, 2.7\ M) \mid Ag(s)$

A) +1.01 V B) –0.83 V C) +1.31 V D) +0.01 V E) –0.66 V

Answer: A
Diff: 3 *Page Ref: 18.6*

43) Calculate the cell potential for the following reaction that takes place in an electrochemical cell at 25°C.

$Mg(s) \mid Mg^{2+}(aq, 2.74\ M) \parallel Cu^{2+}(aq, 0.0033\ M) \mid Cu(s)$

A) –2.80 V B) +2.62 V C) +2.71 V D) +2.12 V E) –1.94 V

Answer: B
Diff: 3 *Page Ref: 18.6*

44) Calculate the cell potential for the following reaction that takes place in an electrochemical cell at 25°C.

$Fe(s) \mid Fe^{3+}(aq, 0.0011\ M) \parallel Fe^{3+}(aq, 2.33\ M) \mid Fe(s)$

A) +0.066 V B) –0.036 V C) 0.00 V D) –0.099 V E) +0.20 V

Answer: A
Diff: 3 *Page Ref: 18.6*

45) Calculate the cell potential for the following reaction that takes place in an electrochemical cell at 25°C.

$$Cu(s) \mid Cu^{2+}(aq, 0.0032\ M) \parallel Cu^{2+}(aq, 4.48\ M) \mid Cu(s)$$

A) 0.00 V B) +0.093 V C) +0.34 V D) +0.186 V E) +0.052 V

Answer: B
Diff: 3 *Page Ref: 18.6*

46) Calculate the cell potential for the following reaction that takes place in an electrochemical cell at 25°C.

$$Al(s) \mid Al^{3+}(aq, 0.115\ M) \parallel Al^{3+}(aq, 3.89\ M) \mid Al(s)$$

A) +1.66 V B) +0.060 V C) 0.00 V D) +0.090 V E) +0.030 V

Answer: E
Diff: 3 *Page Ref: 18.6*

47) Identify the battery that is in most automobiles.
A) dry–cell battery
B) lithium ion battery
C) lead–acid storage battery
D) NiCad battery
E) fuel cell

Answer: C
Diff: 1 *Page Ref: 18.7*

48) Identify the battery that is used as a common flashlight battery.
A) dry–cell battery
B) lithium ion battery
C) lead–acid storage battery
D) NiCad battery
E) fuel cell

Answer: A
Diff: 1 *Page Ref: 18.7*

49) Identify the components of a fuel cell.
A) nickel–metal hydride
B) lithium ion
C) hydrogen–oxygen
D) nickel–cadmium
E) zinc–manganese

Answer: C
Diff: 1 *Page Ref: 18.7*

50) Identify the battery type that has a high overcharge tolerance.
A) NiCad
B) Li ion
C) NiMH
D) lead storage
E) none of the above

Answer: D
Diff: 1 *Page Ref: 18.7*

51) What is the reaction at the anode in a breathalyzer?
A) Ethanol is oxidized to acetic acid.
B) Acetic acid is reduced to ethanol.
C) Oxygen is reduced.
D) Hydrogen is oxidized.
E) Ethanol is oxidized to acetaldehyde.

Answer: A
Diff: 1 *Page Ref: 18.7*

52) What is the reaction at the cathode in a breathalyzer?
A) Ethanol is oxidized to acetic acid.
B) Acetic acid is reduced to ethanol.
C) Oxygen is reduced.
D) Hydrogen is oxidized.
E) Ethanol is oxidized to acetaldehyde.

Answer: C
Diff: 1 *Page Ref: 18.7*

53) Describe how water can be a good conductor of current.
A) use pure water
B) heat the water
C) add salt
D) chill the water
E) vaporize the water

Answer: C
Diff: 1 *Page Ref: 18.8*

54) What mass of silver can be plated onto an object in 33.5 minutes at 8.70 A of current?

$$Ag^+(aq) + e^- \rightarrow Ag(s)$$

A) 19.6 g B) 0.326 g C) 9.78 g D) 3.07 g E) 0.102 g

Answer: A
Diff: 3 *Page Ref: 18.8*

55) What mass of aluminum can be plated onto an object in 755 minutes at 5.80 A of current?
A) 73.5 g B) 24.5 g C) 220. g D) 147 g E) 8.17 g

Answer: B
Diff: 3 *Page Ref: 18.8*

56) Nickel can be plated from aqueous solution according to the following half reaction. How long would it take (in min) to plate 29.6 g of nickel at 4.7 A?

$$Ni^{2+}(aq) + 2\,e^{-} \rightarrow Ni(s)$$

A) 1.7×10^{2} min
B) 5.9×10^{2} min
C) 3.5×10^{2} min
D) 4.8×10^{2} min
E) 6.2×10^{2} min

Answer: C
Diff: 3 *Page Ref: 18.8*

57) Describe the reactions during the electrolysis of water.
A) Oxygen is reduced and hydrogen is oxidized.
B) Oxygen and hydrogen are both oxidized.
C) Oxygen and hydrogen are both reduced.
D) Oxygen is oxidized and hydrogen is reduced.
E) Neither oxygen or hydrogen are oxidized or reduced.

Answer: D, E
Diff: 3 *Page Ref: 18.9*

58) Identify a component in the formation of rust.
A) salt
B) moisture
C) acid
D) all of the above
E) none of the above

Answer: D
Diff: 1 *Page Ref: 18.9*

59) To prevent rust, nails are coated with ________.
A) zinc
B) salt
C) calcium
D) sodium
E) lithium

Answer: A
Diff: 1 *Page Ref: 18.9*

60) Predict the species that will be reduced first if the following mixture of molten salts undergoes electrolysis.

$$Na^{+}, Ca^{2+}, Cl^{-}, Br^{-}, F^{-}$$

A) Na^{+} B) Cl^{-} C) Ca^{2+} D) Br^{-} E) F^{-}

Answer: A
Diff: 1 *Page Ref: 18.9*

61) Predict the species that will be reduced first if the following mixture of molten salts undergoes electrolysis.

Zn^{2+}, Fe^{3+}, Mg^{2+}, Br^{-}, I^{-}

A) Zn^{2+} B) Mg^{2+} C) Br^{-} D) Fe^{3+} E) I^{-}

Answer: D
Diff: 1 *Page Ref: 18.9*

62) Predict the species that will be oxidized first if the following mixture of molten salts undergoes electrolysis.

Cu^{2+}, Mg^{2+}, Cl^{-}, Br^{-}, F^{-}

A) Cl^{-} B) F^{-} C) Cu^{2+} D) Mg^{2+} E) Br^{-}

Answer: E
Diff: 1 *Page Ref: 18.9*

Algorithmic Questions

1) Identify oxidation.
A) increase in oxidation number
B) loss of electrons
C) gain of protons
D) loss of neutrons
E) both A and B

Answer: E
Diff: 1 *Page Ref: 18.2*

2) What element is being oxidized in the following redox reaction?

$$Mg^{2+}(aq) + NH_4^{+}(aq) \rightarrow Mg(s) + NO_3^{-}(aq)$$

A) Mg B) N C) H D) O

Answer: B
Diff: 2 *Page Ref: 18.2*

3) Balance the following redox reaction if it occurs in acidic solution. What are the coefficients in front of Ni and H^{+} in the balanced reaction?

$$Ni^{2+}(aq) + NH_4^{+}(aq) \rightarrow Ni(s) + NO_3^{-}(aq)$$

A) Ni = 1, H^{+} = 8
B) Ni = 1, H^{+} = 4
C) Ni = 4, H^{+} = 10
D) Ni = 2, H^{+} = 4
E) Ni = 3, H^{+} = 5

Answer: C
Diff: 4 *Page Ref: 18.2*

4) Balance the following redox reaction if it occurs in acidic solution. What are the coefficients in front of Sn and Ag^+ in the balanced reaction?

$$Sn(s) + Ag^+(aq) \rightarrow Ag(s) + Sn^{2+}(aq)$$

A) Sn = 1, Ag^+ = 2
B) Sn = 1, Ag^+ = 1
C) Sn = 2, Ag^+ = 1
D) Sn = 2, Ag^+ = 2
E) Sn = 3, Ag^+ = 1

Answer: A
Diff: 4 *Page Ref: 18.2*

5) Balance the following redox reaction if it occurs in basic solution. What are the coefficients in front of Cr and Cl_2 in the balanced reaction?

$$Cr(s) + Cl_2(g) \rightarrow Cr^{3+}(aq) + Cl^-(aq)$$

A) Cr = 2, Cl_2 = 3
B) Cr = 2, Cl_2 = 6
C) Cr = 1, Cl_2 = 1
D) Cr = 2, Cl_2 = 1
E) Cr = 3, Cl_2 = 2

Answer: A
Diff: 4 *Page Ref: 18.2*

6) Determine the redox reaction represented by the following cell notation.

$$Fe(s) \mid Fe^{2+}(aq) \parallel Cu^{2+}(aq) \mid Cu(s)$$

A) $Cu(s) + Fe^{2+}(aq) \rightarrow Fe(s) + Cu^{2+}(aq)$
B) $Fe(s) + Cu^{2+}(aq) \rightarrow Cu(s) + Fe^{2+}(aq)$
C) $2\ Fe(s) + Cu^{2+}(aq) \rightarrow Cu(s) + 2\ Fe^{2+}(aq)$
D) $2\ Cu(s) + Fe^{2+}(aq) \rightarrow Fe(s) + 2\ Cu^{2+}(aq)$
E) $3\ Fe(s) + 2\ Cu^{2+}(aq) \rightarrow 2\ Cu(s) + 3\ Fe^{2+}(aq)$

Answer: B
Diff: 3 *Page Ref: 18.3*

7) Determine the cell notation for the redox reaction given below.

$$Cd(s) + 2\ Ag^{+}(aq) \rightarrow Cd^{2+}(aq) + 2\ Ag(s)$$

A) $Ag^{+}(aq)$ | $Ag(s)$ ‖ $Cd(s)$ | $Cd^{2+}(aq)$
B) $Ag(s)$ | $Ag^{+}(aq)$ ‖ $Cd^{2+}(aq)$ | $Cd(s)$
C) $Cd(s)$ | $Cd^{2+}(aq)$ ‖ $Ag^{+}(aq)$ | $Ag(s)$
D) $Cd^{2+}(aq)$ | $Cd(s)$ ‖ $Ag(s)$ | $Ag^{+}(aq)$
E) $Cd(s)$ | $Ag(s)$ ‖ $Cd^{2+}(aq)$ | $Ag^{+}(aq)$

Answer: C
Diff: 3 *Page Ref: 18.3*

8) Determine the cell notation for the redox reaction given below.

$$Sn(s) + 2\ H^{+}(aq) \rightarrow Sn^{2+}(aq) + H_2(g)$$

A) $H^{+}(aq)$ | $H_2(g)$ | Pt ‖ $Sn(s)$ | $Sn^{2+}(aq)$
B) $H_2(g)$ | $H^{+}(aq)$ | Pt ‖ $Sn^{2+}(aq)$ | $Sn(s)$
C) $Sn^{2+}(aq)$ | $Sn(s)$ ‖ $H_2(g)$ | $H^{+}(aq)$ | Pt
D) $Sn(s)$ | $Sn^{2+}(aq)$ ‖ $H^{+}(aq)$ | $H_2(g)$ | Pt
E) $Sn(s)$ | $H_2(g)$ ‖ $Sn^{2+}(aq)$ | $H^{+}(aq)$ | Pt

Answer: D
Diff: 3 *Page Ref: 18.3*

9) What is the reduction half-reaction for the following overall galvanic cell reaction?

$$Co^{2+}(aq) + 2\ Ag(s) \rightarrow Co(s) + 2\ Ag^{+}(aq)$$

A) $Ag(s) + e^{-} \rightarrow Ag^{+}(aq)$
B) $Ag^{+}(aq) + e^{-} \rightarrow Ag(s)$
C) $Co^{2+}(aq) + 2\ e^{-} \rightarrow Co(s)$
D) $Co^{2+}(aq) + e^{-} \rightarrow Co(s)$

Answer: C
Diff: 3 *Page Ref: 18.3*

10) What is the shorthand notation that represents the following galvanic cell reaction?

$$Pb(s) + Cu(NO_3)_2(aq) \rightarrow Pb(NO_3)_2(aq) + Cu(s)$$

A) $Pb(s)$ | $Pb^{2+}(aq)$ || $Cu^{2+}(aq)$ | $Cu(s)$
B) $Cu(s)$ | $Cu^{2+}(aq)$ || $Pb^{2+}(aq)$ | $Pb(s)$
C) $Pb(s)$ | $NO_3^{-}(aq)$ || $NO_3^{-}(aq)$ | $Cu(s)$
D) $Cu(s)$ | $Cu(NO_3)_2(aq)$ || $Pb(NO_3)_2(aq)$ | $Pb(s)$

Answer: A
Diff: 3 *Page Ref: 18.3*

11) What is the shorthand notation that represents the following galvanic cell reaction?

$2\ Fe^{2+}(aq) + F_2(g) \rightarrow 2\ Fe^{3+}(aq) + 2\ F^-(aq)$

A) $Fe^{2+}(aq) \mid Fe^{3+}(aq) \parallel F_2(g) \mid F^-(aq)$
B) $Fe(s) \mid Fe^{2+}(aq) \parallel Fe^{3+}(aq)\ F_2(g) \mid F^-(aq) \mid C(s)$
C) $Pt(s) \mid Fe^{3+}(aq), Fe^{2+}(aq), F_2(g) \parallel F^-(aq) \mid C(s)$
D) $Pt(s) \mid Fe^{2+}(aq), Fe^{3+}(aq) \parallel F_2(g) \mid F^-(aq) \mid C(s)$

Answer: D
Diff: 3 *Page Ref: 18.3*

12) For the galvanic cell reaction, expressed below using shorthand notation, what half-reaction occurs at the cathode?

$Zn(s) \mid Zn^{2+}(aq) \parallel Ni^{2+}(aq) \mid Ni(s)$

A) $Zn(s) \rightarrow Zn^{2+}(aq) + 2\ e^-$
B) $Zn^{2+}(aq) + 2\ e^- \rightarrow Zn(s)$
C) $Ni(s) \rightarrow Ni^{2+}(aq) + 2\ e^-$
D) $Ni^{2+}(aq) + 2\ e^- \rightarrow Ni(s)$

Answer: D
Diff: 3 *Page Ref: 18.3*

13) What is the balanced chemical equation for the galvanic cell reaction expressed using shorthand notation below?

$Al(s) \mid Al^{3+}(aq) \parallel Fe^{2+}(aq) \mid Fe(s)$

A) $2\ Al(s) + 3\ Fe^{2+}(aq) \rightarrow 2\ Al^{3+}(aq) + 3\ Fe(s)$
B) $3\ Al(s) + 2\ Fe^{2+}(aq) \rightarrow 3\ Al^{3+}(aq) + 2\ Fe(s)$
C) $2\ Fe(s) + 3\ Al^{3+}(aq) \rightarrow 2\ Fe^{2+}(aq) + 3\ Al(s)$
D) $3\ Fe(s) + 2\ Al^{3+}(aq) \rightarrow 3\ Fe^{2+}(aq) + 2\ Al(s)$

Answer: A
Diff: 3 *Page Ref: 18.3*

14) What is the balanced equation for the galvanic cell reaction expressed using shorthand notation below?.

$Ni(s) \mid Ni^{2+}(aq) \parallel Cl_2(g) \mid Cl^-(aq) \mid C(s)$

A) $Ni(s) + 2\ Cl^-(aq) \rightarrow Ni^{2+}(aq) + Cl_2(g)$
B) $Ni(s) + Cl_2(g) \rightarrow Ni^{2+}(aq) + 2\ Cl^-(aq)$
C) $Ni^{2+}(aq) + 2\ Cl^-(aq) \rightarrow Ni(s) + Cl_2(g)$
D) $Ni^{2+}(aq) + 2\ Cl^-(aq) \rightarrow NiCl_2(s)$

Answer: B
Diff: 3 *Page Ref: 18.3*

15) What is the reducing agent in the redox reaction represented by the following cell notation?

$Zn(s) \mid Zn^{2+}(aq) \parallel Ag^+(aq) \mid Ag(s)$

A) Zn(s) B) $Zn^{2+}(aq)$ C) $Ag^+(aq)$ D) Ag(s) E) Pt

Answer: A
Diff: 1 *Page Ref: 18.4*

16) What is the oxidizing agent in the redox reaction represented by the following cell notation?

$Ni(s) \mid Ni^{2+}(aq) \parallel Ag^{+}(aq) \mid Ag(s)$

A) $Ni(s)$ B) $Ni^{2+}(aq)$ C) $Ag^{+}(aq)$ D) $Ag(s)$ E) Pt

Answer: C
Diff: 1 *Page Ref: 18.4*

17) What is the oxidizing agent in the redox reaction represented by the following cell notation?

$Sn(s) \mid Sn^{2+}(aq) \parallel Ag^{+}(aq) \mid Ag(s)$

A) $Sn(s)$ B) $Ag^{+}(aq)$ C) $Sn^{2+}(aq)$ D) $Ag(s)$ E) Pt

Answer: B
Diff: 1 *Page Ref: 18.4*

18) What is the reducing agent in the redox reaction represented by the following cell notation?

$Ca(s) \mid Ca^{2+}(aq) \parallel Ag^{+}(aq) \mid Ag(s)$

A) $Ca(s)$ B) $Ag^{+}(aq)$ C) $Ca^{2+}(aq)$ D) $Ag(s)$ E) Pt

Answer: A
Diff: 1 *Page Ref: 18.4*

19) What is undergoing reduction in the redox reaction represented by the following cell notation?

$Cr(s) \mid Cr^{3+}(aq) \parallel F_2(g) \mid F^{-}(aq) \mid Pt$

A) $Cr(s)$ B) $Cr^{3+}(aq)$ C) $F_2(g)$ D) $F^{-}(aq)$ E) Pt

Answer: C
Diff: 1 *Page Ref: 18.4*

20) What is undergoing oxidation in the redox reaction represented by the following cell notation?

$Fe(s) \mid Fe^{3+}(aq) \parallel Br_2(g) \mid Br^{-}(aq) \mid Pt$

A) $Fe(s)$ B) $Fe^{3+}(aq)$ C) $Br_2(g)$ D) $Br^{-}(aq)$ E) Pt

Answer: A
Diff: 1 *Page Ref: 18.4*

21) What is undergoing oxidation in the redox reaction represented by the following cell notation?

$Fe(s) \mid Fe^{2+}(aq) \parallel H^{+}(aq) \mid H_2(g) \mid Pt$

A) $H_2(g)$ B) $H^{+}(aq)$ C) $Fe^{2+}(aq)$ D) $Fe(s)$ E) Pt

Answer: D
Diff: 1 *Page Ref: 18.4*

22) What is undergoing reduction in the redox reaction represented by the following cell notation?

$Pb(s) | Pb^{2+}(aq) || H^{+}(aq) | H_2(g) | Pt$

A) $H_2(g)$ B) $H^{+}(aq)$ C) $Pb^{2+}(aq)$ D) Pb(s) E) Pt

Answer: B
Diff: 1 *Page Ref: 18.4*

23) Which of the following is the strongest reducing agent?
A) $Cl^{2+}(aq)$ B) $Cr^{3+}(aq)$ C) $Cl^{-}(aq)$ D) Cr(s) E) Cl(s)

Answer: D
Diff: 1 *Page Ref: 18.4*

24) Which of the following is the weakest oxidizing agent?
A) $Sn^{2+}(aq)$ B) $Al^{3+}(aq)$ C) $Sn^{4+}(aq)$ D) $K^{+}(s)$ E) $Cu^{+}(s)$

Answer: D
Diff: 1 *Page Ref: 18.4*

25) Which of the following is the strongest reducing agent?
A) Na(s) B) $Li^{+}(aq)$ C) Ca(s) D) $Ca^{2+}(aq)$ E) K(s)

Answer: E
Diff: 1 *Page Ref: 18.4*

26) Which of the following is the strongest reducing agent?
A) Al(s) B) Mn(s) C) Li(s) D) $Al^{3+}(aq)$ E) $Mg^{2+}(aq)$

Answer: C
Diff: 1 *Page Ref: 18.4*

27) Which of the following is the weakest reducing agent?
A) Cr(s) B) $F^{-}(aq)$ C) K(s) D) $Cr^{3+}(aq)$ E) $Ca^{2+}(aq)$

Answer: B
Diff: 1 *Page Ref: 18.4*

28) Which of the following is the strongest oxidizing agent?
A) $BrO_2(s)$
B) $Br^{-}(aq)$
C) $Cu^{+}(aq)$
D) $SO_4^{2-}(aq)$
E) $MnO_4^{-}(aq)$

Answer: E
Diff: 1 *Page Ref: 18.4*

29) Which of the following is the strongest oxidizing agent?
A) $Br_2(l)$ B) $Au^{3+}(aq)$ C) Li(s) D) $I^{-}(aq)$ E) Li(s)

Answer: B
Diff: 1 *Page Ref: 18.4*

30) Which of the following is the weakest reducing agent?
A) $Cl_2(l)$
B) $ClO_2^-(aq)$
C) Ag(s)
D) $Br^-(aq)$
E) Au(s)
Answer: E
Diff: 1 *Page Ref: 18.4*

31) Which of the following is the strongest oxidizing agent?
A) $H_2O_2(aq)$
B) $Fe^{3+}(aq)$
C) $ClO_2(g)$
D) $I_2(s)$
E) Fe(s)
Answer: A
Diff: 1 *Page Ref: 18.4*

32) Determine which of the following pairs of reactants will result in a spontaneous reaction at 25°C.
A) $Pb^{2+}(aq) + Cu(s)$
B) $Ag^+(aq) + Br^-(aq)$
C) $Li^+(aq) + Cr(s)$
D) $Fe^{3+}(aq) + Cd(s)$
E) None of the above pairs will react.
Answer: D
Diff: 2 *Page Ref: 18.4*

33) Determine which of the following pairs of reactants will result in a spontaneous reaction at 25°C.
A) $Sn^{4+}(aq) + Mn(s)$
B) $Cr^{3+}(aq) + Pb(s)$
C) $Zn(s) + K^+(aq)$
D) $Fe(s) + Ba^{2+}(aq)$
E) None of the above pairs will react.
Answer: A
Diff: 2 *Page Ref: 18.4*

34) Determine which of the following pairs of reactants will result in a spontaneous reaction at 25°C.
A) $I^-(aq) + Fe^{2+}(aq)$
B) $Ca(s) + Mg^{2+}(aq)$
C) $H_2(g) + Sn^{2+}(aq)$
D) $Ag(s) + Sn^{2+}(aq)$
E) All of the above pairs will react.
Answer: B
Diff: 2 *Page Ref: 18.4*

35) Which of the following metals will dissolve in nitric acid but not hydrochloric?

A) Li B) Li C) Cu D) Sn E) Ni

Answer: C

Diff: 2 *Page Ref: 18.4*

36) Which of the following metals will dissolve in nitric acid but not hydrochloric?

A) Fe B) Zn C) Mg D) Ag E) Al

Answer: D

Diff: 2 *Page Ref: 18.4*

37) Which of the following metals will dissolve in HCl?

A) Ba
B) Na
C) Mn
D) Al
E) all of the above

Answer: E

Diff: 2 *Page Ref: 18.4*

38) Which of the following metals will not dissolve in nitric acid or hydrochloric acid?

A) Pb
B) Cu
C) Ni
D) Au
E) all of the above

Answer: D

Diff: 2 *Page Ref: 18.4*

39) Which of the following metals will not dissolve in nitric acid or hydrochloric acid?

A) Al
B) Pb
C) Mg
D) K
E) none of the above

Answer: E

Diff: 2 *Page Ref: 18.4*

40) A galvanic cell consists of a Ni^{2+}/Ni half-cell and a standard hydrogen electrode. If the Ni^{2+}/Ni half-cell standard cell functions as the anode, and the standard cell potential is 0.26 V, what is the standard reduction potential for the Ni^{2+}/Ni half-cell?

A) -0.26 V B) -0.13 V C) +0.13 V D) +0.26 V

Answer: A

Diff: 2 *Page Ref: 18.4*

41) A galvanic cell consists of one half-cell that contains Ag(*s*) and Ag^{+}(*aq*), and one half-cell that contains Cu(*s*) and Cu^{2+}(*aq*). What species are produced at the electrodes under standard conditions?

$Ag^{+}(aq) + e^{-} \rightarrow Ag(s)$ $E° = +0.80$ V

$Cu^{2+}(aq) + 2\ e^{-} \rightarrow Cu(s)$ $E° = +0.34$ V

A) Ag(*aq*) is formed at the cathode and, Cu(*s*) is formed at the anode.
B) Ag(*s*) is formed at the cathode, and Cu^{2+}(*aq*) is formed at the anode.
C) Cu(*s*) is formed at the cathode, and Ag^{+}(*aq*) is formed at the anode.
D) Cu^{2+}(*aq*) is formed at the cathode, and Cu(*s*) is formed at the anode.

Answer: B
Diff: 2 *Page Ref:* 18.4

42) Consider the following standard reduction potentials,

$Ni^{2+}(aq) + 2\ e^{-} \rightarrow Ni(s)$ $E° = -0.26$ V

$I_2(s) + 2\ e^{-} \rightarrow 2\ I^{-}(aq)$ $E° = +0.54$ V

Under standard conditions,

A) Ni^{2+}(*aq*) is a stronger oxidizing agent than I_2(*s*) and I^{-}(*aq*) is a stronger reducing agent than Ni(*s*).
B) I_2(*s*) is a stronger oxidizing agent than Ni^{2+}(*aq*) and Ni(*s*) is a stronger reducing agent than I^{-}(*aq*).
C) Ni(*s*) is a stronger oxidizing agent than I^{-}(*aq*) and Ni^{2+}(*aq*) is a stronger reducing agent than I_2(*s*).
D) I^{-}(*aq*) is a stronger oxidizing agent than Ni(*s*) and I_2(*s*) is a stronger reducing agent than Ni^{2+}(*aq*).

Answer: B
Diff: 2 *Page Ref:* 18.4

43) Consider the galvanic cell, Zn(*s*) | Zn^{2+}(*aq*) || Pb^{2+}(*aq*) | Pb(*s*). Which one of the following changes to the cell would cause the cell potential to increase (i.e., become more positive)?

A) increase the $[Zn^{2+}]$ concentration
B) increase the $[Pb^{2+}]$ concentration
C) increase the mass of Zn(*s*)
D) decrease the mass of Zn(*s*)

Answer: B
Diff: 3 *Page Ref:* 18.4

44) Based on the following information,

$Cl_2(g) + 2\ e^{-} \rightarrow 2\ Cl^{-}(aq)$ $E° = +1.36$ V

$Mg^{2+}(aq) + 2\ e^{-} \rightarrow 2\ Mg(s)$ $E° = -2.37$ V

which of the following chemical species is the strongest reducing agent?

A) Cl_2(*g*)
B) Mg^{2+}(*aq*)
C) Cl^{-}(*aq*)
D) Mg(*s*)

Answer: D
Diff: 3 *Page Ref:* 18.4

45) Using the following standard reduction potentials,

$Fe^{3+}(aq) + e^- \rightarrow Fe^{2+}(aq)$ $E° = +0.77$ V

$Ni^{2+}(aq) + 2 e^- \rightarrow Ni(s)$ $E° = -0.23$ V

calculate the standard cell potential for the galvanic cell reaction given below, and determine whether or not this reaction is spontaneous under standard conditions.

$Ni^{2+}(aq) + 2 Fe^{2+}(aq) \rightarrow 2 Fe^{3+}(aq) + Ni(s)$

A) $E° = -1.00$ V, nonspontaneous
B) $E° = -1.00$ V, spontaneous
C) $E° = +1.00$ V, nonspontaneous
D) $E° = +1.00$ V, spontaneous

Answer: A
Diff: 3 *Page Ref: 18.4*

46) How many electrons are transferred in the following reaction? (The reaction is unbalanced.)

$Ca(s) + Cr^{3+}(aq) \rightarrow Cr(s) + Ca^{2+}(aq)$

A) 6 B) 2 C) 3 D) 1 E) 4

Answer: A
Diff: 2 *Page Ref: 18.5*

47) How many electrons are transferred in the following reaction? (The reaction is unbalanced.)

$Ca^{2+}(aq) + Li(s) \rightarrow Ca(s) + Li^+(aq)$

A) 1 B) 2 C) 3 D) 4 E) 6

Answer: B
Diff: 2 *Page Ref: 18.5*

48) The gas OF_2 can be produced from the electrolysis of an aqueous solution of KF, as shown in the equation below.

$OF_2(g) + 2 H^+(aq) + 4 e^- \rightarrow H_2O(l) + 2 F^-(aq)$ $E° = +2.15$ V

Using the given standard reduction potential, calculate the amount of OF_2 that is produced, and the electrode at which the OF_2 is produced, upon the passage of 0.480 faradays through an aqueous KF solution.

A) 6.48 g of OF_2 at the anode.
B) 26.0 g of OF_2 at the anode.
C) 6.48 g of OF_2 at the cathode.
D) 26.0 g of OF_2 at the cathode.

Answer: A
Diff: 3 *Page Ref: 18.5*

49) Given that $E°_{red} = -0.40$ V for Cd^{2+}/Cd at 25°C, find $E°$ and E for the concentration cell expressed using shorthand notation below.

$Cd(s) \mid Cd^{2+}(1.0 \times 10^{-5}\ M) \parallel Cd^{2+}(0.100\ M) \mid Cd(s)$

A) $E° = 0.00$ V and $E = +0.24$ V
B) $E° = 0.00$ V and $E = +0.12$ V
C) $E° = -0.40$ V and $E = -0.16$ V
D) $E° = -0.40$ V and $E = -0.28$ V

Answer: B
Diff: 3 *Page Ref: 18.6*

50) The standard cell potential (E°) of a voltaic cell constructed using the cell reaction below is 0.76 V:

$$Zn(s) + 2H^+(aq) \rightarrow Zn^{2+}(aq) + H_2(g)$$

With P_{H_2} = 1.0 atm and $[Zn^{2+}]$ = 1.0 M, the cell potential is 0.66 V. The concentration of H^+ in the cathode compartment is ________ M.

A) 2.0×10^{-2}
B) 4.2×10^{-4}
C) 1.4×10^{-1}
D) 4.9×10^{1}
E) 1.0×10^{-12}

Answer: A
Diff: 3 *Page Ref: 18.6*

51) A voltaic cell is constructed with two silver-silver chloride electrodes, where the half-reaction is

$$AgCl(s) + e^- \rightarrow Ag(s) + Cl^-(aq) \qquad E° = +0.222 \text{ V}$$

The concentrations of chloride ion in the two compartments are 0.0222 M and 2.22 M, respectively. The cell emf is ________ V.

A) 0.212 B) 0.118 C) 0.00222 D) 22.2 E) 0.232

Answer: B
Diff: 3 *Page Ref: 18.6*

52) A voltaic cell is constructed with two Zn^{2+}-Zn electrodes, where the half-reaction is

$$Zn^{2+} + 2e^- \rightarrow Zn(s) \qquad E° = -0.763 \text{ V}$$

The concentrations of zinc ion in the two compartments are 5.50 M and 1.11×10^{-2} M, respectively. The cell emf is ________ V.

A) -1.54×10^{-3}
B) -378
C) 0.0798
D) 0.160
E) -0.761

Answer: C
Diff: 3 *Page Ref: 18.6*

53) The standard emf for the cell using the overall cell reaction below is +2.20 V:

$$2Al(s) + 3I_2(s) \rightarrow 2Al^{3+}(aq) + 6I^-(aq)$$

The emf generated by the cell when $[Al^{3+}] = 4.5 \times 10^{-3}$ M and $[I^-] = 0.15$ M is ________ V.

A) 2.20 B) 2.32 C) 2.10 D) 2.39 E) 2.23

Answer: B
Diff: 3 *Page Ref: 18.6*

54) The standard emf for the cell using the overall cell reaction below is +0.48 V:

$$Zn(s) + Ni^{2+}(aq) \rightarrow Zn^{2+}(aq) + Ni(s)$$

The emf generated by the cell when $[Ni^{2+}] = 2.50$ M and $[Zn^{2+}] = 0.100$ M is ________ V.

A) 0.40 B) 0.50 C) 0.52 D) 0.56 E) 0.44

Answer: C
Diff: 3 *Page Ref: 18.6*

55) The electrolysis of molten $AlCl_3$ for 3.25 hr with an electrical current of 15.0 A produces ________ g of aluminum metal.

A) 147
B) 0.606
C) 4.55×10^{-3}
D) 16.4
E) 49.1

Answer: D
Diff: 1 *Page Ref: 18.8*

56) For a galvanic cell that uses the following two half-reactions,

$$Cr_2O_7^{2-}(aq) + 14\ H^+(aq) + 6\ e^- \rightarrow 2\ Cr^{3+}(aq) + 7\ H_2O(l)$$

$$Pb(s) \rightarrow Pb^{2+}(aq) + 2\ e^-$$

how many moles of Pb(s) are oxidized by three moles of $Cr_2O_7^{2-}$?

A) 3 B) 6 C) 9 D) 18

Answer: C
Diff: 2 *Page Ref: 18.8*

57) How many kilowatt-hours of electricity are used to produce 3.00 kg of magnesium in the electrolysis of molten $MgCl_2$ with an applied emf of 4.50 V?

A) 0.0336 B) 0.0298 C) 7.4 D) 29.8 E) 14.9

Answer: D
Diff: 2 *Page Ref: 18.8*

58) The most useful ore of aluminum is bauxite, in which Al is present as hydrated oxides, $Al_2O_3 \cdot xH_2O$. The number of kilowatt-hours of electricity required to produce 4.00 kg of aluminum from electrolysis of compounds from bauxite is ________ when the applied emf is 5.00 V.

A) 0.0168 B) 0.0596 C) 39.7 D) 19.9 E) 59.6

Answer: E
Diff: 2 *Page Ref: 18.8*

59) How many seconds are required to produce 4.00 g of aluminum metal from the electrolysis of molten $AlCl_3$ with an electrical current of 12.0 A?

A) 27.0
B) 9.00
C) 1.19×10^3
D) 2.90×10^5
E) 3.57×10^3

Answer: E
Diff: 2 *Page Ref: 18.8*

60) How many grams of chromium metal are plated out when a constant current of 8.00 A is passed through an aqueous solution containing Cr^{3+} ions for 320. minutes?

A) 27.6 g B) 49.2 g C) 82.4 g D) 248 g

Answer: A
Diff: 2 *Page Ref: 18.8*

61) How many grams of nickel metal are plated out when a constant current of 15.0 A is passed through aqueous $NiCl_2$ for 60.0 minutes?

A) 10.9 g B) 16.4 g C) 32.8 g D) 36.3 g

Answer: B
Diff: 2 *Page Ref: 18.8*

62) How long must a constant current of 50.0 A be passed through an electrolytic cell containing aqueous Cu^{2+} ions to produce 5.00 moles of copper metal?

A) 0.187 hours B) 0.373 hours C) 2.68 hours D) 5.36 hours

Answer: D
Diff: 2 *Page Ref: 18.8*

63) The town of Natrium, West Virginia, derives its name from the sodium produced in the electrolysis of molten sodium chloride (NaCl) mined from ancient salt deposits. The number of kilowatt-hours of electricity required to produce 4.60 kg of metallic sodium from the electrolysis of molten NaCl(s) is ________ when the applied emf is 4.50 V.

A) 24.1 B) 0.0414 C) 0.0241 D) 48.3 E) 12.1

Answer: A
Diff: 3 *Page Ref: 18.8*

Matching Questions

Match the following.

1) $Q = 1$
Diff: 1 *Page Ref: 18.6*

2) $Q > K$
Diff: 1 *Page Ref: 18.6*

3) $Q < K$
Diff: 1 *Page Ref: 18.6*

4) $\Delta G° < 0$
Diff: 1 *Page Ref: 18.6*

5) $\Delta G° > 0$
Diff: 1 *Page Ref: 18.6*

6) $Q = K$
Diff: 1 *Page Ref: 18.6*

A) $E°_{cell} > 0$

B) $E_{cell} = 0$

C) $E_{cell} = E°_{cell}$

D) $E°_{cell} < 0$

E) $E_{cell} < 0$

F) $E_{cell} > 0$

Answers: 1) C 2) E 3) F 4) A 5) D
6) B

Short Answer Questions

1) What is the difference between a voltaic cell and an electrolytic cell?

Answer: A voltaic cell contains a redox reaction that is spontaneous. An electrolytic cell contains a nonspontaneous redox reaction where electrical current must be supplied to the cell in order to drive the reaction.
Diff: 1 *Page Ref: 18.3*

2) Sketch a voltaic cell that contains the following half reactions and label all relevant components.

$Zn^{2+}(aq) + 2e^- \rightarrow Zn(s)$ $E° = -0.76$ V
$Cu^{2+}(aq) + 2e^- \rightarrow Cu(s)$ $E° = +0.34$ V

Answer: The sketch should be very similar to Figure 18.2 in the text.
Diff: 2 *Page Ref: 18.3*

3) Give an example of an inert electrode.

Answer: graphite or platinum
Diff: 2 *Page Ref: 18.3*

4) Explain the significance of the standard hydrogen electrode (SHE) in the tabulation of standard reduction potentials of other species.

Answer: The SHE is used to construct an electrochemical cell with another species. The standard cell potential for the cell is then recorded. The SHE is set to 0, so it is then possible to use $E°_{cell} = E°_{SHE} + E°_{unknown}$ to determine a standard reduction potential for the other species present. The entire table of standard reduction potentials is constructed in this fashion. Any species that is more easily reduced than SHE has a positive $E°_{red}$. Any species that is more easily oxidized than SHE has a negative $E°_{red}$.

Diff: 2 *Page Ref: 18.4*

5) Why, if we multiply a reaction by 2, don't we multiply its $E°_{red}$ by 2?

Answer: Half-cell potentials ($E°_{red}$) are a measure of potential difference per unit charge. It is similar to a density, so it is an intensive property and does not depend on the amount of substance.

Diff: 2 *Page Ref: 18.4*

6) What is electrolysis?

Answer: An electrical current is used to drive an otherwise nonspontaneous redox reaction.

Diff: 1 *Page Ref: 18.8*

7) Why is sugar water not a good conductor of current?

Answer: Sugar water has no ions.

Diff: 1 *Page Ref: 18.8*

8) Why are iron nails coated with zinc?

Answer: Zinc is more easily oxidized than iron, so it will react instead of the iron to maintain the structural integrity of the nail.

Diff: 1 *Page Ref: 18.9*

Chapter 19 Radioactivity and Nuclear Chemistry

Multiple Choice Questions

1) Identify the scientist(s) that were awarded the Nobel Prize in physics for the discovery of radioactivity in 1903.

A) Johannes Geiger, Marie Curie
B) Albert Einstein
C) Antoine–Henri Becquerel, Marie Curie, Pierre Curie
D) Ernest Rutherford, Johannes Geiger
E) Galileo Galilei

Answer: C
Diff: 1 *Page Ref: 19.2*

2) Identify an alpha particle.

A) ${}^{0}_{-1}e$ B) ${}^{1}_{0}n$ C) ${}^{0}_{+1}e$ D) ${}^{4}_{2}He$ E) ${}^{0}_{0}\gamma$

Answer: D
Diff: 3 *Page Ref: 19.3*

3) Which of the following statements is TRUE?

A) Positrons are similar in ionizing power and penetrating power to beta particles.
B) A positron is the antiparticle of the electron.
C) Beta decay occurs when a neutron changes into a proton while emitting an electron.
D) An alpha particle is a helium 2+ ion.
E) All of the above are true.

Answer: E
Diff: 1 *Page Ref: 19.3*

4) Which of the following statements is TRUE?

A) Gamma rays have the lowest ionizing power of any radioactivity.
B) Alpha radiation has the highest penetrating power of any radioactivity.
C) Beta emitters will do more damage than alpha emitters within the body.
D) Beta radiation has the highest ionizing power of any radioactivity.
E) None of the above are true.

Answer: A
Diff: 1 *Page Ref: 19.3*

5) Write a nuclear equation for the alpha decay of ${}^{241}_{95}Am$.

A) ${}^{241}_{95}Am \rightarrow {}^{4}_{2}He + {}^{237}_{93}Np$

B) ${}^{241}_{95}Am \rightarrow {}^{4}_{2}He + {}^{245}_{97}Bk$

C) ${}^{241}_{95}Am \rightarrow {}^{0}_{-1}e + {}^{241}_{96}Cm$

D) ${}^{241}_{95}Am \rightarrow {}^{0}_{+1}e + {}^{241}_{94}Pu$

E) ${}^{241}_{95}Am \rightarrow {}^{1}_{0}n + {}^{240}_{95}Am$

Answer: A
Diff: 2 Page Ref: 19.3

6) Write a nuclear equation for the alpha decay of ${}^{236}_{94}Pu$.

A) ${}^{236}_{94}Pu \rightarrow {}^{0}_{-1}e + {}^{232}_{95}Am$

B) ${}^{236}_{94}Pu \rightarrow {}^{4}_{2}He + {}^{232}_{92}U$

C) ${}^{236}_{94}Pu \rightarrow {}^{0}_{+1}e + {}^{236}_{93}Np$

D) ${}^{236}_{94}Pu \rightarrow {}^{1}_{0}n + {}^{235}_{94}Pu$

E) ${}^{236}_{94}Pu \rightarrow {}^{0}_{-1}e + {}^{236}_{93}Np$

Answer: B
Diff: 2 Page Ref: 19.3

7) Write a nuclear equation for the alpha decay of ${}^{238}_{92}U$.

A) ${}^{238}_{92}U \rightarrow {}^{1}_{0}n + {}^{237}_{92}U$

B) ${}^{238}_{92}U \rightarrow {}^{0}_{-1}e + {}^{238}_{93}Np$

C) ${}^{238}_{92}U \rightarrow {}^{4}_{2}He + {}^{234}_{90}Th$

D) ${}^{238}_{92}U \rightarrow {}^{0}_{+1}e + {}^{238}_{91}Pa$

E) ${}^{238}_{92}U \rightarrow {}^{0}_{-1}e + {}^{238}_{91}Pa$

Answer: C
Diff: 2 Page Ref: 19.3

8) Write the nuclear equation for the alpha decay of ${}^{232}_{90}Th$.

A) ${}^{4}_{2}He + {}^{232}_{90}Th \rightarrow {}^{236}_{92}U$

B) ${}^{1}_{0}n + {}^{232}_{90}Th \rightarrow {}^{233}_{90}Th$

C) ${}^{232}_{90}Th \rightarrow {}^{0}_{+1}e + {}^{232}_{89}Ac$

D) ${}^{232}_{90}Th \rightarrow {}^{4}_{2}He + {}^{228}_{88}Ra$

E) ${}^{232}_{90}Th \rightarrow {}^{0}_{-1}e + {}^{232}_{91}Pa$

Answer: D
Diff: 2 *Page Ref: 19.3*

9) Write the nuclear equation for the alpha decay of ${}^{226}_{88}Ra$.

A) ${}^{226}_{88}Ra + {}^{4}_{2}He \rightarrow {}^{230}_{90}Th$

B) ${}^{226}_{88}Ra \rightarrow {}^{1}_{0}n + {}^{225}_{88}Ra$

C) ${}^{226}_{88}Ra \rightarrow {}^{0}_{-1}e + {}^{226}_{89}Ac$

D) ${}^{226}_{88}Ra + {}^{0}_{+1}e \rightarrow {}^{226}_{89}Ac$

E) ${}^{226}_{88}Ra \rightarrow {}^{4}_{2}He + {}^{222}_{86}Rn$

Answer: E
Diff: 2 *Page Ref: 19.3*

10) Determine the identity of the daughter nuclide from the alpha decay of ${}^{222}_{86}Rn$.

A) ${}^{218}_{84}Po$ B) ${}^{226}_{88}Ra$ C) ${}^{224}_{90}Th$ D) ${}^{223}_{86}Rn$ E) ${}^{221}_{85}At$

Answer: A
Diff: 2 *Page Ref: 19.3*

11) Determine the identity of the daughter nuclide from the alpha decay of ${}^{216}_{84}Po$.

A) ${}^{218}_{84}Po$ B) ${}^{218}_{80}Hg$ C) ${}^{217}_{85}At$ D) ${}^{212}_{82}Pb$ E) ${}^{220}_{86}Rn$

Answer: D
Diff: 2 *Page Ref: 19.3*

12) Determine the identity of the daughter nuclide from the alpha decay of $^{224}_{88}Ra$.

A) $^{223}_{87}Fr$ B) $^{225}_{89}Ac$ C) $^{222}_{84}Po$ D) $^{228}_{90}Th$ E) $^{220}_{86}Rn$

Answer: E
Diff: 2 *Page Ref: 19.3*

13) Determine the identity of the daughter nuclide from the beta decay of $^{89}_{38}Sr$.

A) $^{90}_{38}Sr$ B) $^{90}_{39}Y$ C) $^{89}_{39}Y$ D) $^{85}_{36}Kr$ E) $^{87}_{34}Se$

Answer: C
Diff: 2 *Page Ref: 19.3*

14) Determine the identity of the daughter nuclide from the beta decay of $^{210}_{82}Pb$.

A) $^{208}_{78}Pt$ B) $^{209}_{81}Tl$ C) $^{206}_{80}Hg$ D) $^{210}_{83}Bi$ E) $^{211}_{82}Pb$

Answer: D
Diff: 2 *Page Ref: 19.3*

15) Determine the identity of the daughter nuclide from the beta decay of $^{99}_{43}Tc$.

A) $^{100}_{44}Ru$ B) $^{103}_{45}Rh$ C) $^{95}_{41}Nb$ D) $^{99}_{42}Mo$ E) $^{99}_{44}Ru$

Answer: E
Diff: 2 *Page Ref: 19.3*

16) Determine the identity of the daughter nuclide from the positron emission of $^{11}_{6}C$.

A) $^{11}_{5}B$ B) $^{11}_{7}N$ C) $^{12}_{6}C$ D) $^{10}_{5}B$ E) $^{12}_{7}N$

Answer: A
Diff: 2 *Page Ref: 19.3*

17) Determine the identity of the daughter nuclide from the positron emission of $^{13}_{7}N$.

A) $^{13}_{8}O$ B) $^{13}_{6}C$ C) $^{14}_{8}O$ D) $^{9}_{5}B$ E) $^{17}_{9}F$

Answer: B
Diff: 2 *Page Ref: 19.3*

18) Determine the identity of the daughter nuclide from the positron emission of $^{15}_{8}O$.

A) $^{15}_{6}C$ B) $^{16}_{9}F$ C) $^{15}_{7}N$ D) $^{14}_{7}N$ E) $^{11}_{6}C$

Answer: C
Diff: 2 *Page Ref: 19.3*

19) Determine the identity of the daughter nuclide from the electron capture by $^{81}_{37}Rb$.

A) $^{81}_{36}Kr$ B) $^{81}_{38}Sr$ C) $^{77}_{35}Br$ D) $^{85}_{39}Y$ E) $^{80}_{36}Kr$

Answer: A
Diff: 2 *Page Ref: 19.3*

20) Determine the identity of the daughter nuclide from the electron capture by $^{55}_{26}Fe$.

A) $^{55}_{27}Co$ B) $^{55}_{25}Mn$ C) $^{56}_{27}Co$ D) $^{54}_{25}Mn$ E) $^{51}_{24}Cr$

Answer: B
Diff: 2 *Page Ref: 19.3*

21) Determine the identity of the daughter nuclide from the electron capture by $^{230}_{91}Pa$.

A) $^{229}_{90}Th$ B) $^{234}_{93}Np$ C) $^{226}_{89}Ac$ D) $^{230}_{92}U$ E) $^{230}_{90}Th$

Answer: E
Diff: 2 *Page Ref: 19.3*

22) The following reaction represents what nuclear process?

$$^{241}_{95}Am \rightarrow \, ^{4}_{2}He + \, ^{327}_{93}Np$$

A) beta emission
B) neutron bombardment
C) alpha emission
D) electron capture
E) positron emission

Answer: C
Diff: 2 *Page Ref: 19.3*

23) The following reaction represents what nuclear process?

$$^{137}_{55}\text{Cs} + {}^{0}_{-1}\text{e} \rightarrow {}^{137}_{54}\text{Xe}$$

A) beta emission
B) positron emission
C) gamma emission
D) electron capture
E) alpha capture

Answer: D
Diff: 2 *Page Ref: 19.3*

24) The following reaction represents what nuclear process?

$$^{214}_{82}\text{Pb} \rightarrow {}^{0}_{-1}\text{e} + {}^{214}_{83}\text{Bi}$$

A) alpha emission
B) gamma emission
C) electron capture
D) neutron bombardment
E) beta emission

Answer: E
Diff: 2 *Page Ref: 19.3*

25) Identify the missing particle in the following nuclear equation:

$$^{228}_{90}\text{Th} \rightarrow {}^{4}_{2}\text{He} + ?$$

A) $^{232}_{92}\text{U}$ B) $^{232}_{89}\text{Ac}$ C) $^{228}_{89}\text{Ac}$ D) $^{228}_{88}\text{Ra}$ E) $^{224}_{88}\text{Ra}$

Answer: E
Diff: 3 *Page Ref: 19.3*

26) Identify the missing particle in the following nuclear equation:

$$^{214}_{82}\text{Pb} \rightarrow {}^{0}_{-1}\text{e} + ?$$

A) $^{214}_{83}\text{Bi}$ B) $^{214}_{81}\text{Tl}$ C) $^{213}_{82}\text{Pb}$ D) $^{215}_{82}\text{Pb}$ E) $^{215}_{81}\text{Tl}$

Answer: A
Diff: 3 *Page Ref: 19.3*

27) Which of the following statements is TRUE?
A) If N/Z ratio is too high, there are too many protons and the nuclide will undergo positron emission or electron capture.
B) If N/Z ratio lies somewhere below 1, the nuclide is stable.
C) If N/Z ratio is too low, there are too many neutrons and the nuclide will undergo beta decay.
D) The valley of stability is the geographic location where many of the known nuclides were first discovered.
E) None of the above are true.

Answer: E
Diff: 1 *Page Ref: 19.4*

28) Nuclides above the valley of stability can become more stable through which of the following processes?
A) beta emission
B) positron emission
C) gamma emission
D) electron capture
E) neutron bombardment

Answer: A
Diff: 1 *Page Ref: 19.4*

29) Nuclides below the valley of stability can become more stable through which of the following processes?
A) gamma emission
B) beta emission
C) positron emission
D) neutron emission
E) neutron bombardment

Answer: C
Diff: 1 *Page Ref: 19.4*

30) Which of the following nuclides are most likely to decay via beta decay?
A) I–131 B) Ar–40 C) F–18 D) Zr–90 E) Pb–206

Answer: A
Diff: 2 *Page Ref: 19.4*

31) Which of the following nuclides are most likely to decay via beta decay?
A) I–126 B) Al–24 C) N–13 D) Cs–137 E) Na–20

Answer: D
Diff: 2 *Page Ref: 19.4*

32) Which of the following nuclides are most likely to decay via positron emission?
A) Cs–137 B) I–131 C) Al–24 D) K–42 E) N–14

Answer: C
Diff: 2 *Page Ref: 19.4*

33) Which of the following nuclides are most likely to decay via positron emission?
A) Na–26 B) I–121 C) Ca–42 D) S–30 E) Sb–122

Answer: D
Diff: 2 *Page Ref: 19.4*

34) Identify the instrument(s) used to detect radiation.
A) film–badge dosimeter
B) Geiger–Muller counter
C) scintillation counter
D) all of the above
E) none of the above

Answer: D
Diff: 1 *Page Ref: 19.5*

35) Identify the nuclide that has the longest half–life.
A) ${}^{235}_{92}U$ B) ${}^{14}_{6}C$ C) ${}^{220}_{86}Rn$ D) ${}^{219}_{90}Th$ E) ${}^{232}_{90}Th$

Answer: E
Diff: 1 *Page Ref: 19.6*

36) Identify the nuclide that has the shortest half–life.
A) ${}^{235}_{92}U$ B) ${}^{14}_{6}C$ C) ${}^{220}_{86}Rn$ D) ${}^{219}_{90}Th$ E) ${}^{232}_{90}Th$

Answer: C
Diff: 1 *Page Ref: 19.6*

37) Identify the technique used to predict the age of the Shroud of Turin.
A) uranium–238 to lead–206
B) potassium–40 to argon–40
C) carbon–14 to nitrogen–14
D) none of the above
E) all of the above

Answer: C
Diff: 1 *Page Ref: 19.6*

38) Identify the elements used in radiometric dating.
A) carbon–14 to nitrogen–14
B) uranium–238 to lead–206
C) potassium–40 to argon–40
D) none of the above
E) all of the above

Answer: E
Diff: 2 *Page Ref: 19.6*

39) Fluorine–18 undergoes positron emission with a half–life of 1.10×10^2 minutes. If a patient is given a 248 mg dose for a PET scan, how long will it take for the amount of fluorine–18 to drop to 83 mg? (Assume that none of the fluorine is excreted from the body.)

A) 99 minutes
B) 1.74×10^2 minutes
C) 1.32×10^2 minutes
D) 3.00×10^2 minutes
E) 2.11×10^2 minutes

Answer: B
Diff: 3 *Page Ref: 19.6*

40) The age of an ancient tree trunk is estimated using radiocarbon dating. If the trunk has a C–14 decay rate that is 34% of what it is in living plants, how old is the trunk? The half–life of C–14 is 5730 years.

A) 2.92×10^4 years
B) 1.94×10^4 years
C) 8.92×10^3 years
D) 5.31×10^3 years
E) 1.74×10^2 years

Answer: C
Diff: 3 *Page Ref: 19.6*

41) An archaeologist graduate student found a leg bone of a large animal during the building of a new science building. The bone had a carbon–14 decay rate of 14.8 disintegrations per minute per gram of carbon. Living organisms have a decay rate of 15.3 disintegrations per minute. How old is the bone?

A) 53.3 years
B) 275 years
C) 111 years
D) 25 years
E) 83 years

Answer: C
Diff: 3 *Page Ref: 19.6*

42) The nuclide As–76 has a half–life of 26.0 hours. If a sample of As–76 weighs 344 g, what mass of As–76 remains after 538 minutes?

A) 67.8 g B) 271 g C) 144 g D) 437 g E) 251 g

Answer: B
Diff: 3 *Page Ref: 19.6*

43) A geological sample is found to have a Pb–206/U–238 mass ratio of 0.337/1.00. Assuming there was no Pb–206 present when the sample was formed, how old is it? The half–life of U–238 is 4.5×10^9 years.

A) 7.3×10^{11} years
B) 1.4×10^{10} years
C) 2.4×10^{10} years
D) 2.1×10^9 years
E) 7.1×10^9 years

Answer: D
Diff: 4 Page Ref: 19.6

44) The splitting of a heavy nucleus to form two or more lighter ones is called ________.

A) radioactive cleavage
B) nuclear fission
C) nuclear fusion
D) radioactive merge
E) half life

Answer: B
Diff: 1 Page Ref: 19.7

45) The splitting of the uranium atom is an example of ________.

A) radioactive cleavage
B) nuclear fission
C) nuclear fusion
D) radioactive merge
E) half life

Answer: B
Diff: 1 Page Ref: 19.7

46) Give the goal of the Manhattan project.

A) to build an hydrogen bomb
B) to build the first particle accelerator at Cornell
C) to build an atomic bomb
D) to build the first nuclear reactor to generate electricity for Manhattan
E) to set up the electrification of Manhattan using DC current

Answer: C
Diff: 1 Page Ref: 19.7

47) The following reaction represents what nuclear process?

$$^{235}_{92}U + ^{1}_{0}n \rightarrow ^{139}_{56}Ba + ^{94}_{36}Kr + 3\ ^{1}_{0}n$$

A) nuclear fission
B) nuclear fusion
C) electron capture
D) alpha decay
E) beta emission

Answer: A
Diff: 2 *Page Ref: 19.7*

48) Write a nuclear equation to describe the spontaneous fission of $^{244}_{95}Am$ to form I–134 and Mo–107. Determine how many neutrons are produced in the reaction.

A) 0 B) 1 C) 2 D) 3 E) 4

Answer: D
Diff: 2 *Page Ref: 19.7*

49) Write a nuclear equation to describe the neutron induced fission of U–235 to form Xe–134 and Sr–100. Determine how many neutrons are produced in the reaction.

A) 4 B) 3 C) 1 D) 0 E) 2

Answer: E
Diff: 2 *Page Ref: 19.7*

50) Write a nuclear equation to describe the neutron induced fission of $^{239}_{94}Pu$ to form $^{89}_{36}Kr$ and $^{149}_{58}Ce$. Determine how many neutrons are produced in the reaction.

A) 2 B) 0 C) 3 D) 1 E) 4

Answer: A
Diff: 2 *Page Ref: 19.7*

51) Define mass defect.
A) the difference in mass between an atom and the sum of its separate components
B) an atom with too many neutrons
C) the difference in mass between a radioactive atom and a nonradioactive atom
D) energy released in a radioactive reaction
E) energy absorbed in a radioactive reaction

Answer: A
Diff: 1 *Page Ref: 19.8*

52) Calculate the mass defect in Fe–56 if the mass of an Fe–56 nucleus is 55.921 amu. The mass of a proton is 1.00728 amu and the mass of a neutron is 1.008665 amu.

A) 0.528 amu
B) 3.507 amu
C) 0.564 amu
D) 1.056 amu
E) 0.079 amu

Answer: A
Diff: 2 *Page Ref: 19.8*

53) Calculate the mass defect in Mo–96 if the mass of a Mo–96 nucleus is 95.962 amu. The mass of a proton is 1.00728 amu and the mass of a neutron is 1.008665 amu.

A) 0.197 amu
B) 0.795 amu
C) 0.212 amu
D) 0.812 amu
E) 0.188 amu

Answer: D
Diff: 2 *Page Ref: 19.8*

54) Calculate the mass defect in Ni–59 if the mass of a Ni–59 nucleus is 58.69344 amu. The mass of a proton is 1.00728 amu and the mass of a neutron is 1.008665 amu.

A) 0.232 amu
B) 0.779 amu
C) 0.230 amu
D) 0.775 amu
E) 0.221 amu

Answer: B
Diff: 2 *Page Ref: 19.8*

55) Determine the binding energy of an O–16 nucleus. The O–16 nucleus has a mass of 15.9905 amu. A proton has a mass of 1.00728 amu, a neutron has a mass of 1.008665 amu, and 1 amu is equivalent to 931 MeV of energy.

A) 8.84 MeV B) 128 MeV C) 138 MeV D) 78.1 MeV E) 38.2 MeV

Answer: B
Diff: 3 *Page Ref: 19.8*

56) Determine the binding energy of an F–19 nucleus. The F–19 nucleus has a mass of 18.99840325 amu. A proton has a mass of 1.00728 amu, a neutron has a mass of 1.008665 amu, and 1 amu is equivalent to 931 MeV of energy.

A) 142 MeV
B) 796 MeV
C) 1080 MeV
D) 143 MeV
E) 145 MeV

Answer: D
Diff: 3 *Page Ref: 19.8*

57) Determine the binding energy per nucleon of an Mg–24 nucleus. The Mg–24 nucleus has a mass of 24.30506. A proton has a mass of 1.00728 amu, a neutron has a mass of 1.008665 amu, and 1 amu is equivalent to 931 MeV of energy.
A) 0.3050 MeV
B) 8.83 MeV
C) 0.113 MeV
D) 106 MeV
E) 4.41 MeV

Answer: D
Diff: 3 Page Ref: 19.8

58) The combination of two light nuclei to form a heavier nuclei is called ________.
A) radioactive cleavage
B) nuclear fission
C) nuclear fusion
D) radioactive merge
E) half life

Answer: C
Diff: 1 Page Ref: 19.9

59) The following reaction represents what nuclear process?

$$^{2}_{1}H + ^{2}_{1}H \rightarrow ^{3}_{2}He + ^{1}_{0}n$$

A) nuclear fusion
B) alpha emission
C) beta emission
D) nuclear fission
E) neutron capture

Answer: A
Diff: 1 Page Ref: 19.9

60) Define transmutation.
A) the transformation of one element into another
B) the loss of neutrons from an atom
C) the loss of electrons from an atom
D) the loss of protons from an atom
E) the gain of neutrons to an atom

Answer: A
Diff: 1 Page Ref: 19.10

61) Identify the lowest natural radiation.
A) cosmic radiation from outer space
B) terrestrial radiation
C) natural radionuclides in the body
D) a five–hour jet airplane ride
E) radon gas

Answer: D
Diff: 1 Page Ref: 19.11

62) Identify the highest natural radiation.
A) cosmic radiation from outer space
B) terrestrial radiation
C) natural radionuclides in the body
D) a five-hour jet airplane ride
E) radon gas

Answer: E
Diff: 1 *Page Ref: 19.11*

63) Identify the part of the body that can be studied with radiotracers.
A) spleen
B) brain
C) heart
D) tumors
E) all of the above

Answer: E
Diff: 1 *Page Ref: 19.12*

64) Identify the common radiotracers used in the diagnosis of medical problems
A) fluorine-18
B) iodine-131
C) thallium-201
D) iron-59
E) all of the above

Answer: E
Diff: 1 *Page Ref: 19.12*

Algorithmic Questions

1) Identify the radioactive green light that glows in the dark.
A) phenol red
B) radioactivity
C) phosphorescence
D) desensitivity
E) argon

Answer: C
Diff: 1 *Page Ref: 19.2*

2) Which particle has the highest penetrating power?
A) alpha particle
B) proton particle
C) gamma particle
D) positron capture
E) electron emission

Answer: C
Diff: 1 *Page Ref: 19.3*

3) Which particle has the lowest penetrating power?
   A) alpha particle
   B) neutron particle
   C) gamma particle
   D) positron capture
   E) electron emission

Answer: A
Diff: 1 *Page Ref: 19.3*

4) Describe what changes occur during alpha decay.
   A) The mass number and atomic number decreases.
   B) The mass number and atomic number increases.
   C) The mass number increases and the atomic number decreases.
   D) The mass number is unchanged and the atomic number increases.
   E) The mass number and atomic number do not change.

Answer: A
Diff: 1 *Page Ref: 19.3*

5) Describe what changes occur during beta decay.
   A) The mass number and atomic number decreases.
   B) The mass number and atomic number increases.
   C) The mass number increases and the atomic number is unchanged.
   D) The mass number is unchanged and the atomic number increases.
   E) The mass number and atomic number do not change.

Answer: D
Diff: 1 *Page Ref: 19.3*

6) Describe what changes occur during gamma ray emission.
   A) The mass number and atomic number decreases.
   B) The mass number and atomic number increases.
   C) The mass number is unchanged and the atomic number decreases.
   D) The mass number increases and the atomic number decreases.
   E) The mass number and atomic number do not change.

Answer: E
Diff: 1 *Page Ref: 19.3*

7) Describe what changes occur during positron emission.
   A) The mass number and atomic number decreases.
   B) The mass number and atomic number increases.
   C) The mass number is unchanged and the atomic number decreases.
   D) The mass number decreases and the atomic number is unchanged.
   E) The mass number and atomic number do not change.

Answer: C
Diff: 1 *Page Ref: 19.3*

8) Describe what changes occur during electron capture.
A) The mass number and atomic number decreases.
B) The mass number and atomic number increases.
C) The mass number is unchanged and the atomic number decreases.
D) The mass number decreases and the atomic number is unchanged.
E) The mass number and atomic number do not change.

Answer: C
Diff: 1 *Page Ref: 19.3*

9) Determine the identity of the daughter nuclide from the alpha decay of ${}^{210}_{84}Po$.

A) ${}^{206}_{86}Rn$ B) ${}^{206}_{82}Pb$ C) ${}^{212}_{88}Ra$ D) ${}^{208}_{80}Hg$ E) ${}^{211}_{85}At$

Answer: B
Diff: 2 *Page Ref: 19.3*

10) Determine the identity of the daughter nuclide from the alpha decay of ${}^{228}_{90}Th$.

A) ${}^{232}_{92}U$ B) ${}^{229}_{91}Pa$ C) ${}^{224}_{88}Ra$ D) ${}^{232}_{89}Ac$ E) ${}^{220}_{90}Th$

Answer: C
Diff: 2 *Page Ref: 19.3*

11) Determine the identity of the daughter nuclide from the beta decay of ${}^{14}_{6}C$.

A) ${}^{14}_{7}N$ B) ${}^{9}_{4}Be$ C) ${}^{17}_{7}N$ D) ${}^{14}_{6}C$ E) ${}^{13}_{5}B$

Answer: A
Diff: 2 *Page Ref: 19.3*

12) Determine the identity of the daughter nuclide from the beta decay of ${}^{32}_{15}P$.

A) ${}^{33}_{15}P$ B) ${}^{32}_{16}S$ C) ${}^{35}_{16}S$ D) ${}^{32}_{14}Si$ E) ${}^{30}_{13}Al$

Answer: B
Diff: 2 *Page Ref: 19.3*

13) Determine the identity of the daughter nuclide from the positron emission of ${}^{18}_{9}F$.

A) ${}^{22}_{11}Na$ B) ${}^{20}_{9}F$ C) ${}^{16}_{7}N$ D) ${}^{18}_{8}O$ E) ${}^{18}_{10}Ne$

Answer: D
Diff: 2 *Page Ref: 19.3*

14) Determine the identity of the daughter nuclide from the positron emission of $^{68}_{32}Ge$.

A) $^{67}_{31}Ga$ B) $^{69}_{33}As$ C) $^{68}_{30}Zn$ D) $^{68}_{33}As$ E) $^{68}_{31}Ga$

Answer: E
Diff: 2 *Page Ref: 19.3*

15) Determine the identity of the daughter nuclide from the electron capture by $^{36}_{17}Cl$.

A) $^{37}_{18}Ar$ B) $^{40}_{19}K$ C) $^{36}_{16}S$ D) $^{36}_{15}P$ E) $^{32}_{18}Ar$

Answer: C
Diff: 2 *Page Ref: 19.3*

16) Determine the identity of the daughter nuclide from the electron capture by $^{7}_{4}Be$.

A) $^{11}_{6}C$ B) $^{3}_{2}He$ C) $^{6}_{5}B$ D) $^{7}_{3}Li$ E) $^{8}_{5}B$

Answer: D
Diff: 2 *Page Ref: 19.3*

17) Identify the missing particle in the following nuclear equation:

$$^{3}_{1}H + ^{2}_{1}H \rightarrow ^{4}_{2}He + ? + ^{0}_{0}\gamma$$

A) $^{0}_{-1}n$ B) $^{1}_{0}n$ C) $^{0}_{+1}H$ D) $^{1}_{1}n$ E) $^{0}_{0}\gamma$

Answer: B
Diff: 3 *Page Ref: 19.3*

18) Identify the missing particle in the following nuclear equation:

$$^{235}_{92}U \rightarrow ^{90}_{38}Sr + ? + 2\,^{1}_{0}n + 4\,^{0}_{0}\gamma$$

A) $^{142}_{52}Te$ B) $^{145}_{54}Xe$ C) $^{143}_{54}Xe$ D) $^{144}_{52}Te$ E) $^{92}_{38}Sr$

Answer: C
Diff: 3 *Page Ref: 19.3*

19) Identify the missing particle in the following nuclear equation:

$$^{238}_{92}U \rightarrow ? + ^{4}_{2}He + 2\,^{0}_{0}\gamma$$

A) $^{238}_{90}Th$ B) $^{232}_{88}Ra$ C) $^{236}_{94}Pu$ D) $^{234}_{90}Th$ E) $^{242}_{88}Ra$

Answer: D
Diff: 3 *Page Ref: 19.3*

20) Beta decay of ^{60}Co produces a beta particle and

A) ^{56}Mn. B) ^{59}Co. C) ^{60}Fe. D) ^{60}Ni.

Answer: D
Diff: 1 *Page Ref: 19.3*

21) In addition to a beta particle, what is the other product of beta decay of $^{214}_{82}Pb$?

A) $^{200}_{80}Hg$ B) $^{214}_{81}Tl$ C) $^{214}_{83}Bi$ D) $^{218}_{84}Po$

Answer: C
Diff: 2 *Page Ref: 19.3*

22) The number of nucleons in a $^{60}_{27}Co$ nucleus is

A) 27. B) 33. C) 60. D) 87.

Answer: C
Diff: 2 *Page Ref: 19.3*

23) Which of the following elements would be expected to be particularly stable?

A) $^{12}_{6}C$ B) $^{10}_{5}B$ C) $^{11}_{4}Be$ D) $^{9}_{3}Li$

Answer: A
Diff: 1 *Page Ref: 19.4*

24) Which of the following nuclides is most likely to undergo beta decay?

A) $^{188}_{78}Pt$ B) $^{192}_{78}Pt$ C) $^{196}_{78}Pt$ D) $^{200}_{78}Pt$

Answer: D
Diff: 2 *Page Ref: 19.4*

25) Which nuclide below is most likely to decay by electron capture?

A) $^{183}_{75}Re$ B) $^{185}_{75}Re$ C) $^{187}_{75}Re$ D) $^{189}_{75}Re$

Answer: A
Diff: 2 *Page Ref: 19.4*

26) Above what atomic number are there no stable isotopes of any element?
A) 20 B) 92 C) 83 D) 70 E) 89

Answer: C
Diff: 1 *Page Ref: 19.4*

27) Stable isotopes, with low atomic numbers, have a N/Z ratio of 1. What does that imply?
A) The number of neutrons equals the number of protons.
B) The number of neutrons equals the number of electrons plus protons.
C) The number of protons equals the number of electrons plus neutrons.
D) The atomic number equals the atomic mass.
E) The number of protons equals the number of electrons plus neutrons.

Answer: A
Diff: 1 *Page Ref: 19.4*

28) Atoms with Z > ________ are radioactive and decay in one or more steps involving mostly alpha and beta decay.
A) 60 B) 110 C) 83 D) 160 E) 50

Answer: C
Diff: 1 *Page Ref: 19.4*

29) Neptunium-239 has a half-life of 2.35 days. How many days must elapse for a sample of ^{239}Np to decay to 0.100% of its original quantity?
A) 0.0427 days B) 1.491 days C) 2.04 days D) 23.4 days

Answer: D
Diff: 3 *Page Ref: 19.6*

30) What percentage of a radioactive substance remains after 7.00 half-lives have elapsed?
A) 0.391% B) 0.78% C) 1.56% D) 3.12%

Answer: B
Diff: 3 *Page Ref: 19.6*

31) The half-life of cobalt-60 is 5.20 yr. How many milligrams of a 2.000-mg sample remains after 6.55 years?
A) 0.837
B) 3.23×10^{-15}
C) 4.779
D) 1.588
E) 1.163

Answer: A
Diff: 3 *Page Ref: 19.6*

32) Strontium-90 is a byproduct in nuclear reactors fueled by the radioisotope uranium-235. The half-life of strontium-90 is 28.8 yr. What percentage of a strontium-90 sample remains after 75.0 yr?
A) 68.1 B) 16.5 C) 7.40 D) 38.4 E) 2.60

Answer: B
Diff: 3 *Page Ref: 19.6*

33) Determine the half-life of a nuclide that loses 38.0% of its mass in 407 hours.
A) 204 hours B) 568 hour C) 586 hours D) 281 hours E) 291 hours

Answer: C
Diff: 4 *Page Ref: 19.6*

34) Carbon-11 is used in medical imaging. The half-life of this radioisotope is 20.4 min. What percentage of a sample remains after 60.0 min?
A) 71.2 B) 5.28 C) 13.0 D) 34.0 E) 2.94

Answer: C
Diff: 3 *Page Ref: 19.6*

35) A rock contains 0.313 mg of lead-206 for each milligram of uranium-238. The half-life for the decay of uranium-238 to lead-206 is 4.5×10^9 yr. The rock was formed ________ yr ago.
A) 1.41×10^9
B) 1.08×10^9
C) 1.39×10^9
D) 2.00×10^9
E) 1.56×10^9

Answer: D
Diff: 4 *Page Ref: 19.6*

36) Potassium-40 decays to argon-40 with a half-life of 1.27×10^9 yr. The age of a mineral sample that has a mass ratio of ^{40}Ar to ^{40}K of 0.812 is ________ yr.
A) 1.56×10^9
B) 1.02×10^9
C) 1.47×10^9
D) 7.55×10^8
E) 1.09×10^9

Answer: E
Diff: 4 *Page Ref: 19.6*

37) If we start with 1.000 g of strontium-90, 0.908 g will remain after 4.00 yr. This means that the half-life of strontium-90 is ________ yr.
A) 3.05 B) 4.40 C) 28.8 D) 3.63 E) 41.6

Answer: C
Diff: 4 *Page Ref: 19.6*

38) If we start with 1.000 g of cobalt-60, 0.675 g will remain after 3.00 yr. This means that the half-life of cobalt-60 is ________ yr.
A) 3.08 B) 4.44 C) 2.03 D) 5.30 E) 7.65

Answer: D
Diff: 4 *Page Ref: 19.6*

39) A freshly prepared sample of curium-243 undergoes 3312 disintegrations per second. After 6.00 yr, the activity of the sample declines to 2755 disintegrations per second. The half-life of curium-243 is ________ yr.

A) 4.99 B) 32.6 C) 7.21 D) 0.765 E) 22.6

Answer: E
Diff: 4 *Page Ref: 19.6*

40) The radioactive decay of ________ is the single greatest source of human exposure to radiation.

A) radon B) uranium C) ozone D) hydrogen E) thorium

Answer: A
Diff: 1 *Page Ref: 19.6*

41) Give the maximum age that can be estimated from radiocarbon dating.

A) 100,000 years.
B) 1,000,000 years
C) 50,000 years.
D) 5,000 years.
E) 30,000 years

Answer: C
Diff: 2 *Page Ref: 19.6*

42) Complete the following equation of nuclear fusion.

$$^{2}_{1}H + ^{3}_{1}H \rightarrow ^{4}_{2}He + \_\_\_\_\_$$

A) $^{0}_{+1}H$ B) $^{0}_{0}\gamma$ C) $^{0}_{-1}n$ D) $^{1}_{1}H$ E) $^{1}_{0}n$

Answer: E
Diff: 1 *Page Ref: 19.9*

43) Complete the following equation of transmutation.

$$^{14}_{7}N + ^{4}_{2}He \rightarrow ^{17}_{8}O + \_\_\_\_\_$$

A) $^{0}_{+1}e$ B) $^{0}_{0}\gamma$ C) $^{0}_{-1}n$ D) $^{1}_{1}H$ E) $^{1}_{0}n$

Answer: D
Diff: 1 *Page Ref: 19.10*

44) Complete the following equation of transmutation.

$$^{27}_{13}Al + ^{4}_{2}He \rightarrow ^{30}_{15}P + \_\_\_\_\_$$

A) $^{0}_{+1}H$ B) $^{0}_{0}\gamma$ C) $^{0}_{-1}e$ D) $^{1}_{1}H$ E) $^{1}_{0}n$

Answer: E
Diff: 1 *Page Ref: 19.10*

45) Complete the following equation of nuclear transmutation.

$$^{238}_{92}U + ^{12}_{6}C \rightarrow ^{244}_{98}Cf + 6 \_\_\_\_\_$$

A) $^{0}_{+1}H$ B) $^{0}_{0}\gamma$ C) $^{0}_{-1}e$ D) $^{1}_{1}e$ E) $^{1}_{0}n$

Answer: E
Diff: 1 *Page Ref: 19.10*

46) Identify the following diagnostic procedure that gives the lowest dose of radiation.
A) dental x-ray -- panoramic
B) upper gastrointestinal tract x-ray
C) chest x-ray
D) mammogram
E) x-ray after barium enema

Answer: C
Diff: 1 *Page Ref: 19.11*

47) Identify the following diagnostic procedure that gives the highest dose of radiation.
A) dental x-ray -- two bitewings
B) upper gastrointestinal tract x-ray
C) chest x-ray
D) mammogram
E) thallium heart scan

Answer: E
Diff: 1 *Page Ref: 19.11*

48) Identify the symptom that is not from radiation exposure.
A) measles
B) increased cancer risk
C) death
D) genetic effects
E) weaker immune systems

Answer: A
Diff: 1 *Page Ref: 19.11*

Matching Questions

Match the following.

1) positron
Diff: 1 *Page Ref: 19.3*

2) gamma ray
Diff: 1 *Page Ref: 19.3*

3) alpha particle
Diff: 1 *Page Ref: 19.3*

4) beta particle
Diff: 1 *Page Ref: 19.3*

5) proton
Diff: 1 *Page Ref: 19.3*

6) neutron
Diff: 1 *Page Ref: 19.3*

7) electron
Diff: 1 *Page Ref: 19.3*

A) ${}^{1}_{0}n$

B) ${}^{0}_{+1}e$

C) ${}^{4}_{2}He$

D) ${}^{0}_{0}\gamma$

E) ${}^{0}_{-1}e$

F) ${}^{1}_{1}p$

Answers: 1) B 2) D 3) C 4) E 5) F
6) A 7) E

Short Answer Questions

1) Define radioactivity.

Answer: Radioactivity is the emission of subatomic particles or high-energy electromagnetic radiation by the nuclei of certain atoms.
Diff: 1 *Page Ref: 19.1*

2) Why is an alpha emitter much more harmful if is ingested than when applied to the skin?

Answer: Alpha emitters have the highest ionizing power which make them damaging, but are so large that clothing stops them from attacking cells. If the alpha emitter is ingested however, it has direct access to cells.
Diff: 1 *Page Ref: 19.3*

3) Describe what is meant by the term "valley of stability"?

Answer: If a plot of number of neutrons vs. number of protons is constructed, there are particularly stable isotopes that cluster together amid all of the other isotopes in the plot. These particularly stable isotopes are in the "valley of stability".
Diff: 1 *Page Ref: 19.4*

4) Explain the concept of "magic numbers".

Answer: They are even numbers of protons or neutrons that imply a special nuclear stability.
Diff: 1 *Page Ref: 19.4*

5) How does a dosimeter measure exposure to radioactivity?

Answer: A dosimeter is usually a piece of photographic film in a lightproof container worn on the clothing of anyone who works near radiation. Radioactive particles that strike the film "expose" it during a certain period of time. When the film is developed, the more exposed the film is, the higher the dose of radiation the person was exposed to.
Diff: 1 *Page Ref: 19.5*

6) Define chain reaction in terms of the fission of uranium nucleus.

Answer: Neutrons, produced by the fission of one uranium nucleus, induce fission in other uranium nuclei.
Diff: 1 *Page Ref: 19.7*

7) List two problems associated with nuclear power.

Answer: overheating and waste disposal
Diff: 1 *Page Ref: 19.7*

8) What is the "mass defect"?

Answer: The mass of a nucleus is less than the sum of the masses of its protons and neutrons. The difference between the calculated sum of the protons and neutrons and the actual mass is called the mass defect. This mass difference is related to the binding energy that holds the protons and neutrons together in the nucleus.
Diff: 1 *Page Ref: 19.8*

Chapter 20 Organic Chemistry

Multiple Choice Questions

1) Give the number of covalent bonds that a carbon atom can form.

A) 1 B) 2 C) 3 D) 4 E) 5

Answer: D
Diff: 1 *Page Ref: 20.2*

2) Identify the formula for an alkene.

A) C_nH_{2n+2}
B) C_nH_{2n-2}
C) C_nH_{2n}
D) C_nH_{2n-4}
E) C_nH_{2n+4}

Answer: C
Diff: 1 *Page Ref: 20.3*

3) Identify the formula for an alkane.

A) C_nH_{2n+2}
B) C_nH_{2n-2}
C) C_nH_{2n}
D) C_nH_{2n-4}
E) C_nH_{2n+4}

Answer: A
Diff: 1 *Page Ref: 20.3*

4) Identify the formula for an alkyne.

A) C_nH_{2n+2}
B) C_nH_{2n-2}
C) C_nH_{2n}
D) C_nH_{2n-4}
E) C_nH_{2n+4}

Answer: B
Diff: 1 *Page Ref: 20.3*

5) Nonsuperimposable mirror images are called ________.

A) structural isomers
B) achiral
C) diastereomers
D) enantiomers
E) racemic mixture

Answer: D
Diff: 1 *Page Ref: 20.3*

6) Molecules with the same formula but different structures are called ________.
   A) structural isomers
   B) achiral
   C) diastereomers
   D) enantiomers
   E) racemic mixture

Answer: A
Diff: 1 *Page Ref: 20.3*

7) An equimolar mixture of two optical isomers is called a ________.
   A) structural isomer
   B) achiral
   C) diastereomer
   D) enantiomer
   E) racemic mixture

Answer: A
Diff: 1 *Page Ref: 20.3*

8) How many of the carbons in the following compound are chiral center(s)?

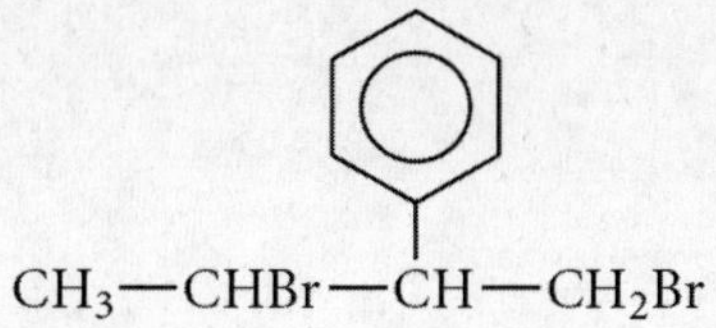

A) 0 B) 1 C) 2 D) 3 E) 4 or more

Answer: C
Diff: 3 *Page Ref: 20.3*

9) How many of the carbons in the following compound are chiral center(s)?

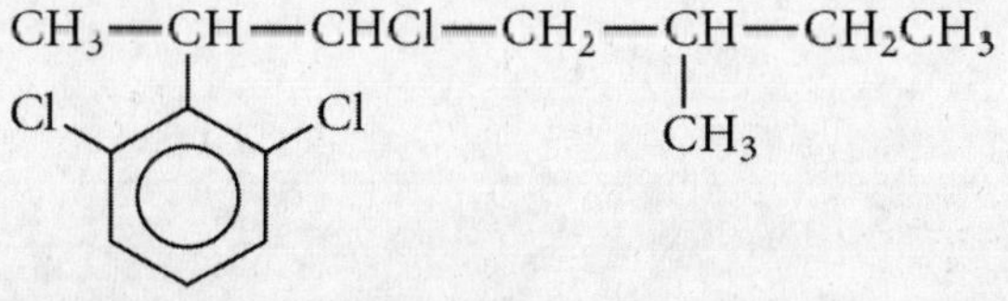

A) 0 B) 1 C) 2 D) 3 E) 4 or more

Answer: D
Diff: 3 *Page Ref: 20.3*

10) Name the following compound.

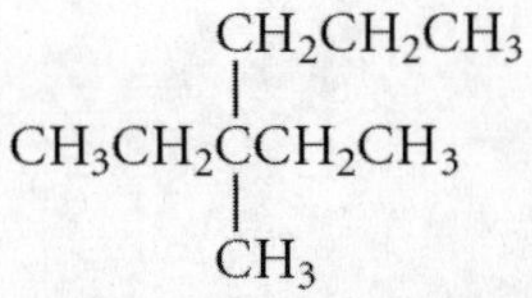

A) 3-ethyl-3-methylhexane
B) 3-methyl-3-propylpentane
C) 3-ethyl-3-propylbutane
D) nonane
E) 2-ethylheptane

Answer: A
Diff: 3 *Page Ref: 20.4*

11) Name the following compound.

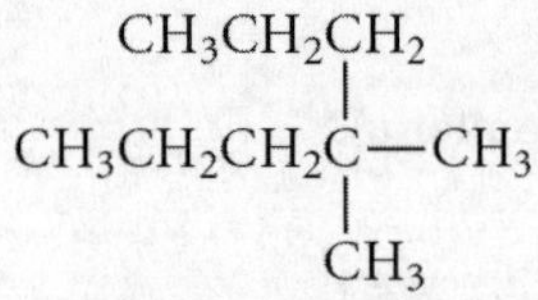

A) 4-methyl-4-propylpentane
B) 4,4-dimethylheptane
C) nonane
D) 4-propyl-4-methylpentane
E) 2-methyl-2-propylpentane

Answer: B
Diff: 3 *Page Ref: 20.4*

12) Name the following compound.

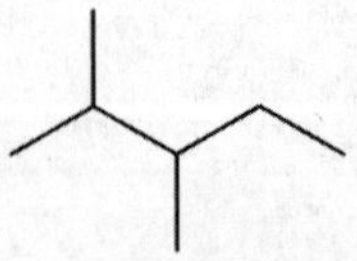

A) 2-methyl-3-methylhexane
B) heptane
C) 2,3-dimethylpentane
D) 3,4-dimethylpentane
E) 2,4-dimethylpropane

Answer: C
Diff: 3 *Page Ref: 20.4*

13) Which of the following names are correct?
   A) 3-butyl-4-pentene
   B) 3-ethyl-1-heptene
   C) 3-etheneheptane
   D) 5-ethyl-6-heptene
   E) All of the above are correct.

Answer: B
Diff: 3 *Page Ref: 20.5*

14) Name the following compound.

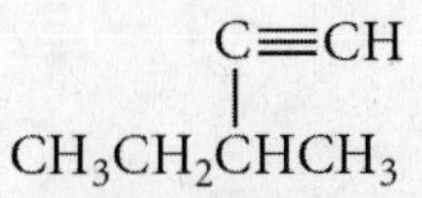

   A) 3-methyl-1-pentyne
   B) 3-methyl-4-pentyne
   C) 2-ethynebutane
   D) 1-hexyne
   E) 3-ethyl-1-butyne

Answer: A
Diff: 3 *Page Ref: 20.5*

15) Name the following compound.

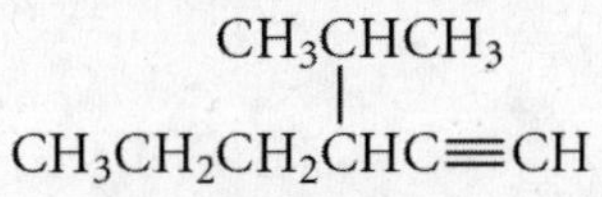

   A) 4-propyl-5-hexyne
   B) 3-isopropyl-1-hexyne
   C) 1-nonyne
   D) 4-methyl-3-propyl-1-pentyne
   E) 2-methyl-4-pentyne

Answer: B
Diff: 3 *Page Ref: 20.5*

16) Name the following compound.

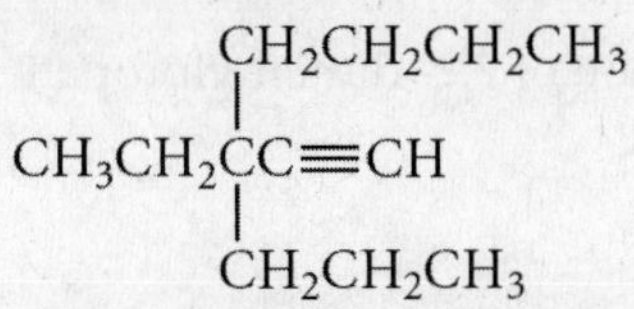

   A) 3-butyl-3-propyl-1-pentyne
   B) 3-butyl-3-propyl-4-pentyne
   C) 3-ethyl-3-propyl-1-heptyne
   D) 5-ethyl-5-propyl-6-heptyne
   E) 3-ethyl-3-butyl-1-hexyne

Answer: C
Diff: 3 *Page Ref: 20.5*

17) Which of the following names are correct?
A) 2-methyl-4-pentyne
B) 3-ethyl-1-butyne
C) 3-butyl-3-propyl-1-pentyne
D) 3-methyl-1-butyne
E) 3-ethyl-3-butyl-1-hexyne

Answer: D
Diff: 3 *Page Ref: 20.5*

18) Which of the following names are correct?
A) 3-methyl-5-heptyne
B) 2-ethyl-4-hexyne
C) 5-ethyl-2-hexyne
D) 4-butyl-1-pentyne
E) 5-methyl-2-heptyne

Answer: E
Diff: 3 *Page Ref: 20.5*

19) Which of the following compounds exhibit geometric isomerism?
A) $CH_2=CH-CH_3$
B) $CCl_2=CBr_2$
C) $CH_3-CH=CH-CH_3$
D) $CCl_2=CHBr$
E) All of the above exhibit geometric isomerism.

Answer: C
Diff: 3 *Page Ref: 20.5*

20) Which of the following compounds exhibit geometric isomerism?
A) $CH_2=CH_2$
B) $CH_2=CCl_2$
C) $CBr_2=CHBr$
D) $CHCl=CHCl$
E) $(CH_3)_2C=CH-CH_3$

Answer: D
Diff: 3 *Page Ref: 20.5*

21) Write a balanced chemical reaction to represent the combustion of 2,2-dimethylpropane.
A) $C_5H_{12} + 8\ O_2 \rightarrow 5\ CO_2 + 6\ H_2O$
B) $C_3H_8 + 5\ O_2 \rightarrow 3\ CO_2 + 4\ H_2O$
C) $C_5H_{12} + H_2 \rightarrow CH_4 + 2\ C_2H_6$
D) $C_3H_8 + H_2 \rightarrow CH_4 + C_2H_6$
E) $2\ C_3H_8 + O_2 \rightarrow 3\ CH_4 + 2\ H_2O$

Answer: A
Diff: 3 *Page Ref: 20.6*

22) Write a balanced chemical equation to represent the reaction of 3-ethyl-2-methylhexane with Br_2.

A) $C_7H_{14} + Br_2 \rightarrow C_7H_{12}Br + HBr$
B) $C_9H_{20} + Br_2 \rightarrow C_9H_{19}Br + HBr$
C) $C_6H_{12} + Br_2 \rightarrow C_6H_{10}Br_2 + H_2$
D) $C_9H_{18} + Br_2 \rightarrow C_9H_{16}Br_2 + H_2$
E) $C_6H_{12} + Br_2 \rightarrow C_6H_{11}Br + HBr$

Answer: B
Diff: 3 *Page Ref: 20.6*

23) Write the balanced chemical equation that represents the addition of Cl_2 to $CH_3CH{=}CH_2$.

A) $CH_3CH{=}CH_2 + Cl_2 \rightarrow CH_3CHClCH_3 + HCl$
B) $CH_3CH{=}CH_2 + Cl_2 \rightarrow CH_3CCl{=}CHCl + H_2$
C) $CH_3CH{=}CH_2 + Cl_2 \rightarrow CH_3CHClCH_2Cl$
D) $CH_3CH{=}CH_2 + 2\ Cl_2 \rightarrow CH_3CHCl_2 + CH_2Cl_2$
E) $CH_3CH{=}CH_2 + 2\ Cl_2 \rightarrow CH_3CCl_2CHCl_2 + H_2$

Answer: C
Diff: 3 *Page Ref: 20.6*

24) Write the balanced chemical equation for the addition of 2 Br_2 to $CH_3C{\equiv}CH$.

A) $CH_3C{\equiv}CH + 2\ Br_2 \rightarrow CH_3CBr_2CHBr_2$
B) $CH_3C{\equiv}CH + 2\ Br_2 \rightarrow CH_3CBr{=}CHBr$
C) $CH_3C{\equiv}CH + 2\ Br_2 \rightarrow CH_3CH_2CHBr_2$
D) $CH_3C{\equiv}CH + 2\ Br_2 \rightarrow CH_3CBr_2CH_3$
E) $CH_3C{\equiv}CH + 2\ Br_2 \rightarrow CH_3CHBrCH_2Br$

Answer: A
Diff: 3 *Page Ref: 20.6*

25) Complete and balance the following addition reaction.

$$CH_3C{\equiv}CCH_3 + 2\ Cl_2 \rightarrow ?$$

A) $CH_3C{\equiv}CCH_3 + 2\ Cl_2 \rightarrow CH_3CCl_2CCl_2CH_3$
B) $CH_3C{\equiv}CCH_3 + 3\ Cl_2 \rightarrow 2\ CH_3CCl_3$
C) $CH_3C{\equiv}CCH_3 + 2\ Cl_2 \rightarrow 4\ CH_3Cl$
D) $CH_3C{\equiv}CCH_3 + 4\ Cl_2 \rightarrow 4\ CH_2Cl_2$
E) $CH_3C{\equiv}CCH_3 + Cl_2 \rightarrow CH_3CHClCHClCH_3$

Answer: A
Diff: 3 *Page Ref: 20.6*

26) Write the balanced chemical equation for the catalytic hydrogenation (addition of H_2) to $CH_3CH{=}CHCH_3$.

A) $CH_3CH{=}CHCH_3 + 2\ H_2 \rightarrow 2\ CH_3CH_3$
B) $CH_3CH{=}CHCH_3 + 3\ H_2 \rightarrow CH_3CH_3 + 2\ CH_4$
C) $CH_3CH{=}CHCH_3 + 2\ H_2 \rightarrow CH_3CH_2CH_3 + CH_4$
D) $CH_3CH{=}CHCH_3 + H_2 \rightarrow CH_3CH_2CH_2CH_3$
E) $CH_3CH{=}CHCH_3 + 4\ H_2 \rightarrow 4\ CH_4$

Answer: D
Diff: 3 *Page Ref: 20.6*

27) Write the balanced chemical equation for the catalytic hydrogenation (addition of H_2) to $CH_3C{\equiv}CH$.

A) $CH_3C{\equiv}CH + H_2 \rightarrow CH_3CH{=}CH_2$
B) $CH_3C{\equiv}CH + 2\ H_2 \rightarrow CH_3CH{=}CH_2$
C) $CH_3C{\equiv}CH + H_2 \rightarrow CH_3CH_2CH_3$
D) $CH_3C{\equiv}CH + 2\ H_2 \rightarrow CH_3CH_2CH_3$
E) $CH_3C{\equiv}CH + H_2 \rightarrow CH_2{=}C{=}CH_2$

Answer: D
Diff: 3 *Page Ref: 20.6*

28) Complete and balance the following complete hydrogenation reaction.

$$CH_3CH_2C{\equiv}CCH_3 + 2\ H_2 \rightarrow ?$$

A) $CH_3CH_2C{\equiv}CCH_3 + 3\ H_2 \rightarrow CH_3CH_2CH_3 + CH_3CH_3$
B) $CH_3CH_2C{\equiv}CCH_3 + 2\ H_2 \rightarrow CH_3CH_2CH_2CH_2CH_3$
C) $CH_3CH_2C{\equiv}CCH_3 + 6\ H_2 \rightarrow 5\ CH_4$
D) $2\ CH_3CH_2C{\equiv}CCH_3 + 2\ H_2 \rightarrow CH_3CH_2CH_2CH_2CH_2CH_2CH_2CH_2CH_3$
E) $CH_3CH_2C{\equiv}CCH_3 + 2\ H_2 \rightarrow CH_3CH_2CH_2{\equiv}CH_2CH_3$

Answer: B
Diff: 3 *Page Ref: 20.6*

29) Write the balanced chemical equation for the addition of HBr to $CH_2{=}CHCH_2CH_3$.

A) $CH_2{=}CHCH_2CH_3 + HBr \rightarrow CH_3BrCH_2CH_2CH_3$
B) $CH_2{=}CHCH_2CH_3 + 2\ HBr \rightarrow 2\ CH_2BrCH_3$
C) $CH_2{=}CHCH_2CH_3 + 2\ HBr \rightarrow CH_3Br + CH_2BrCH_2CH_3$
D) $CH_2{=}CHCH_2CH_3 + 4\ HBr \rightarrow 4\ CH_3Br$
E) $CH_2{=}CHCH_2CH_3 + HBr \rightarrow CH_3CHBrCH_2CH_3$

Answer: E
Diff: 3 *Page Ref: 20.6*

30) Name the following compound.

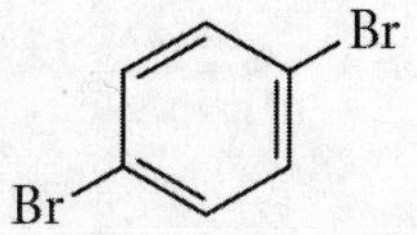

A) 2,5-dibromobenzene
B) 2,5-dibromocyclohexene
C) 1,4-dibromobenzene
D) 1,4-bromocyclohexene
E) 3,6-dibromobenzene

Answer: C
Diff: 3 *Page Ref: 20.7*

31) Name the following compound.

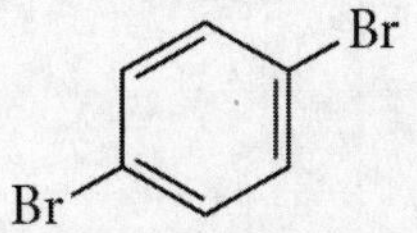

A) o-dibromobenzene
B) m-dibromobenzene
C) p-dibromobenzene
D) p-bromobenzene
E) none of the above

Answer: C
Diff: 3 *Page Ref: 20.7*

32) Name the following compound.

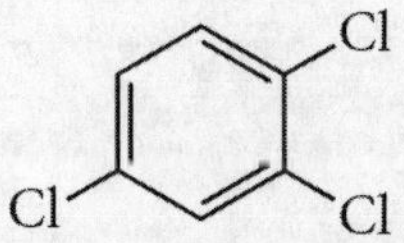

A) 2,3,5-trichlorobenzene
B) 1,3,4-trichlorobenzene
C) trichlorostyrene
D) 1,2,4-trichlorobenzene
E) 1,3,4-trichlorohexene

Answer: D
Diff: 3 *Page Ref: 20.7*

33) Which of the following compounds is an alcohol?
A) $CH_3CH_2CH_2CO_2H$
B) CH_3-O-CH_3
C) $CH_3CO_2CH_3$
D) CH_3CH_2CH=O
E) CH_2OH-CH_2OH

Answer: E
Diff: 2 *Page Ref: 20.8*

34) Which of the following compounds is an aldehyde?
A) $CH_3CH_2CH{=}O$
B) $CH_3CH_2CH_2CO_2H$
C) $CH_3CO_2CH_3$
D) CH_3-O-CH_3
E) CH_2OH-CH_2OH

Answer: A
Diff: 2 *Page Ref:* 20.8

35) Which of the following compounds is an ketone?
A)
$CH_3CH_2CH_2CO_2H$
B)
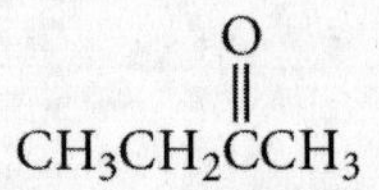
C)
$CH_3CH_2NH_2$
D)
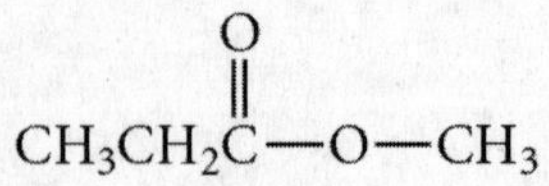
E)
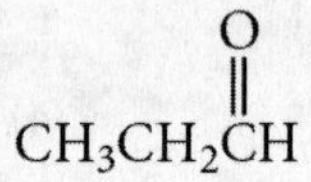

Answer: B
Diff: 2 *Page Ref:* 20.8

36) Which of the following compounds is an carboxylic acid?

A)
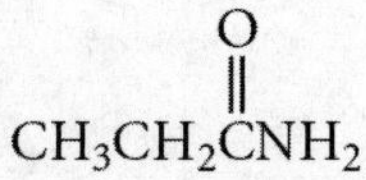

B)
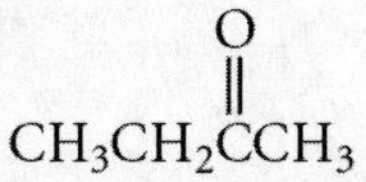

C)
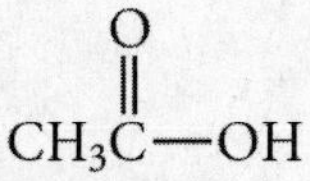

D)
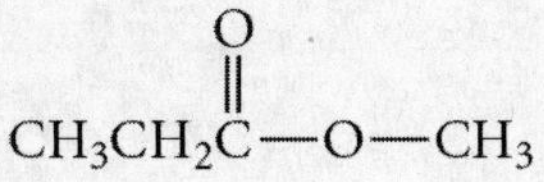

E)
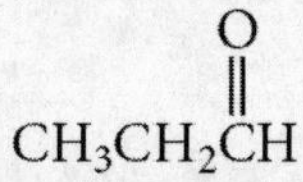

Answer: C
Diff: 2 *Page Ref:* 20.8

37) Which of the following compounds is an ester?

A)
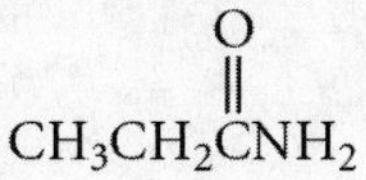

B)
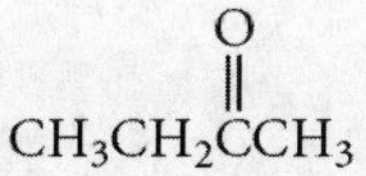

C)
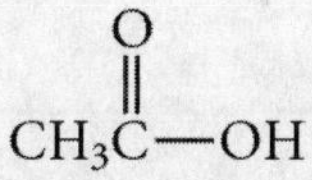

D)
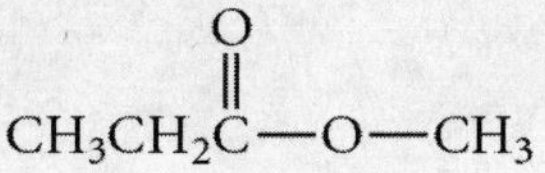

E)
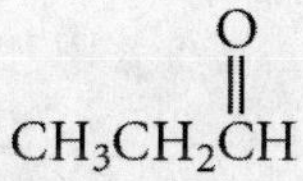

Answer: D
Diff: 2 *Page Ref:* 20.8

38) Which of the following compounds is an ether?

A)

$$CH_3CH_2\overset{O}{\overset{\|}{C}}NH_2$$

B)

$$CH_3CH_2\overset{O}{\overset{\|}{C}}CH_3$$

C)

$$CH_3\overset{O}{\overset{\|}{C}}{-}OH$$

D)

$$CH_3CH_2\overset{O}{\overset{\|}{C}}{-}O{-}CH_3$$

E)

$$CH_3{-}CH_2{-}O{-}CH_3$$

Answer: E

Diff: 2 *Page Ref: 20.8*

39) Which of the following compounds is ethanol?

A) C_2H_6

B) C_2H_5OH

C) CH_3CO_2H

D) $CH_3CO_2CH_3$

E) CH_3-O-CH_3

Answer: B

Diff: 1 *Page Ref: 20.9*

40) Arrange the following in order from most oxidized to least oxidized.

I. CH_3CH_3

II. $CH_3CH_2{-}OH$

III. $CH_3\overset{O}{\overset{\|}{C}}H$

A) II > I > III

B) III > II > I

C) III > I > III

D) I > II > III

E) II > I > III

Answer: B

Diff: 2 *Page Ref: 20.9*

41) Arrange the following in order from least oxidized to most oxidized.

I. $CH_3CH_2\overset{\overset{\displaystyle O}{\|}}{C}H$

II. $CH_3CH_2\overset{\overset{\displaystyle O}{\|}}{C}OCH_3$

III. $CH_3CH_2CH_2{-}OH$

A) II < I < III
B) III < I = II
C) III < I < II
D) I < II < III
E) I < III < II

Answer: C
Diff: 2 *Page Ref: 20.9*

42) Identify the most oxidized compound.
A) CH_3CH_2COOH
B) CH_3CH_2CHO
C) $CH_3CH_2CH_3$
D) $CH_3CH_2OCH_3$
E) $CH_3CH_2CH_2OH$

Answer: A
Diff: 2 *Page Ref: 20.9*

43) Determine the products of the following reaction:

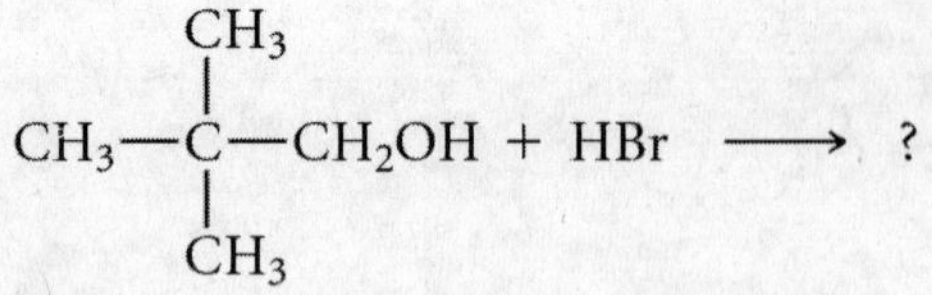

A)

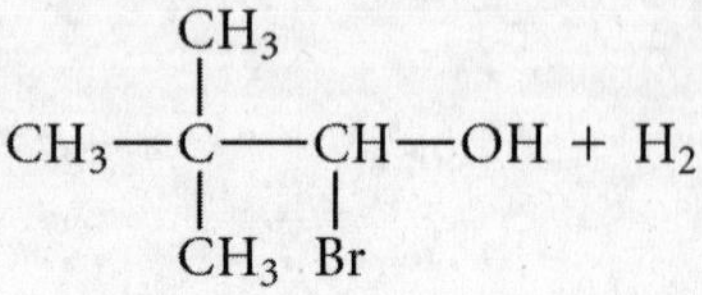

B)

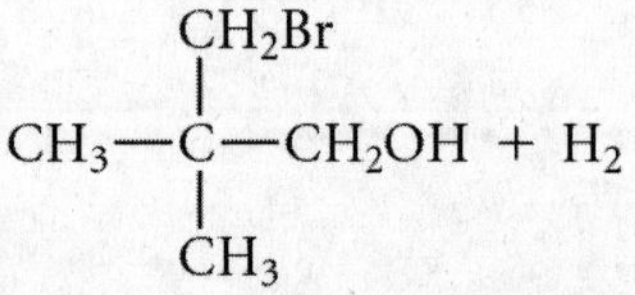

C)

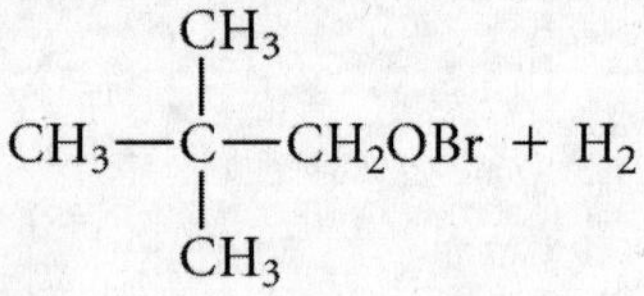

D)

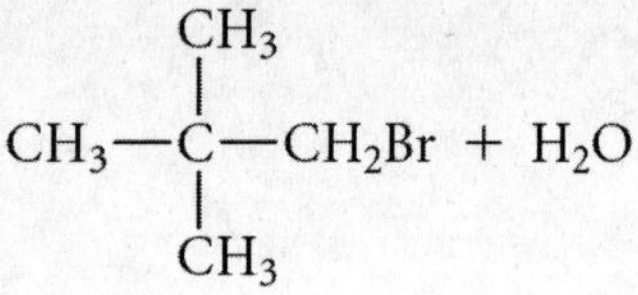

E)

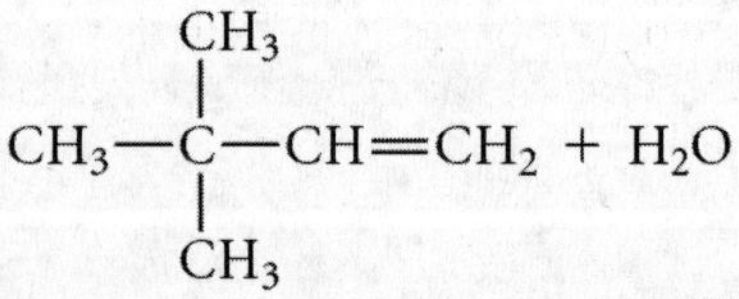

Answer: D
Diff: 3 *Page Ref: 20.9*

44) Give the product for the dehydration of $CH_3CH_2CH_2OH$.
A) $CH_3CH_2CH_2OCH_2CH_2CH_3$
B) $CH_3CH=CH_2$
C) $CH_3CH=CHCH_3$
D) $CH_3CH_2CH=CH_2$
E) $CH_2=C=CH_2$

Answer: B
Diff: 3 *Page Ref: 20.9*

45) Determine the products of the following reaction:

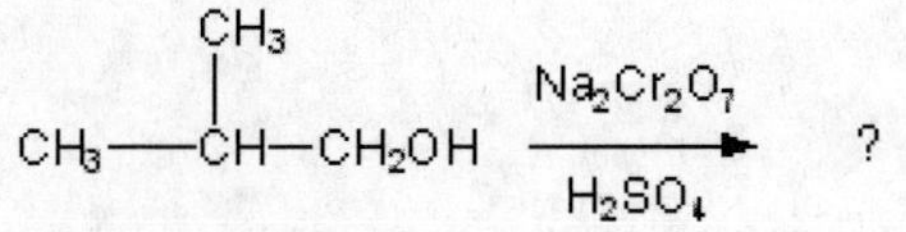

A)

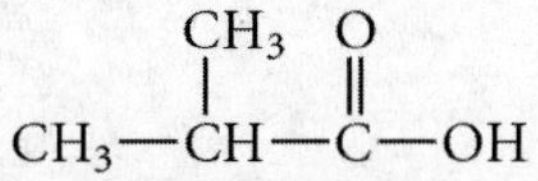

B)

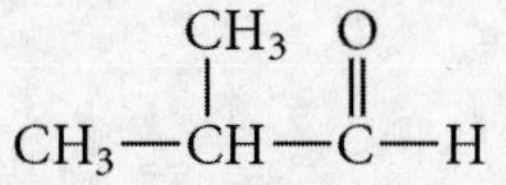

C)

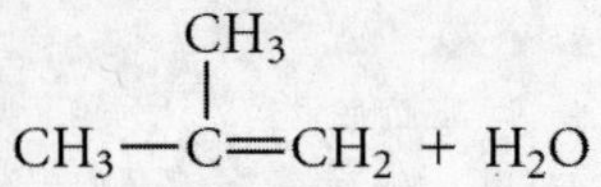

D)

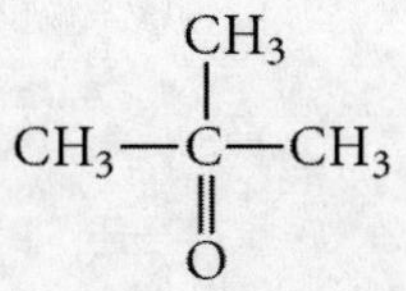

E)

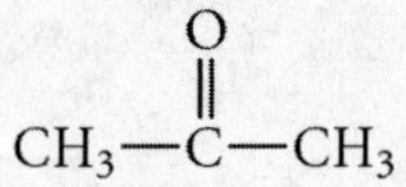

Answer: A

Diff: 3 *Page Ref: 20.9*

46) Determine the products of the following reaction:

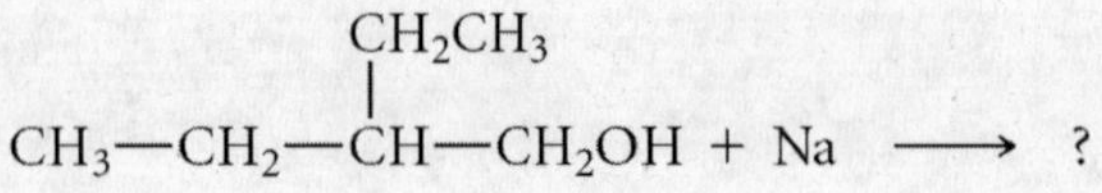

A)

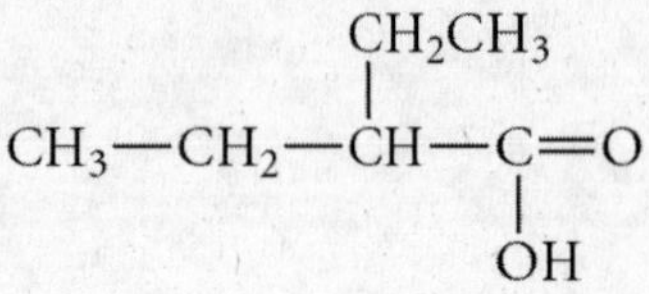

B)

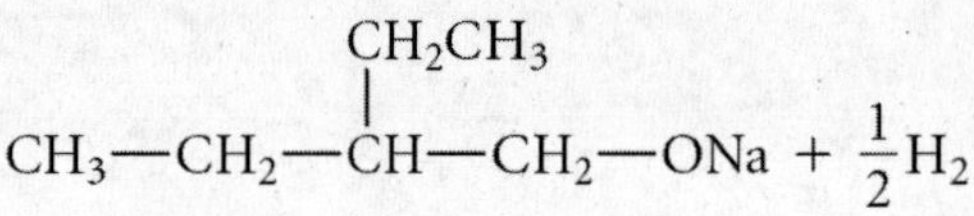

C)

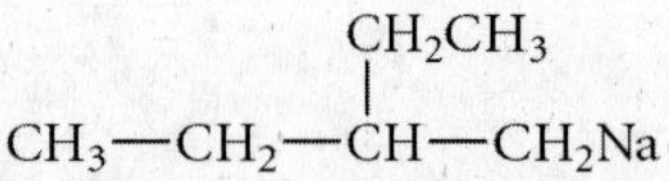

D)

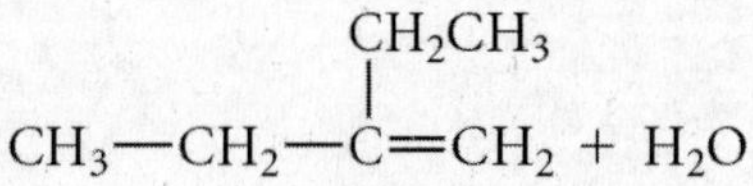

E)

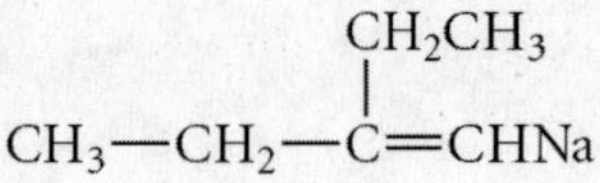

Answer: B
Diff: 3 *Page Ref: 20.9*

47) Determine the products of the following reaction:

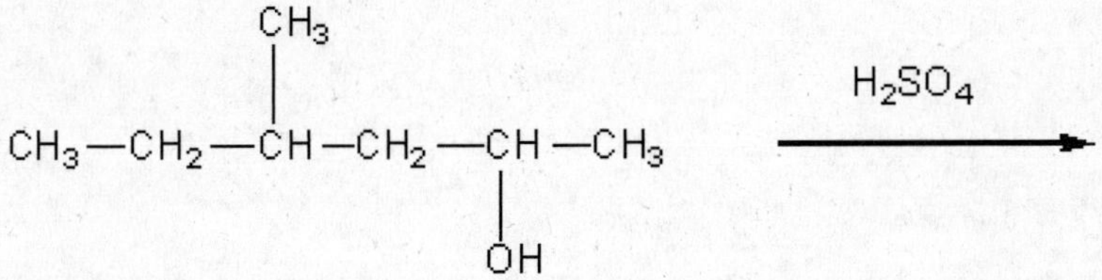

A)

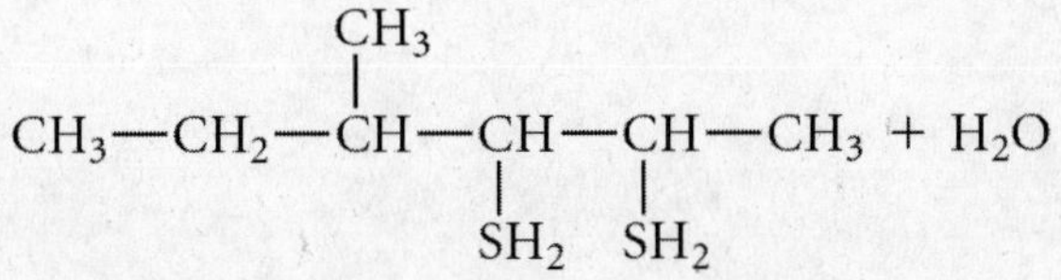

B)

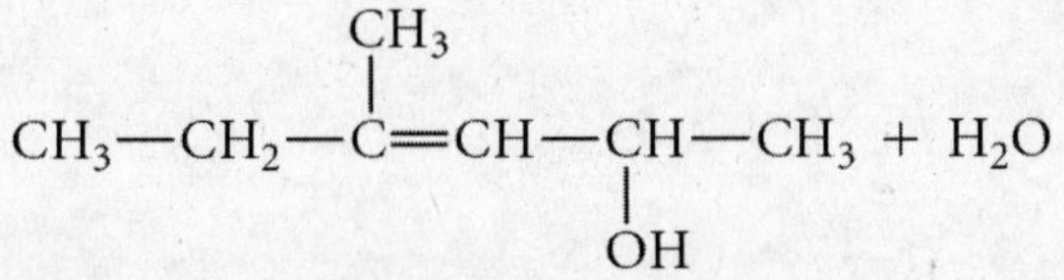

C)

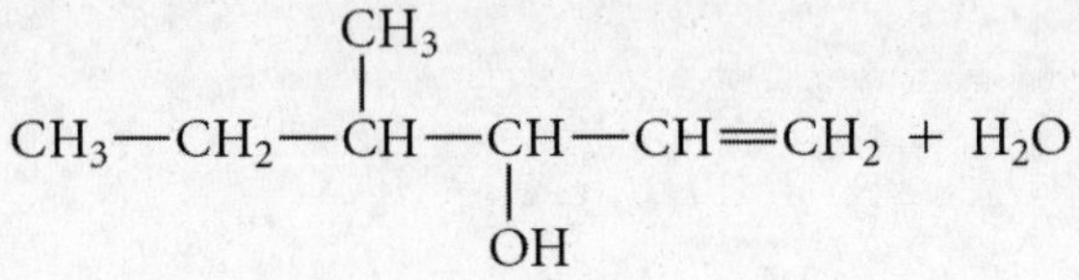

D)

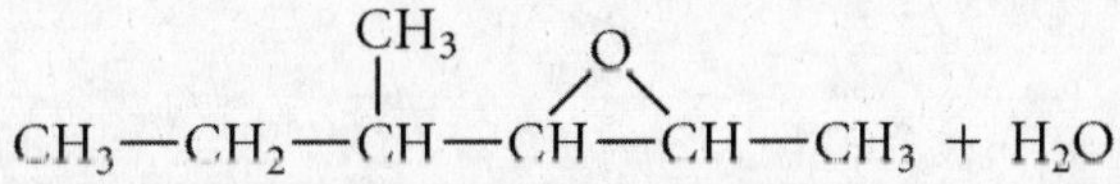

E)

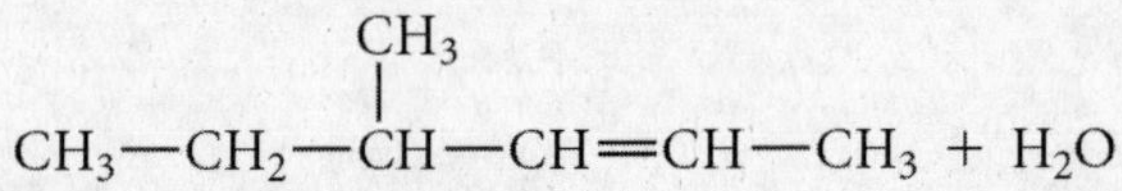

Answer: E
Diff: 3 *Page Ref: 20.9*

48) Give the product of the oxidation of the following compound.

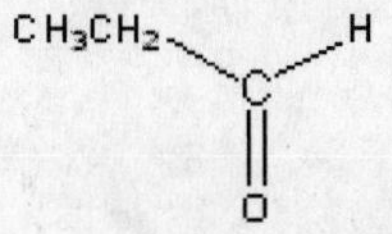

A) CH_3CH_2CHO
B) $CH_3CH_2CH_2OH$
C) $CH_3CH_2CH_3$
D) CH_3CH_2COOH
E) $CH_3CH_2OCH_3$

Answer: D
Diff: 3 *Page Ref: 20.10*

49) Give the product of the reduction of the following compound.

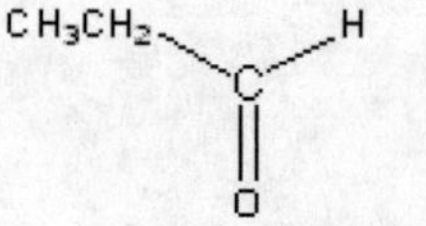

A) CH_3CH_2CHO
B) $CH_3CH_2CH_2OH$
C) $CH_3CH_2CH_3$
D) CH_3CH_2COOH
E) $CH_3CH_2OCH_3$

Answer: B
Diff: 3 *Page Ref: 20.10*

50) Determine the product(s) of the reduction of the following compound:

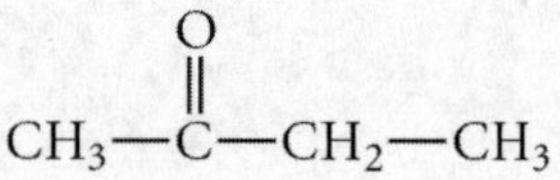

A)
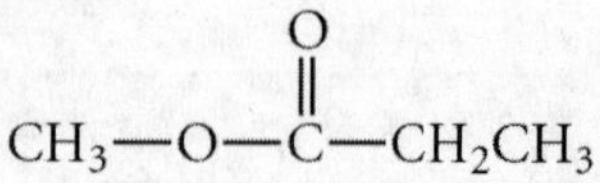

B)
$CH_3-CH=CH-CH_3$

C)
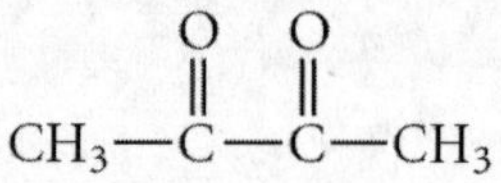

D)
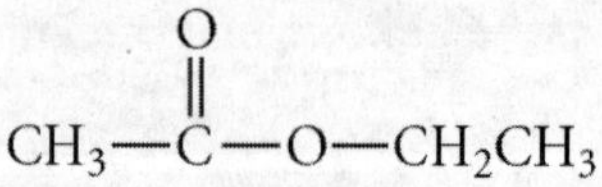

E)
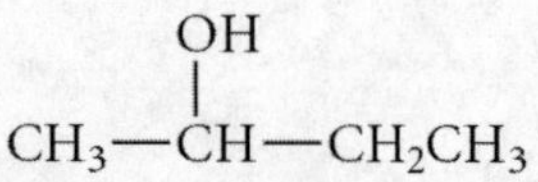

Answer: E
Diff: 3 *Page Ref: 20.10*

51) Determine the product(s) of the following reaction:

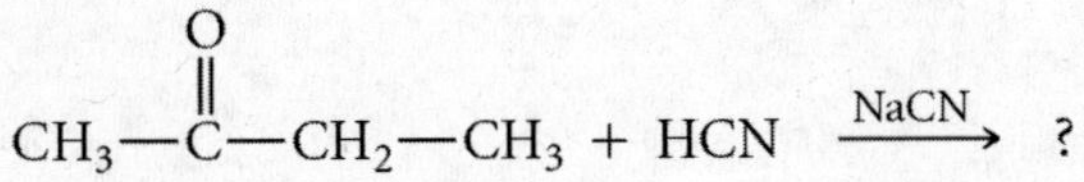

A)

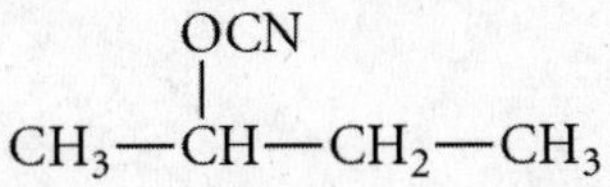

B)

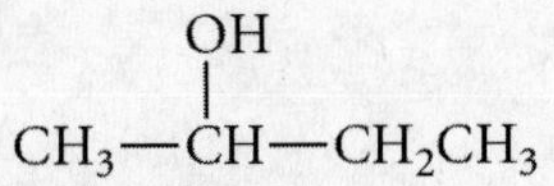

C) $CH_3-CH=CH-CH_3$

D)

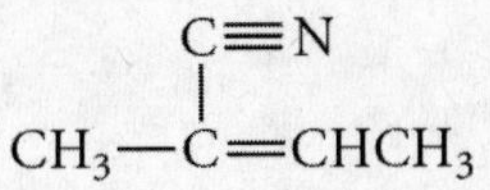

E)

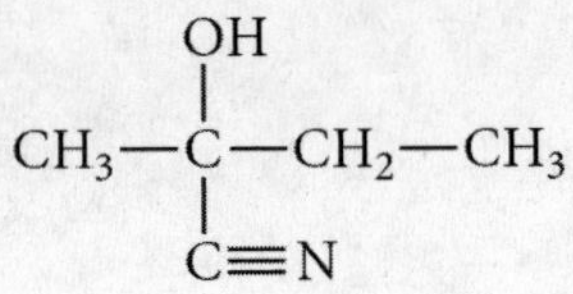

Answer: E

Diff: 3 *Page Ref:* 20.10

52) Name the following compound.

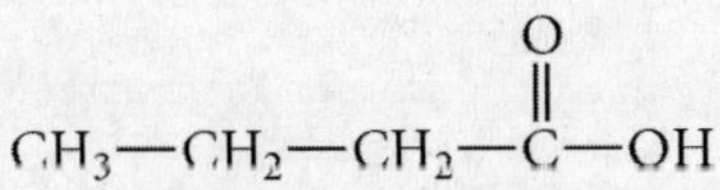

A) propyl butanoate
B) butanoic acid
C) 1-butanal
D) 1-butanoate
E) propyl methanoate

Answer: B

Diff: 3 *Page Ref:* 20.11

53) Give the organic product for the following reaction in the presence of sulfuric acid.

$CH_3CH_2COOH + CH_3CH_2CH_2OH \rightarrow$

A) $CH_3CH_2COOCH_2CH_3$
B) $CH_3CH_2COOCH_2CH_2CH_3$
C) $CH_3CH_2CH_2COOCH_2CH_3$
D) $CH_3CH_2COCH_2CH_2CH_3$
E) $CH_3CH_2OCH_2CH_2CH_3$

Answer: B
Diff: 3 *Page Ref: 20.11*

54) A carboxylic acid reacts an alcohol to form ________.
A) an ether
B) an amine
C) an amide
D) an ester
E) an anhydride

Answer: D
Diff: 3 *Page Ref: 20.11*

55) A carboxylic acid reacts with itself to form ________.
A) an ether
B) an amine
C) an amide
D) an ester
E) an anhydride

Answer: E
Diff: 3 *Page Ref: 20.11*

56) Give the organic product for the following reaction.

$CH_3CH_2CH_2NH_2 + HCl \rightarrow$

A) $ClCH_2CH_2CH_2NH_2$
B) $CH_3CHClCH_2NH_2$
C) $CH_3CH_2CHClNH_2$
D) $CH_3CH_2CH_2NHCl$
E) $CH_3CH_2CH_2NH_3^+ \quad Cl^-$

Answer: E
Diff: 3 *Page Ref: 20.13*

57) Give the organic product for the following reaction.

$CH_3CH_2COOH + CH_3CH_2CH_2NH_2 \rightarrow$

A) $CH_3CH_2NHCH_2CH_2CH_3$
B) $CH_3CH_2CON(CH_2CH_2CH_3)_2$
C) $CH_3CH_2CH_2CONHCH_2CH_3$
D) $CH_3CH_2CONHCH_2CH_2CH_3$
E) $CH_3CH_2N(CH_2CH_2CH_3)_2$

Answer: D
Diff: 3 *Page Ref: 20.13*

58) Which of the following represent the addition polymer formed from the compound below?

$CH_2{=}CH{-}CH_3$

A) $\left(CH{=}C(CH_3) \right)_n$

B) $\left(CH{=}CH \right)_n$

C) $\left(CH_2{-}CH(CH_3) \right)_n$

D) $\left(CH_2{-}CH_2 \right)_n$

E) $\left(CH_2{-}CH_2{-}CH_2 \right)_n$

Answer: C
Diff: 1 *Page Ref: 20.14*

59) Which of the following represent the addition polymer formed from the compound below?

$CH_2{=}CHCl$

A)

$-\!(CH{=}CCl)\!-_n$

B)

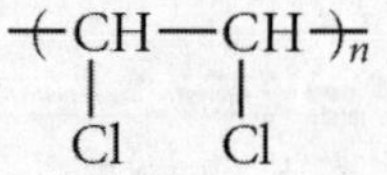

C)

$-\!(CCl{=}C{-}Cl)\!-_n$

D)

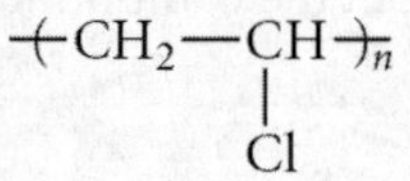

E)

$-\!(CCl_2{-}CCl_2)\!-_n$

Answer: D
Diff: 1 *Page Ref: 20.14*

60) A common use of polyvinyl chloride is ________.
A) plumbing
B) plastic wrap
C) plastic bottles
D) styrofoam
E) nylon

Answer: A
Diff: 1 *Page Ref: 20.14*

61) Identify the copolymer
A) nylon 6,6
B) polyvinyl chloride
C) polyethylene
D) polypropylene
E) polystyrene

Answer: A
Diff: 1 *Page Ref: 20.14*

Algorithmic Questions

1) Which compound is a saturated hydrocarbon?
A) 1-butyne
B) cycloheptene
C) 3-methylheptane
D) 2-heptyne

Answer: C
Diff: 2 *Page Ref: 20.3*

2) How many isomers are there for C_5H_{12} ?

A) 3 B) 4 C) 5 D) 6

Answer: A
Diff: 3 *Page Ref: 20.3*

3) Which of the following compounds exhibit optical isomerism?

A) sec-butylbenzene
B) CH_2Br_2
C) CF_2BrI
D) CHF_2I
E) tert-butylbenzene

Answer: A
Diff: 3 *Page Ref: 20.3*

4) Which of the following compounds exhibits optical isomerism?

A) $CH_3-CH_2-CH_3$
B) $CH_3-CH_2-CHF-CH_3$
C) $CH_3-CHBr-CH_3$
D) $CH_3-CH_2-CH_2F$
E) $CH_3-CH_2-CBr_2-CH_3$

Answer: B
Diff: 3 *Page Ref: 20.3*

5) Which of the following alkane names is correct?

A) 1,2-dimethylhexane
B) 1-propylnonane
C) 4,5-dipropylpentane
D) 2-methyl-3-ethylhexane
E) All of the above names are correct.

Answer: D
Diff: 3 *Page Ref: 20.4*

6) Which of the following alkane names is correct?

A) 2,3-dibutylpropane
B) 3-ethyl-4-methylbutane
C) 4-ethyl-5-butylheptane
D) 3,3-dibutylpentane
E) None of the above names are correct.

Answer: E
Diff: 3 *Page Ref: 20.4*

7) Name the following compound.

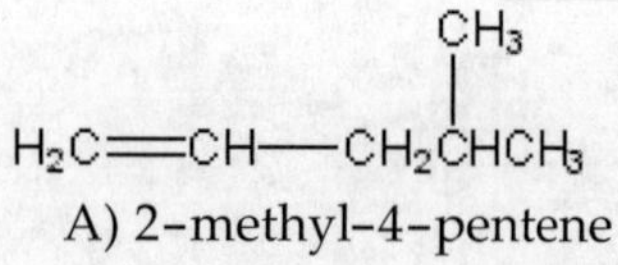

A) 2-methyl-4-pentene
B) 1,1-dimethyl-3-butane
C) 4-methyl-1-pentene
D) hexene
E) 2-methylpentene

Answer: C
Diff: 3 *Page Ref: 20.5*

8) Name the following compound.

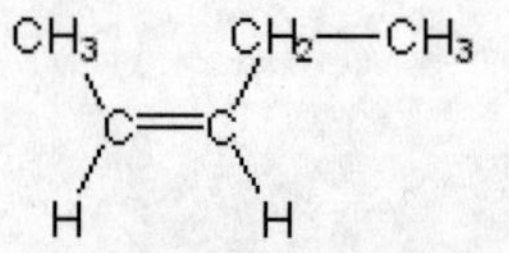

A) trans-2-butane
B) cis-2-butene
C) cis-2-pentene
D) trans-3-pentene
E) 2-pentene

Answer: C
Diff: 3 *Page Ref: 20.5*

9) Name the following compound.

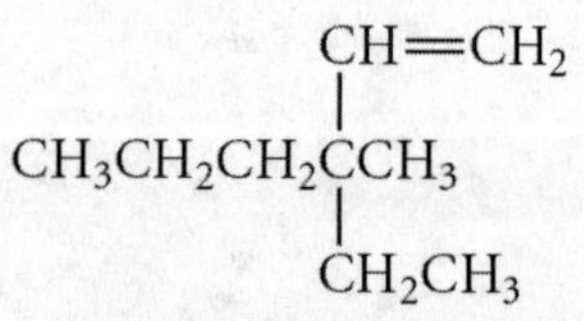

A) 2,2-diethylhexene
B) 2,2-diethyl1-hexene
C) 4-ethyl-4-methyl-5-hexane
D) 3-ethyl-3-methyl-1-hexene
E) 4-ethyl-4-methylhexene

Answer: D
Diff: 3 *Page Ref: 20.5*

10) Name the following compound.

$$\begin{array}{l} \quad CH_2CH_3 \\ \quad | \\ CH_3CHCH_2CH{=}CHCH_3 \end{array}$$

A) 2-ethyl-4-hexane
B) 4-isobutyl-2-butene
C) 3-methyl-5-hexane
D) 4-isopropyl-2-butene
E) 5-methyl-2-heptene

Answer: E
Diff: 3 *Page Ref: 20.5*

11) Which of the following names is correct?
A) 2-methyl-1-butene
B) 3,3-diethyl-4-propene
C) 2-ethyl-3-pentene
D) 1-propyl-2-pentene
E) None of the above are correct.

Answer: A
Diff: 3 *Page Ref: 20.5*

12) Name the following compound.

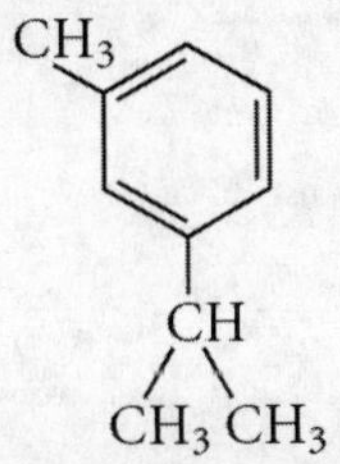

A) 1-methyl-3-isopropylhexene
B) 4-propyl-6-methylhexene
C) 4-isopropyl-6-methylbenzene
D) 1-butylcyclohexene
E) 1-isopropyl-3-methylbenzene

Answer: E
Diff: 3 *Page Ref: 20.7*

13) Name the following compound.

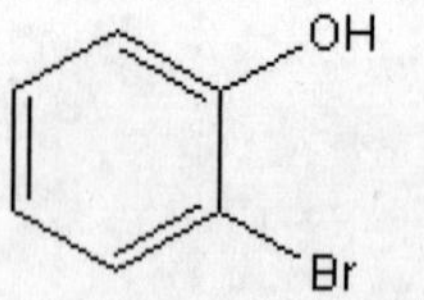

A) 2-bromophenol
B) 2-bromobenzaldehyde
C) 2-bromoanthracene
D) 1-bromotoluene
E) 1-bromoanisole

Answer: A
Diff: 3 *Page Ref: 20.7*

14) Name the following compound.

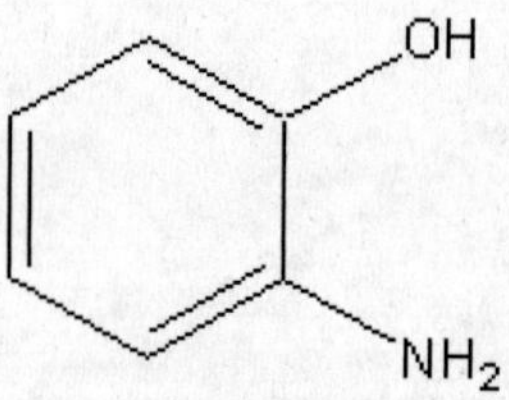

A) 2-aminophenol
B) 2-aminobenzoic acid
C) 2-aminoaniline
D) 2-methylphenol
E) 2-hydroxybenzaldehyde

Answer: A
Diff: 3 *Page Ref: 20.7*

15) Which of the following names is correct?
A) 1,2,4-trifluorobenzene
B) 1-methyl-3-isopropylhexene
C) 1,3,4-fluorobenzene
D) 4-isopropyl-6-methylbenzene
E) 2,5-dibromocyclohexene

Answer: A
Diff: 3 *Page Ref: 20.7*

16) Which of the following names are correct?
A) 4-bromo-2-methylbenzene
B) 1-chloro-3-propylbenzene
C) 3-bromo-4-chlorobenzene
D) 2,2-dibbromobenzene
E) 4-bromo-2-iodobenzene

Answer: B
Diff: 3 *Page Ref: 20.7*

17) Determine the products of the following reaction:

$$\text{Benzene} + Br_2 \xrightarrow{FeBr_3} ?$$

A) 1,2-dibromobenzene and Br_2
B) 1,4-dibromobenzene and HBr
C) ferrobenzene and H_2
D) perbromobenzene and HBr
E) bromobenzene and HBr

Answer: E
Diff: 3 *Page Ref: 20.7*

18) Determine the products of the following reaction:

$$\text{Benzene} + CH_3CH_2Cl \xrightarrow{AlCl_3} ?$$

A) ethylbenzene and HCl
B) 1,2-dimethylbenzene and Cl_2
C) dichlorobenzene and CH_3CH_2Cl
D) trichlorobenzene and CH_3CH_3
E) 1,4-diethylbenzene and Cl_2

Answer: A
Diff: 3 *Page Ref: 20.7*

19) Which one of the following molecules is the most polar?
A) butane B) butanoic acid C) cyclohexane D) ethanol

Answer: B
Diff: 1 *Page Ref: 20.8*

20) Which of the following compounds is an amine?
A) $(CH_3CH_2)_2NH$
B) $CH_3CH_2CH_2CH_2CO_2CH_3$
C) $CH_3CH_2CH_2CH_2Br$
D) CH_3COOH
E) CH_3OCH_3

Answer: A
Diff: 2 *Page Ref: 20.8*

21) Name the following compound.

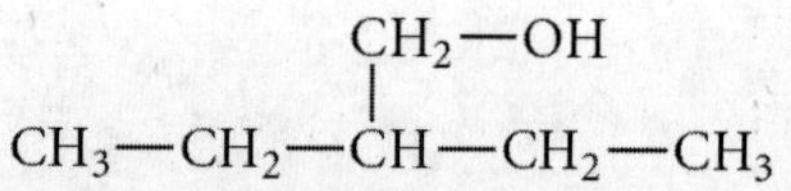

A) 3-pentenol
B) 2-ethyl-1-butanol
C) 3-methyl-3-pentenol
D) 2-propyl-1-propanol
E) hexenal

Answer: B
Diff: 3 *Page Ref: 20.9*

22) Name the following compound.

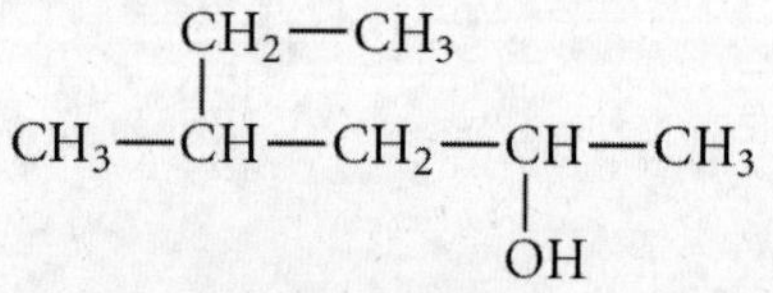

A) 3-methyl-5-hexanol
B) 2-heptenol
C) 4-methyl-2-hexanol
D) 2-ethyl-4-pentanol
E) 4-ethyl-2-pentenal

Answer: C
Diff: 3 *Page Ref: 20.9*

23) Name the following compound.

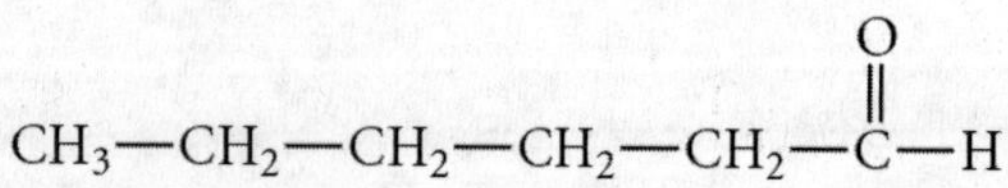

A) pentenel
B) 1-pentanal
C) hexanal
D) 1-pentanone
E) 1-hexenol

Answer: C
Diff: 3 *Page Ref: 20.10*

24) Name the following compound.

$$CH_3-CH_2-\overset{\overset{\displaystyle O}{\|}}{C}-CH_3$$

A) 2-butenal
B) 3-butenal
C) 2-methyl-2-ethylbutanone
D) 2-butanone
E) 2-methylbutanone

Answer: D
Diff: 3 *Page Ref: 20.10*

25) Name the following compound.

$$CH_3-CH_2-CH_2-\overset{\overset{\displaystyle O}{\|}}{C}-O-CH_2-CH_3$$

A) ethyl butyl ether
B) propyl propanoate
C) ethyl butanoate
D) hexanoic acid
E) 4-hexanone

Answer: C
Diff: 3 *Page Ref: 20.11*

26) Name the following compound.

$$CH_3-O-CH_2-CH_2-CH_3$$

A) ethyl propanoate
B) ethyl ethanoate
C) 2 butanal
D) methyl propyl ether
E) 1-butanoic acid

Answer: D
Diff: 3 *Page Ref: 20.12*

27) Choose the weak base from the compounds below.
A) $(CH_3)_2NH$
B) CH_3COOH
C) CH_3SCH_3
D) CH_3CH_3
E) CH_3CH_2I

Answer: A
Diff: 2 *Page Ref: 20.13*

28) Choose the weak acid from the compounds below.

A) $CH_3CH_2NH_2$
B) CH_3CH_2COOH
C) $CH_3CH_2OCH_3$
D) CH_3CH_2I
E) CH_3SeH

Answer: B
Diff: 2 *Page Ref: 20.13*

29) Name the following compound.

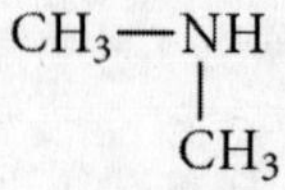

A) methyl amine
B) diethyl sulfide
C) propanol
D) 1-methanoate
E) dimethyl amine

Answer: E
Diff: 3 *Page Ref: 20.13*

Matching Questions

Match the following.

1)

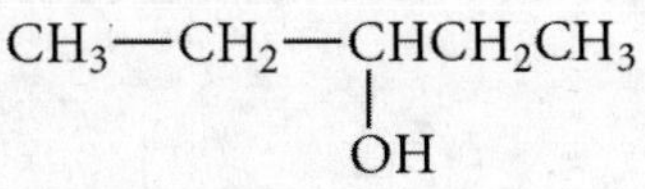

Diff: 1 *Page Ref: 20.8*

2)

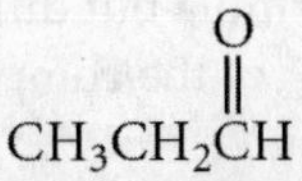

Diff: 1 *Page Ref: 20.8*

3)

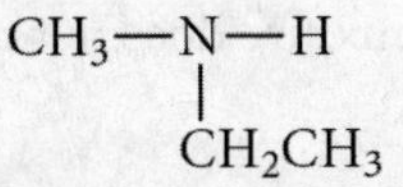

Diff: 1 *Page Ref: 20.8*

4)

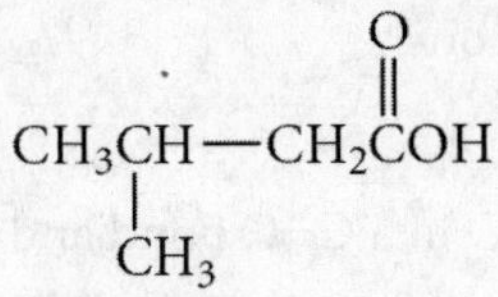

Diff: 1 *Page Ref: 20.8*

5)

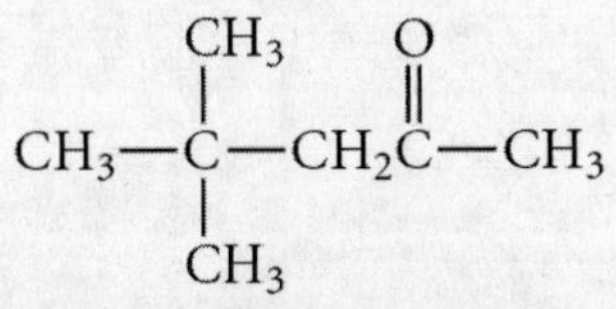

Diff: 1 *Page Ref: 20.8*

6)

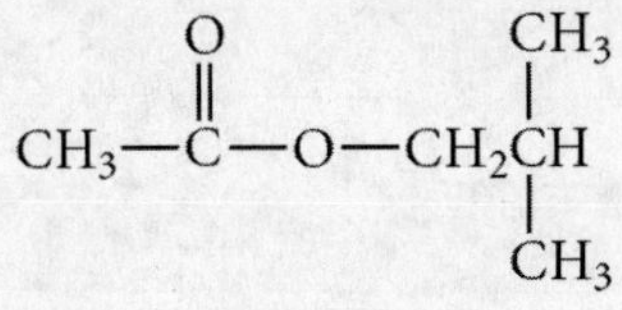

Diff: 1 *Page Ref: 20.8*

A) ester

B) aldehyde

C) ketone

D) alcohol

E) carboxylic acid

F) amine

Answers: 1) D 2) B 3) F 4) E 5) C
6) A

Short Answer Questions

1) Why are there so many more carbon compounds than the number of compounds made up of all the rest of the elements combined?

Answer: Carbon has the ability to form 4 bonds and is able to form a huge number of unique compounds with many other elements.

Diff: 1 *Page Ref: 20.2*

2) What is meant by the term "structural isomer"? Draw a structural isomer of CH_3-CH_2-CH_2-CH_3.

Answer: Structural isomers are compounds that have the same chemical formula but different atomic connectivity. The student should sketch a structural isomer of the compound given. One example would be $(CH_3)_3CH$.

Diff: 2 *Page Ref: 20.3*

3) What does the term "racemic mixture" mean?

Answer: A mixture (equal in mole ratio) of two optical isomers. This mixture does not rotate plane polarized light like each separate optical isomer would.

Diff: 1 *Page Ref: 20.3*

4) Why doesn't benzene typically undergo addition reactions like the alkenes do?

Answer: Benzene's pi electrons are delocalized over the entire structure. This makes benzene particularly stable and unlikely to undergo addition reactions.

Diff: 2 *Page Ref: 20.7*

5) Which part of HCN, the hydrogen or the cyano group, adds to the O in a C=O bond and why?

Answer: The more negative CN attacks the more positive carbon of the C=O. Then the negative oxygen attacks the hydrogen from the HCN.

Diff: 2 *Page Ref: 20.10*

6) Give the organic product of the following reaction.

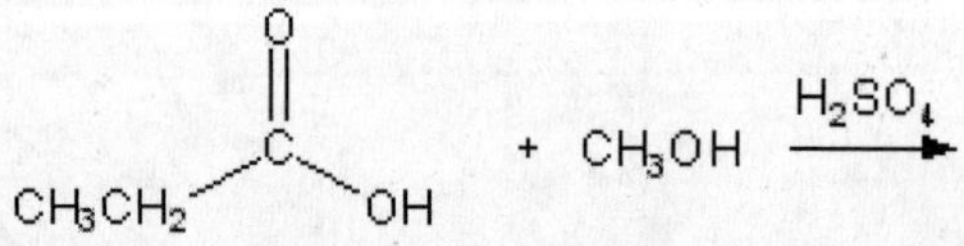

Answer: $CH_3CH_2CO_2CH_3$

Diff: 3 *Page Ref: 20.11*

Chapter 21 Biochemistry

Multiple Choice Questions

1) Type I diabetes develops when the pancreas does not make enough ________.

A) cholesterol
B) sucrose
C) blood
D) insulin
E) iron

Answer: D
Diff: 1 *Page Ref: 21.1*

2) Identify the compound that is a steroid.

A) glucose
B) cholesterol
C) sucrose
D) tristearin
E) glycerin

Answer: B
Diff: 1 *Page Ref: 12.2*

3) What type of lipid is the following compound?

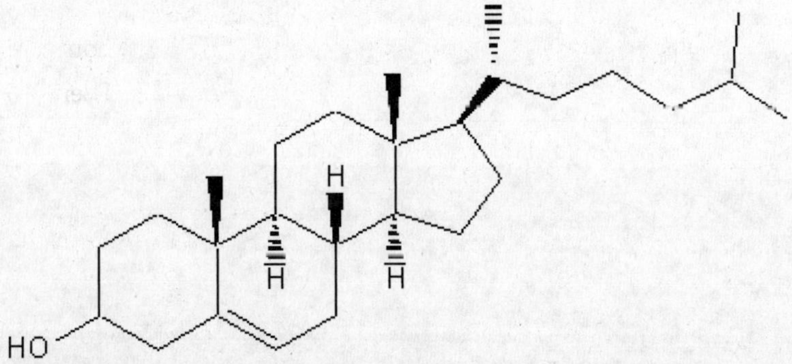

A) steroid
B) phospholipid
C) glycolipid
D) fatty acid
E) triglyceride

Answer: A
Diff: 1 *Page Ref: 21.2*

4) What type of lipid is the following compound?

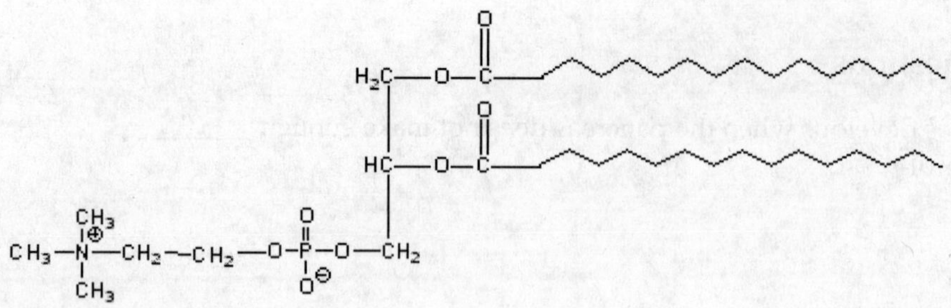

A) steroid
B) phospholipid
C) glycolipid
D) fatty acid
E) triglyceride

Answer: B
Diff: 1 *Page Ref: 21.2*

5) What type of lipid is the following compound?

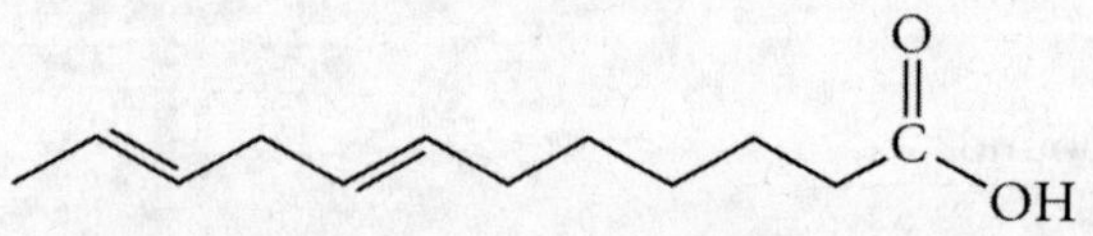

A) steroid
B) phospholipid
C) glycolipid
D) fatty acid
E) triglyceride

Answer: D
Diff: 1 *Page Ref: 21.2*

6) Choose the polyunsaturated fatty acid from the compounds below.

A)

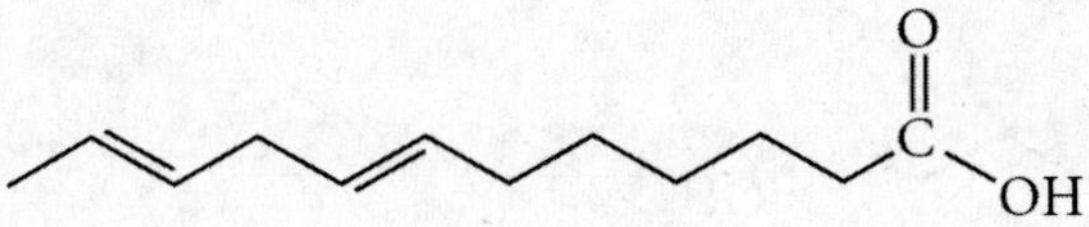

B)

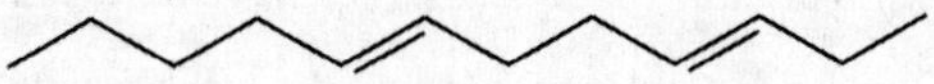

C)

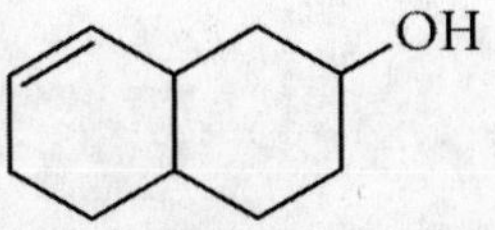

D)

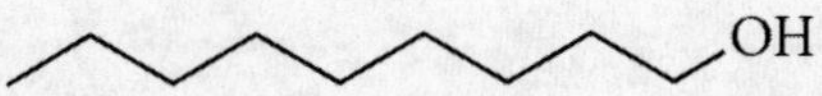

E)

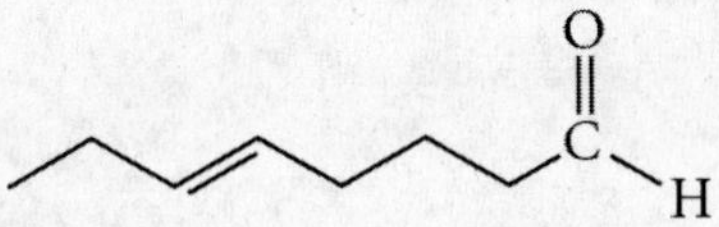

Answer: A
Diff: 1 *Page Ref: 21.2*

7) Choose the polyunsaturated triglyceride from the compounds below.

A)

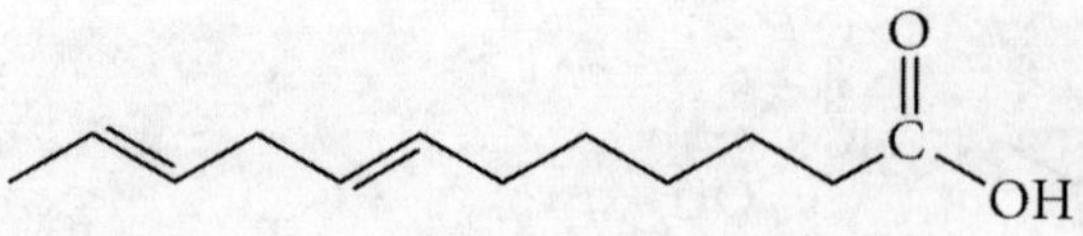

B)

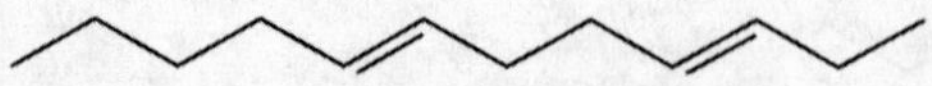

C)

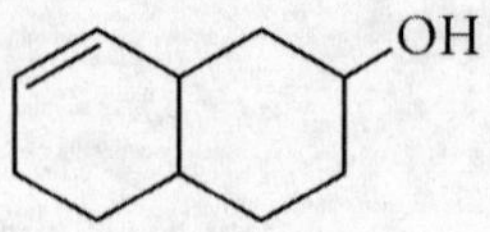

D)

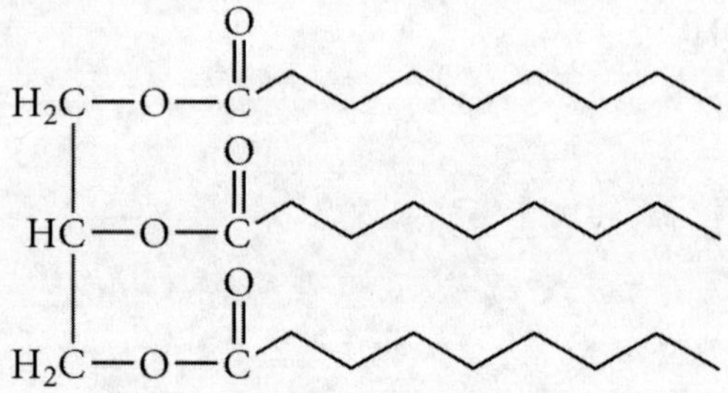

E)

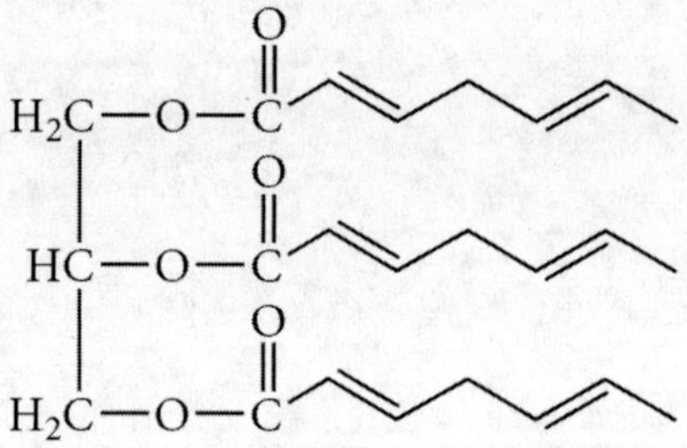

Answer: E

Diff: 1 *Page Ref: 21.2*

8) Choose the saturated fatty acid from the compounds below.

A)

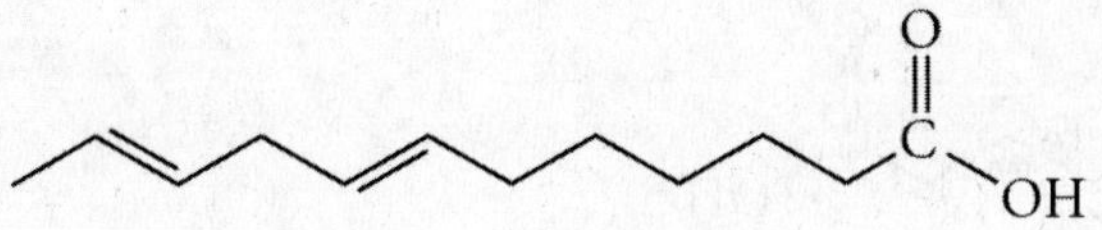

B)

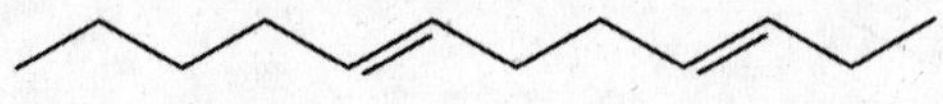

C)

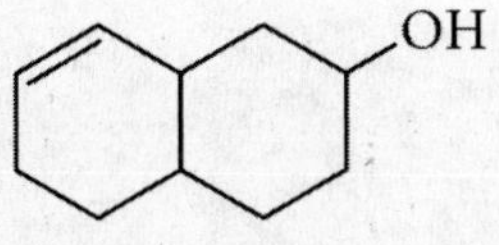

D)

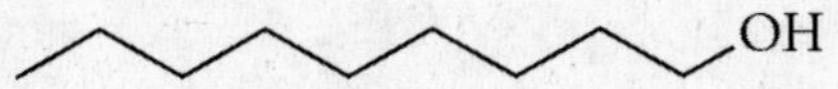

E)

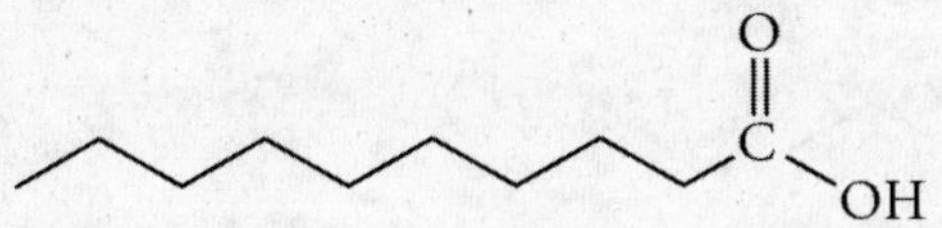

Answer: E

Diff: 1 *Page Ref: 21.2*

9) Choose the saturated triglyceride from the compounds below.

A)

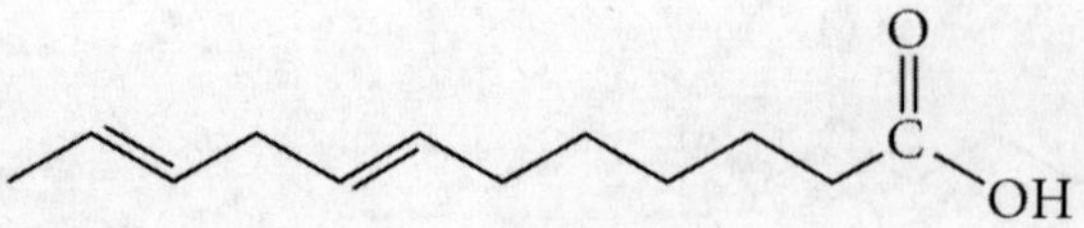

B)

C)

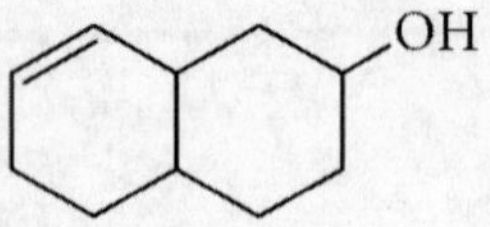

D)

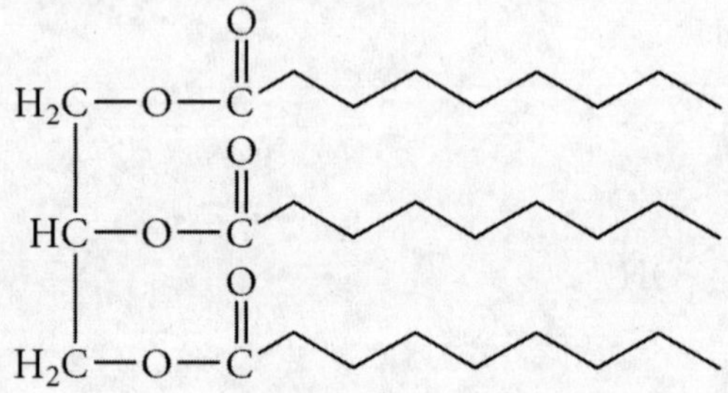

E)

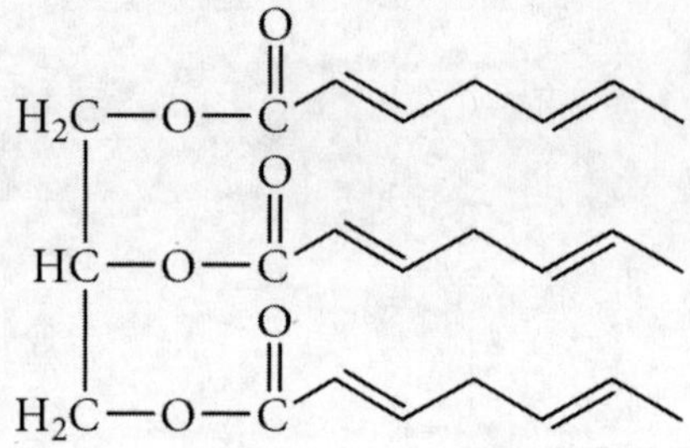

Answer: D

Diff: 1 *Page Ref: 21.2*

10) Which of the following fatty acids is most likely to be a solid at room temperature?

A)

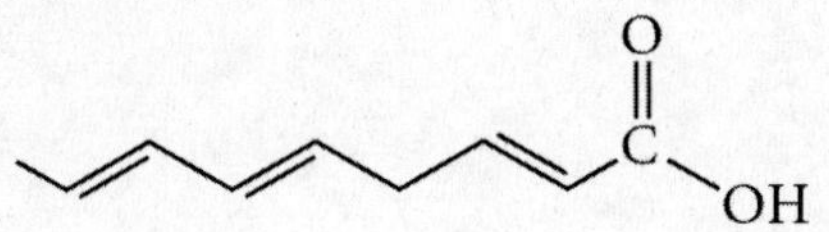

B)

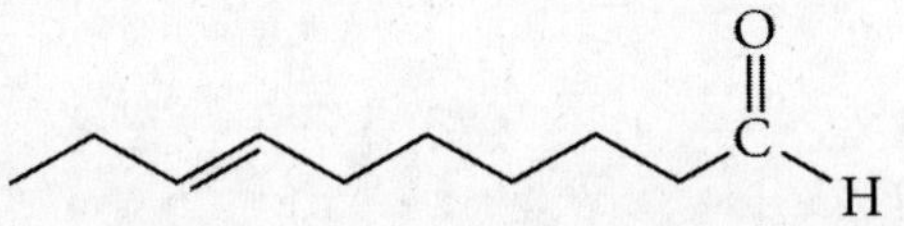

C)

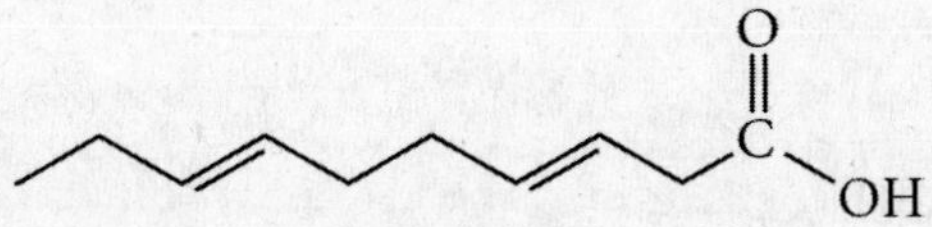

D)

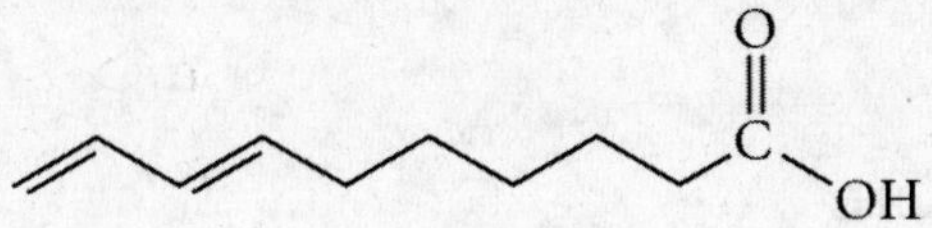

E)

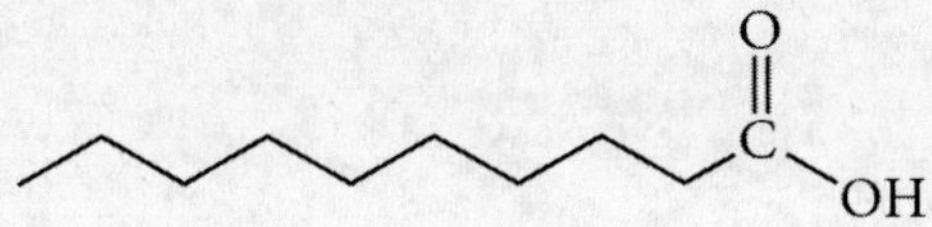

Answer: E

Diff: 1 *Page Ref: 21.2*

11) Which of the following fatty acids is most likely to be a liquid at room temperature?

A)

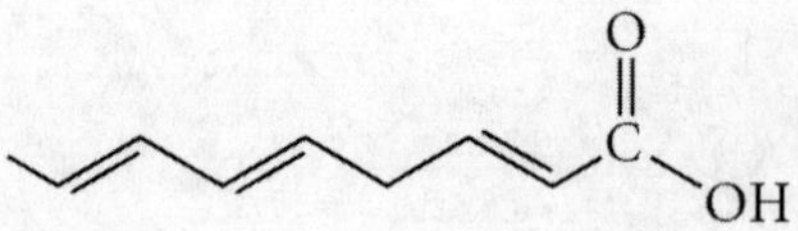

B)

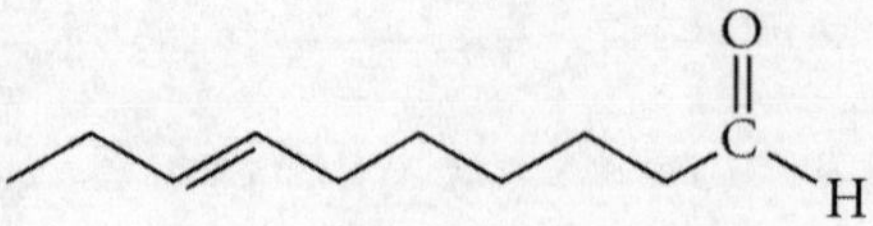

C)

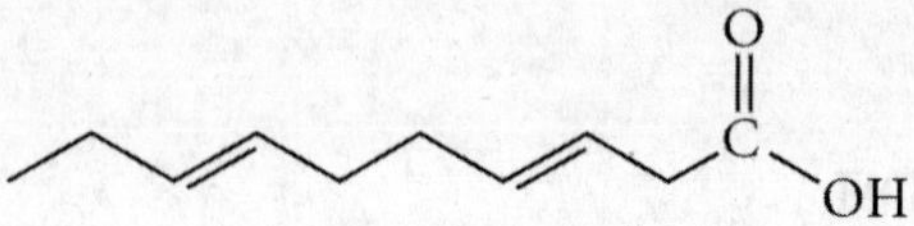

D)

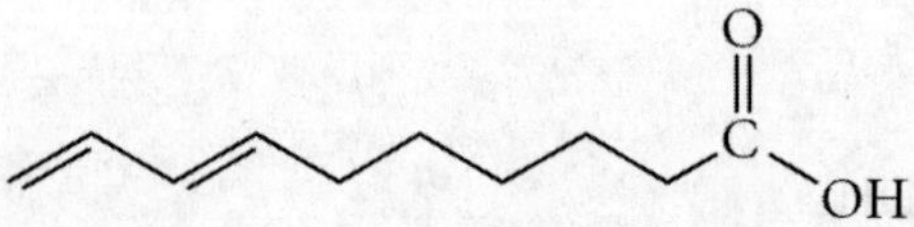

E)

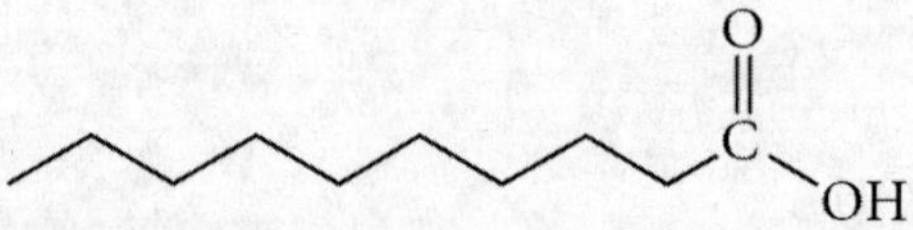

Answer: A

Diff: 1 *Page Ref: 21.2*

12) Identify the most abundant organic substance on earth.

A) cellulose B) fructose C) glucose D) surcrose E) maltose

Answer: A

Diff: 1 *Page Ref: 21.3*

13) Which of the following is an example of an aldopentose?

A)

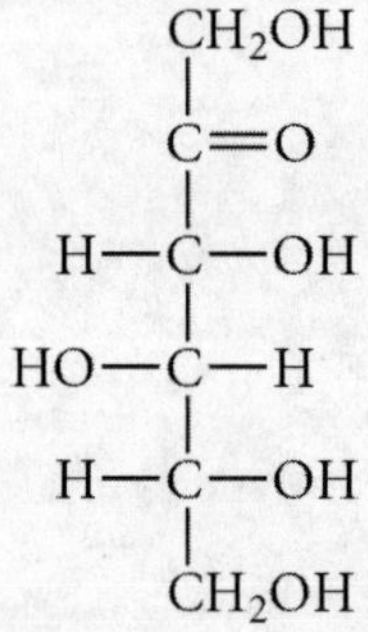

B)

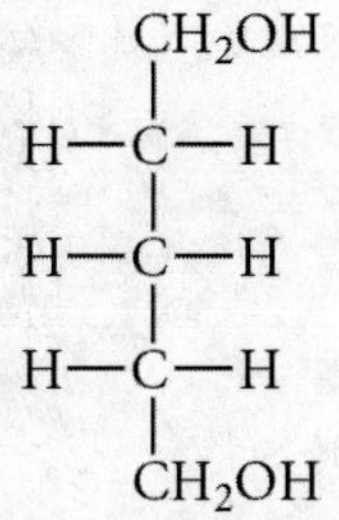

C)

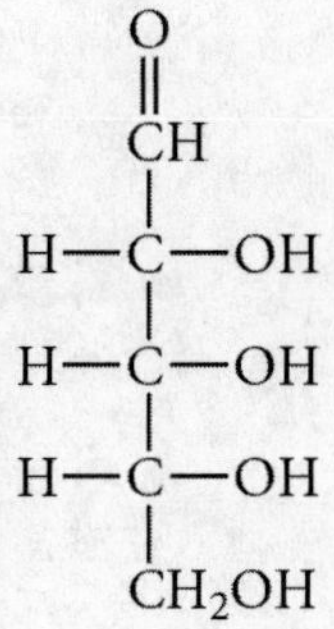

D)

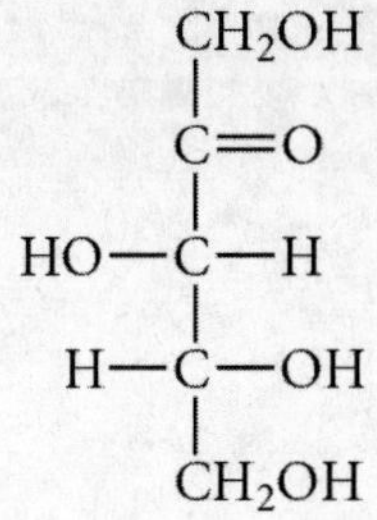

E)

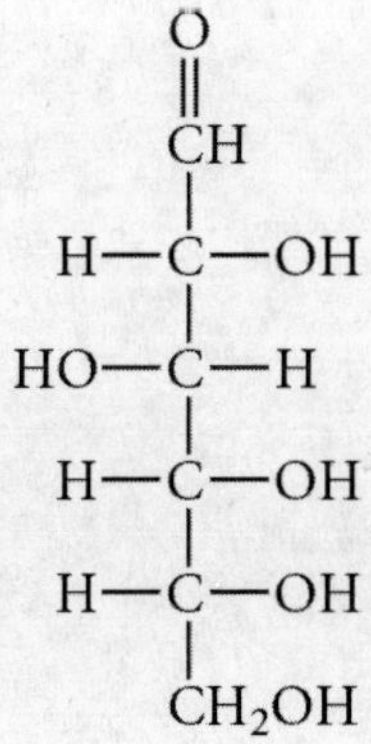

Answer: C
Diff: 1 *Page Ref: 21.3*

14) Which of the following is an example of a ketopentose?

A)

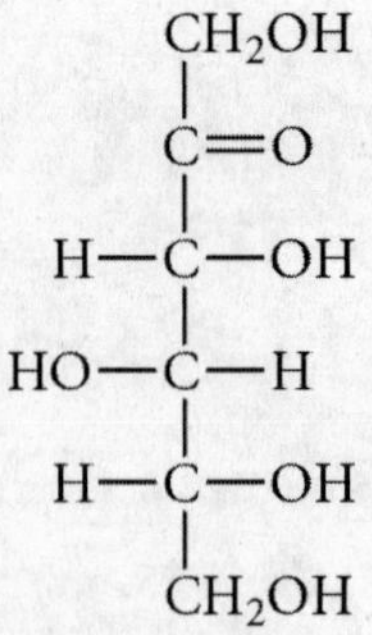

B)

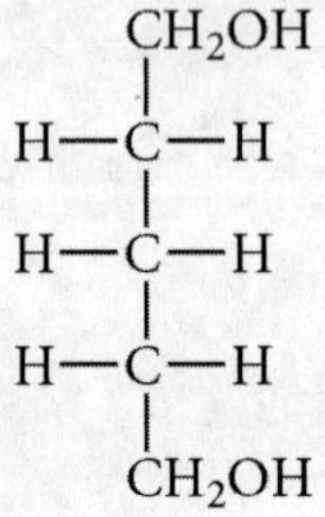

C)

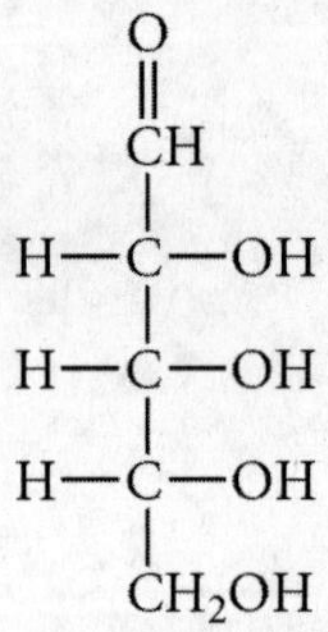

D)

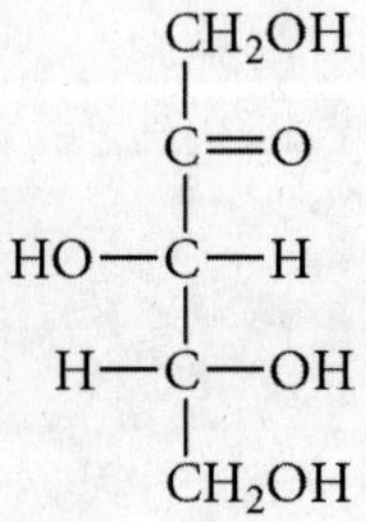

E)

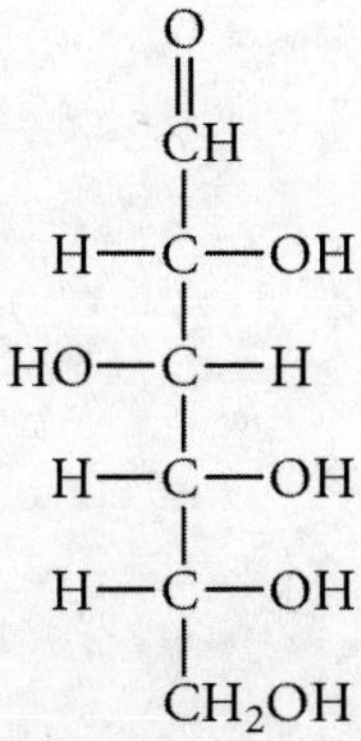

Answer: D

Diff: 1 *Page Ref: 21.3*

15) Which of the following is an example of an aldohexose?

A)

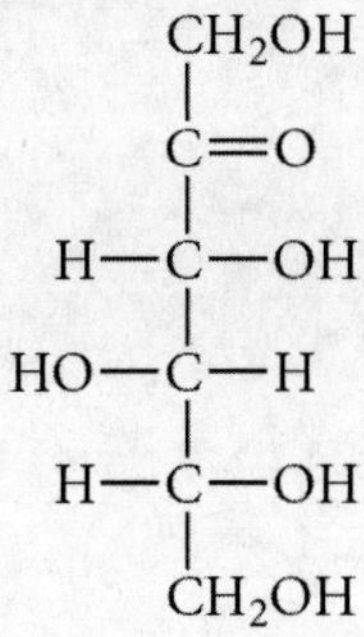

B)

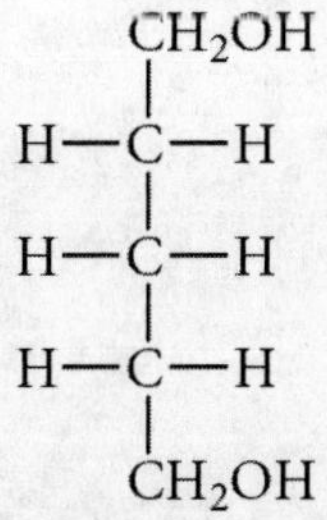

C)

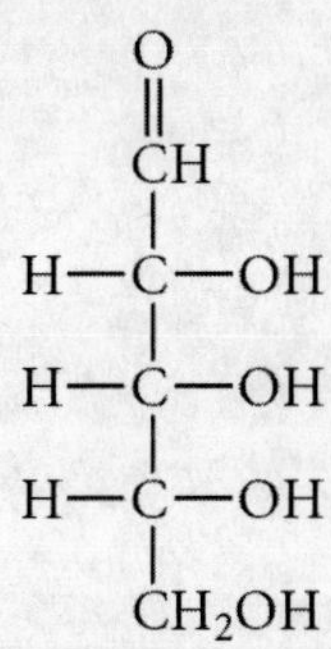

D)

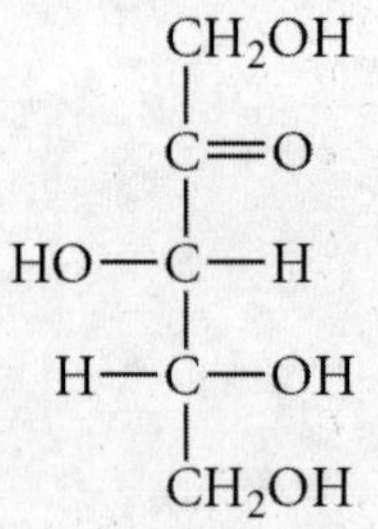

E)

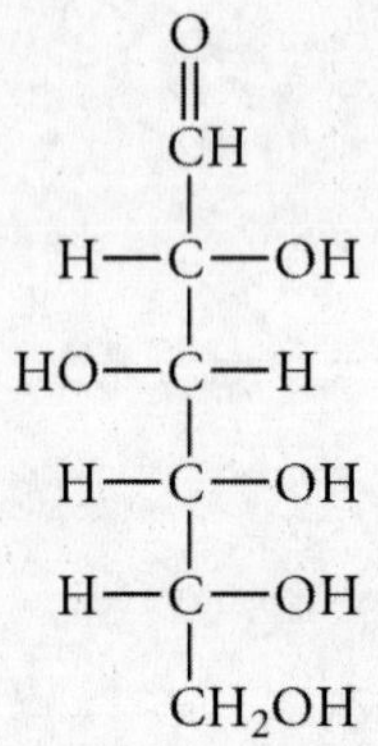

Answer: E
Diff: 1 *Page Ref: 21.3*

16) Which of the following is an example of a ketohexose?

A)

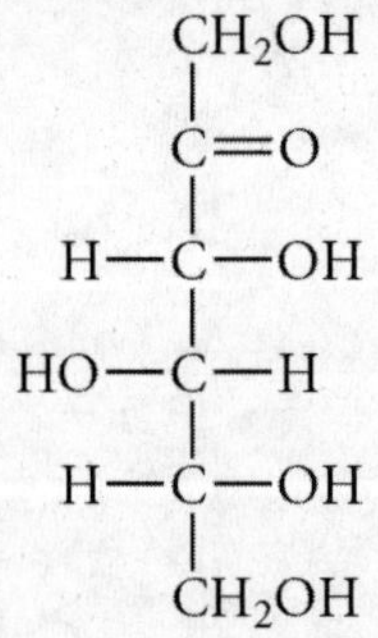

B)

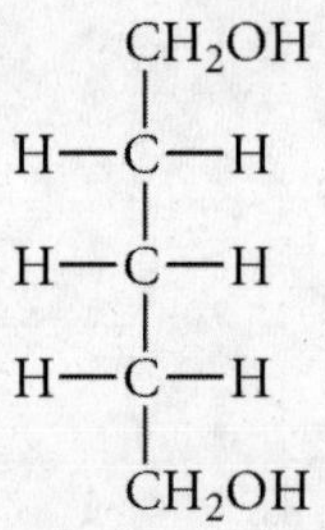

C)

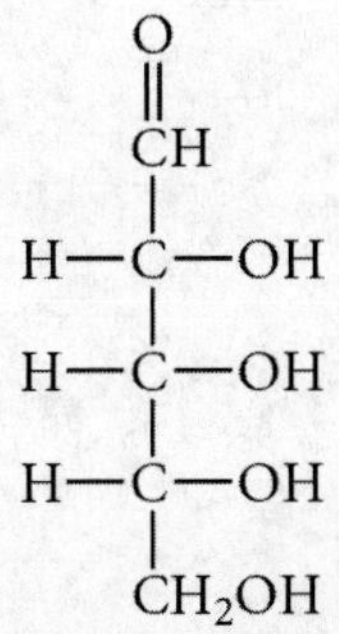

D)

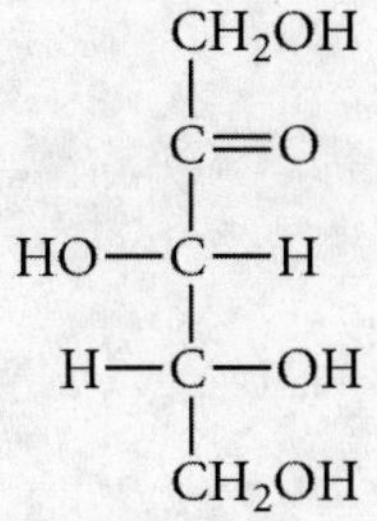

E)

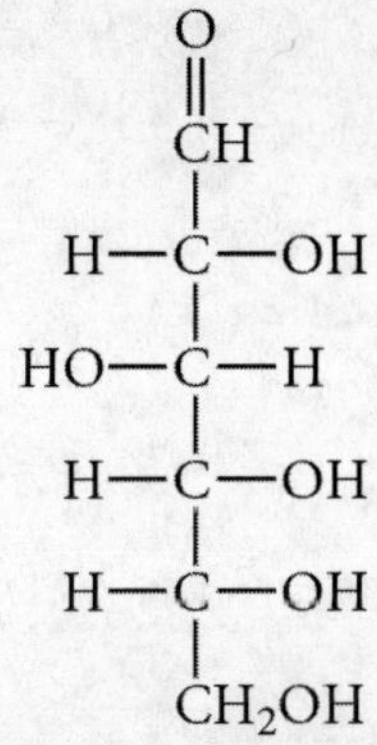

Answer: A
Diff: 1 *Page Ref: 21.3*

17) Identify glucose.

A)

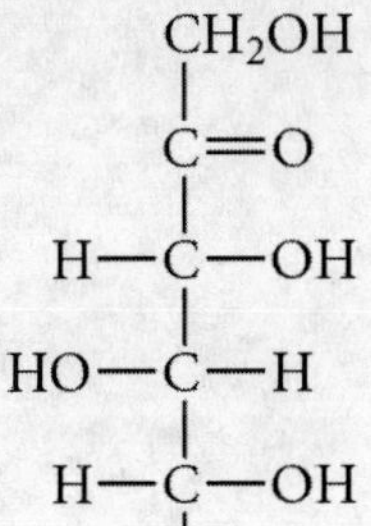

B)

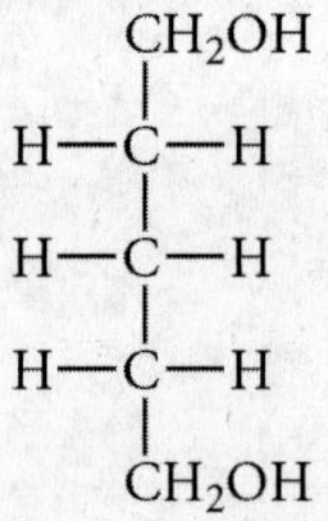

C)

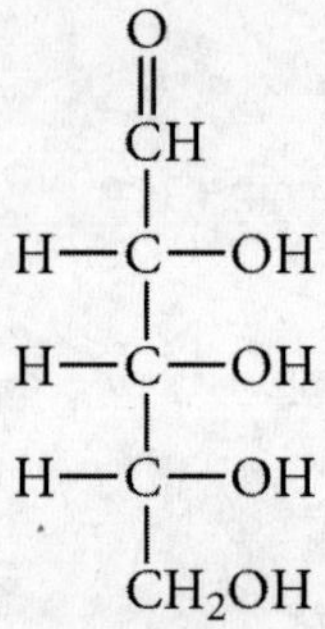

D)

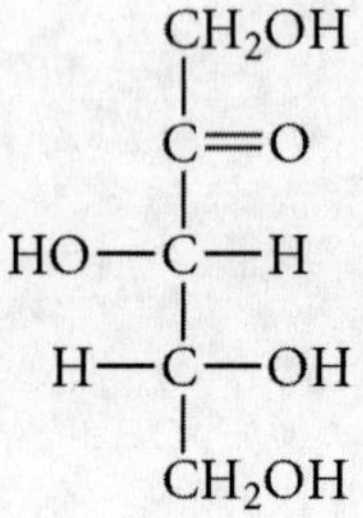

E)

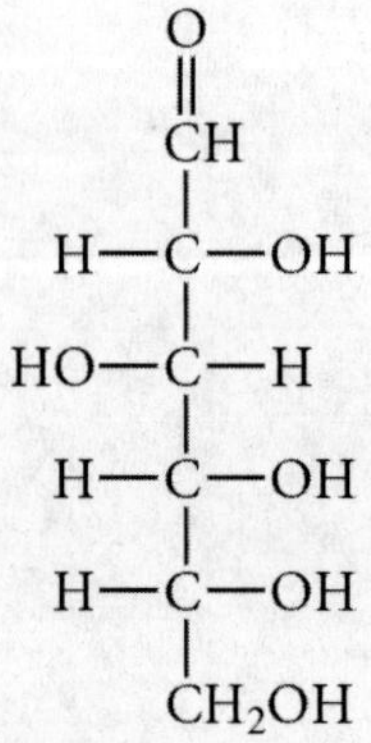

Answer: E

Diff: 1 *Page Ref: 21.3*

18) Identify fructose.

A)

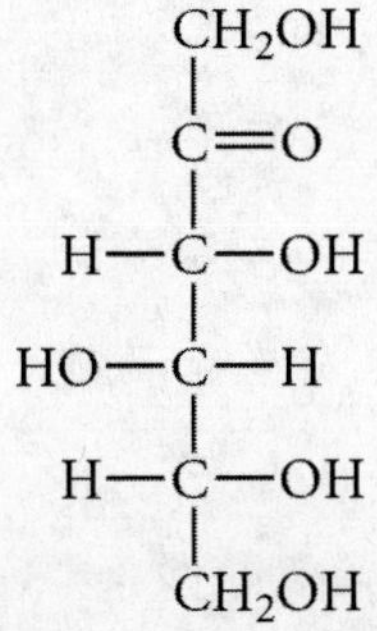

B)

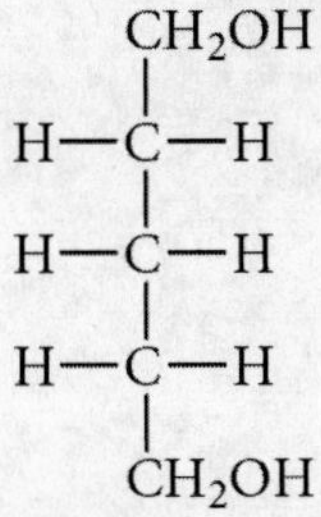

C)

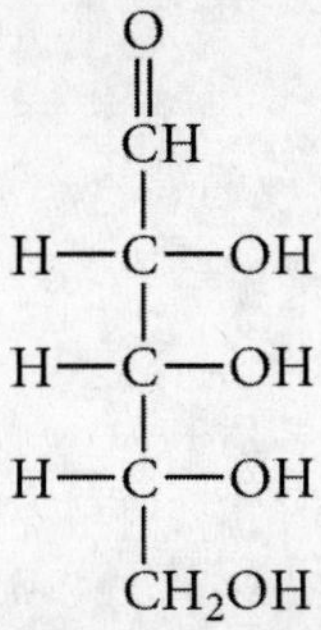

D)

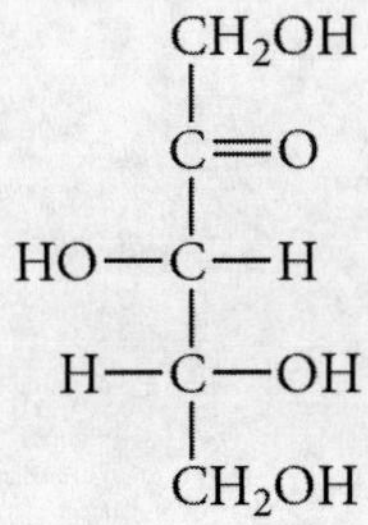

E)

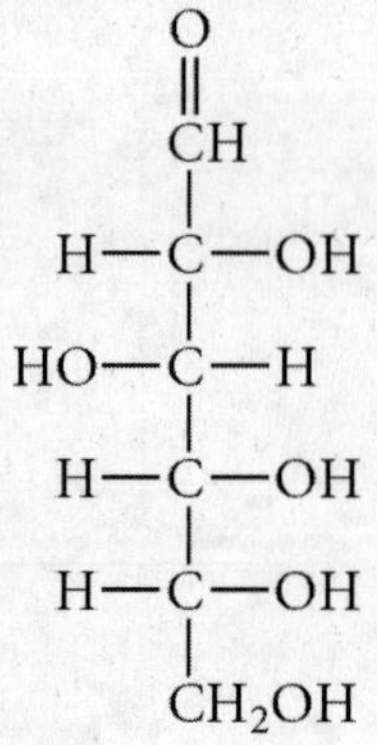

Answer: A
Diff: 1 *Page Ref: 21.3*

19) Choose the monosaccharide from the compounds below.

A)

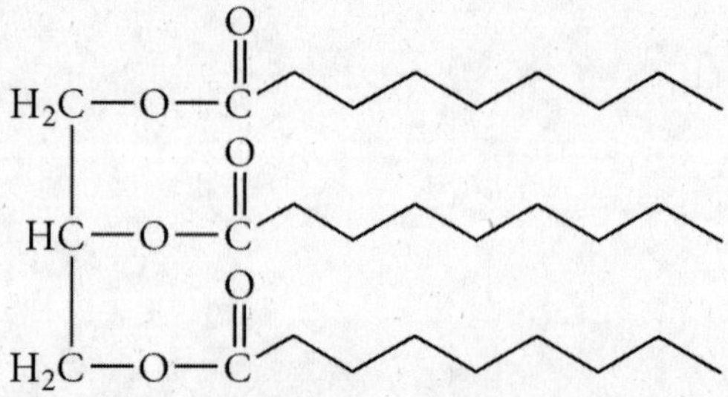

B)

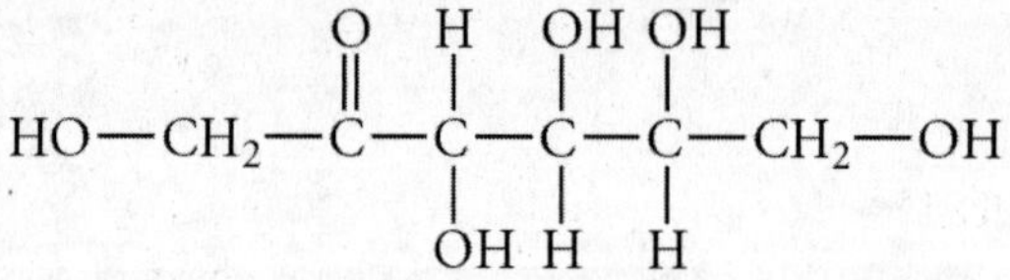

C)

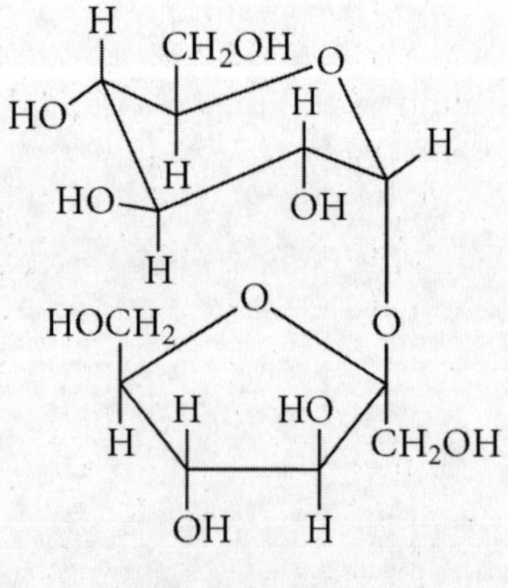

D)

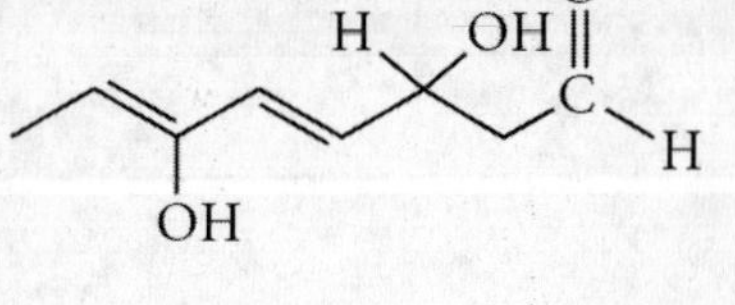

E)

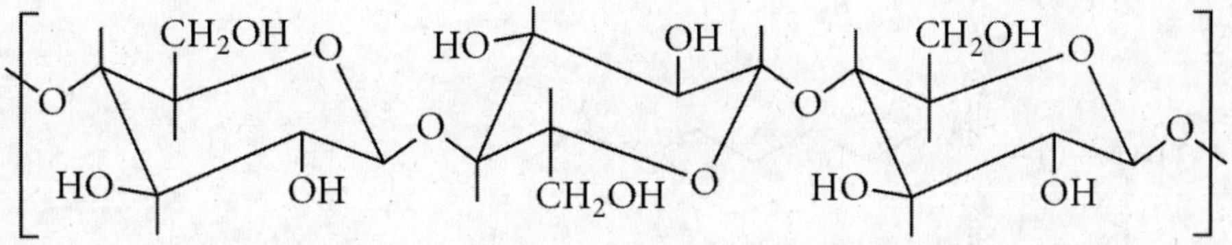

Answer: B
Diff: 1 *Page Ref: 21.3*

20) Choose the disaccharide from the compounds below.

A)

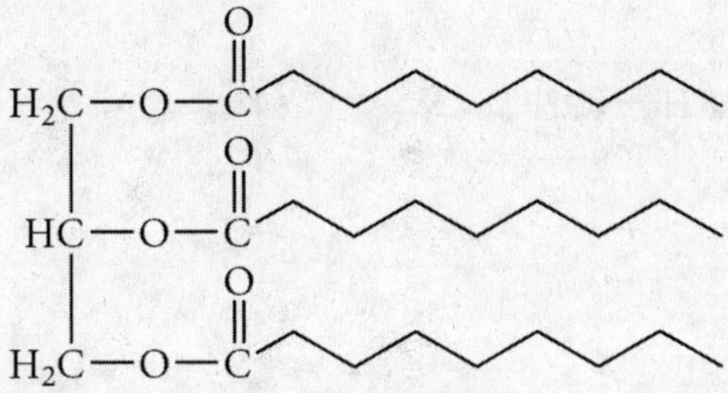

B)

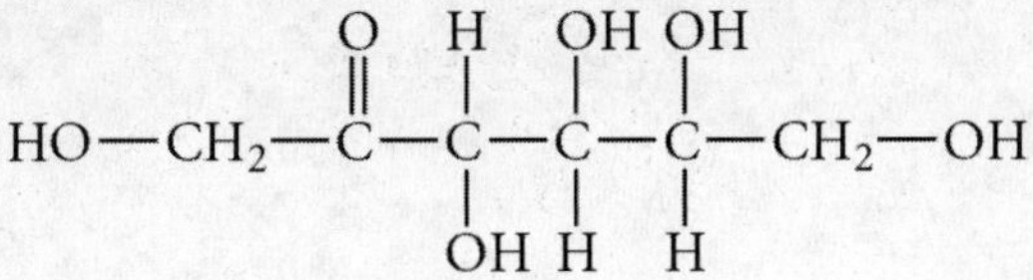

C)

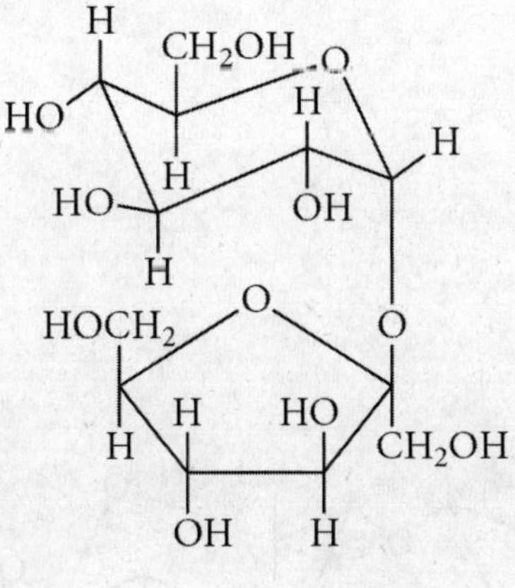

D)

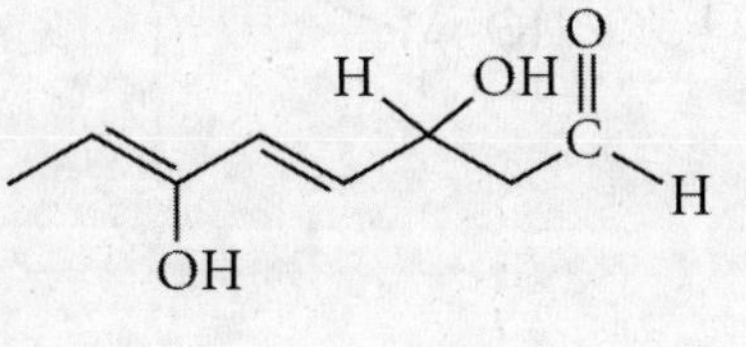

E)

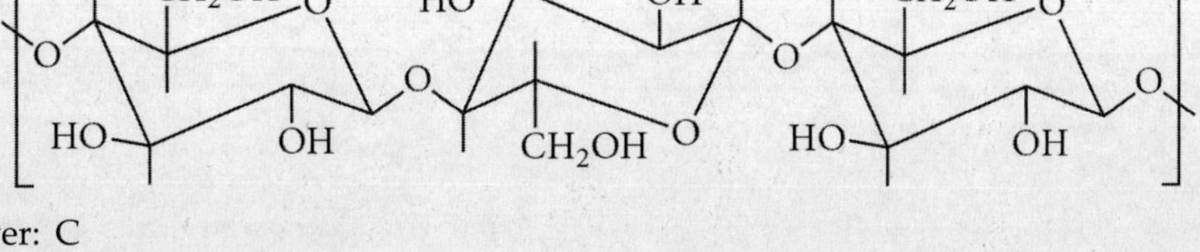

Answer: C
Diff: 1 *Page Ref: 21.3*

21) Choose the polysaccharide from the compounds below.

A)

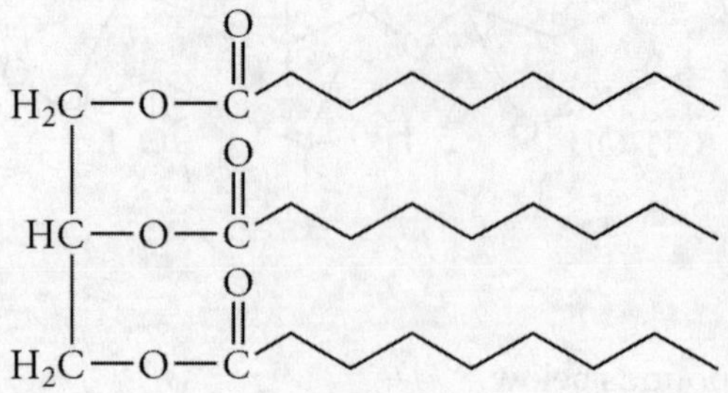

B)

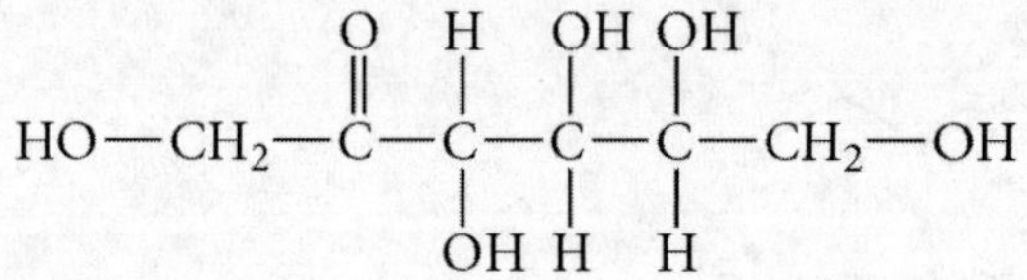

C)

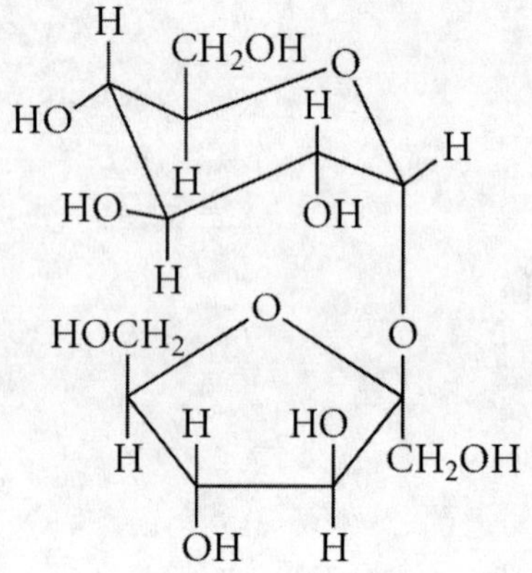

D)

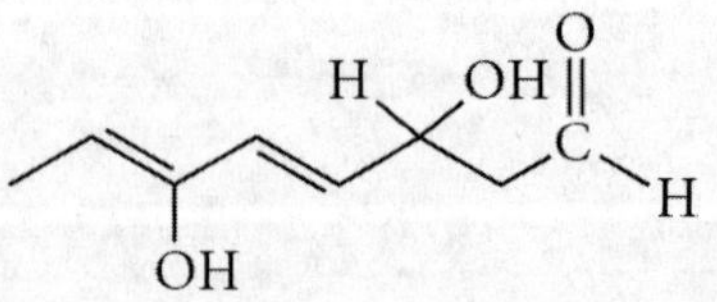

E)

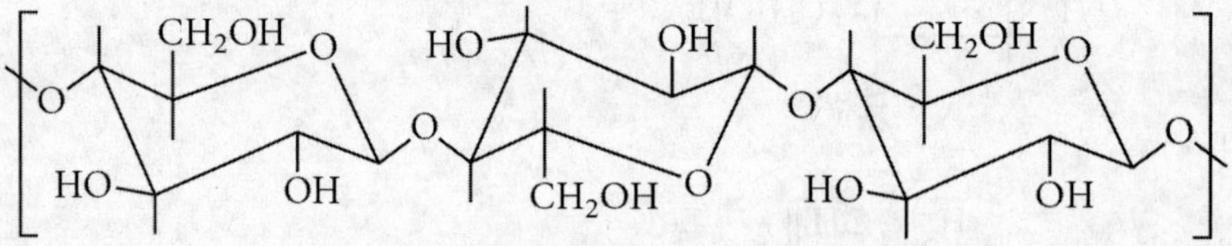

Answer: E

Diff: 1 *Page Ref: 21.3*

22) How many of the carbons in the following structure are chiral?

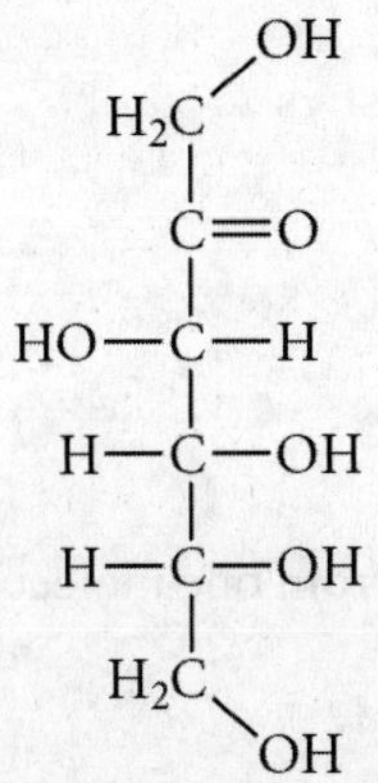

A) 3 B) 0 C) 2 D) 1 E) 4 or more

Answer: A

Diff: 2 *Page Ref: 21.3*

23) Identify the smallest amino acid.

A) glycine
B) cysteine
C) phenylalanine
D) tyrosine
E) aspartic acid

Answer: A

Diff: 2 *Page Ref: 21.4*

24) Identify the compound that is not an amino acid.

A) cysteine
B) threonine
C) serine
D) guanine
E) tryptophan

Answer: D

Diff: 2 *Page Ref: 21.4*

25) Which of the following shows all of the tripeptides that can be formed from one molecule each of glycine (Gly), valine (Val), and leucine (Leu)?

A) GlyValLeu, GlyLeuVal, ValLeuGly, ValGlyLeu, LeuGlyVal, LeuValGly
B) GlyValLeu, GlyLeuVal, LeuGlyVal
C) ValGlyLeu, GlyValLeu, GlyLeuVal, LeuGlyVal
D) ValGlyLeu and GlyLeuVal
E) GlyValLeu

Answer: A

Diff: 2 *Page Ref: 21.4*

26) Which of the following shows all of the tripeptides that can be formed from one molecule each of lysine (Lys), threonine (Thr), and hisitidine (His)?
A) LysThrHis
B) LysThrHis, LysHisThr, HisThrLys, HisLysThr, ThrLysHis, ThrHisLys
C) LysThrHis, LysHisThr, HisLysThr, ThrHisLys
D) LysThrHis, LysHisThr, HisLysThr
E) LysThrHis and LysHisThr

Answer: B
Diff: 2 *Page Ref:* 21.4

27) Which of the following shows all of the tripeptides that can be formed from one molecule each of alanine (Ala), methionine (Met), and proline (Pro)?
A) AlaMetPro
B) AlaMetPro and AlaProMet
C) AlaMetPro, AlaProMet, ProMetAla, ProAlaMet, MetProAla, MetAlaPro
D) AlaMetPro, AlaProMet, ProAlaMet
E) AlaMetPro, AlaProMet, ProAlaMet, MetProAla

Answer: C
Diff: 2 *Page Ref:* 21.4

28) Which of the following describes a primary protein structure?
A) protein structure maintained by disulfide linkages
B) amino acid sequence maintained by peptide bonds
C) protein chains maintained by interactions of peptide backbones
D) arrangement of multiple protein subunits
E) protein structure maintained through multiple hydrogen bonds

Answer: B
Diff: 1 *Page Ref:* 21.5

29) Identify the sugar in RNA.
A) ribose
B) deoxyribose
C) fructose
D) glucose
E) galactose

Answer: A
Diff: 1 *Page Ref:* 21.6

30) Identify the sugar in DNA.
A) ribose
B) deoxyribose
C) fructose
D) glucose
E) galactose

Answer: B
Diff: 1 *Page Ref:* 21.6

31) Which of the following statements is TRUE?
A) A nucleotide is composed of a sugar, a base and a phosphate group.
B) A codon is a sequence of 3 bases that codes for a single amino acid.
C) A gene is a portion of DNA that codes for a single protein
D) A chromosome is a structure within a cell nucleus that houses DNA.
E) All of the above are true.

Answer: E
Diff: 1 *Page Ref: 21.6*

32) The following is part of a DNA sequence. What is its complementary sequence?

AGTTCGAGCCT

A) AGTTCGAGCCT
B) TCCGAGCTTGA
C) CAGTCCA
D) AGGCTCGAACT
E) TCAAGCTCGGA

Answer: E
Diff: 2 *Page Ref: 21.6*

33) The following is part of a DNA sequence. What is its complementary sequence?

TCGTAAGCTTGCG

A) AGCATTCGAACGC
B) CGCAAGCTTACGA
C) TCATTCAC
D) GCGTTCGAATGCT
E) TCGTAAGCTTGCG

Answer: A
Diff: 2 *Page Ref: 21.6*

34) The ability for DNA to replicate itself was discovered by ________.
A) Curie
B) Einstein
C) Watson and Crick
D) Galileo
E) Edison

Answer: C
Diff: 2 *Page Ref: 21.7*

35) What is between the two strands of DNA, holding them together?
A) hydrogen bonds
B) London dispersion forces
C) dipole-dipole bonds
D) ionic bonds
E) covalent bonds

Answer: A
Diff: 2 *Page Ref: 21.7*

Algorithmic Questions

1) Which of the following form between glycerol and fatty acids to form triglycerides?
A) alcohol bonds
B) disulfide linkages
C) amide bonds
D) glycosidic linkages
E) ester linkages

Answer: E
Diff: 1 *Page Ref: 21.2*

2) What type of lipid is the following compound?

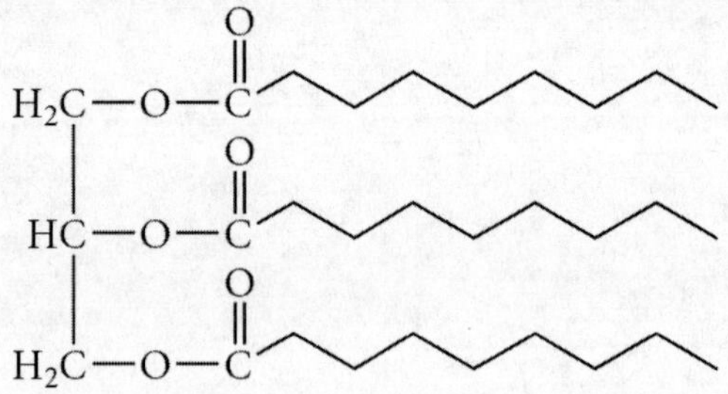

A) steroid
B) phospholipid
C) glycolipid
D) fatty acid
E) triglyceride

Answer: E
Diff: 1 *Page Ref: 21.2*

3) Identify the disaccharide.
A) glucose
B) cholesterol
C) fructose
D) cellulose
E) sucrose

Answer: E
Diff: 1 *Page Ref: 21.3*

4) Identify the polysaccharide.
A) maltose
B) cholesterol
C) galactose
D) cellulose
E) sucrose

Answer: D
Diff: 1 *Page Ref: 21.3*

5) Which of the following bond monosaccharide units together into polysaccharides?
A) hydrogen bonds
B) glycosidic linkages
C) disulfide linkages
D) hydrogen bonds
E) ester linkages

Answer: B
Diff: 1 *Page Ref: 21.3*

6) Which of the following statements is TRUE?
A) Monosaccharides are the building blocks of peptides.
B) Nucleotides are the building blocks of saccharides.
C) Fatty acids are responsible for DNA replication.
D) Peptides are the building blocks of amino acids.
E) Carbohydrates are water soluble and responsible for short term energy storage.

Answer: E
Diff: 2 *Page Ref: 21.3*

7) Which one of the following amino acids does **not** contain a basic side chain?
A) arginine B) histidine C) lysine D) threonine

Answer: D
Diff: 1 *Page Ref: 21.4*

8) Which one of the following amino acids contains a polar side chain?
A) alanine B) glycine C) threonine D) valine

Answer: C
Diff: 1 *Page Ref: 21.4*

9) Which one of the following amino acids contains a hydrophobic side chain?
A) valine B) histidine C) glutamine D) glutamic acid

Answer: A
Diff: 1 *Page Ref: 21.4*

10) In the protein, His–Tyr–Val–Pro, which amino acid contains the N-terminal group?
A) valine B) tyrosine C) proline D) histidine

Answer: D
Diff: 1 *Page Ref: 21.4*

11) In the protein, Arg–Pro–Leu–Gln, which amino acid contains the C-terminal group?
A) proline B) leucine C) glutamine D) arginine

Answer: C
Diff: 1 *Page Ref: 21.4*

12) Which of the following link together amino acid units?
A) glycosidic linkages
B) sulfide linkages
C) peptide bonds
D) hydrogen bonds
E) ester linkages

Answer: C
Diff: 1 *Page Ref: 21.4*

13) Which of the following hold two strands of DNA in an α-helix?
A) ester linkages
B) sulfide linkages
C) peptide bonds
D) hydrogen bonds
E) glycosidic linkages

Answer: D
Diff: 1 *Page Ref: 21.5*

14) Which of the following is an example of a secondary protein structure?
A) tripeptide
B) glyceride
C) α-helix
D) amino acid
E) fatty acid

Answer: C
Diff: 1 *Page Ref: 21.5*

15) Which of the following is an example of a secondary protein structure?
A) fatty acid
B) triglyceride
C) lipid
D) β-pleated sheet
E) dipeptide

Answer: D
Diff: 1 *Page Ref: 21.5*

16) Which of the following has a tertiary protein structure?
A) amino acid
B) ester
C) nucleotide
D) fatty acid
E) fibrous protein

Answer: E
Diff: 1 *Page Ref: 21.5*

17) Which of the following is a type of nucleic acid?
A) DNA
B) amino acid
C) dipeptide
D) carbohydrate
E) lipid

Answer: A
Diff: 1 *Page Ref: 21.6*

18) Which of the following statements is TRUE?
A) Cytosine and Thymine are a complementary base pair that forms 3 hydrogen bonds.
B) Adenine and Guanine are a complementary base pair that forms 3 hydrogen bonds.
C) Cytosine and Guanine are a complementary base pair that forms 2 hydrogen bonds.
D) Thymine and Adenine are a complementary base pair that forms 2 hydrogen bonds.
E) Thymine and Guanine are a complementary base pair that forms 2 hydrogen bonds.

Answer: D
Diff: 2 *Page Ref: 21.6*

19) How many chromosomes are in the nuclei of human cells?
A) 41 B) 44 C) 46 D) 48 E) 52

Answer: C
Diff: 2 *Page Ref: 21.6*

Matching Questions

Match the following.

1)

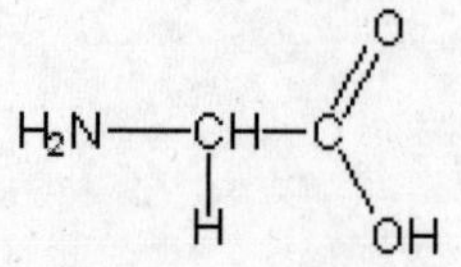

Diff: 1 *Page Ref: 21.4*

2)

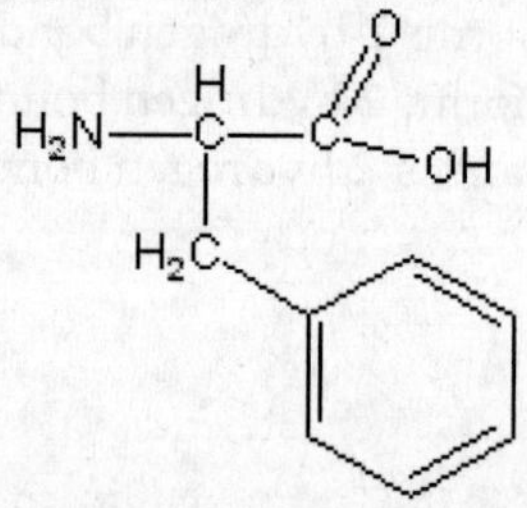

Diff: 1 *Page Ref: 21.4*

3)

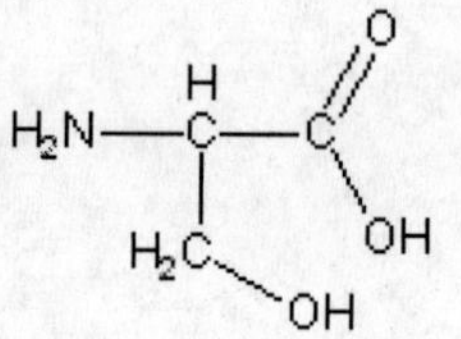

Diff: 1 *Page Ref: 21.4*

4)

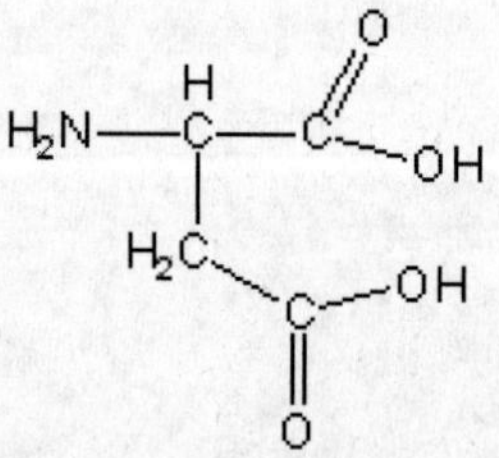

Diff: 1 *Page Ref: 21.4*

5)

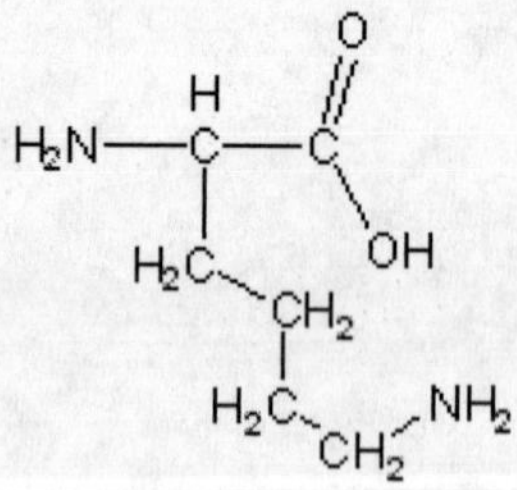

Diff: 1 *Page Ref: 21.4*

A) aspartic acid

B) phenylalanine

C) lysine

D) glycine

E) alanine

F) serine

6)

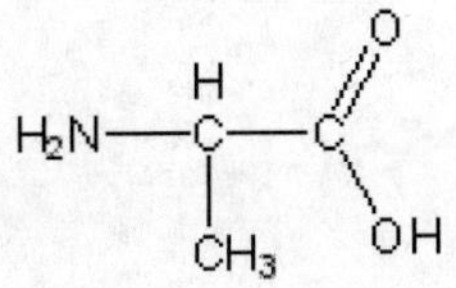

Diff: 1 *Page Ref: 21.4*

Answers: 1) D 2) B 3) F 4) A 5) C
6) E

Short Answer Questions

1) Why are lipids well-suited as structural components of cell membranes?

Answer: Lipids are hydrophobic molecules that are insoluble in aqueous environments. They tend to aggregate and serve as a "wall" between the inner aqueous contents of the cell and the outer aqueous environment.

Diff: 2 *Page Ref: 21.2*

2) Why does the melting point of fatty acids decrease with the number of double bonds present?

Answer: The melting points of fatty acids are determined primarily by the strength of the dispersion forces between the nonpolar portion of the molecules. The more double bonds that are present, the more kinked the molecules and the less strongly they can interact with one another. This weakens the dispersion forces and makes it easier to melt the compound.

Diff: 2 *Page Ref: 21.2*

3) How do cis- and trans-fats differ?

Answer: Since trans double bond containing molecules are more linear than cis double bond containing molecules they will have stronger dispersion forces. Therefore, the trans fats will have a higher melting point than cis fats and similar to saturated fats.

Diff: 2 *Page Ref: 21.2*

4) What is the difference between an α-glycosidic linkage and a β-glycosidic linkage? What implication does this have for the strength of the intermolecular forces between molecules?

Answer: The α-glycosidic linkage in starches cause the oxygen linkages to lie out of the plane of the rest of the molecule. The β-glycosidic linkages in cellulose have the connecting oxygens lying in the plane of the rest of the molecule. Therefore, the β-glycosidic linkages do not disrupt the intermolecular interactions between molecules and form a stronger structure than α-glycosidic linkages.

Diff: 2 *Page Ref: 21.3*

5) What is a codon?

Answer: A codon is a series of 3 bases that codes for one of the 20 amino acids.

Diff: 2 *Page Ref: 21.6*

Chapter 22 Chemistry of the Nonmetals

Multiple Choice Questions

1) Which of the following statements is TRUE?
   A) Silicon atoms are too large to form pi bonds between them.
   B) Silicates are covalent atomic solids that contain silicon and oxygen.
   C) Silicon is the most abundant element in the Earth's crust.
   D) Silicon is one of the strongest oxidizing agents.
   E) All of the above are true.

Answer: B
Diff: 1 *Page Ref:* 22.3

2) Which of the following statements is TRUE?
   A) All silicates are composed of SiO_4 tetrahedra that bond together through an oxygen to form single chains.
   B) The fibrous nature of asbestos comes from its silicate sheet structure.
   C) In some silicates, ionic bonding between a metal ion and an $Si_xO_y^{z-}$ ion is what holds the structure together.
   D) The flakiness of mica comes from its single silicate chain structure.
   E) None of the above are true.

Answer: C
Diff: 1 *Page Ref:* 22.3

3) Determine the formula for an aluminosilicate where one third of the silicons are replaced by an aluminum atom. The charge is balanced by potassium ions.
   A) $K_2(AlO_2)(SiO_2)_2$
   B) $K(AlO_2)(SiO_2)$
   C) $K(AlO_2)_2(SiO_2)_2$
   D) $K(AlO_2)(SiO_2)_2$
   E) $K_2(AlO_2)_2(SiO_2)_2$

Answer: D
Diff: 3 *Page Ref:* 22.3

4) Determine the formula for an aluminosilicate where one fourth of the silicons are replaced by an aluminum atom. The charge is balanced by sodium ions.
   A) $Na_2(AlO_2)_2(SiO_2)$
   B) $Na_2(AlO_2)_2(SiO_2)_3$
   C) $Na(AlO_2)_2(SiO_2)$
   D) $Na_2(AlO_2)_2(SiO_2)_2$
   E) $Na(AlO_2)(SiO_2)_3$

Answer: E
Diff: 3 *Page Ref:* 22.3

5) Determine the formula for an aluminosilicate where one half of the silicons are replaced by an aluminum atom. The charge is balanced by magnesium ions.

A) $Mg(AlO_2)_2(SiO_2)_2$
B) $Mg_2(AlO_2)_2(SiO_2)_2$
C) $Mg_2(AlO_2)_2(SiO_2)$
D) $Mg_2(AlO_2)(SiO_2)_2$
E) $Mg(AlO_2)(SiO_2)_2$

Answer: A
Diff: 3 *Page Ref: 22.3*

6) Give the shape of a boron trihalide.

A) trigonal planar
B) tetrahedral
C) linear
D) trigonal pyramidal
E) bent

Answer: A
Diff: 1 *Page Ref: 22.4*

7) Give the bond angle of a boron trihalide.

A) 60° B) 45° C) 120° D) 109.5° E) 105°

Answer: C
Diff: 1 *Page Ref: 22.4*

8) Which of the following statements is TRUE?

A) Boron is a large component of Earth's crust.
B) Boron is usually found in its elemental state in the Earth's crust.
C) Due to borons small size and low electronegativity, it behaves as a semimetal instead of a metal.
D) Boron atoms don't typically bond to one another due to its small atomic radius.
E) There are at least 5 allotropes of elemental boron.

Answer: E
Diff: 1 *Page Ref: 22.4*

9) Which of the following statements is TRUE?

A) Boron oxide, when added to silicon oxide glass, forms a more heat stable glass called Pyrex.
B) Boron most commonly forms compounds that fulfill the octet rule.
C) Boron forms fairly weak bonds with oxygen, due to their large electronegativity difference.
D) Boron trichloride is an example of a strong Lewis base.
E) None of the above are TRUE.

Answer: A
Diff: 1 *Page Ref: 22.4*

10) Determine the number of vertices and faces in the *closo*–Borane $B_7H_7^{2-}$.

A) vertices = 4, faces = 10
B) vertices = 7, faces = 14
C) vertices = 3, faces = 6
D) vertices = 14, faces = 24
E) vertices = 7, faces = 10

Answer: E
Diff: 2 *Page Ref:* 22.4

11) Determine the number of vertices and faces in the *closo*–Borane $B_{10}H_{10}^{2-}$.

A) vertices = 10, faces = 16
B) vertices = 5, faces = 6
C) vertices = 5, faces = 10
D) vertices = 10, faces = 20
E) vertices = 20, faces = 36

Answer: A
Diff: 2 *Page Ref:* 22.4

12) Determine the number of vertices and faces in the *closo*–Borane $B_{12}H_{12}^{2-}$.

A) vertices = 12, faces = 24
B) vertices = 12, faces = 20
C) vertices = 8, faces = 16
D) vertices = 6, faces = 8
E) vertices = 8, faces = 12

Answer: B
Diff: 2 *Page Ref:* 22.4

13) What is the major detraction from using bituminous coal as an energy source?

A) It has a very low carbon content.
B) It contains a high amount of sulfur, which when burned contributes to acid rain.
C) It has an extremely high surface area which makes it difficult to burn.
D) Bituminous coal is noncrystalline, which gives it a low energy/gram ratio.
E) There are no detractions from using bituminous coal.

Answer: B
Diff: 1 *Page Ref:* 22.5

14) List the name of powdered form of carbon, which is a component of soot.

A) diamond
B) graphite
C) fullerenes
D) bucky balls
E) carbon black

Answer: E
Diff: 1 *Page Ref:* 22.5

15) Identify the form of carbon that is a good lubricant.
A) nanotube
B) buckyball
C) graphite
D) diamond
E) fullerene

Answer: C
Diff: 1 *Page Ref: 22.5*

16) Identify the form of carbon that is used commercially for bike frames and golf clubs.
A) nanotube
B) buckyball
C) graphite
D) diamond
E) fullerene

Answer: A
Diff: 1 *Page Ref: 22.5*

17) Identify the form of carbon that is used commercially for cutting tools and for gems.
A) nanotube
B) buckyball
C) graphite
D) diamond
E) fullerene

Answer: D
Diff: 1 *Page Ref: 22.5*

18) Which of the following statements is TRUE?
A) Sheets of graphitic carbon bond strongly to one another.
B) The graphite form of carbon is more dense than diamond.
C) Carbon atoms sometimes cluster together into substances called fullerenes.
D) The diamond form of carbon conducts electricity.
E) None of the above are true.

Answer: C
Diff: 1 *Page Ref: 22.5*

19) Which of the following statements is TRUE?
A) Ionic carbides form between carbon and another high electronegativity element.
B) Carbides are soft materials with low melting points.
C) Metallic carbides are more malleable than the pure metal.
D) Covalent carbides are formed between carbon and a low electronegativity nonmetal.
E) None of the above are true.

Answer: D
Diff: 1 *Page Ref: 22.5*

20) Determine the oxidation state of carbon in the ionic carbide, CaC_2.
A) -2 B) +4 C) -1 D) +1 E) -4

Answer: C
Diff: 2 *Page Ref: 22.5*

21) Determine the oxidation state of carbon in the carbonate ion, CO_3^{2-}.

A) +2 B) +4 C) +3 D) +6 E) +8

Answer: B
Diff: 2 *Page Ref: 22.5*

22) Determine the oxidation state of carbon in the carbon monoxide, CO.

A) +2 B) +4 C) +3 D) +6 E) +8

Answer: A
Diff: 2 *Page Ref: 22.5*

23) Determine the oxidation state of nitrogen in NO_2^-.

A) −1 B) −2 C) +5 D) E) +3

Answer: E
Diff: 2 *Page Ref: 22.6*

24) Identify an example of a nitrate.

A) NaN_3 B) $NaNO_2$ C) $NaNO_3$ D) NO_2 E) NO

Answer: C
Diff: 2 *Page Ref: 22.6*

25) Determine the oxidation state of phosphorus in $P_2O_7^{4-}$.

A) +5 B) +4 C) +3 D) +7 E) +2

Answer: A
Diff: 2 *Page Ref: 22.6*

26) Which of the following statements is TRUE?

A) Ammonium nitrate is a good explosive.
B) Sodium nitrite is used as a meat preservative.
C) NO_2 reacts with water to form nitric acid.
D) Hydrazine is a good reducing agent.
E) All of the above are true.

Answer: E
Diff: 1 *Page Ref: 22.6*

27) Which of the following is TRUE?

A) Phosphoric acid is used to produce the "fizziness" in soda.
B) Sodium phosphate is an additive in baked goods to give them a longer shelf life.
C) Phosphorus can range in oxidation state from −3 to +5.
D) The phosphorus in the phosphite ion has an oxidation state of − 3
E) None of the above are true.

Answer: C
Diff: 1 *Page Ref: 22.6*

28) Tetraphosphorus decoxide reacts with water to produce phosphoric acid. Write a balanced reaction for this process.

A) $P_4O_{10} + 6\ H_2O \rightarrow 4\ H_3PO_4$
B) $P_2O_5 + 3\ H_2O \rightarrow 2\ H_3PO_3$
C) $P_4 + 3\ O_2 + 6\ H_2O \rightarrow 4\ H_3PO_4$
D) $P_4O_{10} + 4\ H_2O \rightarrow 4\ H_2PO_3$
E) $P_4 + 3\ O_2 + 6\ H_2O \rightarrow 4\ H_3PO_3$

Answer: A
Diff: 3 *Page Ref: 22.6*

29) Identify the oxygen compound that kills bacteria and is used as a disinfectant.

A) O_3 B) O_2 C) SO_2 D) SO_3 E) NO_2

Answer: A
Diff: 2 *Page Ref: 22.7*

30) Which of the following statements is TRUE?

A) Water and H_2S have the same bond angles.
B) Hydrofluoric acid is a strong reducing agent.
C) Sulfur is a stronger oxidizing agent than oxygen.
D) Water is less stable than H_2S.
E) Sulfuric acid is a strong oxidizing agent.

Answer: E
Diff: 1 *Page Ref: 22.8*

31) Identify the compound that smells like a rotten egg.

A) SO_2 B) H_2SO_3 C) H_2SO_4 D) S_4 E) H_2S

Answer: E
Diff: 1 *Page Ref: 22.8*

32) Identify the compound that is formed by volcanos and is a coal pollutant.

A) SO_2 B) H_2SO_3 C) H_2SO_4 D) S_4 E) H_2S

Answer: A
Diff: 1 *Page Ref: 22.8*

33) Determine the oxidation state of sulfur in $S_2O_3^{2-}$.

A) +3 B) +6 C) +2 D) +4 E) 0

Answer: C
Diff: 2 *Page Ref: 22.8*

34) Determine the oxidation state of sulfur in SO_2.

A) +3 B) +6 C) +2 D) +4 E) 0

Answer: D
Diff: 2 *Page Ref: 22.8*

35) Write the balanced equation that represents the formation of hydrofluoric acid from the reaction of calcium fluoride with sulfuric acid.

A) $CaF_2 + H_2S \rightarrow CaS + 2\ HF$
B) $CF_4 + 2\ H_2S \rightarrow 4\ HF + CS_2$
C) $CF_4 + 2\ H_2SO_4 \rightarrow 4\ HF + CS_2 + CO_2$
D) $CaF_2 + H_2SO_4 \rightarrow 2\ HF + CaSO_4$
E) $CaF_2 + H_2SO_3 \rightarrow CaSO_3 + 2\ HF$

Answer: D
Diff: 2 *Page Ref:* 22.9

Algorithmic Questions

1) Which of the following are MAJOR components of Earth's crust?

A) oxygen and silicon
B) calcium and silicon
C) oxygen and hydrogen
D) oxygen, magnesium and carbon
E) oxygen, hydrogen and carbon

Answer: A
Diff: 1 *Page Ref:* 22.3

2) Give the formula of the compound that is in common glass.

A) SiO_2 B) Si_2O_3 C) SiO_3 D) Si_2O_8 E) SiO_4

Answer: A
Diff: 2 *Page Ref:* 22.3

TABLE 22.2 Types of Silicate Structures

Tetrahedrons	Shared Vertices	Formula Unit	Si : O Ratio	Class Name	Example
Single tetrahedron	0	SiO_4^{4-}	1:4	Orthosilicates, nesosilicates	Olivine, Mg_2SiO_4
Double tetrahedron	1	$Si_2O_7^{6-}$	2:7	Pyrosilicates, sorosilicates	Hardystonite, $Ca_2ZnSi_2O_7$
Single chain	2	SiO_3^{2-}	1:3	Pyroxenes, inosilicates	Jadeite, $NaAl(SiO_3)_2$
Double chain	2 and 3	$Si_4O_{11}^{6-}$	4:11	Amphiboles	Tremolite, $Ca_2(OH)_2Mg_5(Si_4O_{11})_2$
Sheet	3	$Si_2O_5^{2-}$	2:5	Phyllosilicates	Talc, $Mg_3(Si_2O_5)_2(OH)_2$
Network covalent	4	SiO_2	1:2	Silicas, tectosilicates	Quartz, SiO_2
Network covalent	4	$AlSi_3O_8^-$ or $Al_2Si_2O_8^{2-}$	Variable	Feldspars	Albite, $NaAlSi_3O_8$

3) Refer to the above table. Use charge balance to determine the value of x in the phyllosilicate Zn $_3(Si_2O_5)_2(OH)_x$.

A) 6 B) 4 C) 3 D) 1 E) 2

Answer: E
Diff: 3 *Page Ref:* 22.3

4) Refer to the above table. Use charge balance to determine the value of x in the pyroxene $Sr_x(SiO_3)_2$.

A) 2 B) 1 C) 4 D) 6 E) 3

Answer: A
Diff: 3 *Page Ref:* 22.3

5) Refer to the above table. Use charge balance to determine the value of x in the pyrosilicate $Ba_xSi_2O_7$.

A) 1 B) 3 C) 6 D) 4 E) 2

Answer: B
Diff: 3 *Page Ref:* 22.3

6) Refer to the above table. Use charge balance to determine the value of x in the orthosilicate Ba_xSiO_4.

A) 3 B) 1 C) 2 D) 4 E) 6

Answer: C
Diff: 3 *Page Ref:* 22.3

7) Refer to the above table. Use charge balance to determine the value of x in the amphibole $Ca_2(OH)_2Zn_x(Si_4O_{11})_2$.

A) 1 B) 4 C) 2 D) 5 E) 3

Answer: D
Diff: 3 *Page Ref:* 22.3

8) Refer to the above table. Predict the silicate structure for $ZnBa_2(Si_2O_7)$.

A) orthosilicate
B) pyrosilicate
C) pyroxene
D) amphibole
E) phyllosilicate

Answer: B
Diff: 3 *Page Ref:* 22.3

9) Refer to the above table. Predict the silicate structure for $CaZn(SiO_4)$.

A) pyrosilicate
B) amphibole
C) orthosilicate
D) phyllosilicate
E) pyroxene

Answer: C
Diff: 3 *Page Ref:* 22.3

10) Refer to the above table. Predict the silicate structure for $NaAl(SiO_3)_2$.

A) phyllosilicate
B) orthosilicate
C) amphibole
D) pyroxene
E) pyrosilicate

Answer: D
Diff: 3 *Page Ref: 22.3*

11) Determine the oxidation state of nitrogen in KN_3.

A) +1 B) +3 C) -2 D) $-\frac{1}{3}$ E) -3

Answer: D
Diff: 2 *Page Ref: 22.6*

12) Determine the oxidation state of nitrogen in NO_3^-.

A) -1 B) -6 C) +2 D) +4 E) +5

Answer: E
Diff: 2 *Page Ref: 22.6*

13) Identify an example of a nitrite.

A) $NaNO_2$ B) NO C) N_2O D) $LiNO_3$ E) NO_3

Answer: A
Diff: 2 *Page Ref: 22.6*

14) What is the oxidation number of oxygen in K_2O_2?

A) -4 B) +4 C) -3 D) -1 E) +1

Answer: D
Diff: 2 *Page Ref: 22.7*

15) What is the oxidation number of oxygen in K_2O?

A) -4 B) +4 C) -2 D) -3 E) +1

Answer: C
Diff: 2 *Page Ref: 22.7*

16) Determine the oxidation state of bromine in BrO^-.

A) 0 B) +1 C) -3 D) -1 E) +2

Answer: B
Diff: 2 *Page Ref: 22.9*

Matching Questions

Match the following.

1) orthosilicate
Diff: 2 *Page Ref: 22.3*

2) pyroxene
Diff: 2 *Page Ref: 22.3*

3) pyrosilicate
Diff: 2 *Page Ref: 22.3*

4) amphibole
Diff: 2 *Page Ref: 22.3*

5) phyllosilicate
Diff: 2 *Page Ref: 22.3*

A) $Si_2O_7^{6-}$

B) SiO_3^{2-}

C) $Si_2O_5^{2-}$

D) SiO_4^{4-}

E) $Si_4O_{11}^{6-}$

Answers: 1) D 2) B 3) A 4) E 5) C

Short Answer Questions

1) Why are phosphate compounds added to detergents?

Answer: They have the ability to remove Ca^{2+} and Mg^{2+} from hard water, making the detergent work more efficiently.
Diff: 2 *Page Ref: 22.6*

2) Why are the chemical properties of nitrogen and phosphorus so different when they are in the same family?

Answer: Phosphorus is much larger and has d orbitals available where nitrogen is smaller with no d orbitals. Phosphorus is also much less electronegative.
Diff: 2 *Page Ref: 22.6*

3) Describe the major production method for obtaining oxygen.

Answer: The most cost effective method is called air fractionation. Air is cooled down to liquify all components. Then it is slowly warmed and each component boils off separately, leaving oxygen behind.
Diff: 2 *Page Ref: 22.7*

4) Describe the Frasch process for obtaining sulfur.

Answer: Since sulfur deposits are typically found underground, superheated water is piped into the deposit where it liquifies the sulfur. Then air is pumped into the deposit where the pressure forces the liquified sulfur to the surface through pipes.
Diff: 2 *Page Ref: 22.8*

5) Draw the Lewis structure for IF_4^- and determine the geometry of the ion.

Answer: Student should show the Lewis structure with 4 bonds to the I and 2 lone pairs on the I. The geometry is square planar.

Diff: 2 *Page Ref:* 22.9

6) Write the balanced reaction that shows the formation of ICl_3 from its elements.

Answer: $I_2(s) + 3\ Cl_2(g) \rightarrow 2\ ICl_3(g)$

Diff: 3 *Page Ref:* 22.9

Chapter 23 Metals and Metallurgy

Multiple Choice Questions

1) Which of the following is NOT generally a property of a metal?
   A) high resistivity
   B) good conductor of heat
   C) malleable
   D) ductile
   E) opaque

Answer: A
Diff: 1 *Page Ref: 23.2*

2) Identify the substances which are homogeneous, naturally occurring, crystalline inorganic solids.
   A) slag B) minerals C) ores D) gangue E) alloy

Answer: B
Diff: 1 *Page Ref: 23.2*

3) Identify the substances which contains a high concentration of a specific mineral.
   A) slag B) minerals C) ores D) gangue E) alloy

Answer: C
Diff: 1 *Page Ref: 23.2*

4) Which of the following describes gangue?
   A) The undesirable product formed during smelting.
   B) The undesirable portion of a metal-containing ore.
   C) The volatile product formed during calcination.
   D) A waste material that is the product of smelting a metal ore.
   E) None of the above describe gangue.

Answer: B
Diff: 1 *Page Ref: 23.3*

5) Which of the following describes flux?
   A) A waste material that is the product of smelting a metal ore.
   B) An aqueous portion of the gangue separated from a metal ore.
   C) A low density waste product, formed from the reaction of gangue with another substance, that can be easily separated from the metal ore of interest.
   D) A substance that reacts with the gangue to make a low density liquid compound that can be easily separated from the metal of interest.
   E) None of the above describe flux.

Answer: D
Diff: 1 *Page Ref: 23.3*

6) Which of the following describes slag?
   A) A waste material that is the product of smelting a metal ore.
   B) An aqueous portion of the gangue separated from a metal ore.
   C) A low density waste product, formed from the reaction of gangue with another substance, that can be easily separated from the metal ore of interest.
   D) A substance that reacts with the gangue to make a low density liquid compound that can be easily separated from the metal of interest.
   E) None of the above describe slag.

Answer: C
Diff: 1 *Page Ref: 23.3*

7) What method is used to obtain pure aluminum metal from Al_2O_3?
   A) smelting
   B) powder metallurgy
   C) Hall process
   D) leaching
   E) calcination

Answer: C
Diff: 1 *Page Ref: 23.3*

8) Which of the following is TRUE of powder metallurgy?
   A) Machining is required after forming the component.
   B) There is more waste than with casting the molten metal.
   C) Iron powder from mill scrap makes a denser component than iron particles formed through water atomization.
   D) It is used to separate gangue from the rest of the metal ore.
   E) It usually requires lower temperatures than casting the molten metal.

Answer: E
Diff: 1 *Page Ref: 23.3*

9) Which of the following is an example of calcination?
   A) $2\ CuS(s) + 3\ O_2 \xrightarrow{heat} 2\ CuO(s) + 2\ SO_2(g)$
   B) $SnO_2(s) + 2\ C(s) \xrightarrow{heat} Sn(l) + 2\ CO(g)$
   C) $CaCO_3(s) + SiO_2(s) \xrightarrow{heat} CaSiO_3(l) + CO_2(g)$
   D) $MgCO_3(s) \xrightarrow{heat} MgO(s) + CO_2(g)$
   E) $PbO(s) + CO(g) \xrightarrow{heat} Pb(l) + CO_2(g)$

Answer: D
Diff: 2 *Page Ref: 23.3*

10) Which of the following is an example of roasting?

A) $CaCO_3(s) \rightarrow CaO(s) + CO_2(g)$

B) $PbO(s) + CO(g) \rightarrow Pb(l) + CO_2(g)$

C) $SnO_2(s) + 2\ C(s) \xrightarrow{heat} Sn(l) + 2\ CO(g)$

D) $Ni(OH)_2(s) \xrightarrow{heat} NiO(s) + H_2O(g)$

E) $2\ CuS(s) + 3\ O_2 \xrightarrow{heat} 2\ CuO(s) + 2\ SO_2(g)$

Answer: E
Diff: 2 *Page Ref:* 23.3

11) Which of the following is an example of smelting?

A) $PbO(s) + CO(g) \rightarrow Pb(l) + CO_2(g)$

B) $2\ CuS(s) + 3\ O_2 \xrightarrow{heat} 2\ CuO(s) + 2\ SO_2(g)$

C) $CaCO_3(s) \rightarrow CaO(s) + CO_2(g)$

D) $Ni(OH)_2(s) \xrightarrow{heat} NiO(s) + H_2O(g)$

E) $2\ Hg(l) + O_2(g) \xrightarrow{heat} 2\ HgO(s)$

Answer: A
Diff: 2 *Page Ref:* 23.3

12) Which compound in the following reaction is the slag?

$$2\ Ca_3(PO_4)_2(s) + 6\ SiO_2(s) + 10\ C(s) \xrightarrow{heat} P_4(g) + 6\ CaSiO_3(l) + 10\ CO(s)$$

A) $CaSiO_3$
B) CO
C) P_4
D) SiO_2
E) $Ca_3(PO_4)_2$

Answer: A
Diff: 2 *Page Ref:* 23.3

13) Which of the following is an example of a hydrometallurgical process?

A) $Al(OH)_3(s) + SiO_2(s) \xrightarrow{heat} AlSiO_3(s) + 3\ H_2O(g)$

B) $Ni(OH)_2(s) \xrightarrow{heat} NiO(s) + H_2O(g)$

C) $2\ Al(OH)_3(s) \xrightarrow{heat} Al_2O_3(s) + 3\ H_2O(g)$

D) $CuSO_4{\cdot}5H_2O(s) + H_2O(l) \rightarrow CuS(s) + H_2SO_4(aq) + 5\ OH^-(aq)$

E) $Al_2O_3{\cdot}nH_2O(s) + 2\ OH^-(aq) + 2\ H_2O(l) \rightarrow 2\ Al(OH)_4{}^-(aq)$

Answer: E
Diff: 2 *Page Ref:* 23.3

14) Write a balanced reaction for the smelting of SnO_2 with carbon.

A) $SnO_2(s) + CO_2(g) \xrightarrow{\text{heat}} SnCO_3(s)$

B) $SnO_2(s) + CaCO_3(s) \xrightarrow{\text{heat}} SnC(s) + CaO(s) + O_2(g)$

C) $SnO_2(s) + 2\ C(s) \xrightarrow{\text{heat}} Sn(l) + 2\ CO(g)$

D) $2\ SnO_2(s) + 2\ CaC_2(s) \xrightarrow{\text{heat}} 2\ SnC_2(s) + 2\ CaO(s) + O_2(g)$

E) None of the above represent smelting.

Answer: C
Diff: 3 *Page Ref: 23.3*

15) Write a balanced equation for the calcination of $Al(OH)_3$.

A) $2\ Al(OH)_3(s) \xrightarrow{\text{heat}} Al_2O_3(s) + 3\ H_2O(g)$

B) $Al(OH)_3(s) + SiO_2(s) \xrightarrow{\text{heat}} AlSiO_3(s) + 3\ H_2O(g)$

C) $2\ Al(OH)_3(s) + O_2(g) \xrightarrow{\text{heat}} Al_2O_3(s) + 3\ H_2O(g)$

D) $Al(OH)_3(s) + C(s) \xrightarrow{\text{heat}} Al_2O_3(s) + CO_2(g) + H_2O(g)$

E) None of the above represent calcination.

Answer: A
Diff: 3 *Page Ref: 23.3*

16) Write a balanced equation for the calcination of $NiCO_3$.

A) $NiCO_3(s) + CaO(s) \xrightarrow{\text{heat}} NiO(s) + CaCO_3(s)$

B) $NiCO_3(s) \xrightarrow{\text{heat}} NiO(s) + CO_2(g)$

C) $NiCO_3(s) + C(s) \xrightarrow{\text{heat}} NiO(s) + 2\ CO(g)$

D) $NiCO_3(s) + SiO_2(s) \xrightarrow{\text{heat}} NiSiO_3(s) + CO_2(g)$

E) None of the above represent calcination.

Answer: B
Diff: 3 *Page Ref: 23.3*

17) In a substitutional alloy,

A) one nonmetal substitutes for another nonmetal in a crystal structure.
B) small, usually nonmetallic atoms fit in between the metallic atoms of a crystal.
C) small metallic atoms fit in between the nonmetallic atoms of a crystal
D) one metal substitutes for another metal in a crystal structure.
E) one metal substitutes for a nonmetal in a crystal structure.

Answer: D
Diff: 1 *Page Ref: 23.4*

18) In an interstitial alloy,
A) one nonmetal substitutes for another nonmetal in a crystal structure.
B) small, usually nonmetallic atoms fit in between the metallic atoms of a crystal.
C) small metallic atoms fit in between the nonmetallic atoms of a crystal
D) one metal substitutes for another metal in a crystal structure.
E) one metal substitutes for a nonmetal in a crystal structure.

Answer: B
Diff: 1 *Page Ref: 23.4*

19) Which of the following is TRUE concerning substitutional alloys?
A) The metals involved must have similar radii.
B) One type of metal atom is inserted in between the metal atoms in the initial crystal structure of the other metal.
C) The two metals must be miscible over the entire composition range.
D) Titanium carbide is an example of a substitutional alloy.
E) None of the above are true.

Answer: A
Diff: 1 *Page Ref: 23.4*

20) Why can't chromium and nickel form a miscible solid solution over the entire composition range?
A) The lever rule says that these two metals can't coexist in an alloy.
B) The two metals crystallize into different cubic structures in their pure forms.
C) The tetrahedral holes in the nickel crystal are not large enough to accommodate the chromium.
D) The octahedral holes in the chromium crystal are not large enough to accommodate the nickel.
E) None of the above are true.

Answer: B
Diff: 1 *Page Ref: 23.4*

21) Which of the following is TRUE concerning point "D" on the figure?

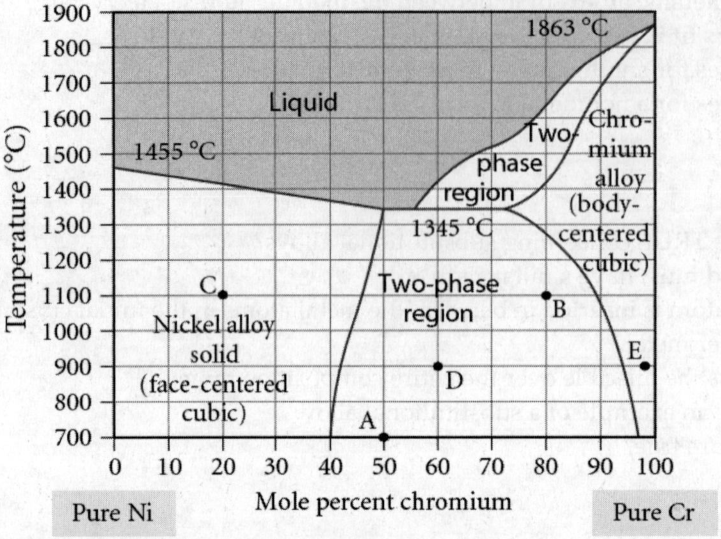

A) The alloy is composed of the BCC structure.
B) The alloy is composed of the FCC structure.
C) The alloy is composed of both FCC and BCC phases, with a larger portion of FCC present.
D) The alloy is composed of both FCC and BCC phases, with a larger portion of BCC present.
E) None of the above can be determined from the information given.

Answer: C
Diff: 1 *Page Ref: 23.4*

22) Which of the following is TRUE concerning point "B" on the figure?

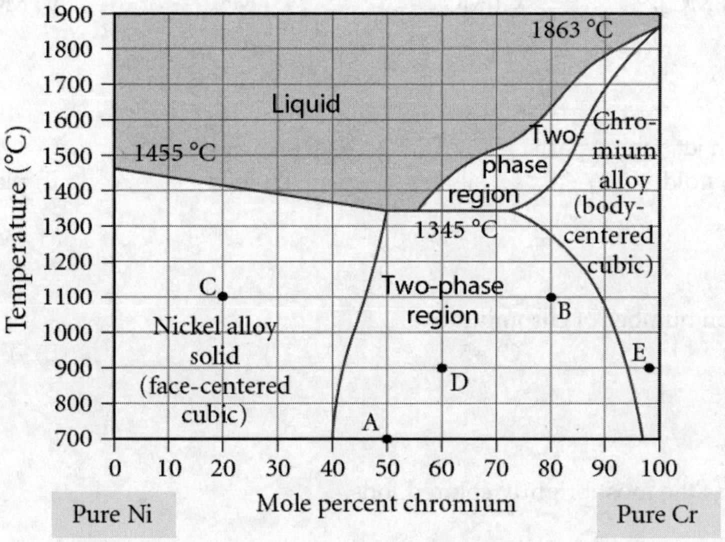

A) The alloy is composed of the BCC structure.
B) The alloy is composed of the FCC structure.
C) The alloy is composed of both FCC and BCC phases, with a larger portion of BCC present.
D) The alloy is composed of both FCC and BCC phases, with a larger portion of FCC present.
E) None of the above can be determined from the information given.

Answer: C
Diff: 1 *Page Ref: 23.4*

23) Which of the following is TRUE concerning interstitial alloys?
A) The metal takes on the properties of the foreign atoms once they are present.
B) Small atoms fit in between the metal atoms in the crystal structure.
C) Brass is an example of an interstitial alloy.
D) The foreign atoms present weaken the metallic structure.
E) None of the above are true.

Answer: B
Diff: 1 *Page Ref: 23.4*

24) An interstitial alloy contains nitrogen in half of the octahedral holes of a closest packed metal, M. What is the formula of this alloy?
A) MN B) MN_2 C) M_4N D) M_2N E) MN_4

Answer: D
Diff: 2 *Page Ref: 23.4*

25) An interstitial alloy contains hydrogen in half of the tetrahedral holes of a closest packed metal, M. What is the formula of this alloy?
A) MH B) MH_4 C) MH_2 D) M_2H E) M_4H

Answer: E
Diff: 2 *Page Ref: 23.4*

26) An interstitial alloy contains carbon in one-fourth of the octahedral holes of a closest packed metal, M. What is the formula of this alloy?
A) M_4C B) MC_4 C) MC D) M_2C E) MC_2

Answer: A
Diff: 2 *Page Ref:* 23.4

27) Identify the metal used in jet engine parts.
A) titanium B) gold C) silver D) lead E) platinum

Answer: A
Diff: 2 *Page Ref:* 23.5

28) Give the highest oxidation number of chromium.
A) +3 B) +4 C) +5 D) +7 E) +6

Answer: E
Diff: 1 *Page Ref:* 23.5

29) Identify the metal that has the most brightly colored ions.
A) gold
B) silver
C) tin
D) chromium
E) lead

Answer: D
Diff: 1 *Page Ref:* 23.5

30) Give the highest oxidation number of manganese.
A) +3 B) +4 C) +5 D) +7 E) +6

Answer: D
Diff: 1 *Page Ref:* 23.5

31) Identify the ferromagnetic elements
A) cobalt
B) iron
C) nickel
D) all of the above
E) none of the above

Answer: D
Diff: 1 *Page Ref:* 23.5

Algorithmic Questions

1) Which substance is the best conductor of electricity?
A) arsenic B) boron C) silver D) tellurium

Answer: C
Diff: 1 *Page Ref:* 23.2

2) Identify the metal that is a noble metal.
A) platinum
B) potassium
C) calcium
D) carbon
E) aluminum

Answer: A
Diff: 1 *Page Ref: 23.2*

3) Pyrometallurgy is the
A) refining of metal ores using reactions with aqueous solutions.
B) refining of metal ores using heat.
C) refining of metal ores using reduction reactions.
D) forming metal parts using heat and small crystals of metal.
E) forming metal parts using electricity.

Answer: B
Diff: 1 *Page Ref: 23.3*

4) Calcination is
A) heating an ore, in the presence of argon or another substance, in order to cause a chemical reaction that drives off newly formed volatile compounds.
B) heating an ore, in the presence of helium or another substance, in order to purify the ore and obtain the liquid metal.
C) selectively dissolving a metal in solution to separate it from its ore.
D) the process of heating an ore to drive off volatile compounds.
E) forming metal parts using heat and small crystals of metal.

Answer: D
Diff: 1 *Page Ref: 23.3*

5) Roasting is
A) forming metal parts using heat and small crystals of metal.
B) the process of heating an ore to drive off volatile compounds.
C) selectively dissolving a metal in strong acid to separate it from its ore.
D) heating an ore, in the presence of helium or another substance, in order to purify the ore and obtain the liquid metal.
E) heating an ore, in the presence of oxygen or another substance, in order to cause a chemical reaction that drives off newly formed volatile compounds.

Answer: E
Diff: 1 *Page Ref: 23.3*

6) Smelting is
A) heating an ore, in the presence of oxygen or another substance, in order to purify the ore and obtain the liquid metal.
B) selectively dissolving a metal in strong acid to separate it from its ore.
C) the process of cooling an ore to drive off volatile compounds.
D) heating an ore, in the presence of oxygen or another substance, in order to cause a chemical reaction that drives off newly formed volatile compounds.
E) forming metal parts using heat and small crystals of metal.

Answer: A
Diff: 1 *Page Ref: 23.3*

7) Hydrometallurgy is
   A) refining of metal ores using oxidation reactions.
   B) refining of metal ores using heat.
   C) refining of metal ores using reactions with aqueous solutions.
   D) forming metal parts using heat and small crystals of metal.
   E) forming tiny metal crystals using heat and water spray.

Answer: C
Diff: 1   *Page Ref: 23.3*

8) Electrometallurgy is
   A) refining of metal ores using cold.
   B) forming metal parts using steam.
   C) forming metal parts using cold and small crystals of metal.
   D) refining of metal ores using oxidation/reduction reactions.
   E) refining of metal ores using reactions with aqueous solutions.

Answer: D
Diff: 1   *Page Ref: 23.3*

9) Leaching is
   A) melting an ore and separating it from the lower density gangue.
   B) selectively dissolving a metal in solution to separate it from its ore.
   C) the process of heating an ore to drive off volatile compounds.
   D) heating an ore, in the presence of argon or another substance, in order to purify the ore and obtain the liquid metal.
   E) heating an ore, in the presence of argon or another substance, in order to cause a chemical reaction that drives off newly formed volatile compounds.

Answer: B
Diff: 1   *Page Ref: 23.3*

10) Powder metallurgy is
   A) refining of metal ores using reduction reactions.
   B) refining of metal ores using cold.
   C) forming metal parts using electricity.
   D) refining of metal ores using reactions with basic solutions.
   E) forming metal parts using heat and small crystals of metal.

Answer: E
Diff: 1   *Page Ref: 23.3*

11) Identify the normal crystal structure at room temperature for Cu.
   A) face-centered cubic
   B) trigonal planar
   C) tetrahedral
   D) hexagonal
   E) hexagonal closest packed

Answer: A
Diff: 1   *Page Ref: 23.4*

12) Identify the normal crystal structure at room temperature for Co.
A) face-centered cubic
B) trigonal planar
C) octahedral
D) hexagonal
E) hexagonal closest packed

Answer: E
Diff: 1 *Page Ref: 23.4*

13) Identify the metal most commonly used for wires in the electrical industry.
A) copper B) silver C) zinc D) platinum E) lithium

Answer: A
Diff: 1 *Page Ref: 23.5*

14) Give the components of bronze.
A) copper and tin
B) lead and aluminum
C) nickel and zinc
D) copper and lead
E) zinc and lead

Answer: A
Diff: 1 *Page Ref: 23.5*

15) Give the components of brass.
A) copper and tin
B) lead and tin
C) nickel and gold
D) copper and zinc
E) zinc and lead

Answer: D
Diff: 1 *Page Ref: 23.5*

16) Identify the metal used as a coating for steel objects such as nails.
A) gold
B) silver
C) zinc
D) palladium
E) manganese

Answer: C
Diff: 1 *Page Ref: 23.5*

Matching Questions

Match the following.

1) Hg
Diff: 1 *Page Ref: 23.2*

2) Pb
Diff: 1 *Page Ref: 23.2*

3) V
Diff: 1 *Page Ref: 23.2*

4) Ti
Diff: 1 *Page Ref: 23.2*

5) Zn
Diff: 1 *Page Ref: 23.2*

6) Mn
Diff: 1 *Page Ref: 23.5*

7) Cu
Diff: 1 *Page Ref: 23.5*

A) rhodochrosite

B) galena

C) rutile

D) carnotite

E) sphalerite

F) cinnabar

G) malachite

Answers: 1) F 2) B 3) D 4) C 5) E 6) A 7) G

Short Answer Questions

1) What is an advantage of hydrometallurgical processes over pyrometallurgical processes?

Answer: Hydrometallurgy generally doesn't require elevated temperatures like pyrometallurgy. This results in a more economical processing of a metal ore.
Diff: 2 *Page Ref: 23.3*

2) Why does a "two-phase" structure occur in a substitutional alloy?

Answer: When a substitutional alloy is formed from two metals with differing crystal structures, there are some compositions (ratios of metal 1 to metal 2) that result in the presence of both crystal structures.
Diff: 2 *Page Ref: 23.4*

3) Describe the difference between a substitutional alloy and an interstitial alloy.

Answer: A substitutional alloy just "swaps" one metal atom for another of similar size. In an interstitial alloy, an additional element is added to the empty spaces between metal atoms in their crystalline structure.
Diff: 2 *Page Ref: 23.4*

4) What is the difference between ferromagnetism and paramagnetism?

Answer: When a substance is paramagnetic, it contains unpaired electrons that can interact with a magnetic field. When a substance is ferromagnetic, all of the unpaired electrons are aligned in a particular direction, resulting in a permanent magnetic field.

Diff: 2 *Page Ref:* 23.5

5) Why is zinc used to coat steel objects?

Answer: Zinc is a stronger reducing agent than iron. As a result, it will react with oxygen in the atmosphere to form a protective zinc oxide coating that prevents corrosion of the steel.

Diff: 2 *Page Ref:* 23.5

Chapter 24 Transition Metals and Coordination Compounds

Multiple Choice Questions

1) Identify the ion that is responsible for the red color of rubies.

A) Cr^{3+} B) Cr^{4+} C) Cr^{5+} D) Cr^{6+} E) Cr^{7+}

Answer: A

Diff: 1 *Page Ref: 24.1*

2) Identify the ion that is responsible for the green color in emeralds.

A) Cr^{3+} B) Cr^{4+} C) Cr^{5+} D) Cr^{6+} E) Cr^{7+}

Answer: A

Diff: 1 *Page Ref: 24.1*

3) Identify the ion that is responsible for the blue of turquoise.

A) Cr^{2+} B) Cu^{+} C) Cu^{2+} D) Cr^{3+} E) Fe^{2+}

Answer: C

Diff: 1 *Page Ref: 24.1*

4) Identify the ion that is responsible for the red in garnet and the yellow-green of peridot.

A) Cr^{2+} B) Cu^{+} C) Cu^{2+} D) Cr^{3+} E) Fe^{2+}

Answer: E

Diff: 1 *Page Ref: 24.1*

5) Choose the electron configuration for Fe^{3+}.

A) $[Ar]4s^23d^3$
B) $[Ar]3d^5$
C) $[Ar]4s^23d^9$
D) $[Ar]4s^13d^4$
E) $[Ar]4s^23d^6$

Answer: B

Diff: 2 *Page Ref: 24.2*

6) Name the following: $Fe[AlF_6]$

A) ironaluminumhexafluoride
B) iron(II) hexafluoroaluminum
C) iron (III) hexafluoroaluminate
D) iron(I) aluminumhexafluoride
E) aluminumhexafluoroferrate

Answer: C

Diff: 3 *Page Ref: 24.3*

7) Name the following: $[Fe(H_2O)_4Br_2]Cl$

A) tetraaquadibromoiron(II) chloride
B) dibromotetrahydroironchloride
C) dibromotetrahydroiron(II)chloride
D) tetraaquadibromoiron(III) chloride
E) iron(II)dibromotetraquachloride

Answer: D
Diff: 3 *Page Ref: 24.3*

8) Name the following: $[Ni(NH_3)_3(H_2O)_3]Cl_2$

A) nickel(II)trihydrotriamminechloride
B) nickel(I)triamminetrihydrochloride
C) dichloronickel(II)triamminetrihydride
D) nickel(III)chloride
E) triamminetriaquanickel(II)chloride

Answer: E
Diff: 3 *Page Ref: 24.3*

9) Name the following: $[Pt(H_2O)_4F_2]Br_2$

A) tetraaquadifluoroplatinum(IV) bromide
B) tetraaquadibromodifluoroplatinate
C) platinum(II)bromide
D) platinum(III)tetraaquadifluorobromide
E) platinum (II) dibromodifluorotetrahydride

Answer: A
Diff: 3 *Page Ref: 24.3*

10) Name the following: $[Fe(CO)_6][Ag(CN)_2]_2$

A) iron(III)hexacarbonylsilverdicyanide
B) hexacarbonyliron(II) dicyanoargentate
C) iron(II) carbonyl silvercyanide
D) hexacarbonyliron(III) dicyanosilvate
E) iron(II)hexacarbonyl silvercyanide

Answer: B
Diff: 3 *Page Ref: 24.3*

11) Determine the chemical formula for the compound, diamminetetraaquairon(II)chloride.

A) $[Fe(NH_3)_2(H_2O)_4Cl]$
B) $[Fe(NH_3)_2][(H_2O)_4Cl]$
C) $[Fe(NH_3)_2(H_2O)_4]Cl_2$
D) $[Fe(H_2O)_4][(NH_3)_2Cl]$
E) $[Fe(NH_3)_2(H_2O)_4]Cl_3$

Answer: C
Diff: 3 *Page Ref: 24.3*

12) Determine the chemical formula for the compound, hexaammineiron(III) hexacyanoferrate(III).

A) $[Fe_2(NH_3)_6(CN)_6]^{3+}$
B) $[Fe(NH_3)_6][Fe(CN)_6]_2$
C) $[Fe(NH_3)_6]_3[Fe(CN)_6]_2$
D) $[Fe(NH_3)_6][Fe(CN)_6]$
E) $[Fe_2(NH_3)_3(CN)_3]^{6+}$

Answer: D
Diff: 3 *Page Ref: 24.3*

13) Determine the chemical formula for the compound, diaquadicarbonylzinc tetrabromopalladate (IV).

A) $[PdZn(H_2O)_2(CO)_2]Br_4$
B) $[Zn(H_2O)_2(CO)_2]_2[PdBr_4]$
C) $[Pd(H_2O)_2][Zn(CO)_2]Br_4$
D) $[Pd(H_2O)_2]_2[Zn(CO)_2]_3Br_4$
E) $[Zn(H_2O)_2(CO)_2][PdBr_4]$

Answer: E
Diff: 3 *Page Ref: 24.3*

14) Determine the chemical formula for the compound, potassiumtrisethylenediamine chromate(III).

A) $K_3[Cr(NH_2CH_2CH_2NH_2)_3]$
B) $[K(NH_2CH_2CH_2NH_2)_3]Cr^{3+}$
C) $[KCr(NH_2CH_2CH_2NH_2)_3]^{3+}$
D) $K_2[Cr(NH_2CH_2CH_2NH_2)_4]$
E) $[Cr(NH_2CH_2CH_2NH_2)_2]K_2$

Answer: A
Diff: 3 *Page Ref: 24.3*

15) Determine the chemical formula for the compound, tetracarbonylplatinum(IV) chloride.

A) $[Pt(CO)_4]Cl_2$
B) $[Pt(CO)_4]Cl_4$
C) $[Pt(CO)_4Cl_4]^{2-}$
D) $[PtCl_4](CO)_4$
E) $[Pt(CO)_4Cl_4]^{4-}$

Answer: B
Diff: 3 *Page Ref: 24.3*

16) Identify the isomers that have ligands with different spatial arrangements about the metal ions.
   A) linkage isomers
   B) geometric isomers
   C) coordination isomers
   D) optical isomers
   E) structural isomers

Answer: B
Diff: 1 *Page Ref: 24.4*

17) Identify the isomers that have ligands which coordinates to metal in different ways.
   A) linkage isomers
   B) geometric isomers
   C) coordination isomers.
   D) optical isomers
   E) stereoisomers.

Answer: A
Diff: 1 *Page Ref: 24.4*

18) Which of the following pairs of coordination compounds or complex ions are examples of coordination isomers?
   A) $[Fe(CO)_5NO_2]^{2+}$ and $[Fe(CO)_5ONO]^{2+}$
   B) $[Fe(NH_3)_2(H_2O)_4]Cl_2$ and $[Fe(NH_3)_2(H_2O)_4]Br_2$
   C) $[Fe(NH_3)_2(H_2O)_4]Cl_2$ and $[Fe(NH_3)_4(H_2O)_2]Cl_2$
   D) $[Cu(CO)_5Br]Cl$ and $[Cu(CO)_5Cl]Br$
   E) $[MnCl_3Br]^{2-}$ and $[MnClBr_3]^{2-}$

Answer: D
Diff: 3 *Page Ref: 24.4*

19) Which of the following pairs of coordination compounds or complex ions are examples of linkage isomers?
   A) $[Cu(NH_3)_5Br]Cl$ and $[Cu(NH_3)_5Cl]Br$
   B) $[Fe(NH_3)_2(H_2O)_4]Cl_2$ and $[Fe(NH_3)_4(H_2O)_2]Cl_2$
   C) $[Fe(CO)_5NO_2]^{2+}$ and $[Fe(CO)_5ONO]^{2+}$
   D) $[Fe(NH_3)_2(H_2O)_4]Cl_2$ and $[Fe(NH_3)_2(H_2O)_4]Br_2$
   E) $[Cr(H_2O)_6]^{3+}$ and $[Cr(NH_3)_6]^{3+}$

Answer: C
Diff: 3 *Page Ref: 24.4*

20) Which of the following compounds can exhibit cis–trans isomerism?

A) $[Fe(CO)_5NO_2]^{2+}$

B) $[Cu(CO)_5Cl]^+$

C) $[MnClBr_3]^{2-}$

D) $[Cr(H_2O)_6]^{3+}$

E) $[Ni(CO)_2(NH_3)_2]^{2+}$

Answer: E

Diff: 3 *Page Ref: 24.4*

21) Which of the following compounds can exhibit cis–trans isomerism?

A) $[Cr(H_2O)_4Br_2]^+$

B) $[Cu(CO)_5Br]^+$

C) $[Fe(CO)_3(NH_3)_3]^{3+}$

D) $[Fe(CO)_5ONO]^{2+}$

E) $[Fe(NH_3)_6]^{2+}$

Answer: A

Diff: 3 *Page Ref: 24.4*

22) Which of the following compounds can exhibit fac–mer isomerism?

A) $[Cu(CO)_5Cl]^+$

B) $[Co(H_2O)_3(CO)_3]^{3+}$

C) $[Fe(CO)_5NO_2]^{2+}$

D) $[Fe(NH_3)_2(H_2O)_4]^{2+}$

E) $[Cr(H_2O)_4Br_2]^+$

Answer: B

Diff: 3 *Page Ref: 24.4*

23) Which of the following compounds can exhibit fac–mer isomerism?

A) $[Cr(H_2O)_4Br_2]^+$

B) $[Fe(CO)_5ONO]^{2+}$

C) $[Fe(CO)_3(NH_3)_3]^{3+}$

D) $[Cu(CO)_5Br]^+$

E) $[Fe(NH_3)_4(H_2O)_2]^{2+}$

Answer: C

Diff: 3 *Page Ref: 24.4*

24) Which of the following complex ions absorbs light of the longest wavelength?

A) $[Cr(CN)_6]^{3-}$

B) $[CrCl_6]^{3-}$

C) $[Cr(en)_3]^{3+}$

D) $[Cr(NH_3)_6]^{3+}$

E) $[Cr(NO_2)_6]^{3-}$

Answer: B

Diff: 2 *Page Ref: 24.5*

25) Which of the following complex ions absorbs light of the shortest wavelength?

A) $[FeCl_6]^{3-}$

B) $[Fe(OH)_6]^{3-}$

C) $[Fe(CN)_6]^{3-}$

D) $[Fe(H_2O)_6]^{3+}$

E) $[FeI_6]^{3-}$

Answer: C

Diff: 2 *Page Ref: 24.5*

26) The complex ion, $[Ni(NH_3)_6]^{2+}$, has a maximum absorption near 580 nm. Calculate the crystal field splitting energy (in kJ/mol) for this ion.

A) 114 kJ/mol

B) 292 kJ/mol

C) 343 kJ/mol

D) 206 kJ/mol

E) 485 kJ/mol

Answer: D

Diff: 3 *Page Ref: 24.5*

27) How many unpaired electrons would you expect for the complex ion: $[Co(OH)_6]^{3-}$?

A) 1 B) 2 C) 4 D) 0 E) 3

Answer: C

Diff: 3 *Page Ref: 24.5*

28) How many unpaired electrons would you expect for the complex ion: $[Fe(NH_3)_6]^{2+}$?

A) 5 B) 2 C) 6 D) 4 E) 0

Answer: D

Diff: 3 *Page Ref: 24.5*

29) How many unpaired electrons would you expect for the complex ion: $[Cr(CN)_6]^{4-}$?

A) 0 B) 5 C) 3 D) 1 E) 2

Answer: E

Diff: 3 *Page Ref: 24.5*

30) How many unpaired electrons would you expect for the complex ion: $[FeCl_6]^{4-}$?

A) 2 B) 6 C) 4 D) 3 E) 0

Answer: E
Diff: 3 *Page Ref: 24.5*

31) Identify the shape of the hemoglobin complex.

A) octahedral
B) tetrahedral
C) pentadral
D) hexadral
E) heptadral

Answer: A
Diff: 3 *Page Ref: 24.6*

32) Identify the transition metal that is the coordinated metal in chlorophyll.

A) magnesium
B) manganese
C) selenium
D) iron
E) silver

Answer: A
Diff: 1 *Page Ref: 24.6*

Algorithmic Questions

1) What is the highest possible oxidation state for manganese?

A) +2 B) +5 C) +6 D) +7

Answer: D
Diff: 1 *Page Ref: 24.2*

2) Choose the electron configuration for Nb^{3+}.

A) $[Kr]5s^24d^2$
B) $[Kr]5s^2$
C) $[Kr]4d^2$
D) $[Ar]4s^23d^{10}4p^6$
E) $[Kr]5s^24d^4$

Answer: C
Diff: 2 *Page Ref: 24.2*

3) Choose the electron configuration for Ru^{7+}.

A) $[Kr]5s^14d^3$
B) $[Kr]5s^24d^2$
C) $[Kr]5s^24d^1$
D) $[Kr]4d^3$
E) $[Kr]5s^24d^6$

Answer: D
Diff: 2 *Page Ref: 24.2*

4) Choose the electron configuration for Cr^{3+}.

A) $[Ne]3s^23p^6$
B) $[Ar]4s^13d^2$
C) $[Ar]4s^23d^1$
D) $[Ar]4s^23d^7$
E) $[Ar]3d^3$

Answer: E
Diff: 2 *Page Ref:* 24.2

5) Choose the electron configuration for V^{3+}.

A) $[Ar]3d^2$
B) $[Ar]4s^23d^2$
C) $[Ar]4s^23d^4$
D) $[Ar]4s^2$
E) $[Ne]3s^23p^6$

Answer: A
Diff: 2 *Page Ref:* 24.2

6) What is the ground-state electron configuration for the element chromium (Z = 24)?

A) $[Ne]\ 4s^2\ 3d^4$ B) $[Ar]\ 4s^2\ 3d^4$ C) $[Ar]\ 4s^1\ 3d^5$ D) $[Ar]\ 3d^6$

Answer: C
Diff: 2 *Page Ref:* 24.2

7) What is the ground-state electron configuration for Cr^{3+} (Z = 24)?

A) $[Ar]\ 4s^2\ 3d^7$ B) $[Ar]\ 4s^2\ 3d^1$ C) $[Ar]\ 3d^3$ D) $[Ar]\ 4s^1\ 3d^2$

Answer: C
Diff: 2 *Page Ref:* 24.2

8) Which of the following species has the electron configuration $[Ar]3d^6$?

A) Cr B) Fe^{3+} C) Co^{3+} D) Ni^{3+}

Answer: C
Diff: 2 *Page Ref:* 24.2

9) How many *d* electrons are there in WO_4^{2-}?

A) 0 B) 1 C) 2 D) 3

Answer: A
Diff: 2 *Page Ref:* 24.2

10) Which of the following can function as a bidentate ligand?

A) CO B) OH^- C) NH_3 D) $C_2O_4^{\ 2-}$

Answer: D
Diff: 1 *Page Ref:* 24.3

11) Which of the following can function as a chelating agent?

A) SH^-

B) H_2O

C) $H_2NCH_2CO_2^-$

D) SCN^-

Answer: C
Diff: 1 *Page Ref: 24.3*

12) Choose the bidentate ligand from the substances below.
A) ethylenediamine
B) EDTA
C) chloride ion
D) ammonia
E) carbon monoxide

Answer: A
Diff: 1 *Page Ref: 24.3*

13) Choose the polydentate ligand from the substances below.
A) oxalate ion
B) EDTA
C) nitrite ion
D) hydroxide ion
E) carbon monoxide

Answer: B
Diff: 1 *Page Ref: 24.3*

14) Identify the geometry of $[PdI_4]^{2-}$.
A) tetrahedral
B) square planar
C) linear
D) octahedral
E) trigonal planar

Answer: B
Diff: 2 *Page Ref: 24.3*

15) Identify the geometry of $[Ag(NH_3)_2]^+$.
A) tetrahedral
B) square planar
C) linear
D) bent
E) trigonal pyramidal

Answer: C
Diff: 2 *Page Ref: 24.3*

16) Identify the geometry of $[Fe(H_2O)_6]^{3+}$.

A) tetrahedral
B) square planar
C) linear
D) octahedral
E) trigonal planar

Answer: D
Diff: 2 *Page Ref: 24.3*

17) How many moles of aqueous ions will be produced from the dissolution of 1.0 mole of $K_3[Fe(I)_6]$ in water?

A) 4 B) 10 C) 2 D) 9 E) 1

Answer: A
Diff: 2 *Page Ref: 24.3*

18) Which ion would you expect to have the largest crystal field splitting Δ?

A) $[Rh(CN)_6]^{4-}$ B) $[Rh(CN)_6]^{3-}$ C) $[Rh(H_2O)_6]^{2+}$ D) $[Rh(H_2O)_6]^{3+}$

Answer: B
Diff: 2 *Page Ref: 24.5*

19) Which of the following species is diamagnetic?

A) an isolated, gas-phase Cr^{3+} ion
B) a high-spin octahedral Cr^{2+} complex
C) an isolated, gas-phase Cu^{2+} ion
D) a low-spin octahedral Co^{3+} complex

Answer: D
Diff: 3 *Page Ref: 24.5*

20) What type of hybridization (according to valence bond theory) does Fe exhibit in the complex ion, $[Fe(H_2O)_4(Br)_2]^+$?

A) dsp^3 B) $sp3$ C) d^2sp^2 D) d^2sp^3 E) d^2sp

Answer: D
Diff: 3 *Page Ref: 24.5*

21) What type of geometry (according to valence bond theory) does Fe exhibit in the complex ion, $[Fe(H_2O)_4(Cl)_2]^+$?

A) linear
B) trigonal planar
C) square planar
D) octahedral
E) tetrahedral

Answer: D
Diff: 3 *Page Ref: 24.5*

22) What type of hybridization (according to valence bond theory) does Pt exhibit in the complex ion, $[Pt(NH_3)_4]^{2+}$?

A) dsp^3 B) sp C) d^2sp^3 D) d^2sp E) dsp^2

Answer: E
Diff: 3 *Page Ref: 24.5*

23) What type of geometry (according to valence bond theory) does Pt exhibit in the complex ion, $[Pt(NH_3)_4]^{2+}$?
A) square pyramidal
B) see-saw
C) trigonal planar
D) tetrahedral
E) square planar

Answer: E
Diff: 3 *Page Ref:* 24.5

24) How many unpaired electrons would you expect for the complex ion: $[Co(NH_3)_6]^{4+}$?
A) 1 B) 0 C) 5 D) 2 E) 3

Answer: C
Diff: 3 *Page Ref:* 24.5

25) Identify the transition metal that is used in oxygen transport in the human body.
A) chromium
B) iron
C) sodium
D) copper
E) silver

Answer: B
Diff: 1 *Page Ref:* 24.6

26) Identify the transition metal that is used in hemoglobin synthesis in the human body.
A) potassium
B) zinc
C) copper
D) manganese
E) chromium

Answer: C
Diff: 1 *Page Ref:* 24.6

27) Identify the transition metal that is used infat and carbohydrate synthesis in the human body.
A) potassium
B) zinc
C) copper
D) manganese
E) chromium

Answer: D
Diff: 1 *Page Ref:* 24.6

Matching Questions

Match the following.

1) linkage
Diff: 2 *Page Ref:* 24.4

2) coordination
Diff: 2 *Page Ref:* 24.4

3) cis–trans
Diff: 2 *Page Ref:* 24.4

4) mer–fac
Diff: 2 *Page Ref:* 24.4

5) optical
Diff: 2 *Page Ref:* 24.4

A) $[Cr(en)_3]^{3+}$

B) $[Cu(SCN)_4]^{3-}$ and $[Cu(NCS)_4]^{3-}$

C) $[Ni(CN)_2(OH)_2]^{2-}$

D) $[Zn(NH_3)_2Cl_2]Br_2$ and $[Zn(NH_3)_2Br_2]Cl_2$

E) $[Cr(NH_3)_3(CO)_3]^{3+}$

Answers: 1) B 2) D 3) C 4) E 5) A

Short Answer Questions

1) What is the Lanthanide contraction?

Answer: It is the explanation for why the 6th period transition elements are about the same size as the 5th period transition elements. The 6th period transition elements contain an extra 14 electrons compared to the elements above them. These extra electrons have been added into inner f orbitals. These electrons do not shield outer electrons from the nucleus very well and the atoms end up smaller than predicted because of the higher nuclear charge and lower shielding.

Diff: 2 *Page Ref:* 24.2

2) Why is +2 a common oxidation state for transition elements?

Answer: The transition metals lose their outer s electrons before losing any of their d electrons. Since the outer s orbital holds 2 electrons, the loss of these electrons when forming an ion results in a +2 oxidation state.

Diff: 2 *Page Ref:* 24.2

3) What is a coordinate covalent bond?

Answer: A covalent bond is formed through the sharing of a pair of electrons, typically one electron from each atom in the bond. A coordinate covalent bond forms when one atom in the bond donates a pair of electrons to an electron deficient atom.

Diff: 2 *Page Ref:* 24.5

4) What is the difference between a weak-field complex and a strong-field complex?

Answer: The type of ligand attached to the central atom in a complex ion determines how the degenerate d-orbital energies will be split. In a weak-field complex, the d-orbitals split with a small energy difference. In a strong-field complex, the d-orbitals split with a large energy difference. These energy differences then determine the color of the complex from the transitions of electrons between the split d-orbitals.

Diff: 3 *Page Ref:* 24.5

5) Explain how EDTA is used to treat lead poisoning.

Answer: A calcium complex of EDTA is administered and since the lead complex of EDTA has a much higher K_f than the calcium complex, the lead displaces the calcium. The lead/EDTA complex is then excreted.

Diff: 2 *Page Ref:* 24.6